D1285555

Optimum Systems Control

second edition

OPTIMUM SYSTEMS CONTROL

second edition

Andrew P. Sage
School of Engineering and Applied Science
University of Virginia

Chelsea C. White, III
Department of Applied Math and Computer Science
University of Virginia

Prentice-Hall, Inc.
Englewood Cliffs, New Jersey 07632

Library of Congress Cataloging in Publication Data

Sage, Andrew P
Optimum systems control.

Includes bibliographies and index.
1. Control theory. 2. Mathematical optimization.
I. White, Chelsea C., joint author. II. Title.
QA402.3.S24 1977 629.8'312 76-49473
ISBN 0-13-638296-7

Englewood Cliffs, New Jersey 07632

Printed in the United States of America

10 9 8 7 6 5 4 3 2 1

PRENTICE-HALL INTERNATIONAL, INC., *London*
PRENTICE-HALL OF AUSTRALIA PTY. LIMITED, *Sydney*
PRENTICE-HALL OF CANADA, LTD., *Toronto*
PRENTICE-HALL OF INDIA PRIVATE LIMITED, *New Delhi*
PRENTICE-HALL OF JAPAN, INC., *Tokyo*
PRENTICE-HALL OF SOUTHEAST ASIA PTE. LTD., *Singapore*
WHITEHALL BOOKS LIMITED, *Wellington, New Zealand*

Contents

3 Variational calculus and continuous optimal control

4 The maximum principle and Hamilton-Jacobi theory

5
Optimum systems control examples *87*

6
Discrete variational calculus and the discrete maximum principle *125*

7
Systems concepts *145*

8 Optimum state estimation 191

9 Combined estimation and control—the linear quadratic Gaussian problem 263

10 Computational methods in optimum systems control 287

Appendices

C
Random variables and stochastic processes 397

D
Proof of the matrix inversion lemma 405

Index 407

Preface

In the last several years, strong interest has continued in the study of optimization theory as applied to the control of systems. The purpose of this text is to provide a reasonably comprehensive treatment of this optimum systems control field at a level comparable to that of a beginning graduate student. In this regard, the book does not require prior background in state space techniques, calculus of variations, or probability theory, although some exposure, particularly to the first and third topics, would be of value. The text has been written strictly from the point of view of an engineer with interest in the study of systems. Consequently, we emphasize the basic concepts of various techniques and the relations, similarities, and limitations of these basic concepts at the expense of mathematical rigor. As befits an introductory text, the level of presentation is generally monotone increasing from chapter to chapter.

Structurally, the text is divided into four areas although overlap certainly exists. These are:

1. Optimal control with deterministic inputs (Chapters 2, 3, 4, 5, 6).
2. Systems concepts including controllability, observability, sensitivity, and stability (Chapter 7).
3. State estimation and combined estimation and control (Chapters 8, 9).
4. Computational techniques in systems control (Chapter 10).

There are several ways in which the text can be used. There is undoubtedly too much material covered for a one three-semester credit hour course, although the material can easily be covered in two three-quarter hour courses. For a single three-semester hour course, we suggest that the instructor consider eliminating either Chapters 8 and 9 or Chapter 10, as best fits following courses in mathematical system theory. If a course using this text has been preceded by a graduate level course in state space techniques, then Chapter 7 may be eliminated, and the more advanced backgrounds of the students may well allow completion of the remainder of the text in one semester.

This edition of *Optimum Systems Control* is considerably revised and we hope much improved over the first edition. Every chapter in the original text has been subject to this revision. Several new derivations and examples have been included as have developments in optimum systems control that were unknown during the writing of the first edition. The senior author considers it his personal good fortune that he was able to obtain the full and complete collaboration of an outstanding young professional who has contributed mightily to this updating. The authors wish to acknowledge the helpful assistance of many former students, including those mentioned in the first edition, who have offered many helpful comments when reading through earlier versions of the text.

University of Virginia

ANDREW P. SAGE
CHELSEA C. WHITE, III

Optimum Systems Control

second edition

Introduction

1

In recent years much attention has been focused upon optimizing the behavior of systems. A particular problem may concern maximizing the range of a rocket, maximizing the profit of a business, minimizing the error in estimation of position of an object, minimizing the energy or cost required to achieve some required terminal state, or any of a vast variety of similar statements. The search for the control which attains the desired objective while minimizing (or maximizing) a defined system criterion constitutes the fundamental problem of optimization theory.

The fundamental problem of optimization theory may be subdivided into four interrelated parts:

1. Definition of a goal.
2. Knowledge of our current position with respect to the goal.
3. Knowledge of all environmental factors influencing the past, present, and future.
4. Determination of the best policy from the goal definition (1) and knowledge of our current state (2) and environment (3).

To solve an optimization problem, we must first define a goal or a cost function for the process we are attempting to optimize. This requires an adequate definition of the problem in physical terms and a translation of this physical description into mathematical terms. To effectively control a process, we must know the current state of the process. This we will call the problem of state estimation. Also, we must be able to characterize the process by an effective model which will depend upon various environmental factors. This we will call system identification. With a knowledge of the cost function, and the system states and parameters, we then determine the best control which minimizes (or maximizes) the cost function. Thus we may define five problems, which are again interrelated, and which we must solve in order to determine the best, or optimum, system:

1. The Control Problem. We are given a known system with relation between system states and input control. We desire to find the control which changes the state $x(t)$ so as to accomplish some desirable objective. Figure 1.1 illustrates the salient features of the control problem. This may be an open- or closed-loop problem, depending upon whether or not the control is a function of the state.

Fig. 1.1 Deterministic optimum control problem.

2. The State Estimation Problem. We are given a known system with a random input and measurement noise such that we measure an output $z(t)$ which is a corrupted version of $x(t)$ as indicated in Fig. 1.2. We know the statistics of the plant noise $w(t)$ and the measurement noise $v(t)$, and we desire to determine a "best" estimate $\hat{x}(t)$ of the true system state $x(t)$ from a knowledge of $z(t)$.

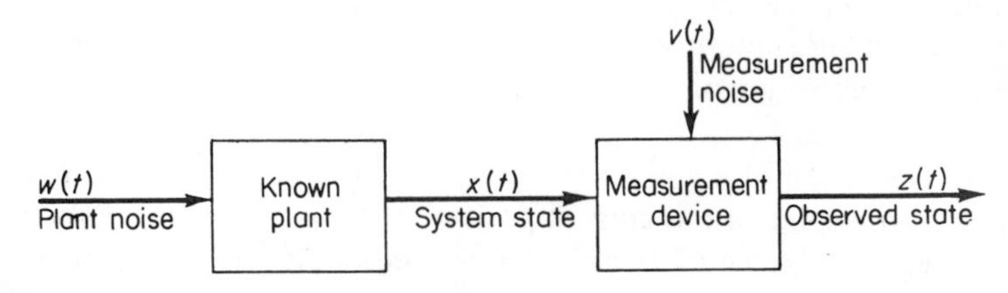

Fig. 1.2 State estimation problem.

3. The Stochastic Control Problem. We may combine problems 1 and 2 to form a stochastic control problem as depicted in Fig. 1.3. We desire to determine a control $u(t)$ such that the output state $x(t)$ is changed in accordance with some desired objective. Plant noise $w(t)$ and measurement noise $v(t)$ are present. We know the statistics of these noises and must of course determine a best estimate, $\hat{x}(t)$, of $x(t)$ from a knowledge of the output $z(t)$ before we may discern the "best" control which may be open- or closed-loop.

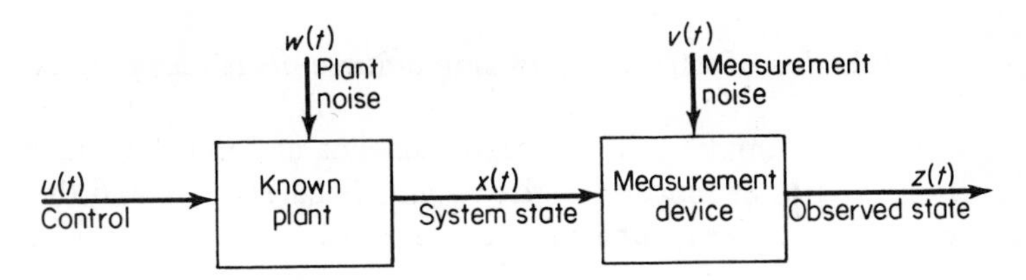

Fig. 1.3 Stochastic control problem.

4. The Parameter Estimation Problem. In many systems we must incorporate some method of identification of system parameters which may vary as a function of the environment. We are given a system such as that shown in Fig. 1.4, where we again know the statistical characteristics of the plant and the measurement noise, and we wish to determine the best estimate of certain plant parameters based upon a knowledge of the deterministic input $u(t)$, the measured output $z(t)$, and possibly some a priori knowledge of the system plant structure. As we shall see, we often must accomplish state estimation in order to obtain parameter estimation.

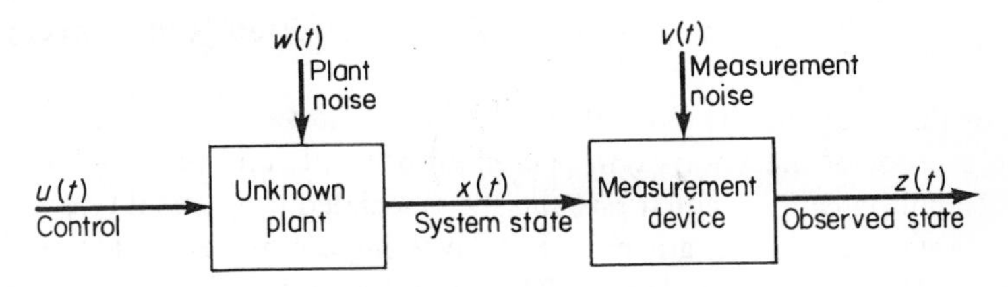

Fig. 1.4 Parameter estimation problem.

5. The Adaptive Control Problem. We may combine problems 1 through 4 to form an adaptive control problem. We are given the statistical characteristics of $w(t)$ and $v(t)$ or some method of determining these characteristics. Plant parameters are random. We desire to determine a control $u(t)$ to best accomplish some desired objective in terms of the measurement noise and plant noise as well as the uncertainty in system dynamics. If the control $u(t)$ is determined as a function of the measured output $z(t)$, we have a closed-loop adaptive system.

We will divide our efforts in optimum systems control into ten chapters. These chapters and their respective purposes and contents will now be described briefly. Each chapter will contain several examples to illustrate our developed theory. Many problems, of varying complexity, will be posed at the end of each chapter for the interested reader.

2
Calculus of extrema and single-stage decision processes

This chapter examines ordinary scalar maxima and minima and extrema of functions of two or more variables. Constrained extrema and the vector formulation of extrema problems are presented for single-stage decision processes.

3
Variational calculus and continuous optimal control

In this chapter, we introduce the subject of the variational calculus for continuous decision processes through a derivation of the Euler-Lagrange equations and associated transversality conditions. We discuss the use of Lagrange multipliers to treat equality constraints and briefly mention the inequality constraint problem. Several very simple optimal control problems are considered.

4
The maximum principle and Hamilton-Jacobi theory

In this chapter, the Bolza formulation of the variational calculus leads into a proof of the Pontryagin maximum principle and the development of the Hamilton canonic equations and the associated transversality conditions. We discuss at some length problems involving control and state, and state variable inequality constraints. The Hamilton-Jacobi equations are then developed and modified to produce Bellman's equations of continuous dynamic programming.

5
Optimum systems control examples

This chapter formulates and solves numerous optimal control problems of interest; among those solved are:

1. Minimum time problems.
2. Linear regulator problems.
3. Servomechanism problems.
4. Minimum fuel problems.
5. Minimum energy problems.

6. Singular solution problems.
7. Distributed parameter problems.

6
Discrete variational calculus and the discrete maximum principle

In this chapter, we develop a simplified discrete maximum principle for cases in which control and state variable inequality constraints are absent. We give a meaningful comparison of the discrete maximum principle and the discretized results of application of the continuous maximum principle for a rather general optimization problem. We conclude our discussion with a brief presentation of the relationship between discrete time optimal control and mathematical programming.

7
Systems concepts

After having established and solved many state estimation problems and optimal control problems, we now inquire into the conditions which must be established in order for many of these problems to have meaningful solutions. First we examine the manner in which the output of a system is constrained with respect to the ability to observe system states. Then we examine the dual requirement and find the characterization of the manner in which a system is constrained with respect to control of system states or system outputs.

Also presented are various methods for studying the parameter sensitivity problem in continuous systems. The use of sensitivity concepts in optimal and optimal adaptive systems are presented. A brief introduction to system stability concepts and a discussion of stability-optimality relations for linear systems concludes the chapter.

8
Optimum state estimation

Chapter 8 introduces the subject of optimum filtering. The state transition approach is used, which allows us to develop the celebrated Kalman-Wiener computational algorithms for nonstationary filtering. The dual relations between the filter and the regulator problems are observed, and the difference between optimum smoothing and optimum filtering is discussed.

9
Combined estimation and control—the linear quadratic Gaussian problem

A very fundamental problem in optimum system control results when we desire to obtain the closed-loop control for a linear stochastic system with noise-corrupted observations of the system states. The cost function to be minimized is the expected value of an integral of a quadratic function of the system states and controls. We present a rather thorough discussion of this problem and derive a separation theorem that, in linear estimation and control problems, allows us to separate the problem of estimation from that of control. We also present several pitfalls associated with application of the separation theorem and discuss several situations in which separation is not optimal.

10
Computational methods in optimum systems control

This chapter discusses several methods for solving optimal control problems. We begin with a discussion of discrete dynamic programming. The gradient method or method of steepest descent is presented along with modifications to allow us to consider terminal manifold equality constraints and inequality constraints. The second variation method is developed and applied to several illustrative examples to indicate the improved speed of convergence compared with the gradient method. We conclude the chapter with a modification of the Newton-Raphson method for the indirect solution of two-point boundary value problems. Problems of optimal control and those of parameter estimation are considered.

A
The algebra, calculus, and differential equations of vectors and matrices

In this appendix, we summarize many of the matrix and vector calculus operations used throughout the text.

B
Abstract spaces

A deeper and fuller understanding of the more mathematical aspects of optimum systems control requires an appreciation for abstract spaces. Although not required as a preliminary to understanding this text, this appendix presents the salient results in abstract spaces associated with our developments in the text.

C
Random variables and stochastic processes

Our developments in Chapters 8 and 9 presume a beginning knowledge of probability and stochastic processes. For the convenience of the reader, we summarize results in probability and stochastic processes useful in these two chapters. We also present a discussion of stochastic dynamic programming, results of which are used to derive the separation theorem in Chapter 9.

D
Proof of the matrix inversion lemma

A matrix inversion lemma is needed to resolve certain problems associated with linear optimal control in Chapters 5 and 6, estimation in Chapter 8, and stochastic control in Chapter 9. We derive this matrix inversion lemma here.

Calculus of extrema and single-stage decision processes 2

Many problems in modern system theory may be simply stated as extreme value problems. These can be resolved via the calculus of extrema which is the natural solution method whenever one desires to find parameter values which minimize or maximize a quantity dependent upon them. In this chapter, we consider several such problems, starting with simple scalar problems and concluding with a discussion of the vector case. The method of Lagrange multipliers is introduced and used to solve constrained extrema problems for single-stage decision processes. A brief discussion of linear and nonlinear programming is presented. Multistage decision processes, which can be treated by the calculus of extrema, are reserved for a variational treatment which will result in a discrete maximum principle.

2.1 Unconstrained extrema

Let D be a subset of a normed linear vector space having norm $\|\cdot\|$ (see Appendix B), and let f be a real-valued function defined on D, i.e., $f: D \longrightarrow R$, where R^n is Euclidean n-space and $R = R^1$. The function f has a relative maximum (minimum) at $\alpha \in D$ if and only if there is a positive real number δ such that for any $x \in D$ satisfying $\|x - \alpha\| < \delta$, $f(x) \leq f(\alpha)$ $[f(\alpha) \leq f(x)]$. If the inequalities hold for all $x \in D$, then f has an absolute minimum or maximum. We use the word *extremum* to refer to either a maximum or a minimum.

A point $\alpha \in D$ is an interior point of D if and only if there is a positive real number δ such that if x satisfies $\|x - \alpha\| < \delta$, then $x \in D$. Let D be the real line, i.e., $D = R$, and assume $df(x)/dx$ exists and is continuous at $x = \alpha$. Then, a necessary condition for a point α to be an interior maximum or minimum is that

$$\left.\frac{df(x)}{dx}\right|_{x=\alpha} = 0. \tag{2.1-1}$$

Assume that f has a continuous second derivative at $x = \alpha$. It is well-known that if Eq. (2.1-1) is satisfied, and if $[d^2f(x)/dx^2]|_{x=\alpha}$ is positive (negative), then α is a relative minimum (maximum). Minima and maxima for various functions are illustrated in Fig. 2.1-1.

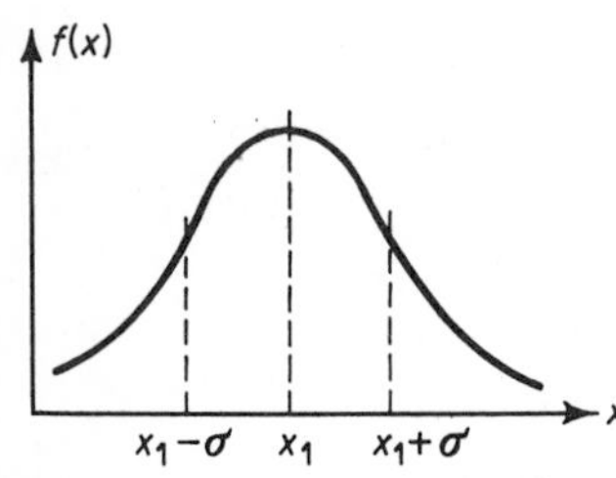

(a)
$f(x) = \frac{1}{\sqrt{2\pi}\,\sigma} \exp\left[-(x-x_1)^2/2\sigma^2\right]$
For x in the interval: $(-\infty, \infty)$
$f(x)$ has an absolute maximum at $x = x_1$.

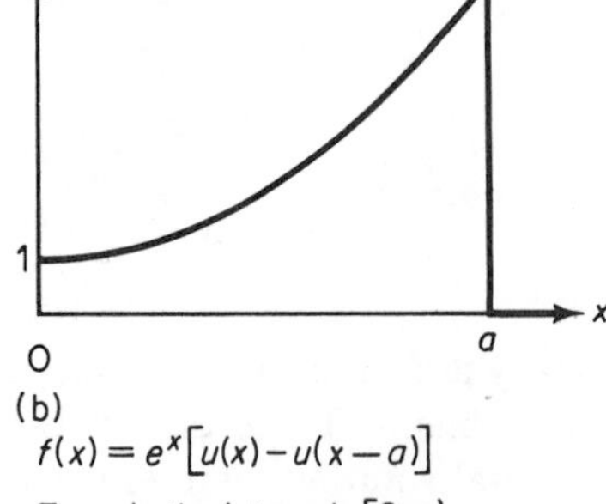

(b)
$f(x) = e^x[u(x) - u(x-a)]$
For x in the interval: $[0, a)$
$f(x)$ has an absolute minimum at $x = 0$, and an absolute maximum at $x = a - \delta$ where δ is an arbitrarily small positive number.

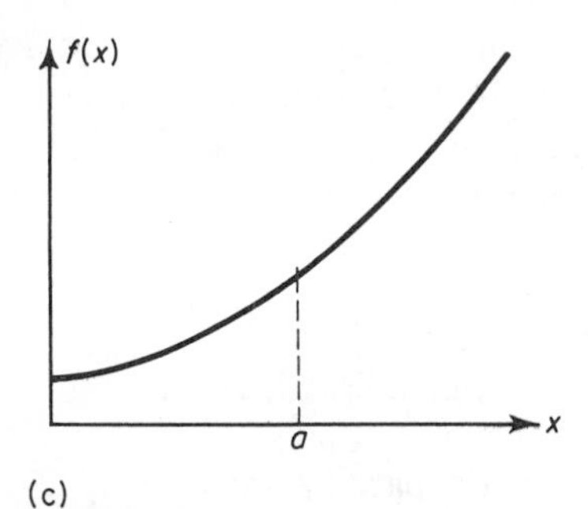

(c)
$f(x) = e^x u(x)$
For x in the interval: $[0, a]$
$f(x)$ has an absolute minimum at $x = 0$, and an absolute maximum at $x = a$.
For x in the interval: $[0, +\infty]$
$f(x)$ has an absolute minimum at $x = 0$.

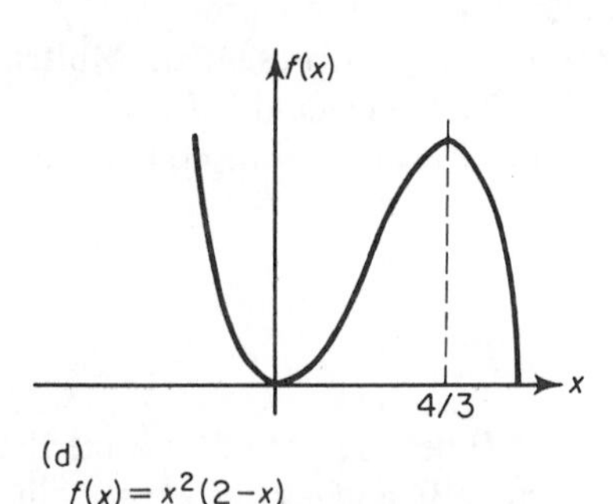

(d)
$f(x) = x^2(2-x)$
For x in the interval: $(-\infty, +\infty)$
$f(x)$ has a relative minimum at $x = 0$, and a relative maximum at $x = 4/3$.

Fig. 2.1-1 Illustrations of extrema.

The extrema-finding technique can be extended to include functions of more than one scalar variable, using partial derivatives rather than total derivatives. Let $f\colon R^n \longrightarrow R$. Analogous to Eq. (2.1-1), a necessary condition for a vector $\boldsymbol{\alpha}$ to be an interior maximum or minimum is that

$$\left.\frac{\partial f(\mathbf{x})}{\partial \mathbf{x}}\right|_{\mathbf{x}=\boldsymbol{\alpha}} = \mathbf{0}, \tag{2.1-2}$$

(see Appendix A, Sec. 2.2) where all partial derivatives exist and are continuous at $\mathbf{x} = \boldsymbol{\alpha}$. Assuming the appropriate second partial derivatives exist and are continuous at $\mathbf{x} = \boldsymbol{\alpha}$, a sufficient condition that $\boldsymbol{\alpha}$ be a relative maximum (minimum) is that Eq. (2.1-2) holds and that the matrix having as its ijth element $[\partial^2 f(\mathbf{x})/\partial x_i\, \partial x_j]|_{\mathbf{x}=\boldsymbol{\alpha}}$ is negative (positive) definite (see Appendix A, Secs. A.1-23 and A.2-4. A simple example will illustrate the procedure to be followed.

Example 2.1-1. Let us consider the maximization of

$$f(\mathbf{x}) = \frac{1}{(x_1 - 1)^2 + (x_2 - 1)^2 + 1}, \qquad \mathbf{x}^T = [x_1, x_2],$$

where $\mathbf{x}^T$ is used to indicate transpose of the column vector $\mathbf{x}$.† Following an extended version of the foregoing scalar procedure, we take the partial derivatives of f with respect to x_1 and x_2 and set them equal to zero to obtain:

$$\frac{\partial f}{\partial x_1} = \frac{(-1)(2x_1 - 2)}{[(x_1 - 1)^2 + (x_2 - 1)^2 + 1]^2} = 0, \qquad \alpha_1 = 1$$

$$\frac{\partial f}{\partial x_2} = \frac{(-1)(2x_2 - 2)}{[(x_1 - 1)^2 + (x_2 - 1)^2 + 1]^2} = 0, \qquad \alpha_2 = 1.$$

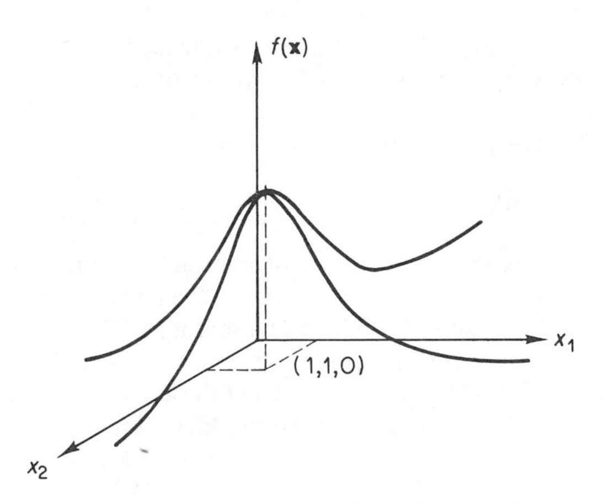

Fig. 2.1-2 $f(x) = 1/[(x_1 - 1)^2 + (x_2 - 1)^2]$.

†Appendix A contains a brief presentation of vector matrix notations and vector matrix calculus.

Thus, since $\alpha_1 = \alpha_2 = 1$ are the only extrema, and since a simple computation shows that the second derivatives are nonpositive at this extrema, we see that we have a maximum at the point $\mathbf{x}^T = [1, 1]$. Figure 2.1-2 presents a sketch of $f(\mathbf{x})$.

2.2 Extrema of functions with equality constraints

We now consider extremizing the function $\theta: R^n \longrightarrow R$, subject to the equality constraint $\mathbf{f}(\mathbf{x}) = \mathbf{0}$, where $\mathbf{f}: R^n \longrightarrow R^m$. Several approaches are applicable to the solution of this problem, two of which are the direct substitution method and the Lagrange multiplier method. Direct substitution is a particularly straightforward procedure; however, it is often more cumbersome than the Lagrange multiplier approach. For simplicity of illustration, assume $n = 2$ and $m = 1$.

The method of direct substitution can be described as follows. Assume f is such that there exists an $h: R \longrightarrow R$ satisfying $h(x_2) = x_1$. Then, $\theta(x_1, x_2) = \theta[h(x_2), x_2]$ is a function of x_2 only, and techniques for the unconstrained problem are then applicable. We further illustrate this approach with two examples.

Example 2.2-1. Consider extremizing the function defined in Example 2.1-1 subject to the constraint that the Euclidean norm of $\mathbf{x}$ equals one. Symbolically, this means that $\|\mathbf{x}\|^2 = \mathbf{x}^T\mathbf{x} = x_1^2 + x_2^2 + \ldots + x_n^2 = \langle \mathbf{x}, \mathbf{x} \rangle$. Since the dimension of the example we are considering is two, the Euclidean norm squared becomes $\|\mathbf{x}\|^2 = x_1^2 + x_2^2$.

One approach to the problem is to solve for x_1 in terms of x_2, then solve for $\Theta = f(\mathbf{x})$ in terms of x_2 alone. This will then allow us to use the standard scalar procedure. From the given constraint on the length of the Euclidean norm, we have $x_1 = (1 - x_2^2)^{1/2}$. Substituting this into the expression for $\Theta(\mathbf{x})$ of Example 2.1-1, we find that

$$\Theta(x_2) = \frac{1}{(\pm\sqrt{1 - x_2^2} - 1)^2 + (x_2 - 1)^2 + 1}$$

where $\Theta(x_2)$ has the given constraint imbedded into it. The next step is to differentiate this expression with respect to the remaining variable x_2 and set the result equal to zero. This yields two solutions. The second-derivative test shows that a maximum (which is easily shown to be an absolute maximum) occurs at $\mathbf{x}^T = [0.707, 0.707]$ and that an (absolute) minimum occurs at $\mathbf{x}^T = [-0.707, -0.707]$. Figure 2.2-1 illustrates $\Theta(x_2)$ for this example.

We note that, in the absence of the equality constraint, this problem has no relative minimum.

Example 2.2-2. A tin can manufacturer wants to maximize the volume of a certain run of cans subject to the constraint that the area of tin used be

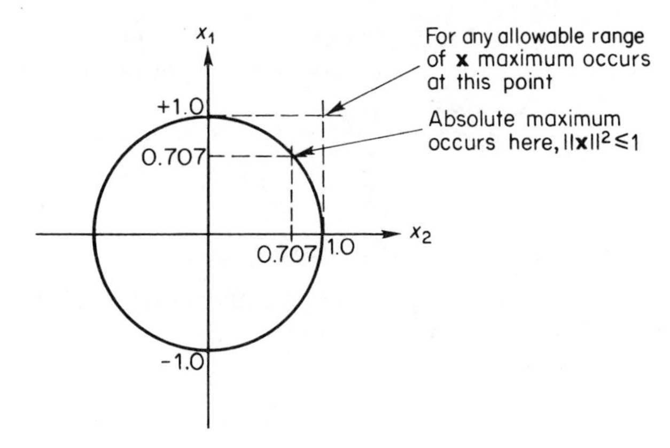

Fig. 2.2-1 Top view of Fig. 2.1-1 showing the region defined in state space by $||x||^2 < 1$.

a given constant. If a fixed metal thickness is assumed, a volume of tin constraint implies that the cross-sectional area is constrained.

The defining equations for this problem are:

$$\text{Volume} = V(r, l) = \pi r^2 l \tag{1}$$

$$\text{Cross-sectional area} = A(r, l) = 2\pi r^2 + 2\pi rl = A_o. \tag{2}$$

Our problem is to maximize $V(r, l)$ subject to keeping $A(r, l) = A_o$, where A_o is a given constant. The same approach can be used here as in Example 2.2-1. We solve for l in terms of r (or if preferred, r in terms of l) and then express the volume as a function of r alone, noting that the constraint on the cross-sectional area is now imbedded into the expression for the volume. We then examine the first and second derivatives to discern the character and location of the extrema.

From Eq. (2) we have

$$l = \frac{A_o - 2\pi r^2}{2\pi r}. \tag{3}$$

By substituting Eq. (3) into Eq. (1), we obtain

$$V(r) = \frac{r}{2} A_o - 2\pi r^2. \tag{4}$$

We differentiate V with respect to r and set the result equal to zero to obtain

$$\frac{dV(r)}{dr} = \frac{A_o}{2} - 3\pi r^2 = 0, \qquad r = \sqrt{\frac{A_o}{6}}. \tag{5}$$

We now substitute Eq. (5) into Eq. (2) and solve for l:

$$l = \sqrt{\frac{2A_o}{3\pi}}. \tag{6}$$

It is interesting to obtain the optimum length-to-radius ratio. In doing this, we see that, to get maximum volume, we make the length of the tin can equal the diameter, keeping cross-sectional area equal to a given constant.

We will now develop the Lagrange multiplier approach. Assuming all appropriate partial derivatives exist, we use the chain rule to obtain

$$\frac{d\theta(\mathbf{x})}{dx_1} = \frac{\partial\theta(\mathbf{x})}{dx_1} + \frac{\partial\theta(\mathbf{x})}{\partial x_2}\frac{dx_2}{dx_1} \tag{2.2-1}$$

and

$$\frac{df(\mathbf{x})}{dx_1} = \frac{\partial f(\mathbf{x})}{\partial x_1} + \frac{\partial f(\mathbf{x})}{\partial x_2}\frac{dx_2}{dx_1}. \tag{2.2-2}$$

Since $\mathbf{x}$ is constrained to satisfy $f(\mathbf{x}) = 0$, it follows from Eq. (2.2-2) that a candidate for a constrained extremal must satisfy

$$\frac{\partial f(\mathbf{x})}{\partial x_1} + \frac{\partial f(\mathbf{x})}{\partial x_2}\frac{dx_2}{dx_1} = 0.$$

It follows from (2.2-1) that a candidate for an extremal must also satisfy

$$\frac{\partial\theta(\mathbf{x})}{\partial x_1} + \frac{\partial\theta(\mathbf{x})}{\partial x_2}\frac{dx_2}{dx_1} = 0.$$

Assuming all of the above partial derivatives are nonzero, it then follows that

$$\frac{dx_2}{dx_1} = \frac{-\partial\theta(\mathbf{x})}{\partial x_1}\bigg/\frac{\partial\theta(\mathbf{x})}{\partial x_2} = \frac{-\partial f(\mathbf{x})}{\partial x_1}\bigg/\frac{\partial f(\mathbf{x})}{\partial x_2}; \tag{2.2-3}$$

that is, the ratios

$$\frac{\partial\theta(\mathbf{x})}{\partial x_1} : \frac{\partial\theta(\mathbf{x})}{\partial x_2} \quad \text{and} \quad \frac{\partial f(\mathbf{x})}{\partial x_1} : \frac{\partial f(\mathbf{x})}{\partial x_2}$$

must be proportional for a vector to be a candidate for an interior extremal. Let λ be the proportionality constant, which we call the Lagrange multiplier. Thus,

$$\lambda = \frac{-\partial\theta(\mathbf{x})}{\partial x_1}\bigg/\frac{\partial\theta(\mathbf{x})}{\partial x_2} = \frac{-\partial f(\mathbf{x})}{\partial x_1}\bigg/\frac{\partial f(\mathbf{x})}{\partial x_2}. \tag{2.2-4}$$

We define the Lagrangian

$$L(\mathbf{x}, \lambda) = \theta(\mathbf{x}) + \lambda f(\mathbf{x}).$$

Then, (2.2-4) is equivalent to $\partial L/\partial \mathbf{x} = \mathbf{0}$, and $f(\mathbf{x}) = 0$ is equivalent to $\partial L/\partial \lambda = 0$.

Example 2.2-3. Consider the problem stated in Example 2.2-2. By using the Lagrange multiplier, we again want to extremize (maximize) the volume $V(r, l)$ subject to the constraint $A(r, l) = A_o$. First we form the adjoined function

$$V'(r, l) = V(r, l) + \lambda[A(r, l) - A_o]$$

where λ is the Lagrange multiplier. In terms of the parameters of the tin can, this expression becomes

$$V'(r, l) = \pi r^2 l + \lambda[2\pi r^2 + 2\pi r l - A_o].$$

We take the first partial derivative with respect to each of the variables and set each result equal to zero. Thus we obtain

$$\frac{\partial V'(r, l)}{\partial l} = \pi r^2 + \lambda 2\pi r = 0, \qquad r = -2\lambda$$

$$\frac{\partial V'(r, l)}{\partial r} = 2\pi r l + \lambda[4\pi r + 2\pi l] = 0, \qquad l = 2r.$$

We now evaluate λ subject to given constraint, $A(r, l) = A_o$ or

$$A_o = 2\pi r^2 + 2\pi r l.$$

In terms of the obtained values of r and l, this becomes

$$A_o = 2\pi(4\lambda^2) + 2\pi(-2\lambda)(-4\lambda)$$

so

$$\lambda = \pm\sqrt{\frac{A_o}{24\pi}}.$$

Thus, we have

$$r = 2\sqrt{\frac{A_o}{24\pi}}, \qquad l = 4\sqrt{\frac{A_o}{24\pi}}.$$

We note that the negative square root is selected for λ to make r and l physically realizable quantities. We further note that the length-to-radius ratio is the same as obtained by direct substitution, as it well should be.

The Lagrange multiplier approach is easily generalized to consider the case where m and n are arbitrary, finite, positive integers. The Lagrangian is again defined as $L(\mathbf{x}, \boldsymbol{\lambda}) = \theta(\mathbf{x}) + \boldsymbol{\lambda}^T\mathbf{f}(\mathbf{x})$, where now $\boldsymbol{\lambda} \in R^m$. If $\boldsymbol{\alpha}$ is a constrained maximum or minimum, then there is a $\boldsymbol{\lambda} \in R^m$ such that $\partial L/\partial \mathbf{x} = \mathbf{0}$ and $\partial L/\partial \boldsymbol{\lambda} = \mathbf{0}$ evaluated at $\boldsymbol{\alpha}$ and $\boldsymbol{\lambda}$.

For the majority of systems problems, it is convenient to distinguish between control vectors and state vectors. We generally desire to find a control vector $\mathbf{u}$, or $\mathbf{u}(k)$ or $\mathbf{u}(t)$ if we have a multistage or continuous process, which minimizes or maximizes some scalar index of performance of the system. This performance index will be called J. Possibly the simplest single-stage decision process with equality constraints is to minimize or maximize the scalar index of performance

$$J = \theta[\mathbf{x}, \mathbf{u}] \tag{2.2-5}$$

subject to the equality constraint

$$\mathbf{f}(\mathbf{x}, \mathbf{u}) = \mathbf{0} \tag{2.2-6}$$

where $\mathbf{x}$ is an n vector

$$\mathbf{x}^T = [x_1, x_2, \ldots, x_n]$$

$\mathbf{u}$ is an m vector

$$\mathbf{u}^T = [u_1, u_2, \ldots, u_m]$$

$\mathbf{f}$ is an n vector function

$$\mathbf{f}^T(\mathbf{x}, \mathbf{u}) = [f_1(\mathbf{x}, \mathbf{u}), f_2(\mathbf{x}, \mathbf{u}), \ldots, f_n(\mathbf{x}, \mathbf{u})].$$

The solution proceeds as follows. We adjoin Eq. (2.2-6) to Eq. (2.2-5) with a vector Lagrange multiplier in order to form the scalar quantity

$$L(\mathbf{x}, \mathbf{u}, \boldsymbol{\lambda}) = \theta(\mathbf{x}, \mathbf{u}) + \boldsymbol{\lambda}^T \mathbf{f}(\mathbf{x}, \mathbf{u})$$
$$\boldsymbol{\lambda}^T = [\lambda_1, \lambda_2, \ldots, \lambda_n].$$

We now adjust $\mathbf{x}$ and $\mathbf{u}$ such that L is a maximum or minimum. This requires

$$\frac{\partial L}{\partial \mathbf{x}} = \frac{\partial \theta}{\partial \mathbf{x}} + \frac{\partial}{\partial \mathbf{x}} \mathbf{f}^T(\mathbf{x}, \mathbf{u})\boldsymbol{\lambda} = \mathbf{0}$$

$$\frac{\partial L}{\partial \mathbf{u}} = \frac{\partial \theta}{\partial \mathbf{u}} + \frac{\partial}{\partial \mathbf{u}} \mathbf{f}^T(\mathbf{x}, \mathbf{u})\boldsymbol{\lambda} = \mathbf{0}$$

where
$$\left[\frac{\partial L}{\partial \mathbf{u}}\right]^T = \left[\frac{\partial L}{\partial u_1}, \frac{\partial L}{\partial u_2}, \ldots, \frac{\partial L}{\partial u_m}\right].$$

Thus $\partial L/\partial \mathbf{u}$ may be interpreted as the gradient of L with respect to $\mathbf{u}$, which is commonly designated $\nabla_{\mathbf{u}} L$. Also,

$$\frac{\partial}{\partial \mathbf{x}} \mathbf{f}^T(\mathbf{x}, \mathbf{u}) = \begin{bmatrix} \frac{\partial f_1}{\partial x_1} & \frac{\partial f_2}{\partial x_1} & \cdots & \frac{\partial f_n}{\partial x_1} \\ \cdot & & & \cdot \\ \cdot & & & \cdot \\ \cdot & & & \cdot \\ \frac{\partial f_1}{\partial x_n} & & & \frac{\partial f_n}{\partial x_n} \end{bmatrix}. \tag{2.2-7}$$

It should be noted that Eq. (2.2-7) is similar to the transpose of the Jacobian of a vector

$$[J_{\mathbf{x}} \mathbf{f}(\mathbf{x}, \mathbf{u})]^T = \begin{vmatrix} \frac{\partial f_1}{\partial x_1} & \frac{\partial f_2}{\partial x_1} & \cdots & \frac{\partial f_n}{\partial x_1} \\ \cdot & & & \cdot \\ \cdot & & & \cdot \\ \cdot & & & \cdot \\ \frac{\partial f_1}{\partial x_n} & & & \frac{\partial f_n}{\partial x_n} \end{vmatrix}$$

with at least two important differences: $\partial \mathbf{f}(\mathbf{x}, \mathbf{u})/\partial \mathbf{u}$ need not be square and is a matrix rather than a determinant. In order that J be an extremum, not only must

$$\frac{\partial L}{\partial \mathbf{x}} = \mathbf{0}; \qquad \frac{\partial L}{\partial \mathbf{u}} = \mathbf{0},$$

but also the second variation of L must be greater than zero for a minimum or less than zero for a maximum (see second-derivative test, Sec. 2.1) To see what this constraint on the second variation of L means, in terms of the necessary conditions required for making $J(\mathbf{x}, \mathbf{u})$ have an extremum, let us now formulate the second variation of $L(\mathbf{x}, \mathbf{u}, \boldsymbol{\lambda})$. The first variation of $L(\mathbf{x}, \mathbf{u}, \boldsymbol{\lambda})$ is

$$\delta L = \left(\frac{\partial L}{\partial \mathbf{x}}\right)^T \boldsymbol{\delta}\mathbf{x} + \left(\frac{\partial L}{\partial \mathbf{u}}\right)^T \boldsymbol{\delta}\mathbf{u}$$

which is the linear part of

$$\Delta L = L[\mathbf{x} + \boldsymbol{\delta}\mathbf{x}, \mathbf{u} + \boldsymbol{\delta}\mathbf{u}] - L[\mathbf{x}, \mathbf{u}]. \tag{2.2-8}$$

To get the second variation of L, denoted $\delta^2 L$, we take the second-order part of the expansion of Eq. (2.2-8) in a Taylor series about $\boldsymbol{\delta}\mathbf{u} = \mathbf{0}$, $\boldsymbol{\delta}\mathbf{x} = \mathbf{0}$ to obtain

$$\begin{aligned}\delta^2 L = \frac{1}{2}\boldsymbol{\delta}\mathbf{x}^T\left\{\left[\frac{\partial}{\partial \mathbf{x}}\frac{\partial L}{\partial \mathbf{x}}\right]\boldsymbol{\delta}\mathbf{x} + \left[\frac{\partial}{\partial \mathbf{u}}\frac{\partial L}{\partial \mathbf{x}}\right]\boldsymbol{\delta}\mathbf{u}\right\} \\ + \frac{1}{2}\boldsymbol{\delta}\mathbf{u}^T\left\{\left[\frac{\partial}{\partial \mathbf{u}}\frac{\partial L}{\partial \mathbf{x}}\right]^T\boldsymbol{\delta}\mathbf{x} + \left[\frac{\partial}{\partial \mathbf{u}}\frac{\partial L}{\partial \mathbf{u}}\boldsymbol{\delta}\mathbf{u}\right]\right\}.\end{aligned} \tag{2.2-9}$$

In more compact notation, this becomes

$$\delta^2 L = \frac{1}{2}[\boldsymbol{\delta}\mathbf{x}^T\ \boldsymbol{\delta}\mathbf{u}^T]\begin{bmatrix} \dfrac{\partial}{\partial \mathbf{x}}\dfrac{\partial L}{\partial \mathbf{x}} & \dfrac{\partial}{\partial \mathbf{u}}\dfrac{\partial L}{\partial \mathbf{x}} \\ \left[\dfrac{\partial}{\partial \mathbf{u}}\dfrac{\partial L}{\partial \mathbf{x}}\right]^T & \dfrac{\partial}{\partial \mathbf{u}}\dfrac{\partial L}{\partial \mathbf{u}} \end{bmatrix}\begin{bmatrix}\partial \mathbf{x} \\ \partial \mathbf{u}\end{bmatrix}. \tag{2.2-10}$$

If we define

$$\boldsymbol{\delta}\mathbf{z}^T = [\boldsymbol{\delta}\mathbf{x}^T\ \boldsymbol{\delta}\mathbf{u}^T], \qquad \mathbf{P} = \begin{bmatrix} \dfrac{\partial}{\partial \mathbf{x}}\dfrac{\partial L}{\partial \mathbf{x}} & \dfrac{\partial}{\partial \mathbf{u}}\dfrac{\partial L}{\partial \mathbf{x}} \\ \left[\dfrac{\partial}{\partial \mathbf{u}}\dfrac{\partial L}{\partial \mathbf{x}}\right]^T & \dfrac{\partial}{\partial \mathbf{u}}\dfrac{\partial L}{\partial \mathbf{u}} \end{bmatrix}$$

Eq. (2.2-10) reduces to

$$\delta^2 L = \tfrac{1}{2}\boldsymbol{\delta}\mathbf{z}^T\mathbf{P}\boldsymbol{\delta}\mathbf{z} = \tfrac{1}{2}\,||\boldsymbol{\delta}\mathbf{z}||^2_{\mathbf{P}}$$

which is recognized as the standard quadratic form. A positive definite quadratic form is defined as one for which $\boldsymbol{\delta}\mathbf{z}^T\mathbf{P}\,\boldsymbol{\delta}\mathbf{z} > 0$ for all nonzero $\boldsymbol{\delta}\mathbf{z}$. A positive semidefinite matrix, $\mathbf{P}$, is defined as one which has the property that $\boldsymbol{\delta}\mathbf{z}^T\mathbf{P}\,\boldsymbol{\delta}\mathbf{z} \geq 0$ for all nonzero $\boldsymbol{\delta}\mathbf{z}$. In a similar fashion, negative definite and negative semidefinite quadratic forms and matrices are defined. Section A.1-23 of Appendix A delineates a method which we can use to discern positive definiteness of a square matrix. Thus we can state the two necessary conditions [1] for $J(\mathbf{x}, \mathbf{u})$ to have an extremum in a given interval of $\mathbf{x}$ for convex or concave $J(\mathbf{x}, \mathbf{u})$. If $J(\mathbf{x}, \mathbf{u})$ is not convex or concave, the second condition is only sufficient, and a quantity known as the bordered Hessian must be used to obtain the second necessary condition.

1. The following vectors are zero:

$$\frac{\partial L}{\partial \mathbf{x}} = \mathbf{0}; \qquad \frac{\partial L}{\partial \mathbf{u}} = \mathbf{0}.$$

2. The following matrix

$$\begin{bmatrix} \dfrac{\partial}{\partial \mathbf{x}}\dfrac{\partial L}{\partial \mathbf{x}} & \dfrac{\partial}{\partial \mathbf{u}}\left(\dfrac{\partial L}{\partial \mathbf{x}}\right) \\ \left[\dfrac{\partial}{\partial \mathbf{u}}\dfrac{\partial L}{\partial \mathbf{x}}\right]^T & \dfrac{\partial}{\partial \mathbf{u}}\dfrac{\partial L}{\partial \mathbf{u}} \end{bmatrix}$$

is $\begin{cases}\text{positive semidefinite for a minimum along } \mathbf{f}(\mathbf{x}, \mathbf{u}) = \mathbf{0} \\ \text{negative semidefinite for a maximum along } \mathbf{f}(\mathbf{x}, \mathbf{u}) = \mathbf{0}.\end{cases}$

A sufficient condition for a function to have a minimum (maximum) given that the first variation vanishes is that the second variation be positive (negative) where the first variation vanishes [1]. These conditions are general and need be modified only if the possibility of a singular solution exists.

Example 2.2-4. Suppose that we have a linear system represented by

$$\mathbf{f}(\mathbf{x}, \mathbf{u}) = \mathbf{A}\mathbf{x} + \mathbf{B}\mathbf{u} + \mathbf{c} = \mathbf{0}$$

and wish to find the m vector $\mathbf{u}$ which minimizes

$$J(\mathbf{x}, \mathbf{u}) = \tfrac{1}{2}\|\mathbf{u}\|_{\mathbf{R}}^2 + \tfrac{1}{2}\|\mathbf{x}\|_{\mathbf{Q}}^2$$

where $\mathbf{A}$ is an $n \times n$ matrix, $\mathbf{B}$ is an $n \times m$ matrix, $\mathbf{x}$, $\mathbf{c}$, and $\mathbf{0}$ are n vectors. $\mathbf{R}$ and $\mathbf{Q}$ are positive definite symmetric matrices of dimensionality $m \times m$ and $n \times n$.

The Lagrangian function is formed by adjoining the cost function to the given constraint via the Lagrange multiplier technique which gives us

$$L = \tfrac{1}{2}\mathbf{u}^T\mathbf{R}\mathbf{u} + \tfrac{1}{2}\mathbf{x}^T\mathbf{Q}\mathbf{x} + \boldsymbol{\lambda}^T[\mathbf{A}\mathbf{x} + \mathbf{B}\mathbf{u} + \mathbf{c}].$$

In order to minimize J, it is necessary that

$$\frac{\partial L}{\partial \mathbf{x}} = \mathbf{Q}\mathbf{x} + \mathbf{A}^T\boldsymbol{\lambda} = \mathbf{0}, \qquad \frac{\partial L}{\partial \mathbf{u}} = \mathbf{R}\mathbf{u} + \mathbf{B}^T\boldsymbol{\lambda} = \mathbf{0}$$

where $\boldsymbol{\lambda}$ is to be adjusted so that the given equality constraint is satisfied, or

$$\mathbf{A}\mathbf{x} + \mathbf{B}\mathbf{u} + \mathbf{c} = \mathbf{0}.$$

Thus we find that

$$\begin{aligned}\mathbf{u} &= -(\mathbf{R} + \mathbf{B}^T\mathbf{A}^{-T}\mathbf{Q}\,\mathbf{A}^{-1}\mathbf{B})^{-1}\,\mathbf{B}^T\mathbf{A}^{-T}\mathbf{Q}\,\mathbf{A}^{-1}\mathbf{c} \\ &= -\mathbf{R}^{-1}\mathbf{B}^T(\mathbf{A}\mathbf{Q}^{-1}\,\mathbf{A}^T + \mathbf{B}\mathbf{R}^{-1}\mathbf{B}^T)^{-1}\mathbf{c}\end{aligned}$$

is the optimum $\mathbf{u}$ vector. In order to determine whether this solution does in fact cause $J(\mathbf{x}, \mathbf{u})$ to have a minimum, we find the second variation and check the necessary condition 2 given earlier. From Eq. (2.2-9) and the specifications for this problem, we have

$$\delta^2 J = \tfrac{1}{2}[\boldsymbol{\delta}\mathbf{x}^T\ \boldsymbol{\delta}\mathbf{u}^T]\begin{bmatrix}\mathbf{Q} & \mathbf{0} \\ \mathbf{0} & \mathbf{R}\end{bmatrix}\begin{bmatrix}\boldsymbol{\delta}\mathbf{x} \\ \boldsymbol{\delta}\mathbf{u}\end{bmatrix} = \tfrac{1}{2}\,\boldsymbol{\delta}\mathbf{x}^T\mathbf{Q}\,\boldsymbol{\delta}\mathbf{x} + \tfrac{1}{2}\,\boldsymbol{\delta}\mathbf{u}^T\mathbf{R}\,\boldsymbol{\delta}\mathbf{u}.$$

For $J(\mathbf{x}, \mathbf{u})$ to have a minimum, $\delta^2 J \geq 0$, therefore, it is sufficient that $\mathbf{Q}$ and $\mathbf{R}$ be nonnegative definite. A further requirement is obtained by noting that the first variation of $\mathbf{f}(\mathbf{x}, \mathbf{u}) = \mathbf{0}$ yields $\mathbf{A}\delta\mathbf{x} + \mathbf{B}\delta\mathbf{u} = \mathbf{0}$, and it is therefore only necessary, for $\delta^2 J > 0$, that $\mathbf{R} + \mathbf{B}^T\mathbf{A}^{-T}\mathbf{Q}\mathbf{A}^{-1}\mathbf{B}$ be positive definite.

Example 2.2-5. Suppose that we wish to minimize the cost function

$$J = \tfrac{1}{2}\|\mathbf{x}\|_{\mathbf{Q}}^2$$

subject to the constraint

$$\mathbf{x} + \mathbf{b}u + \mathbf{c} = \mathbf{0}$$

where the scalar control is bounded such that $|u| \leq 1$.

This problem can be solved without the magnitude constraint on the control with the result (from the last example)

$$u = -(\mathbf{b}^T\mathbf{Q}\mathbf{b})^{-1}\mathbf{b}^T\mathbf{Q}\mathbf{c}.$$

If $|u|$ obtained from the foregoing problem is less than 1, we obtain what is called a singular solution. This is so because the L function is linear in the control variable and $\partial L/\partial u = \boldsymbol{\lambda}^T\mathbf{b} = 0$ is the equation for a stationary point which may well be a minimum. If $\mathbf{b}^T\mathbf{Q}\mathbf{b}$ is positive definite, it is at least a local minimum. If the value of u obtained is within the boundary, that value solves our problem. If the value obtained is greater in magnitude than 1, the true solution for u must be on the boundary. This type of problem is of concern in optimal control theory and will be considered in some detail for dynamic processes.

Example 2.2-6. [2] Suppose that observations of a constant vector are taken after being corrupted with noise. Symbolically, we express this as

$$\mathbf{z} = \mathbf{H}\mathbf{x} + \mathbf{v}$$

where $\mathbf{z}$ which is composed of observed numbers is an m vector, $\mathbf{H}$ is an $m \times n$ matrix, $\mathbf{x}$ is an n vector, and $\mathbf{v}$ is an m vector representing measurement noise. It is desired to obtain the best estimate of $\mathbf{x}$, denoted $\hat{\mathbf{x}}$, such that

$$J = \tfrac{1}{2}\|\mathbf{z} - \mathbf{H}\hat{\mathbf{x}}\|_{\mathbf{R}^{-1}}^2$$

is minimum where $\mathbf{R}$ is a symmetric positive definite matrix. We accomplish this by setting

$$\frac{\partial J}{\partial \hat{\mathbf{x}}} = \mathbf{H}^T\mathbf{R}^{-1}(\mathbf{z} - \mathbf{H}\hat{\mathbf{x}}) = \mathbf{0}.$$

Thus to obtain the best least-square error estimate of $\mathbf{x}$ we have

$$\hat{\mathbf{x}} = (\mathbf{H}^T\mathbf{R}^{-1}\mathbf{H})^{-1}\mathbf{H}^T\mathbf{R}^{-1}\mathbf{z}.$$

One of the simplest cases of interest occurs when we take m estimates of a scalar. In that case it is reasonable to take $\mathbf{H}$ as a unit vector of dimension m or, in other words, a column vector of 1's, and $\mathbf{R}$ as the identity matrix. For this simplest case, we have for the "best" estimate of x

$$\hat{\mathbf{x}} = \frac{\mathbf{H}^T\mathbf{z}}{m} = \frac{1}{m}\sum_{i=1}^{m} z_i$$

which is the well-known expression for the average of a number of observations.

Another interesting case occurs when we have computed $\hat{\mathbf{x}}$ for r measurements and someone gives us an additional measurement. A great deal of effort would be involved in multiplying and inverting $\mathbf{H}^T\mathbf{R}^{-1}\mathbf{H}$ if $\mathbf{H}$ is, say, a 1000 by 20 matrix. To repeat this procedure for a new 1001 by 20 matrix would probably be prohibitive of computer time, particularly if "on-line" computation is a requirement. We are thus led to seek a solution which allows us to add the new measurement without repeating the entire calculation. A method which allows us to do this is called a recursive or sequential estimation scheme. Such schemes are of considerable importance in modern system theory and will be explored in much more detail in Chapter 8.

Assume a set of measurements is represented by

$$\mathbf{z} = \mathbf{H}\mathbf{x} + \mathbf{v}$$

$$\mathbf{z} = \begin{bmatrix} z_1 \\ z_2 \\ \cdot \\ \cdot \\ \cdot \\ z_m \end{bmatrix} = \begin{bmatrix} h_{11} & h_{12} & \cdots & h_{1n} \\ h_{21} & & & \\ \cdot & & & \\ \cdot & & & \\ \cdot & & & \\ h_{m1} & & \cdots & h_{mn} \end{bmatrix} \begin{bmatrix} x_1 \\ x_2 \\ \cdot \\ \cdot \\ \cdot \\ x_n \end{bmatrix} + \begin{bmatrix} v_1 \\ v_2 \\ \cdot \\ \cdot \\ \cdot \\ v_m \end{bmatrix}$$

where $\hat{\mathbf{x}}_m$ is given by $(\mathbf{H}^T\mathbf{R}^{-1}\mathbf{H})^{-1}\mathbf{H}^T\mathbf{R}^{-1}\mathbf{z}$. Now suppose that we obtain an additional measurement such that we have

$$\begin{bmatrix} z \\ \hline \mathbf{z}_{m+1} \end{bmatrix} = \begin{bmatrix} \mathbf{H} \\ \hline \mathbf{h}^T \end{bmatrix} [\hat{\mathbf{x}}_m + \Delta\mathbf{x}] + \begin{bmatrix} \mathbf{v} \\ \hline v_{m+1} \end{bmatrix}.$$

The problem now becomes one of obtaining the best estimate of $\mathbf{x}$, $\hat{\mathbf{x}}_{m+1}$, such that

$$J = \frac{1}{2}\left\| \begin{bmatrix} \mathbf{z} \\ \hline z_{m+1} \end{bmatrix} - \begin{bmatrix} \mathbf{H} \\ \hline \mathbf{h}^T \end{bmatrix} \hat{\mathbf{x}}_{m+1} \right\|^2$$

is minimum. Following a procedure similar to the previous one, we find the best estimate of $\mathbf{x}$ is

$$\hat{\mathbf{x}}_{m+1} = \left(\begin{bmatrix} \mathbf{H} \\ \hline \mathbf{h}^T \end{bmatrix}^T \begin{bmatrix} \mathbf{H} \\ \hline \mathbf{h}^T \end{bmatrix} \right)^{-1} \begin{bmatrix} \mathbf{H} \\ \hline \mathbf{h}^T \end{bmatrix}^T \begin{bmatrix} \mathbf{z} \\ \hline z_{m+1} \end{bmatrix}$$

where for convenience we will now assume that the matrix $\mathbf{R}$ is an identity matrix. This amounts to placing equal weight on each measurement. A recursive scheme may be developed by the use of the matrix inversion lemma [2, 3], which is stated and proved in Appendix D. We recall that

$$\left\{ \begin{bmatrix} \mathbf{H} \\ \hline \mathbf{h}^T \end{bmatrix}^T \begin{bmatrix} \mathbf{H} \\ \hline \mathbf{h}^T \end{bmatrix} \right\}^{-1} = [\mathbf{H}^T\mathbf{H} + \mathbf{h}\mathbf{h}^T]^{-1}.$$

If we define

$$\mathbf{P}_m^{-1} = \mathbf{H}^T\mathbf{H}, \qquad \mathbf{P}_{m+1}^{-1} = \begin{bmatrix} \mathbf{H} \\ \hline \mathbf{h}^T \end{bmatrix}^T \begin{bmatrix} \mathbf{H} \\ \hline \mathbf{h}^T \end{bmatrix} = \mathbf{P}_m^{-1} + \mathbf{h}\mathbf{h}^T$$

then the matrix inversion lemma implies

$$\mathbf{P}_{m+1} = \mathbf{P}_m - \mathbf{P}_m\mathbf{h}[\mathbf{h}^T\mathbf{P}_m\mathbf{h} + 1]^{-1}\mathbf{h}^T\mathbf{P}_m,$$

which yields the recursion formula

$$\begin{aligned}\hat{\mathbf{x}}_{m+1} &= \mathbf{P}_{m+1}[\mathbf{H}^T\mathbf{z} + \mathbf{h}z_{m+1}] \\ &= \mathbf{P}_m\mathbf{H}^T\mathbf{z} + \mathbf{P}_m\mathbf{h}z_{m+1} - \mathbf{P}_m\mathbf{h}[\mathbf{h}^T\mathbf{P}_m\mathbf{h} + 1]^{-1}\mathbf{h}^T\mathbf{P}_m[\mathbf{H}^T\mathbf{z} + \mathbf{h}z_{m+1}] \\ &= \hat{\mathbf{x}}_m + \mathbf{P}_m\mathbf{h}[\mathbf{h}^T\mathbf{P}_m\mathbf{h} + 1]^{-1}[z_{m+1} - \mathbf{h}^T\hat{\mathbf{x}}_m].\end{aligned}$$

Thus the new estimate is equal to the old plus a linear correction term based on the new data and the old $\mathbf{P}_m$ only. For m estimates of a scalar x with H as a unit vector of dimension m, we have

$$\mathbf{P}_m^{-1} = m, \qquad \mathbf{P}_{m+1} = \frac{1}{m+1}, \qquad \mathbf{x}_m = \frac{1}{m}\sum_{i=1}^{m} z_i$$

$$\hat{\mathbf{x}}_{m+1} = \hat{\mathbf{x}}_m + \frac{1}{m+1}[z_{m+1} - \hat{\mathbf{x}}_m] = \hat{\mathbf{x}}_m\left[\frac{m}{m+1}\right] + \frac{z_{m+1}}{m+1}$$

which is, of course, the expected answer in this simple case.

2.3 Nonlinear programming

In this section, we briefly discuss some of the essential aspects of a type of optimization problem called the *nonlinear programming* (*NLP*) *problem.* Consider the following NLP problem formulation: determine a vector in R^n (if it exists) such that the function $\theta: R^n \longrightarrow R$ is minimized with respect to all vectors in R^n satisfying $\mathbf{\Lambda}(\mathbf{x}) \leq \mathbf{0}$, $\mathbf{\Lambda}: R^n \longrightarrow R^m$. We note that this problem generalizes the problems formulated in Secs. 2.1 and 2.2. Thus, the essential difference between the NLP problem defined above and the classical minimization problem, either unconstrained or having equality constraints, is the presence of inequalities. Such inequalities, illustrated in Fig. 2.3-2, play a fundamental role in nonlinear programming and are studied in detail in the nonlinear programming literature [4]. We now consider an example of an NLP problem.

Example 2.3-1. Again, consider extremizing the function defined in Example 2.1-1, where we now suppose that the allowable range of $\mathbf{x}$ is constrained such that $|x_i| \leq \frac{1}{2}$, $i = 1, 2$. We wish to determine the value of $\mathbf{x}$ that maximizes f with respect to the set of allowable values of $\mathbf{x}$. From Fig. 2.3-1, it is apparent for this simple problem that f has an extremum on the boundary of the allowable set of vectors and the maximum is attained at $x^T = [\frac{1}{2}, \frac{1}{2}]$.

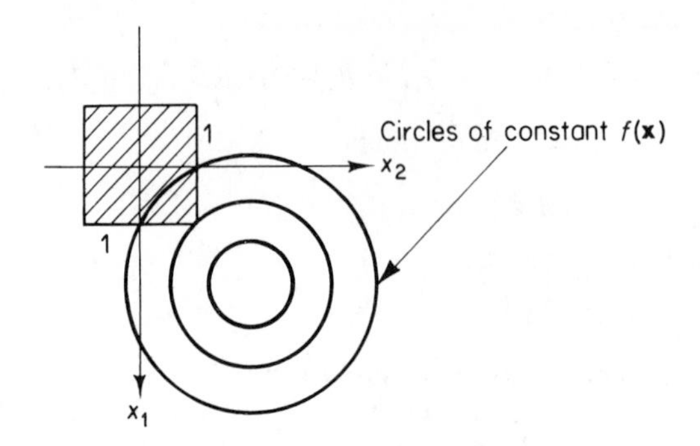

Fig. 2.3-1 Top view of f in Example 2.1-1 showing the region defined by $|x_i| \leq \frac{1}{2}$, $i = 1, 2$.

As we have seen, ordinary calculus methods may be used to find the extremum of unconstrained functions. If ordinary calculus is applied to extremize θ, and if the resulting optimum vector $\mathbf{x}$ lies entirely within the constraint set $\Lambda_i \leq 0$, and if $x_i \geq 0$, then that value of $\mathbf{x}$ solves the optimization problem with the constraint. If the optimum value of $\mathbf{x}$ computed by extremizing θ is outside the constraint set $\Lambda \leq 0$, then the optimum value of $\mathbf{x}$ lies on the boundary of the constraint set. If we knew which one of the m constraints Λ determined the optimum, then we could apply the Lagrange multiplier method, use an equality sign for that particular constraint, and ignore the other constraints since the optimum $\mathbf{x}$ will be on the boundary of one of the known m inequality constraints. In general, we find it necessary to exploit each of the inequality constraints to determine which one to use. It is possible that more than one of the m inequality constraints will determine the optimum $\mathbf{x}$ as illustrated in Fig. (2.3-2). We should remark that, in the typical nonlinear programming problem, the functions Λ are convex, which insures that the possible region for an optimum $\mathbf{x}$ is also convex. Also, θ is convex if minimization is required and concave if maximization is required. This requires that any local optimum is a global optimum of the cost function in the possible region of a constraint Λ_i [1].

A special type of NLP problem is the *linear programming* (*LP*) *problem*, where θ and Λ are linear in $\mathbf{x}$, and $\mathbf{x}$ is constrained to have nonnegative elements. For the LP problem, we are assured by our geometric insight that an optimum admissible value lies on a vertex of the (convex) set of admissible or feasible vectors. The well-known simplex method [5, 6] is an organized technique for proceeding from vertex to vertex to determine an optimizing vertex. By the nature of the problem formulation, the number of vertices to be considered is finite, and hence the simplex algorithm converges in a finite number of steps.

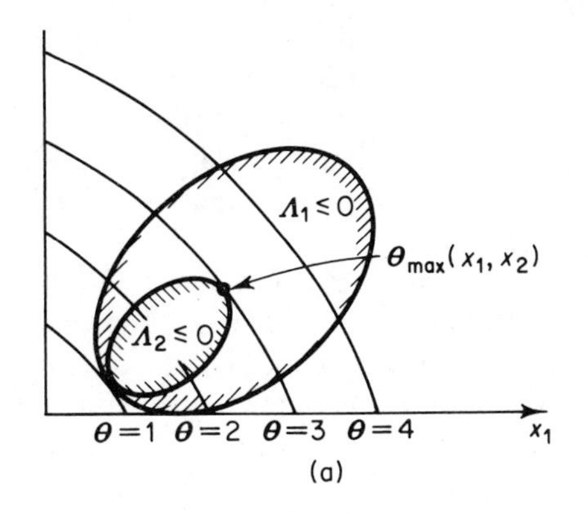

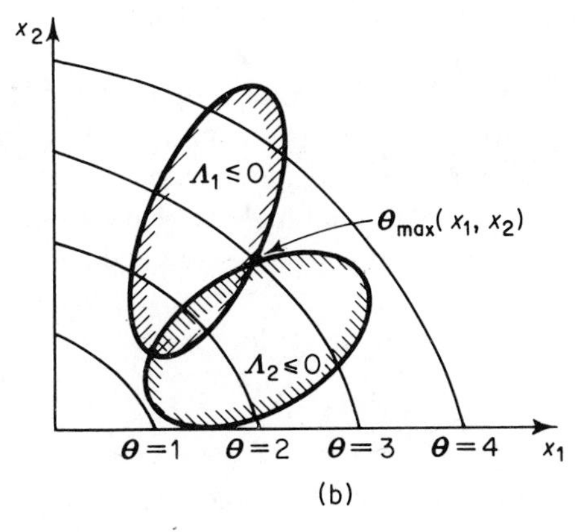

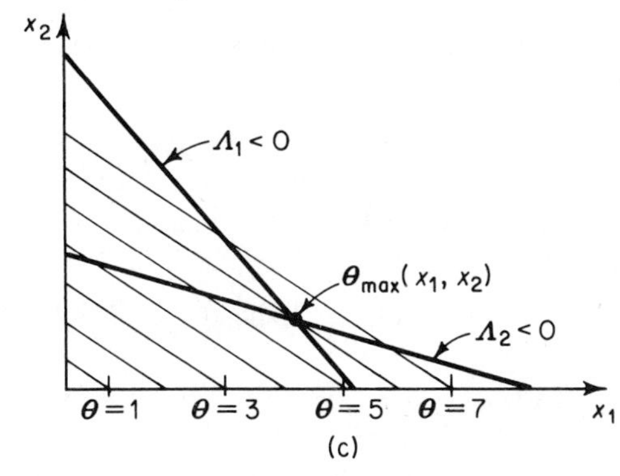

Fig. 2.3-2 Illustrations of nonlinear programming (a, b) and linear programming (c).

Although there may appear at first glance to be only a remote relationship between the NLP problem and the optimal control problems to be formulated and analyzed throughout the remainder of this text, the relationship is quite fundamental and subject to much theoretical and computational exploitation [7, 8, 9]. An examination of the theoretical relationship [10, 11, 12] is beyond the scope of this book; however, the computational relationship will be examined in Chapter 10.

The extrema-finding techniques of this chapter, although quite sufficient for many different situations, will not, in general, allow the solution to many problems associated with control systems. Whereas the previously discussed techniques deal with methods for extremizing functions of one or several independent variables, in control-system design we are typically concerned with extremizing certain types of functions whose independent variables are actually other functions. This type of function is called a *functional*. Although, as we might expect, many of the basic approaches for extremizing functionals are similar to those for extremizing functions, the end results are sometimes quite different. The solution to a given problem in extremizing a given function of one variable is, perhaps, a number associated with a coordinate point, while the analogous solution to a functional problem is a number associated with a function. The body of mathematics developed for extremizing functionals is variational calculus. This subject is at the very heart of optimal control theory and is a subject that we explore in some detail throughout the remainder of this text.

References

1. SAATY, T.L., and J. BRAM, *Nonlinear Mathematics*. McGraw-Hill Book Co., New York, 1964.

2. HO, Y.C., "Method of Least Squares and Optimal Filtering Theory." Rand Memo RM 3329 PR, 1962.

3. LEE, R.C.K., *Optimal Estimation Identification and Control*. Technology Press, Cambridge, Mass., 1964.

4. MANGASARIAN, O.L., *Nonlinear Programming*. McGraw-Hill Book Co., New York, 1969.

5. DANTZIG, G.B., *Linear Programming and Extensions*. Princeton University Press, Princeton, New Jersey, 1963.

6. SPIVEY, W.A., and THRALL, R.M., *Linear Optimization*. Holt, New York, 1970.

7. TABAK, D., and KUO, B.C., *Optimal Control by Mathematical Programming*. Prentice-Hall, Inc., Englewood Cliffs, New Jersey, 1971.

8. CANON, M.D., CULLUM, C.D., JR., and POLAK, E., *Theory of Optimal Control and Mathematical Programming*. McGraw-Hill Book Co., New York, 1970.

9. LUENBERGER, D.G., *Optimization by Vector Space Methods.* Wiley, New York, 1969.

10. MANGASARIAN, O.L., and FROMOVITZ, S., "A Maximum Principle in Mathematical Programming," A.V. Balakrishnan and L.W. Neustadt, eds., *Mathematical Theory of Control.* Academic Press, New York, 1967, pp. 85–95.

11. NEUSTADT, L.W., "An Abstract Variational Theory with Applications to a Broad Class of Optimization Problems. I. General Theory." *SIAM J. Control,* 4, (1966), 505–27.

12. NEUSTADT, L.W., "An Abstract Variational Theory with Applications to a Broad Class of Optimization Problems. II. Applications." *SIAM J. Control,* 5, (1967), 90–137.

Problems

1. Find u such that

$$J = x^2 + u^2$$

is minimized subject to the equation

$$xu = 1.$$

Use the Lagrange multiplier technique as well as the basic method.

2. Discuss the singular solution problem where $\mathbf{x}$ is a two vector.

3. Find $\hat{x}_6$ for a set of measurements where $\mathbf{z} = \mathbf{Hx}$, where

$$\mathbf{z} = \begin{bmatrix} 1.01 \\ 2.03 \\ 3.00 \\ 3.05 \\ 1.95 \\ 0.97 \end{bmatrix} \qquad \mathbf{H} = \begin{bmatrix} 1 & 0 \\ 0 & 1 \\ 1 & 1 \\ 1 & 1 \\ 0 & 1 \\ 1 & 0 \end{bmatrix}.$$

4. Now suppose that an additional measurement

$$z_7 = 3.0; \qquad \mathbf{h}^T = [1, 1]$$

is taken. Compute $\hat{x}_7$ by the smoothing method and the matrix inversion lemma method. Compare the effort involved via each method.

5. Verify the matrix inversion lemma if

$$\mathbf{P}_{r+1}^{-1} = \mathbf{P}_r^{-1} + \mathbf{h}\mathbf{h}^T$$

$$\mathbf{P}_{r+1} = \mathbf{P}_r - \mathbf{P}_r\mathbf{h}(\mathbf{h}^T\mathbf{P}_r\mathbf{h} + 1)^{-1}\mathbf{h}^T\mathbf{P}_r$$

by showing that

$$\mathbf{P}_{r+1}^{-1}\mathbf{P}_{r+1} = \mathbf{I}.$$

6. From Eqs. (2.2-7) and (2.2-8) calculate the third variation of the Lagrangian.

7. Find the maximum value of

$$\theta(\mathbf{x}) = x_1^2 + x_2^2, \qquad x_1 \geq 0, \qquad x_2 \geq 0$$

subject to the inequality constraints

$$(x_1 - 4)^2 + x_2^2 \leq 1$$
$$(x_1 - 1)^2 + x_2^2 \leq 4.$$

8. Find the maximum value of

$$J = x_1 + x_2, \qquad x_1 \geq 0, \qquad x_2 \geq 0$$

subject to the constraints

$$x_1 + \tfrac{1}{2}x_2 \leq 1$$
$$\tfrac{1}{2}x_1 + x_2 \leq 2.$$

9. Two alternate expressions were developed for the optimum $\mathbf{u}$ vector of Eq. (2.2-1). Show that the two expressions are equivalent and that the first solution will be easier to implement computationally if the dimension of $\mathbf{u}$ is lower than that of $\mathbf{x}$.

Variational calculus and continuous optimal control 3

In this chapter we introduce the subject of variational calculus through a derivation of the Euler-Lagrange equations and associated transversality conditions. The existence of the definite integrals defining the cost function is assumed. This chapter deals with most of the basic concepts necessary for solving the types of variational problems commonly classified as control-system problems. Several such examples of continuous control problems are solved. Many of the restrictions posed here are removed in the following chapter. Although there is a vast literature devoted to the subject of variational calculus, we present here those topics most related to the subject of continuous time optimal control. Of the many references available, [1] through [11] are especially pertinent to our efforts.

3.1 Dynamic optimization without constraints

We now consider the problem of selecting a continuously differentiable function $x:[t_o, t_f] \longrightarrow R$ to minimize the cost function

$$J(x) = \int_{t_o}^{t_f} \phi[x(t), \dot{x}(t), t]\, dt \qquad (3.1\text{-}1)$$

with respect to the set of all real-valued, continuously differentiable functions on the interval $[t_o, t_f]$. Such functions will be referred to as *admissible trajec-*

tories. Throughout, we assume ϕ is continuous in x, $\dot{x}$, and t and has continuous partial derivatives with respect to x and $\dot{x}$.

It is of interest to note that the cost function presented in Eq. (3.1-1), called the Lagrange form, is equivalent under certain smoothness assumptions to several other useful cost function descriptions, two of which are the Bolza form

$$J'(x) = \theta[x(t_f), t_f] + \int_{t_o}^{t_f} \Lambda[x(t), \dot{x}(t), t]\, dt$$

and the Mayer form

$$J''(x) = \sigma[x(t_f), t_f].$$

The Mayer form, which can be expressed so that σ can also depend on $x(t_o)$, is of particular theoretical interest due to its inherent simplicity. Clearly, both the Mayer and Lagrange forms are special cases of the Bolza form. Letting $\phi = \dot{\theta} + \Lambda$, we note that the Bolza form can be equivalently expressed in the Lagrange form under the assumption that $\theta[x(t_o), t_o]$ is fixed. To show that the Bolza form can also be expressed in the Mayer form, one can augment the state space of the Bolza problem with the variable $\dot{x}^0(t) = \Lambda\,[x(t), \dot{x}(t), t]$, where $x^0(t_o) = 0$, and define $\sigma[x(t_f), t_f] = \theta[x(t_f), t_f] + x^0(t_f)$.

We now develop necessary conditions, called the Euler-Lagrange equation and the transversality conditions, for an admissible trajectory to be optimal. Let $\hat{x}$ be an admissible optimal trajectory. Consider an admissible, but not necessarily optimal, trajectory x, where for all $t \in [t_o, t_f]$

$$x(t) = \hat{x}(t) + \epsilon\eta(t) \tag{3.1-2}$$

where $\eta(t)$ is a variation in $x(t)$ and ϵ is a small number. A plot of $J\langle x\rangle$ versus ϵ for various choices of $\eta(t)$ might appear as shown in Fig. 3.1-1. It is obvious that as $\epsilon = 0$, all curves are minimum since

$$\hat{x}(t) = x(t)\,|_{\epsilon=0}.$$

Thus on the extremals we have

$$\left.\frac{\partial J\langle x\rangle}{\partial \epsilon}\right|_{\epsilon=0} = 0 \tag{3.1-3}$$

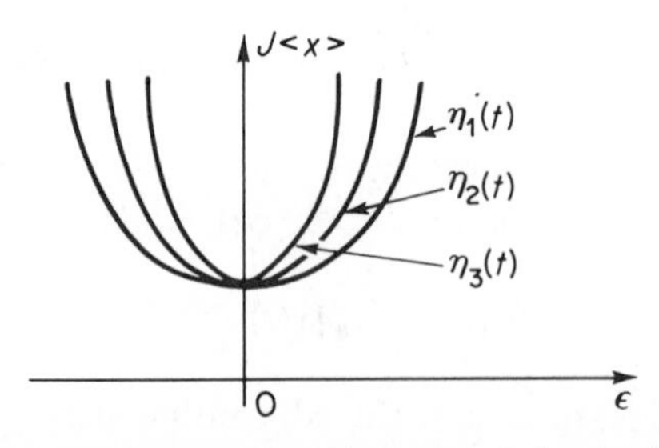

Fig. 3.1-1 Minimization problem of variational calculus.

independent of the value of $\eta(t)$ chosen. Strictly speaking, the solution obtained from Eq. (3.1-2) could cause $J\langle x\rangle$ to have a maximum or minimum or be a stationary point. The condition for a minimum is that $\partial^2 J/\partial\epsilon^2$ be positive at $\epsilon = 0$ independent of $\eta(t)$. However, in most physical problems, it is apparent that if a solution to Eq. (3.1-3) exists, it will be a solution which minimizes (maximizes) the integral, $J\langle x\rangle$, as desired. Now we can extremize Eq. (3.1-1) by using Eqs. (3.1-2) and (3.1-3). By differentiating Eq. (3.1-2) with respect to t, we obtain

$$\dot{x}(t) = \dot{\hat{x}}(t) + e\dot{\eta}(t). \tag{3.1-4}$$

If we substitute Eqs. (3.1-2) and (3.1-4) into the given functional (3.1-1), we then have

$$J\langle x\rangle = \int_{t_0}^{t_f} \phi[\hat{x}(t) + \epsilon\eta(t), \dot{\hat{x}}(t) + \epsilon\dot{\eta}(t), t]\, dt. \tag{3.1-5}$$

We should note that

$$\lim_{\epsilon=0} J(x) = J(\hat{x}), \qquad \lim_{\epsilon=0} x(t) = \hat{x}(t).$$

Therefore, to find the extremals of $J\langle x\rangle$ we now use Eq. (3.1-3)†

$$\left.\frac{\partial J\langle x\rangle}{\partial\epsilon}\right|_{\epsilon=0} = \int_{t_0}^{t_f} \left\{\eta(t)\frac{\partial\phi(\hat{x}, \dot{\hat{x}}, t)}{\partial\hat{x}} + \dot{\eta}(t)\frac{\partial\phi(\hat{x}, \dot{\hat{x}}, t)}{\partial\dot{\hat{x}}}\right\} dt = 0 \tag{3.1-6}$$

or

$$0 = \int_{t_0}^{t_f} \eta(t)\frac{\partial\phi(\hat{x}, \dot{\hat{x}}, t)}{\partial\hat{x}}\, dt + \int_{t_0}^{t_f} \dot{\eta}(t)\frac{\partial\phi(\hat{x}, \dot{\hat{x}}, t)}{\partial\dot{\hat{x}}}\, dt. \tag{3.1-7}$$

After simplification Eq. (3.1-7) becomes,‡

$$0 = \int_{t_0}^{t_f} \eta(t)\left[\frac{\partial\phi}{\partial\hat{x}} - \frac{d}{dt}\frac{\partial\phi}{\partial\dot{\hat{x}}}\right] dt + \left.\frac{\partial\phi}{\partial\dot{\hat{x}}}\eta(t)\right|_{t_0}^{t_f}. \tag{3.1-8}$$

---

†The following is given without proof: If $u = f(x, y, z, \ldots)$ is a function of several variables, each of which is a differentiable function of $r, v, w, \ldots$, then u as a function of these new independent variables, is differentiable, and the following chain rule applies

$$\frac{\partial u}{\partial r} = \frac{\partial u}{\partial x}\frac{\partial x}{\partial r} + \frac{\partial u}{\partial y}\frac{\partial y}{\partial r} + \cdots$$

$$\frac{\partial u}{\partial v} = \frac{\partial u}{\partial x}\frac{\partial x}{\partial v} + \frac{\partial u}{\partial y}\frac{\partial y}{\partial v} + \cdots.$$

‡Applying the formula for integration by parts, which is

$$\int_a^b u\, dv = uv\Big|_a^b - \int_a^b v\, du$$

by letting

$$u = \frac{\partial\phi}{\partial\dot{\hat{x}}}, \qquad dv = \dot{\eta}(t)\, dt$$

$$du = \frac{d}{dt}\frac{\partial\phi}{\partial\dot{\hat{x}}}\, dt, \qquad v = n(t)$$

we have

$$\int_{t_0}^{t_f} \dot{\eta}(t)\frac{\partial\phi}{\partial\dot{\hat{x}}}\, dt = \eta(t)\left.\frac{\partial\phi}{\partial\dot{\hat{x}}}\right|_{t_0}^{t_f} - \int_{t_0}^{t_f} \eta(t)\frac{d}{dt}\frac{\partial\phi}{\partial\dot{\hat{x}}}\, dt.$$

Clearly, if

$$\frac{\partial \phi}{\partial \hat{x}} - \frac{d}{dt}\frac{\partial \phi}{\partial \dot{\hat{x}}} = 0 \tag{3.1-9}$$

$$\frac{\partial \phi}{\partial \dot{\hat{x}}}\eta(t) = 0, \qquad \text{for} \quad t = t_o, t_f \tag{3.1-10}$$

along the optimal trajectory, Eq. (3.1-8) is satisfied. Since Eq. (3.1-8) must hold for all admissible variations, the converse also holds; i.e., if Eq. (3.1-8) holds independent of η, then Eq. (3.1-9) and Eq. (3.1-10) hold. For the case where $x(t_o)$ and $x(t_f)$ are fixed, and hence $\eta(t_o) = \eta(t_f) = 0$, we justify Eq. (3.1-9) by the following lemma.

If $x(t)$ is continuous on the closed interval $t \in [t_1, t_2]$ and if $\int_{t_1}^{t_2} x(t)\eta(t)\,dt = 0$ for every $\eta(t)$ contained in $[t_1, t_2]$ such that $\eta(t_1) = \eta(t_2) = 0$, then $x(t) = 0$ for all t in $[t_1, t_2]$. Proof of this lemma is given in references [1, 9].

These two very important relationships form a good foundation for solving variational problems. Equation (3.1-9) is commonly known as the Euler-Lagrange equation, and Eq. (3.1-10) is the associated transversality condition. These equations specify a two-point boundary value differential equation which, when solved, determines a candidate for an optimal trajectory in terms of a known ϕ.

The above derivation of necessary conditions suffers from the implicit assumption that

$$\frac{d}{dt}\frac{\partial \phi}{\partial \dot{x}}$$

is continuous along the optimal trajectory. An alternative derivation can be given which avoids this drawback [9]. Equation (3.1-9) can also be expressed in the integral form

$$\frac{\partial \phi}{\partial \dot{x}} = \int_{t_o}^{t} \frac{\partial \phi}{\partial x}\,dt + c$$

for constant c [3].

In summary, for the fixed final time optimal control problem defined above, if $\hat{x}$ is an admissible optimal trajectory, then the Euler-Lagrange equation (3.1-9) and the transversality conditions, Eq. (3.1-10), must be satisfied for $\hat{x}$. We note that Eqs. (3.1-9) and (3.1-10) are in general only necessary conditions; therefore, a trajectory satisfying these conditions is not necessarily optimal.

3.2 Transversality conditions

We now examine the transversality conditions in some detail. Equation (3.1-10) must be examined in conjunction with given boundary conditions on the trajectory. Boundary conditions, i.e., restrictions on $x(t)$ at times t_o and t_f, are usually of the following form: the value of $x(t_b)$, where t_b is either t_o or t_f or both, is fixed, totally free, or constrained. If $x(t_b)$ is fixed, then $\hat{x}(t_b) = x(t_b)$ for all admissible trajectories x; therefore, $\eta(t_b) = 0$, and no restriction is placed on the value of $\partial\phi/\partial\dot{x}$ by Eq. (3.1-10). If $x(t_b)$ is unspecified, there is no restriction placed on the independence of $\eta(t_b)$, and hence it follows from Eq. (3.1-10) that $\partial\phi/\partial\dot{x}$ must equal zero at t_b. Combinations of these cases are examined graphically in Fig. 3.2-1. When we examine the vector version of the optimal control problem considered above, we will investigate the case where partial constraints are placed on $\eta(t_b)$. When the

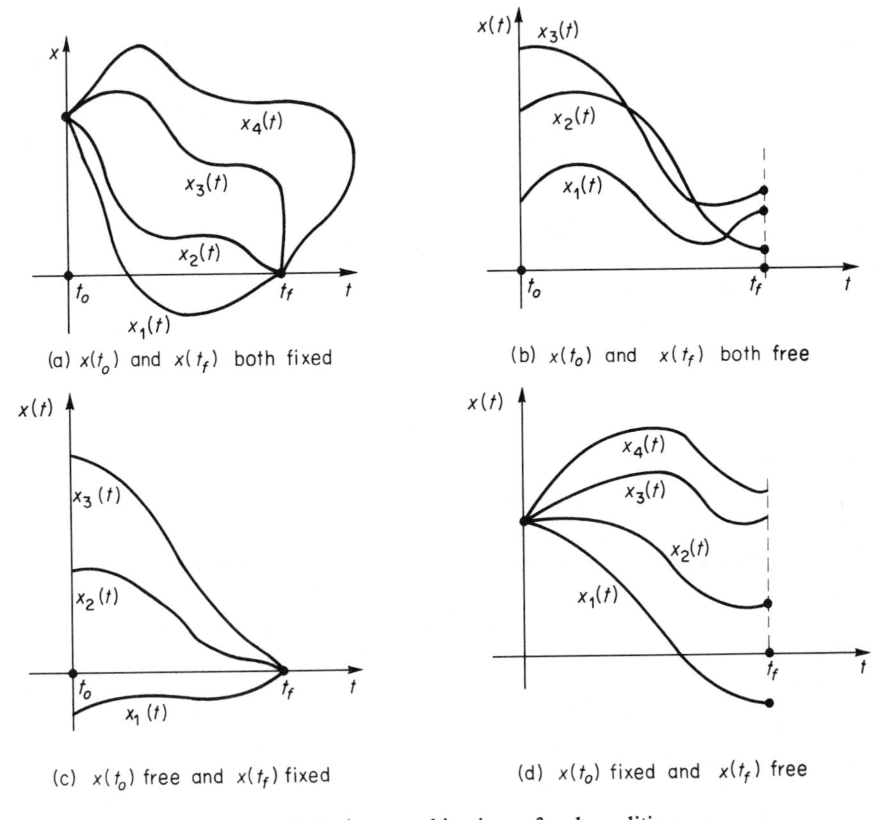

Fig. 3.2-1 Various combinations of end conditions.

values of both boundary conditions $x(t_o)$ and $x(t_f)$ are dependent, e.g., $x(t_o) = x(t_f)$, $\eta(t_o)$ and $\eta(t_f)$ will not necessarily be totally independent and Eq. (3.1-10) will therefore not necessarily hold. (Recall that Eq. (3.1-8) implies Eq. (3.1-10) due to the fact that $\eta(t_o)$ and $\eta(t_f)$ are allowed to be independent of each other.) For example, let $x(t_o)$ and $x(t_f)$ be such that $F[x(t_o), x(t_f)] = F[\hat{x}(t_o) + \epsilon\eta(t_o), \hat{x}(t_f) + \epsilon\eta(t_f)] = 0$. Assuming first partial derivatives of F exist and are continuous at $\hat{x}(t_o)$ and $\hat{x}(t_f)$, it is easily shown that

$$\frac{\partial F}{\partial x(t_o)}[\hat{x}(t_o), \hat{x}(t_f)]\eta(t_o) + \frac{\partial F}{\partial x(t_f)}[\hat{x}(t_o), \hat{x}(t_f)]\eta(t_f) = 0. \tag{3.2-1}$$

Since $\eta(t_o)$ and $\eta(t_f)$ are no longer independent, Eq. (3.1-8) implies

$$\left.\frac{\partial \phi}{\partial \dot{x}}\right|_{t=t_f} \eta(t_f) - \left.\frac{\partial \phi}{\partial \dot{x}}\right|_{t=t_o} \eta(t_o) = 0. \tag{3.2-2}$$

Combining Eqs. (3.2-1) and (3.2-2), we find that

$$\eta(t_f)\left(\left.\frac{\partial \phi}{\partial \dot{x}}\right|_{t=t_f} + \frac{a}{b}\left.\frac{\partial \phi}{\partial \dot{x}}\right|_{t=t_o}\right) = 0 \tag{3.2-3}$$

where

$$a = \frac{\partial F}{\partial x(t_o)}[\hat{x}(t_o), \hat{x}(t_f)] \quad \text{and} \quad b = \frac{\partial F}{\partial x(t_f)}[\hat{x}(t_o), \hat{x}(t_f)].$$

Since $\eta(t_f)$ can now be chosen arbitrarily, Eq. (3.2-3) implies

$$\left.\frac{\partial \phi}{\partial \dot{x}}\right|_{t=t_f} + \frac{a}{b}\left.\frac{\partial \phi}{\partial \dot{x}}\right|_{t=t_o} = 0,$$

which is our transversality condition for a problem having $x(t_o)$ and $x(t_f)$ constrained such that

$$F[x(t_o), x(t_f)] = 0.$$

3.3 Sufficient conditions for (weak) extrema

Thus far, we have only considered one condition of optimality, the condition that the first derivative of J with respect to ϵ for $\epsilon = 0$ must necessarily vanish along an optimal trajectory. In this section we determine sufficient conditions for a trajectory to be optimal by examining the second derivative of J with respect to ϵ for $\epsilon = 0$. Simple examples are used to help illustrate application of the second derivative. Throughout this section we assume that second partial derivatives of ϕ in $\dot{x}$ and x exist and are continuous.

To establish the nature of an extremum, it is necessary to obtain $\partial^2 J/\partial \epsilon^2$ evaluated at $\epsilon = 0$ from Eq. (3.1-1) under the conditions of Eq. (3.1-2).

This is

$$\left.\frac{\partial^2 J\langle x\rangle}{\partial \epsilon^2}\right|_{\epsilon=0} = \int_{t_0}^{t_f} \left\{\eta^2 \frac{\partial^2 \phi(\hat{x}, \dot{\hat{x}}, t)}{\partial \hat{x}^2} + 2\dot{\eta}\eta \frac{\partial^2 \phi(\hat{x}, \dot{\hat{x}}, t)}{\partial \hat{x}\, \partial \dot{\hat{x}}} + \dot{\eta}^2 \frac{\partial^2 \phi(\hat{x}, \dot{\hat{x}}, t)}{\partial \dot{\hat{x}}^2}\right\} dt. \tag{3.3-1}$$

Applying integration by parts and the transversality conditions, Eq. (3.1-10), we have

$$2\int_{t_0}^{t_f} \eta\dot{\eta} \frac{\partial^2 \phi(\hat{x}, \dot{\hat{x}}, t)}{\partial \hat{x}\, \partial \dot{\hat{x}}}\, dt = -\int_{t_0}^{t_f} \left\{\frac{d}{dt} \frac{\partial^2 \phi(\hat{x}, \dot{\hat{x}}, t)}{\partial \hat{x}\, \partial \dot{\hat{x}}}\right\} \eta^2\, dt. \tag{3.3-2}$$

Thus the second variation of J becomes

$$\left.\frac{\partial^2 J\langle x\rangle}{\partial \epsilon^2}\right|_{\epsilon=0} = \int_{t_0}^{t_f} \left\{\eta^2 \left[\frac{\partial^2 (\hat{x}, \dot{\hat{x}}, t)}{\partial \hat{x}^2} - \frac{d}{dt} \frac{\partial^2 \phi(\hat{x}, \dot{\hat{x}}, t)}{\partial \hat{x}\, \partial \dot{\hat{x}}}\right] + \dot{\eta}^2 \frac{\partial^2 \phi(\hat{x}, \dot{\hat{x}}, t)}{\partial \dot{\hat{x}}^2}\right\} dt. \tag{3.3-3}$$

To establish a minimum (maximum) of J, the first necessary condition is that $\partial J/\partial \epsilon = 0$ at $\epsilon = 0$ independently of the variation $\eta(t)$. The second necessary condition for a minimum (maximum) is that the second derivative of J with respect to ϵ, evaluated at $\epsilon = 0$, be equal to or greater than (equal to or less than) zero. Sufficient conditions for a weak minimum (maximum) require that the derivative be positive (negative). All of this must, of course, be true independent of the variation $\eta(t)$ and need only be true along the optimal "trajectory," $\hat{x}(t)$.

We can rewrite Eq. (3.3-1) as the quadratic form integral

$$\left.\frac{\partial^2 J(x)}{\partial \epsilon^2}\right|_{\epsilon=0} = \int_{t_0}^{t_f} [\eta(t)\dot{\eta}(t)] \begin{bmatrix} \frac{\partial^2 \phi(\hat{x}, \dot{\hat{x}}, t)}{\partial \hat{x}^2} & \frac{\partial^2 \phi(\hat{x}, \dot{\hat{x}}, t)}{\partial \hat{x}\, \partial \dot{\hat{x}}} \\ \frac{\partial^2 \phi(\hat{x}, \dot{\hat{x}}, t)}{\partial \hat{x}\, \partial \dot{\hat{x}}} & \frac{\partial^2 \phi(\hat{x}, \dot{\hat{x}}, t)}{\partial \dot{\hat{x}}^2} \end{bmatrix} \begin{bmatrix} \eta(t) \\ \dot{\eta}(t) \end{bmatrix} dt. \tag{3.3-4}$$

If the matrix in this expression is at least positive (negative) semidefinite, we have certainly established a minimum (maximum). Alternately, from Eq. (3.3-3) we are assured that the second derivative is equal to or greater than zero if

$$\frac{\partial^2 \phi(\hat{x}, \dot{\hat{x}}, t)}{\partial \hat{x}^2} - \frac{d}{dt}\left[\frac{\partial^2 \phi(\hat{x}, \dot{\hat{x}}, t)}{\partial \hat{x}\, \partial \dot{\hat{x}}}\right] \geq 0 \tag{3.3-5}$$

and

$$\frac{\partial^2 \phi(\hat{x}, \dot{\hat{x}}, t)}{\partial \dot{\hat{x}}^2} \geq 0.\dagger \tag{3.3-6}$$

For many problems in which we will have interest, the foregoing conditions are fulfilled, and we can establish necessary and sufficient conditions for a minimum. It is still possible, however, for Eq. (3.3-1) or Eq. (3.3-2) to be

†Additionally, (3.3-6) can be shown to be a necessary condition, called the Legendre condition [10]. Other necessary conditions are discussed in [3].

greater than zero even if the requirements of Eqs. (3.3-4), (3.3-5), and (3.3-6) are not satisfied, since $\eta(t)$ and $\dot{\eta}(t)$ are not independent of one another. Complete exploitation of this point is beyond the intent of this chapter. Chapters 5 and 6 of Reference [1] provide an excellent and readable discussion of the necessary and sufficient conditions for a minimum. We will return again to this point in Chapter 4. We must again emphasize here that we are establishing conditions for a relative extremum, sometimes called a weak extremum, which may or may not be an absolute extremum. In Sec. 4.1 we will discuss some requirements for an absolute or strong extremum.

Example 3.3-1. We desire to find the curve with minimum arc length between the point $x(0) = 1$ and the line $t_f = 2$.

The first step toward solving this problem is to formulate the functional $J\langle x\rangle$. If we define the differential arc length as ds, the functional we desire to minimize is

$$J\langle x\rangle = \int_0^2 ds$$

with associated boundary conditions

$$x(t = 0) = 1, \qquad x(t = 2) = \text{open}.$$

Noting that for a differential arc length

$$(ds)^2 = (dx)^2 + (dt)^2$$

we have

$$\frac{ds}{dt} = [1 + \dot{x}^2]^{1/2}.$$

By substituting into the given cost function, we obtain

$$J\langle x\rangle = \int_0^2 [1 + \dot{x}^2]^{1/2}\, dt.$$

Upon referring back to the functional defined in Eq. (3.1-1), we see that

$$\phi(x, \dot{x}, t) = [1 + \dot{x}^2]^{1/2}.$$

The Euler-Lagrange equation for this problem is therefore

$$\frac{\partial \phi}{\partial \hat{x}} - \frac{d}{dt}\frac{\partial \phi}{\partial \dot{\hat{x}}} = 0,$$

and thus we obtain

$$\frac{-d}{dt}\left[\frac{\dot{\hat{x}}}{(1 + \dot{\hat{x}}^2)^{1/2}}\right] = 0.$$

Upon integrating, we obtain

$$\frac{\dot{\hat{x}}}{(1 + \dot{\hat{x}}^2)^{1/2}} = c = \text{constant}, \qquad \dot{\hat{x}}^2 = \frac{c^2}{1 - c^2} = a^2.$$

Thus we see that the extremal curve is given by

$$\hat{x}(t) = at + b.$$

Therefore, the shortest distance between a point and a straight line is another straight line.

We obtain the particular solution by properly applying the transversality equation to the given boundary conditions. We note that this problem has a fixed beginning–variable terminal point. Thus, $x(t_0) = x(0) = 1$ and

$$\frac{\partial \phi}{\partial \dot{x}} = 0 = \frac{\dot{x}}{[1 + \dot{x}^2]^{1/2}}, \qquad \text{at} \quad t = 2$$

or $\dot{x} = 0$ at $t = 2$.

Differentiating the solution for $\hat{x}$ with respect to t, we have $\dot{\hat{x}} = a$, and, using the transversality conditions, we obtain $a = 0$ and $b = 1$. Therefore, the extremal curve satisfying the given boundary condition and minimizing the given arc length is $x = 1$.

To mathematically demonstrate that we have obtained a minimum rather than a maximum or stationary point, it is necessary to show that the second variation, represented by Eq. (3.3-3), is greater than zero. The pertinent terms in Eq. (3.3-3) are, for this example,

$$\frac{\partial^2 \phi}{\partial \hat{x} \partial \dot{\hat{x}}} = 0, \qquad \frac{\partial^2 \phi}{\partial \dot{\hat{x}}^2} = \frac{1}{(1 + \dot{\hat{x}}^2)^{3/2}}.$$

Since $\dot{\hat{x}} = 0$ is the extremal solution, $\partial^2 \phi / \partial \dot{\hat{x}}^2$ is always greater than zero. Thus the second variation is greater than zero, and we have indeed established a minimum. Physically this was, of course, evident from the start.

Example 3.3-2. We desire to find the equation of the curve which minimizes the functional (boundary conditions unspecified)

$$J\langle x \rangle = \int_0^2 [\tfrac{1}{2}\dot{x}^2 + x\dot{x} + \dot{x} + x]\, dt.$$

The Euler-Lagrange equation for this problem is

$$\dot{x} + 1 - \dot{x} - \ddot{x} = 0 = 1 - \ddot{x}.$$

By integrating directly, we obtain the solution to this equation:

$$x(t) = \frac{t^2}{2} + C_1 t + C_2.$$

To determine C_1 and C_2 we must now apply the transversality equation to the given boundary conditions. Since this is a variable beginning–terminal point problem,

$$\frac{\partial \phi}{\partial \dot{x}} = \dot{x} + x + 1 = 0, \qquad \text{for} \quad t = 0, 2.$$

Therefore, from the solution for x and its derivative, we have

$$\frac{\partial \phi}{\partial \dot{x}} = t + C_1 + \frac{t^2}{2} + C_1 t + C_2 + 1 = 0, \qquad \text{for} \quad t = 0, 2.$$

We can now solve for C_1 and C_2 from the simultaneous equations

$$C_1 + C_2 = -1, \qquad 3C_1 + C_2 = -5$$

to obtain $C_1 = -2$ and $C_2 = 1$. Therefore the extremal curve, which satisfies the given boundary conditions, is

$$x(t) = \frac{t^2}{2} - 2t + 1.$$

The actual value of the extremum is obtained when we substitute into the given cost function and carry out the integration to obtain $J_{\min} = \frac{4}{3}$. It follows directly from Eqs. (3.3-5) and (3.3-6) that the determined extremum is indeed a minimum.

3.4 Unspecified terminal time problems

In deriving conditions for optimality, we have thus far assumed that the final time t_f is fixed. We now consider a generalization of the Lagrange problem formulation above, where the final time is defined as the first time after the initial time t_0 that the state trajectory is a member of a target set or terminal manifold. An important example of an unspecified final time Lagrange problem is the minimum time (to, for example, the origin) problem which exhibits an optimal control having a particularly interesting structure. The minimum time problem and other well-known control problems having simple and thus easily implementable optimal control structures will be discussed in detail in Chapter 5. As in the problem defined in Sec. 3.1, the original state may be specified or unspecified. A graphical illustration of a problem where the terminal time is defined in terms of a target set is given in Fig. 3.4-1.

We now wish to determine necessary conditions for the following problem: Minimize the cost function,

$$J\langle x\rangle = \int_{t_0}^{t_f} \phi\,(x, \dot{x}, t)\,dt, \tag{3.4-1}$$

with respect to the set of admissible trajectories, where t_o is known, $x(t_o)$ may or may not be specified, and t_f is the first time the state trajectory intersects the target set $C(t)$; i.e., t_f is the first time $x(t_f) = C(t_f)$. We assume through-

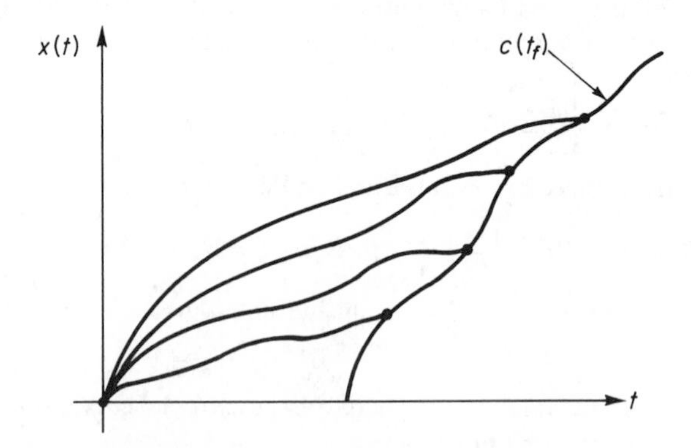

Fig. 3.4-1 Illustration of variable terminal time problem where $x(t_f) = c(t_f)$.

out that, in addition to the smoothness assumptions on ϕ specified in Sec. 3.1, the first derivative of C exists.

In developing the desired necessary conditions, we again assume that $\hat{x}$ is an optimal state trajectory. A family of curves, which includes the optimal trajectory $\hat{x}(t)$, starting at t_o and ending at t_f is given by

$$x(t) = \hat{x}(t) + \epsilon\eta_x(t) \tag{3.4-2}$$

with time derivative

$$\dot{x}(t) = \dot{\hat{x}}(t) + \epsilon\dot{\eta}_x(t) \tag{3.4-3}$$

where $\eta_x(t)$ is a variation in x which depends on t.

Since the terminal time is unspecified, it must be treated as a variable and, therefore, must be examined to see if perhaps there is a final time, $\hat{t}_f$, which is optimal. We will therefore define a family of final times, one of which is the optimal final time $\hat{t}_f$:

$$t_f = \hat{t}_f + \epsilon\eta_t(t_f) \tag{3.4-4}$$

where $\eta_t(t_f)$ is a variation in t_f.

Our first step in minimizing the cost function, Eq. (3.4-1), is to substitute Eqs. (3.4-2), (3.4-3), and (3.4-4) into it, which gives us

$$J\langle x\rangle = \int_{t_o}^{\hat{t}_f+\epsilon\eta_t(t_f)} \phi[\hat{x}(t) + \epsilon\eta_x(t), \dot{\hat{x}}(t) + \epsilon\dot{\eta}_x(t), t]\, dt. \tag{3.4-5}$$

We now set $\partial J/\partial\epsilon = 0$ at $\epsilon = 0$ and obtain

$$\left.\frac{\partial J}{\partial\epsilon}\right|_{\epsilon=0} = 0 = \int_{t_o}^{\hat{t}_f} \left\{\eta_x(t)\frac{\partial\phi}{\partial\hat{x}} + \dot{\eta}_x(t)\frac{\partial\phi}{\partial\dot{\hat{x}}}\right\} dt + \eta_t(\hat{t}_f)\phi[\hat{x}(\hat{t}_f), \dot{\hat{x}}(\hat{t}_f), \hat{t}_f]. \tag{3.4-6}$$

Integrating a portion of Eq. (3.4-6) by parts, we obtain

$$\int_{t_o}^{\hat{t}_f} \eta_x(t)\left[\frac{\partial\phi}{\partial\hat{x}} - \frac{d}{dt}\frac{\partial\phi}{\partial\dot{\hat{x}}}\right] dt + \eta_x\frac{\partial\phi}{\partial\dot{\hat{x}}}\Bigg|_{t=t_o}^{\hat{t}_f} + \eta_t(\hat{t}_f)\phi[\hat{x}(\hat{t}_f), \dot{\hat{x}}(\hat{t}_f), \hat{t}_f] = 0. \tag{3.4-7}$$

We now digress to obtain a relationship between $\eta_t(\hat{t}_f)$ and $\eta_x(\hat{t}_f)$ which will then allow for the determination of necessary conditions. At the terminal time, the terminal line $C(t)$ or, in higher dimensions, terminal manifold, and the optimal trajectory $x(t)$ intersect, as shown in Fig. 3.4-1. Therefore, using Eqs. (3.4-2) and (3.4-4), we have

$$\hat{x}[\hat{t}_f + \epsilon\eta_t(t_f)] + \epsilon\eta_x[\hat{t}_f + \epsilon\eta_t(t_f)] = C[\hat{t}_f + \epsilon\eta_t(t_f)]. \tag{3.4-8}$$

We take the partial derivative of this equation with respect to ϵ and evaluate it at $\epsilon = 0$ to obtain

$$\eta_t(\hat{t}_f)\dot{\hat{x}}(\hat{t}_f) + \eta_x(\hat{t}_f) = \eta_t(\hat{t}_f)\dot{C}(\hat{t}_f) \tag{3.4-9}$$

where $\dot{\hat{x}}(t) = \partial\hat{x}/\partial t$ and $\dot{C}(t) = \partial C/\partial t$ at $t = \hat{t}_f$. Thus

$$\eta_x(\hat{t}_f) = \eta_t(\hat{t}_f)[\dot{C}(\hat{t}_f) - \dot{\hat{x}}(\hat{t}_f)]. \tag{3.4-10}$$

It now follows from the argument that Eq. (3.4-7) must hold independent of η_x and η_t, but subject to Eq. (3.4-10), that

$$\frac{\partial \phi}{\partial \hat{x}} - \frac{d}{dt}\frac{\partial \phi}{\partial \dot{\hat{x}}} = 0 \qquad \text{for} \quad t \in [t_0, \hat{t}_f],$$

$$\eta_x(\hat{t}_f)\frac{\partial \phi}{\partial \dot{\hat{x}}}[\hat{x}(\hat{t}_f), \dot{\hat{x}}(\hat{t}_f), \hat{t}_f] + \eta_t(\hat{t}_f)\phi[\hat{x}(\hat{t}_f), \dot{\hat{x}}(\hat{t}_f), \hat{t}_f] = 0 \tag{3.4-11}$$

and

$$\eta_x(t_0)\frac{\partial \phi}{\partial \dot{\hat{x}}}[\hat{x}(t_0), \dot{\hat{x}}(t_0), t_0] = 0.$$

Substituting Eq. (3.4-10) into Eq. (3.4-11) implies that any optimal trajectory x must necessarily satisfy

$$\frac{\partial \phi}{\partial x} - \frac{d}{dt}\frac{\partial \phi}{\partial \dot{x}} = 0 \tag{3.4-12}$$

$$\eta_t(t)\left[(\dot{C} - \dot{x})\frac{\partial \phi}{\partial \dot{x}} + \phi\right] = 0, \qquad \text{for} \quad t = \hat{t}_f \tag{3.4-13}$$

$$\eta_x(t)\frac{\partial \phi}{\partial \dot{x}} = 0, \qquad \text{for} \quad t = t_o. \tag{3.4-14}$$

Equation (3.4-12) is the familiar Euler-Lagrange equation while Eqs. (3.4-13) and (3.4-14) comprise the transversality conditions for this problem. Conditions useful for the solution of the optimal control problem are once again obtainable from the transversality and boundary conditions in a fashion analogous to that discussed in Sec. 3.2. We now attempt to apply our results to a simple problem.

Example 3.4-1. We wish to minimize

$$J\langle x \rangle = \int_0^{t_f} [1 + \dot{x}^2]^{1/2}\, dt$$

with $x(0) = 1$ such that $x(t_f) = C(t_f) = 2 - t_f$.

We should recognize that the cost function is actually the arc length, which means that the distance between a point and a line is being minimized. Application of the Euler-Lagrange equation yields the optimal trajectory $x = at + b$, as in Example 3.3-1. To evaluate the arbitrary constants a and b, we make proper use of the transversality Eqs. (3.4-13) and (3.4-14). Here we specify $x(0) = 1$; thus $\eta_x(t_o) = 0$. And since t_f is unspecified, Eq. (3.4-13) becomes

$$(\dot{C} - \dot{x})\frac{\partial \phi}{\partial \dot{x}} + \phi = 0, \qquad \text{at} \quad t = t_f.$$

Thus we obtain $\dot{x} = 1$ at the unspecified terminal time t_f. From the solution to the Euler-Lagrange equation and the specified initial condition, we have $x(t = 0) = 1$; so we must have $b = 1$ and $\dot{x}(t = t_f) = a = 1$. Therefore the optimal trajectory is $x(t) = t + 1$, and the final time t_f is $t_f = \frac{1}{2}$. Salient features of this problem are indicated in Fig. 3.4-2. An interesting fact here is that the optimal trajectory intersects the terminal manifold at right angles.

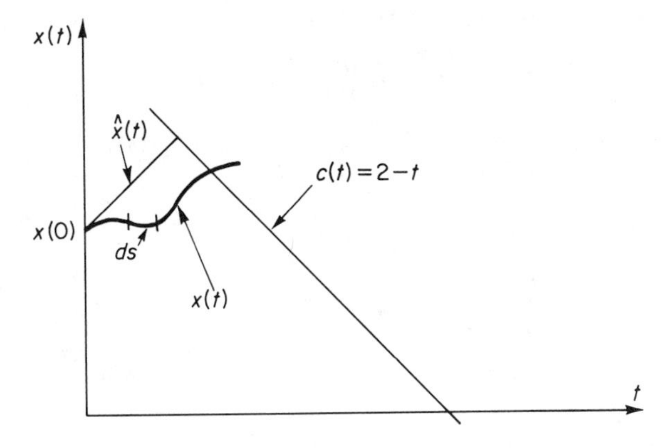

Fig. 3.4-2 Illustration of variable terminal time-variable end point problem, Example (3.4-1).

In general, the optimal trajectory will always be nontangent to the terminal manifold. This nontangency condition is, in fact, called the transversality condition.

3.5 Euler-Lagrange equations and transversality conditions—vector formulation

The previous results can be easily generalized to include scalar cost functions in n-dimensional variables via the state-space approach. That is, we desire to minimize

$$J\langle \mathbf{x} \rangle = \int_{t_o}^{t_f} \phi(\mathbf{x}, \dot{\mathbf{x}}, t)\, dt \tag{3.5-1}$$

where $\mathbf{x}$ is the system state, an n vector such that $\mathbf{x}^T = [x_1, x_2, \ldots, x_n]$; t_o, the starting time, is generally specified (it may not be); $\mathbf{x}(t_o)$ may or may not be specified; $\mathbf{x}(t_f)$ is specified by a given terminal manifold denoted $\mathbf{C}(t_f)$.† As before, the terminal time t_f does not have to be known. After following a procedure quite similar to the scalar one, we have, after setting $\partial J/\partial \epsilon$ at $\epsilon = 0$ and dropping the ^ notation, the requirement that among other things

$$\int_{t_o}^{t_f} \boldsymbol{\eta}^T(t)\left[\frac{\partial \phi}{\partial \mathbf{x}} - \frac{d}{dt}\frac{\partial \phi}{\partial \dot{\mathbf{x}}}\right] dt = 0 \tag{3.5-2}$$

†In general, all the states of $\mathbf{x}(t)$ need not be specified at the terminal time. If this is in fact the case for a given problem, great care must be exercised in applying the equations derived for transversality conditions in this section. This point will again be stressed at an appropriate time in the next chapter.

be true independent of $\boldsymbol{\eta}(t)$. This leads to the requirement that

$$\frac{\partial \phi}{\partial \mathbf{x}} - \frac{d}{dt}\frac{\partial \phi}{\partial \dot{\mathbf{x}}} = \mathbf{0} \tag{3.5-3}$$

which is simply an extended version of the Euler-Lagrange equation. The associated transversality conditions are given by

$$\boldsymbol{\eta}_{\mathbf{x}}^T \frac{\partial \phi}{\partial \dot{\mathbf{x}}} = 0, \qquad \text{at} \quad t = t_o \tag{3.5-4}$$

$$\boldsymbol{\eta}_{\mathbf{x}}^T \frac{\partial \phi}{\partial \dot{\mathbf{x}}} + \eta_t \phi = 0, \qquad \text{at} \quad t = t_f \tag{3.5-5}$$

where η_t can be related $\boldsymbol{\eta}_{\mathbf{x}}$ by an equation that can be obtained exactly as Eq. (3.4-8) was obtained:

$$\eta_t \left[\frac{d\mathbf{x}}{dt} - \frac{d\mathbf{C}}{dt}\right] + \boldsymbol{\eta}_{\mathbf{x}} = \mathbf{0}. \tag{3.5-6}$$

We now could submit Eq. (3.5-6) into Eq. (3.5-5) to obtain the vector version of Eq. (3.4-13); however, it is often more convenient to leave these equations in their present form.

Although the notation of this section may appear somewhat cumbersome, in an actual problem it is not, as the next example shows. Use of the Lagrange multiplier technique, as in Sec. 3.7 will alleviate some of the burdensome notation.

Example 3.5-1. We desire to find the transversality conditions for the minimization of

$$J = \int_{t_o}^{t_f} \phi(\mathbf{x}, \dot{\mathbf{x}}, t)\, dt$$

such that $\mathbf{x}(t_f) = \mathbf{C}(t_f)$, where $\mathbf{C}^T(t) = [c_1(t), 0, 0]$ and $\mathbf{x}^T = [x_1, x_2, x_3]$, $\mathbf{x}(t_o) = \mathbf{x}_o$, with t_o specified and t_f unspecified. The Euler-Lagrange equations are

$$\frac{\partial \phi}{\partial x_1} - \frac{d}{dt}\frac{\partial \phi}{\partial \dot{x}_1} = 0, \qquad \frac{\partial \phi}{\partial x_2} - \frac{d}{dt}\frac{\partial \phi}{\partial \dot{x}_2} = 0, \qquad \frac{\partial \phi}{\partial x_3} - \frac{d}{dt}\frac{\partial \phi}{\partial \dot{x}_3} = 0$$

with associated boundary conditions, $\mathbf{x}(t_o) = \mathbf{x}_o$, which represents the initial condition for the two-point boundary value problem, and

$$\frac{\partial \phi}{\partial \dot{x}_1} + \frac{\phi}{\dot{c}_1 - \dot{x}_1} = 0, \qquad x_2(t) = 0, \qquad x_3(t) = 0, \qquad \text{at} \quad t = t_f.$$

Although it may seem that all unspecified terminal time problems may now be worked by mere substitution into the derived relationships, Eqs. (3.5-3) through (3.5-6), this is not the case. Many problems do not fall precisely into a form which allows direct use of our derived formulas. When this type of problem is encountered, a good procedure to follow is to derive the transversality condition for the particular problem. An example demonstrating this type of approach follows.

Example 3.5-2. We wish to find the transversality conditions for the minimization of

$$J = \int_{t_o}^{t_f} \phi(\mathbf{x}, \dot{\mathbf{x}}, t)\, dt$$

such that $||\mathbf{x}(t_f)||^2 = 1$, where $\mathbf{x}^T = [x_1 x_2]$, with specified starting time t_o and terminal time t_f. Thus, we would like to reach the region of state-space specified by $x_1^2 + x_2^2 = 1$ at a specified terminal time t_f, given the state at the starting time t_o, denoted by $\mathbf{x}(t_o)$.

The transversality conditions are, from Eq. (3.5-4),

$$\left(\frac{\partial \phi}{\partial \dot{\mathbf{x}}}\right)^T \boldsymbol{\eta}_{\mathbf{x}} = 0 = \frac{\partial \phi}{\partial \dot{x}_1}\eta_{x_1} + \frac{\partial \phi}{\partial \dot{x}_2}\eta_{x_2}, \qquad \text{at} \quad t = t_f.$$

As before, we assume that $\mathbf{x}(t) = \hat{\mathbf{x}}(t) + \epsilon \boldsymbol{\eta}_{\mathbf{x}}(t)$ where $\mathbf{x}$ is the optimal trajectory. For this problem, this relation in component form becomes $x_1 = \hat{x}_1 + \epsilon\eta_{x_1}$ and $x_2 = \hat{x}_2 + \epsilon\eta_{x_2}$. Substituting these results into the given terminal manifold, we obtain

$$(\hat{x}_1 + \epsilon\eta_{x_1})^2 + (\hat{x}_2 + \epsilon\eta_{x_2})^2 = 1, \qquad \text{at} \quad t = t_f.$$

Taking the partial derivative of the foregoing equation with respect to ϵ and then setting $\epsilon = 0$, we have

$$\hat{x}_1\eta_{x_1} + \hat{x}_2\eta_{x_2} = 0, \qquad t = t_f.$$

We thus see that the specification of the terminal manifold

$$x_1^2(t_f) + x_2^2(t_f) = 1$$

leads to a linear relationship between η_{x_1} and η_{x_2} at the terminal time. If we combine this relation with the previously stated transversality condition, we obtain for one of the terminal boundary conditions

$$\frac{\partial \phi}{\partial \dot{x}_1}\frac{x_2}{x_1} - \frac{\partial \phi}{\partial \dot{x}_2} = 0, \qquad \text{at} \quad t = t_f.$$

Therefore the two boundary conditions at $t = t_f$ are

$$x_1^2(t_f) + x_2^2(t_f) = 1$$

$$\frac{\partial \phi}{\partial \dot{x}_1(t_f)}\frac{x_2(t_f)}{x_1(t_f)} - \frac{\partial \phi}{\partial \dot{x}_2(t_f)} = 0.$$

Thus for a given $\phi(\mathbf{x}, \dot{\mathbf{x}}, t)$, we can resolve this problem completely by solving for the optimal trajectory through the Euler-Lagrange equations and the appropriate boundary conditions which we have just obtained.

3.6 A variational approach

The approach taken thus far in the development of necessary conditions for a trajectory $\hat{x}$ to be optimal has been to perturb away from x by an amount $\epsilon\eta$ and then to determine necessary conditions from the equation

$$\left.\frac{\partial J}{\partial \epsilon}\right|_{\epsilon=0} = 0,$$

where the equation must hold independent of η. An alternative approach, the variational approach, is to consider the linear part of $\Delta J = J\langle \hat{x} + \epsilon\eta \rangle - J\langle \hat{x} \rangle$, and note that it, too, must equal zero when $\epsilon = 0$, independent of η. The linear part of ΔJ, which we designate as δJ, is called the first variation of J (at $\hat{x}$), the Gateaux differential of J, or the closely related Frechet differential of J [9]. As the last two terminologies explicitly state, the variational approach to necessary conditions is concerned with differentials rather than derivatives.

We now wish to develop necessary conditions, using the variational approach, for the fixed final time problem of minimizing

$$J = \int_{t_0}^{t_f} \phi(x, \dot{x}, t)\, dt. \tag{3.6-1}$$

We assume, as in Sec. 3.1, that both $x(t)$ and $\dot{x}(t)$ are representable by a family of curves

$$x(t) = \hat{x}(t) + \epsilon\eta(t), \qquad \dot{x}(t) = \dot{\hat{x}}(t) + \epsilon\dot{\eta}(t) \tag{3.6-2}$$

where $x(t)$ is the optimal (extremal) curve and $\eta(t)$ is a variation in $x(t)$ depending upon t. We substitute Eq. (3.6-2) into Eq. (3.6-1) and expand $\phi(x, \dot{x}, t)$ in a Taylor series about the point $\epsilon = 0$,

$$\phi[\hat{x}(t) + \epsilon\eta(t), \dot{\hat{x}}(t) + \epsilon\dot{\eta}(t), t] = \phi(\hat{x}, \dot{\hat{x}}, t) + \frac{\partial \phi}{\partial \hat{x}}\epsilon\eta(t) + \frac{\partial \phi}{\partial \dot{\hat{x}}}\epsilon\dot{\eta}(t) + \text{H.O.T.} \tag{3.6-3}$$

where H.O.T. is used to indicate higher-order terms in $\eta(t)$ and $\dot{\eta}(t)$.

We note that

$$\begin{aligned} \Delta J &= \int_{t_0}^{t_f} \{\phi[\hat{x}(t) + \epsilon\eta(t), \dot{\hat{x}} + \epsilon\dot{\eta}, t] - \phi[\hat{x}(t), \dot{\hat{x}}, t]\}\, dt \\ &= \int_{t_0}^{t_f} \left\{\frac{\partial \phi}{\partial \hat{x}}\epsilon\eta(t) + \frac{\partial \phi}{\partial \dot{\hat{x}}}\epsilon\dot{\eta}(t) + \text{H.O.T.}\right\} dt. \end{aligned} \tag{3.6-4}$$

Now we define the first variation of $x(t)$ and $\dot{x}(t)$ as

$$\epsilon\eta(t) = \delta x, \qquad \epsilon\dot{\eta}(t) = \delta\dot{x}. \tag{3.6-5}$$

Thus

$$\Delta J = \int_{t_0}^{t_f} \left[\frac{\partial \phi}{\partial \hat{x}}\delta x + \frac{\partial \phi}{\partial \dot{\hat{x}}}\delta\dot{x} + \text{H.O.T.}\right] dt. \tag{3.6-6}$$

The first variation of J is

$$\delta J = \int_{t_0}^{t_f} \left[\frac{\partial \phi}{\partial \hat{x}}\delta x + \frac{\partial \phi}{\partial \dot{\hat{x}}}\delta\dot{x}\right] dt. \tag{3.6-7}$$

A necessary condition for an extremum at $x(t) = \hat{x}(t)$, i.e., $\epsilon = 0$, is that the first variation of J, δJ, be zero. Applying this to Eq. (3.6-7), along with the minor simplification of integrating by parts and dropping the ^ notation, we obtain

$$\int_{t_o}^{t_f} \left[\frac{\partial \phi}{\partial x} - \frac{d}{dt}\frac{\partial \phi}{\partial \dot{x}}\right] \delta x \, dt + \frac{\partial \phi}{\partial \dot{x}} \delta x \Big|_{t=t_o}^{t=t_f} = 0. \tag{3.6-8}$$

For Eq. (3.6-8) to equal zero independent of the variation δx, we must have

$$\frac{\partial \phi}{\partial x} - \frac{d}{dt}\frac{\partial \phi}{\partial \dot{x}} = 0 \tag{3.6-9}$$

$$\frac{\partial \phi}{\partial \dot{x}} \delta x = 0, \qquad \text{for} \quad t = t_o, t_f. \tag{3.6-10}$$

We note that Eq. (3.6-9) is the Euler-Lagrange equation and Eq. (3.6-10) is its associated transversality condition.

In a similar manner, it is also easy to show that the second variation of Eq. (3.6-1), written $\delta^2 J$, is

$$\delta^2 J = \frac{1}{2} \int_{t_o}^{t_f} \left\{(\delta x)^2 \left[\frac{\partial^2 \phi}{\partial x^2} - \frac{d}{dt}\frac{\partial^2 \phi}{\partial \dot{x}^2}\right] + (\delta \dot{x})^2 \frac{\partial^2 \phi}{\partial \dot{x}^2}\right\} dt \tag{3.6-11}$$

where the second variation is now defined as the quadratic part of Eq. (3.6-6) of twice Eq. (3.3-4). As previously stated, the interpretations of the second variation are that $\delta^2 J \geq 0$ implies a minimum of J, and $\delta^2 J \leq 0$ implies a maximum of J. A quadratic form integral similar to Eq. (3.3-4) also follows directly.

3.7 Dynamic optimization with equality constraints—Lagrange multipliers

The problem examined in the previous section excluded consideration of optimal control problems having constraint relationships between the scalar elements of the state trajectory, a situation which almost always occurs in problems with physical origins. We now examine the following constrained problem: Minimize the cost function,

$$J = \int_{t_o}^{t_f} \phi(\mathbf{x}, \dot{\mathbf{x}}, t) \, dt \tag{3.7-1}$$

subject to the m-vector equality constraint,

$$\mathbf{\Lambda}(\mathbf{x}, \dot{\mathbf{x}}, t) = \mathbf{0}, \quad \text{for all} \quad t \in [t_0, t_f]. \tag{3.7-2}$$

A state trajectory is now said to be admissible if, in addition to the smoothness assumptions stated in Sec. 3.1, Eq. (3.7-2) is satisfied. It can be shown that this constrained problem is equivalent to the problem of minimizing the cost function

$$J' = \int_{t_o}^{t_f} [\phi(\mathbf{x}, \dot{\mathbf{x}}, t) + \boldsymbol{\lambda}^T(t)\mathbf{\Lambda}(\mathbf{x}, \dot{\mathbf{x}}, t)] \, dt \tag{3.7-3}$$

subject to no constraints, where the time-varying m-vector $\boldsymbol{\lambda}(t)$ is the vector equivalent of the Lagrange multiplier discussed in Chapter 2 [5].

To illustrate the development of the Lagrange multiplier, let us consider a special case where $\mathbf{x}$ is a two vector. Suppose that we wish to minimize

$$J = \int_{t_0}^{t_f} \phi(x_1, x_2, \dot{x}_1, \dot{x}_2, t)\, dt \tag{3.7-4}$$

subject to the constraint (with fixed end points)

$$\Lambda(x_1, x_2, t) = 0. \tag{3.7-5}$$

We will use the variational notation just developed to establish a method for treating the given equality constraint. To establish a minimum, it is necessary that the first variation of Eq. (3.7-4) be zero, that is

$$\delta J = \int_{t_0}^{t_f} \left\{ \delta x_1 \left[\frac{\partial \phi}{\partial x_1} - \frac{d}{dt}\frac{\partial \phi}{\partial \dot{x}_1} \right] + \delta x_2 \left[\frac{\partial \phi}{\partial x_2} - \frac{d}{dt}\frac{\partial \phi}{\partial \dot{x}_2} \right] \right\} dt = 0. \tag{3.7-6}$$

If δx_1 were independent of δx_2, we could simply set each term of Eq. (3.7-6) equal to 0. Since the constraint provides a dependence on x_1 and x_2, we must take the given constraint into consideration. Taking the variation of Eq. (3.7-5), we have

$$\delta \Lambda = \frac{\partial \Lambda}{\partial x_1} \delta x_1 + \frac{\partial \Lambda}{\partial x_2} \delta x_2 = 0. \tag{3.7-7}$$

It also follows that, for any $\lambda(t)$, we may multiply Eq. (3.7-7) by $\lambda(t)$ and integrate so that

$$\int_{t_0}^{t_f} \lambda(t) \left[\frac{\partial \Lambda}{\partial x_1} \delta x_1 + \frac{\partial \Lambda}{\partial x_2} \delta x_2 \right] dt = 0. \tag{3.7-8}$$

If we add Eq. (3.7-6) to Eq. (3.7-8) we obtain

$$0 = \int_{t_0}^{t_f} \left\{ \delta x_1 \left[\frac{\partial \phi}{\partial x_1} - \frac{d}{dt}\frac{\partial \phi}{\partial \dot{x}_1} + \lambda \frac{\partial \Lambda}{\partial x_1} \right] + \delta x_2 \left[\frac{\partial \phi}{\partial x_2} - \frac{d}{dt}\frac{\partial \phi}{\partial \dot{x}_2} + \lambda \frac{\partial \Lambda}{\partial x_2} \right] \right\} dt. \tag{3.7-9}$$

We will now adjust λ so that the term within the first brackets under the integral is zero. It also must follow that, since δx_2 is arbitrary, the term in the second brackets under the integral is also equal to zero. It is apparent that we would have obtained the same results had we reformulated the given problem by adjoining to the cost function the constraint, via a Lagrange multiplier as in Eq. (3.7-3), and used the Euler-Lagrange equations on this cost function. The resulting Euler-Lagrange equations would then be solved subject to the equality constraint of Eq. (3.7-2).

Several other types of constraints on the state trajectory other than Eq. (3.7-2) are considered in [2, 4, 5]. We will study in depth a special case of Eq. (3.7-2) in Chapter 4, the case where $\mathbf{g}(\mathbf{x}, \dot{\mathbf{x}}, t) = \dot{\mathbf{x}} - \mathbf{h}(\mathbf{x}, t)$.

Example 3.7-1. We are given the differential system

$$\ddot{\theta} = u(t)$$

which may be interpreted as the moment of inertia of a rocket in free space, and we desire to minimize

$$J = \frac{1}{2}\int_0^2 (\ddot{\theta})^2\, dt$$

such that

$$\theta(t=0) = 1, \qquad \theta(t=2) = 0$$
$$\dot{\theta}(t=0) = 1, \qquad \dot{\theta}(t=2) = 0.$$

To cast this problem in state space notation, we let

$$x_1(t) = \theta(t), \qquad \dot{x}_1 = x_2(t), \qquad \dot{x}_2 = u(t).$$

Now the differential system can be represented by

$$\dot{\mathbf{x}} = \mathbf{A}\mathbf{x}(t) + \mathbf{b}u(t)$$

where

$$\mathbf{x}^T = [x_1 \quad x_2], \qquad \mathbf{A} = \begin{bmatrix} 0 & 1 \\ 0 & 0 \end{bmatrix}, \qquad \mathbf{b}^T = [0 \quad 1].$$

When we apply Eq. (3.7-3) [$u(t)$ is treated as another state variable, x_3], the problem becomes one of minimizing

$$J = \int_0^2 \{\tfrac{1}{2}u^2(t) + \boldsymbol{\lambda}^T(t)[\mathbf{A}\mathbf{x}(t) + \mathbf{b}u(t) - \dot{\mathbf{x}}]\}\, dt$$
$$= \int_0^2 \{\tfrac{1}{2}u^2(t) + \lambda_1(t)[x_2(t) - \dot{x}_1] + \lambda_2(t)[u(t) - \dot{x}_2]\}\, dt.$$

The Euler-Lagrange equations yield

$$\dot{\lambda}_1 = 0, \qquad \dot{\lambda}_2 = -\lambda_1(t), \qquad u(t) = -\lambda_2(t).$$

The final solution, obtained by means of the given differential relationships and boundary conditions, is

$$x_1 = \tfrac{1}{2}t^3 - \tfrac{7}{4}t^2 + t + 1, \qquad x_2 = \tfrac{3}{2}t^2 - \tfrac{7}{2}t + 1, \qquad u = 3t - \tfrac{7}{2}.$$

This system, along with a plot of the system trajectories, is shown in Fig. 3.7-1.

Example 3.7-2. Linear Servomechanism† Suppose that we wish to minimize

$$J = \frac{1}{2}\int_{t_o}^{t_f} \{\|\mathbf{u}(t)\|^2_{\mathbf{R}(t)} + \|\mathbf{x}(t) - \mathbf{r}(t)\|^2_{\mathbf{Q}(t)}\}\, dt$$

for the general time-varying system specified by

$$\dot{\mathbf{x}} = \mathbf{A}(t)\mathbf{x}(t) + \mathbf{B}(t)\mathbf{u}(t)$$

with $\mathbf{x}(t_o) = \mathbf{x}_o$ as the initial condition vector; $\mathbf{r}(t)$ is the desired value of the

†A considerably more detailed treatment of this problem is given in Chapter 5.

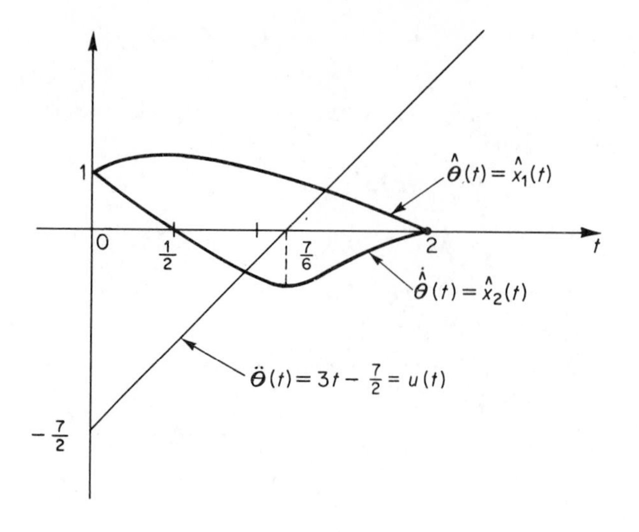

Fig. 3.7-1 Block diagram, optimal control, and state variables for system of Example (3.7-1).

state vector $\mathbf{x}(t)$. As before, it is necessary to assume that all matrices and vectors are of compatible orders. We adjoin the differential system equality constraint to the cost function by the Lagrange multiplier to obtain

$$J' = \int_{t_0}^{t_f} \{\tfrac{1}{2}\|\mathbf{u}(t)\|^2_{\mathbf{R}(t)} + \tfrac{1}{2}\|\mathbf{x}(t) - \mathbf{r}(t)\|^2_{\mathbf{Q}(t)} + \boldsymbol{\lambda}^T(t)[\mathbf{A}(t)\mathbf{x}(t) + \mathbf{B}(t)\mathbf{u}(t) - \dot{\mathbf{x}}]\}\,dt.$$

The exact nature of the cost function used depends upon the particular problem being solved. Therefore $\mathbf{R}(t)$ and $\mathbf{Q}(t)$, both penalty-weighting matrices, are generally chosen with regard to the physical conditions present. We also assume that both $\mathbf{R}(t)$ and $\mathbf{Q}(t)$ are symmetric, since there is no loss in generality by doing so. The control vector, $\mathbf{u}(t)$, is treated just as if it were a state vector. Then we apply the Euler-Lagrange equations, which in this case are

$$\frac{\partial \Phi}{\partial \mathbf{x}} - \frac{d}{dt}\frac{\partial \Phi}{\partial \dot{\mathbf{x}}} = \mathbf{0}, \qquad \frac{\partial \Phi}{\partial \mathbf{u}} - \frac{d}{dt}\frac{\partial \Phi}{\partial \dot{\mathbf{u}}} = \mathbf{0}$$

where

$$\Phi = \tfrac{1}{2}\|\mathbf{u}(t)\|^2_{\mathbf{R}(t)} + \tfrac{1}{2}\|\mathbf{x}(t) - \mathbf{r}(t)\|^2_{\mathbf{Q}(t)} + \boldsymbol{\lambda}^T(t)[\mathbf{A}(t)\mathbf{x}(t) + \mathbf{B}(t)\mathbf{u}(t) - \dot{\mathbf{x}}].$$

Thus

$$\frac{\partial \Phi}{\partial \mathbf{x}} = \mathbf{Q}(t)[\mathbf{x}(t) - \mathbf{r}(t)] + \mathbf{A}^T(t)\boldsymbol{\lambda}(t), \qquad \frac{\partial \Phi}{\partial \dot{\mathbf{x}}} = -\boldsymbol{\lambda}(t)$$

$$\frac{\partial \Phi}{\partial \mathbf{u}} = \mathbf{R}(t)\mathbf{u}(t) + \mathbf{B}^T(t)\boldsymbol{\lambda}(t), \qquad \frac{\partial \Phi}{\partial \dot{\mathbf{u}}} = \mathbf{0}.$$

The Euler-Lagrange equations for this problem become

$$\dot{\boldsymbol{\lambda}} = -\mathbf{A}^T(t)\boldsymbol{\lambda}(t) - \mathbf{Q}(t)[\mathbf{x}(t) - \mathbf{r}(t)], \qquad \mathbf{u}(t) = -\mathbf{R}^{-1}(t)\mathbf{B}^T(t)\boldsymbol{\lambda}(t).$$

Since $\mathbf{x}(t_f)$ is unspecified, the transversality condition at the terminal time yields $\boldsymbol{\lambda}(t_f) = \mathbf{0}$. This solution can be block-diagrammed as in Fig. 3.7-2. We note that the solution for the optimal control requires that $\mathbf{R}(t)$ have an inverse. Also, certain other requirements must be met to insure a minimum of the cost function; specifically, $\mathbf{R}(t)$ and $\mathbf{Q}(t)$ must be nonnegative definite to insure a nonnegative second variation. Thus we see that $\mathbf{R}(t)$ must be positive definite.

Although it appears that we have solved the originally stated problem, there are still some further refinements which are highly desired. Since the state of the system is specified at t_0, we are given $\mathbf{x}(t_0)$, while the adjoint operator $\boldsymbol{\lambda}(t)$ is specified at the terminal time, $\boldsymbol{\lambda}(t_f) = \mathbf{0}$. What we, in fact, have to do is solve a two-point boundary value problem (TPBVP), something which, in general, cannot always be done without recourse to electronic computers. In this particular case, since the differential equations are all linear, superposition can be invoked and a closed-form analytical solution obtained with great difficulty.

If we let $\mathbf{r}(t)$ be either a constant vector or the null vector, the foregoing problem reduces to a regulator problem. The treatment of the servomechanism problem can be made more general if we assume that indirect state observation is made available to us; that is, for the system

$$\dot{\mathbf{x}} = \mathbf{A}(t)\mathbf{x}(t) + \mathbf{B}(t)\mathbf{u}(t)$$

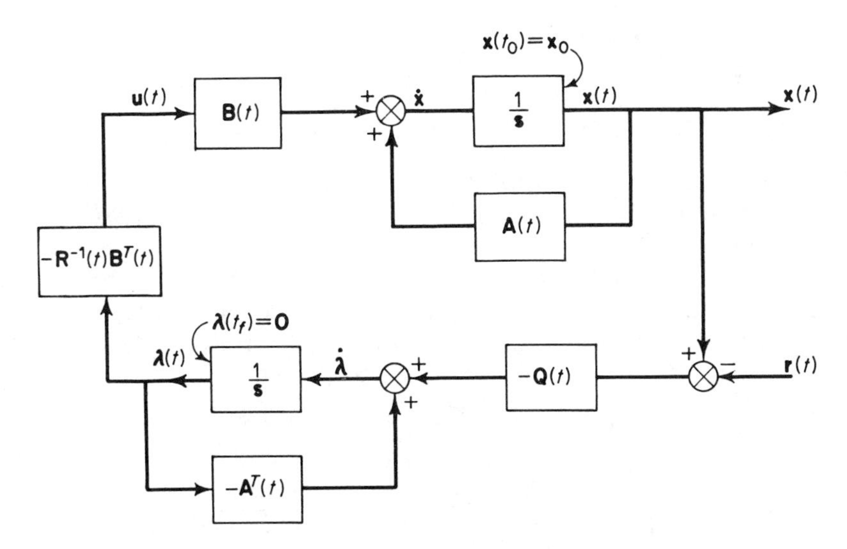

Fig. 3.7-2 Block diagram of a possible solution to the servomechanism problem.

we can obtain directly only

$$\mathbf{z}(t) = \mathbf{C}(t)\mathbf{x}(t) + \mathbf{D}(t)\mathbf{u}(t).$$

The procedure and results are quite similar to the ones obtained in this example except that requirements on observability and controllability, to be discussed in Chapter 7, are present.

To solve this two-point boundary value problem, we must require a knowledge of $\mathbf{r}(t)$ for all time in the closed interval t_o to t_f, or, in shorthand notation, $\forall t \in [t_o, t_f]$. Since a two-point boundary value problem must be solved before we can determine the optimum control for this problem, it is clear that a closed-loop control has not been found. After we have formulated the Hamilton-Jacobi equations and the Pontryagin maximum principle, we will have a great deal more to say about this important problem.

3.8 Dynamic optimization with inequality constraints

In many physical problems of interest to the control engineer there are various inequality constraints on the control vector. For example, the maximum thrust from a reaction jet is physically limited as is the maximum input reactivity in a nuclear reactor. When inequality constraints are present it is necessary that we consider them in determining optimum system design. Thus we are faced with minimizing a cost function of the form

$$J = \int_{t_o}^{t_f} \phi(\mathbf{x}, \dot{\mathbf{x}}, t)\, dt \tag{3.8-1}$$

with equality constraints of the form

$$\mathbf{\Lambda}(\mathbf{x}, \dot{\mathbf{x}}, t) = \mathbf{0} \tag{3.8-2}$$

and inequality constraints of the form

$$\mathbf{\Gamma}_{\min} \leq \mathbf{\Gamma}(\mathbf{x}, \dot{\mathbf{x}}, t) \leq \mathbf{\Gamma}_{\max}. \tag{3.8-3}$$

In order for a trajectory to be admissible, it must now satisfy Eqs. (3.8-2) and (3.8-3) as well as the smoothness conditions specified in Sec. 3.1. Several techniques are available for solving the inequality constraint problem, one of which is the slack-variable method. The slack-variable method converts the inequality constraint, Eq. (3.8-3), to an equality constraint by introducing new state variables γ_i satisfying the equations

$$(\mathbf{\Gamma}_{\max i} - \mathbf{\Gamma}_i)(\mathbf{\Gamma}_i - \mathbf{\Gamma}_{\min i}) = \gamma_i^2, \qquad i = 1, 2, \ldots. \tag{3.8-4}$$

It is easily demonstrated that Eq. (3.8-4) is equivalent to Eq. (3.8-3) since in order for γ_i, $i = 1, 2$, to be real, Eq. (3.8-3) must be satisfied, and vice versa. Lagrange multipliers can now be used to adjoin Eqs. (3.8-2) and (3.8-4) to Eq. (3.8-1), and the usual necessary conditions are then applicable. We illustrate this solution technique with an example.

Example 3.8-1. Let us consider the same plant dynamics as in the previous example (3.7-1)

$$\dot{x}_1 = x_2(t), \qquad \dot{x}_2 = u(t)$$

with the initial conditions $x_1(t_o) = x_o$ and $x_2(t_o) = v_o$. The problem is to find the control which maximizes $x_1(t_f)$, for fixed t_f, subject to the boundary condition equality constraint that $x_2(t_f) = v_f$ and the inequality constraint on the scalar control $u_{\min} \leq u \leq u_{\max}$. We convert the inequality constraint to an equality constraint by introducing a new variable $\alpha(t)$ and replacing the inequality constraint by

$$(u - u_{\min})(u_{\max} - u) - \alpha^2 = 0.$$

Thus the problem may be recast as one of minimizing $J = -x_1(t_f)$ subject to the equality constraints

$$\begin{aligned} &\dot{x}_1 = x_2(t), \qquad x_1(t_o) = x_o, \qquad x_1(t_f) = \text{open} \\ &\dot{x}_2 = u(t), \qquad x_2(t_o) = v_o, \qquad x_2(t_f) = v_f \\ &(u - u_{\min})(u_{\max} - u) - \alpha^2 = 0. \end{aligned}$$

The cost function with the adjoined Lagrange multiplier becomes

$$J' = -x_1(t_o) + \int_{t_o}^{t_f} \{-\dot{x}_1 + \lambda_1[x_2 - \dot{x}_1] + \lambda_2[u - \dot{x}_2] + \lambda_3[(u - u_{\min})(u_{\max} - u) - \alpha^2]\}\, dt.$$

The Euler-Lagrange equation

$$\frac{d}{dt}\frac{\partial \Phi}{\partial \dot{\mathbf{x}}} - \frac{\partial \Phi}{\partial \mathbf{x}} = 0, \qquad \mathbf{x}^T = [x_1, x_2, u]$$

with

$$\Phi = \lambda_1[x_2 - \dot{x}_1] + \lambda_2[u - \dot{x}_2] + \lambda_3[(u - u_{\min})(u_{\max} - u) - \alpha^2] - \dot{x}_1$$

yields

$$\dot{\lambda}_1 = 0, \qquad \dot{\lambda}_2 = -\lambda_1$$

$$0 = -\lambda_2 + \lambda_3[2u - u_{\max} - u_{\min}], \qquad 0 = \alpha\lambda_3.$$

Application of the natural boundary condition equation (transversality condition) to determine the single missing terminal condition on $x_1(t_f)$ yields

$$\left.\frac{\partial \Phi}{\partial \dot{x}_1}\right|_{t=t_f} = 0 = -1 - \lambda_1(t_f).$$

Thus we have arrived at the two-point boundary value problem whose solution determines the optimal state and control variables. This TPBVP is

$$\begin{aligned} &\dot{x}_1 = x_2(t), \qquad x_1(t_o) = x_o \\ &\dot{x}_2 = x(t), \qquad x_2(t_o) = v_o \\ &\dot{\lambda}_1 = 0, \qquad \lambda_1(t_f) = -1 \\ &\dot{\lambda}_2 = -\lambda_1(t), \qquad x_2(t_f) = v_f \\ &\alpha(t)\lambda_3(t) = 0 \\ &\lambda_2(t) = \lambda_3(t)[2u(t) - u_{\max} - u_{\min}] \\ &\alpha^2(t) = [u(t) - u_{\min}][u_{\max} - u(t)]. \end{aligned}$$

This TPBVP is nonlinear because of the last three coupling equations above and is quite difficult to solve without recourse to a computer. In a usual version of this problem, $u_{\min} = -1$ and $u_{\max} = +1$. In that case, it is possible to show that $\alpha(t) = 0$ and

$$u(t) = -\text{sign}\ \lambda_2(t)$$

where

$$\text{sign}\ \lambda_2 = 1 \quad \text{if} \quad \lambda_2 > 0$$
$$\text{sign}\ \lambda_2 = -1 \quad \text{if} \quad \lambda_2 < 0.$$

This does not, however, change the nonlinear nature of the two-point boundary problem. In a later chapter we will devote considerable time to various computational techniques for solving nonlinear two-point (and multi-point) boundary value problems.

References

1. BERKOVITZ, L.D., "Variational Methods in Problems of Control and Programming." *J. Math. Anal. Appl.*, 3, (1961), 145–69.
2. BLISS, G.A., *Lectures on the Calculus of Variations.* University of Chicago Press, Chicago, 1963.
3. DREYFUS, S.E., *Dynamic Programming and the Calculus of Variations.* Academic Press, New York, 1965.
4. ELSGOC, L.E., *Calculus of Variations.* Pergamon Press, Ltd., New York, 1961.
5. GELFAND, I.M., and FOMIN, S.V., *Calculus of Variations.* Prentice-Hall, Inc., Englewood Cliffs, New Jersey, 1963.
6. KALMAN, R.E., "The Theory of Optimal Control and the Calculus of Variations." *Mathematical Optimization Techniques*, R. Bellman (ed.). University of California Press, Berkeley, 1963.
7. LEITMAN, G., *Optimal Control.* McGraw-Hill Book Company, New York, 1966.
8. LEITMAN, G., ed., *Optimization Techniques.* Academic Press, New York, 1962.
9. LUENBERGER, D.G., *Optimization by Vector Space Methods.* Wiley, New York, 1969.
10. PIERRE, D.A., *Optimization Theory with Applications.* Wiley, New York, 1969.
11. TOU, J., *Modern Control Theory.* McGraw-Hill Book Company, New York, 1964.

Problems

1. A linear differential system is described by

$$\dot{\mathbf{x}} = \mathbf{A}\mathbf{x} + \mathbf{B}\mathbf{u}$$

where

$$\mathbf{A} = \begin{bmatrix} 0 & 1 \\ 0 & 0 \end{bmatrix}, \qquad \mathbf{B} = \begin{bmatrix} 1 & 0 \\ 0 & 1 \end{bmatrix}, \quad \mathbf{x}^T = [x_1, x_2], \qquad \mathbf{u}^T = [u_1, u_2].$$

Find $\mathbf{u}(t)$ such that

$$J = \tfrac{1}{2} \int_0^2 ||\mathbf{u}||^2 \, dt$$

is minimum, given $\mathbf{x}^T(0) = [1, 1]$ and $x_1(2) = 0$.

2. Find the conditions necessary for minimizing

$$J = \theta[x(t_f)] + \int_{t_o}^{t_f} \phi(x, \dot{x}, t) \, dt$$

given $x(t_o) = x_0$ and $g(x, \dot{x}, t) = 0$.

3. Use the results of Problem 2 to find the control $u(t)$, which minimizes

$$J = \frac{S}{2} x^2(2) + \tfrac{1}{2} \int_0^2 u^2 \, dt$$

such that $\dot{x} = u(t)$, $x(0) = 1$.

4. A linear system is described by

$$\dot{x} = -x + u, \qquad x(0) = 1.$$

It is desired to minimize

$$J = \tfrac{1}{2} \int_0^2 (x^2 + u^2) \, dt.$$

A feedback law is obtained if we let $u(t) = \alpha x(t)$ where $d\alpha/dt = 0$ such that α is a constant. Find the equations defining the optimum value of α.

5. Find the differential equations and associated boundary conditions whose solutions minimize

$$J = \tfrac{1}{2} \int_0^{t_f} u^2 \, dt$$

for the differential system described by

$$\dot{x}_1 = -x_1 + x_2$$
$$\dot{x}_2 = u$$

with end points given by

$$x_1(0) = x_2(0) = 0$$
$$x_1^2(t_f) + x_2^2(t_f) = t_f^2 + 1.$$

6. Find the value of u which minimizes (for t_f unspecified)

$$J = \int_0^{t_f} [\alpha + u^2(t) + x^2(t)] \, dt$$

for the differential system

$$\dot{x} = -x(t) + u(t), \qquad x(0) = 1, \qquad x(t_f) = 0.$$

7. A linear second-order differential equation is described by

$$\dot{x}_1 = x_2(t), \qquad x_1(0) = 1$$
$$\dot{x}_2 = u \qquad x_2(0) = 1.$$

Find, by use of the Euler-Lagrange equations and transversality conditions, the optimal control $u(t)$ which minimizes:

(a) $$J = \int_0^1 u^2\,dt, \qquad x_1(1) = x_2(1) = 0$$

(b) $$J = \int_0^1 u^2\,dt, \qquad x_1(1) = 0$$

(c) $$J = \int_0^{t_f} u^2\,dt, \qquad x_1(t_f) = c(t_f) = -t_f^2$$

[Also determine t_f and $x_1(t_f)$.]

(d) $$J = \int_0^{t_f} u^2\,dt, \qquad x_1(t_f) = c(t_f) = -t_f^2, \qquad x_2(t_f) = 0$$

(e) $$J = \int_0^1 \{\|\mathbf{x}\|^2 + \|\mathbf{u}\|^2\}\,dt.$$

For all cases, sketch both the optimal system trajectory $\mathbf{x}(t)$ and the optimal system control $u(t)$.

8. For the fixed plant dynamics given by

$$\dot{x} = u$$

determine the optimal closed-loop system which minimizes

$$J = \tfrac{1}{2}\int_0^2 \{u^2 + (x - i)^2\}\,dt$$

where $i(t) = 1 - e^{-t}$.

9. For the fixed plant dynamics given by $\dot{x} = u(t)$, $x(0) = x_0$, determine the optimal closed-loop control which minimizes for fixed t_f

$$J = \tfrac{1}{2}sx^2(t_f) + \tfrac{1}{2}\int_0^{t_f} u^2\,dt$$

where s is an arbitrary constant. Do this by first determining the optimum open-loop control and trajectory and then let $u(t) = k(t)x(t)$.

4 The maximum principle and Hamilton-Jacobi theory

In the previous chapter, we formulated many problems in the classical calculus of variations [1]. A derivation of the Euler-Lagrange equations for both the scalar and vector cases was presented. We discussed the associated transversality conditions and some of the difficulties which we may encounter if inequality constraints are present. Several simple optimal control problems were stated and solved. In this chapter we wish to reexamine many of the problems presented in the previous chapter and obtain more general solutions for some of them. In addition, we will develop methods for handling some problems which could not be conveniently formulated by the methods in the previous chapter.

To these ends, we will present the Bolza formulation of the variational calculus using Hamiltonian methods. This will lead us into a proof of the Pontryagin maximum principle and the associated transversality conditions [2–5]. We will proceed then to a development of the Hamilton-Jacobi equations [12–14], which are equivalent to Bellman's equations of continuous dynamic programming. Finally, we will give brief mention to some limitations of dynamic programming. Examples to illustrate the methods will be presented. We will reserve the next chapter for a discussion of some of the many problems which we can formulate and solve using the maximum principle.

In order to fully develop our approach to optimization theory where the terminal time is not fixed and where the control and state vectors are not necessarily smooth functions, we must consider in more detail the first variation for such problems.

4.1 The variational approach for functions with terminal times not fixed

We now extend the variational approach introduced in Sec. 3.6 to problems having unspecified terminal times. Consider extremizing

$$J = \int_{t_o}^{t_f} \Phi[\mathbf{x}(t), \dot{\mathbf{x}}(t), t]\, dt \tag{4.1-1}$$

with respect to the set of all admissible (see Sec. 2.1) trajectories. Let t_f be the terminal time associated with optimal trajectory $\mathbf{x}$. Associated with each perturbation $\mathbf{h}$ away from the optimal trajectory is a perturbation δt_f in the terminal time. Let the first variation δJ be the part of

$$\Delta J = J(\mathbf{x} + \mathbf{h}, t_f + \delta t_f) - J(\mathbf{x}, t_f) \tag{4.1-2}$$

which is linear in $\mathbf{h}$ and δt_f. Substituting Eq. (4.1-1) into Eq. (4.1-2), taking the linear terms of ΔJ (in $\mathbf{h}$, δt_f, and $\dot{\mathbf{h}}$), and performing the usual integration by parts to reduce terms dependent on $\dot{\mathbf{h}}$ to terms dependent on $\mathbf{h}$, we obtain

$$\delta J = \Phi[\mathbf{x}(t_f), \dot{\mathbf{x}}(t_f), t_f]\, \delta t_f + \mathbf{h}^T(t_f)\frac{\partial \Phi[\mathbf{x}(t_f), \dot{\mathbf{x}}(t_f), t_f]}{\partial \dot{\mathbf{x}}(t_f)} + \int_{t_o}^{t_f} \mathbf{h}^T(t)\left(\frac{\partial \Phi}{\partial \mathbf{x}} - \frac{d}{dt}\frac{\partial \Phi}{\partial \dot{\mathbf{x}}}\right) dt \tag{4.1-3}$$

where for convenience, we have assumed that the initial condition is fixed and hence $\mathbf{h}(t_o) = \mathbf{0}$.

In order to rephrase Eq. (4.1-3) into a convenient form, we introduce the following notation. We define

$$\delta\mathbf{x}(t_f) = \mathbf{h}(t_f) + \dot{\mathbf{x}}(t_f)\, \delta t_f. \tag{4.1-4}$$

From a Taylor series expansion of $\mathbf{x}(t_f + \delta t_f)$, we note that $\delta\mathbf{x}(t_f)$ is a close approximation to that part of $[\mathbf{x}(t_f + \delta t_f) + \mathbf{h}(t_f + \delta t_f)] - \mathbf{x}(t_f)$ which is linear in $\mathbf{h}(t_f)$ and δt_f. Submitting Eq. (4.1-4) into Eq. (4.1-3) and rearranging, the first variation becomes

$$\delta J = \left\{\Phi[\mathbf{x}(t_f), \dot{\mathbf{x}}(t_f), t_f] - \dot{\mathbf{x}}^T(t_f)\frac{\partial \Phi[\mathbf{x}(t_f), \dot{\mathbf{x}}(t_f), t_f]}{\partial \dot{\mathbf{x}}(t_f)}\right\} \delta t_f + \delta\mathbf{x}^T(t_f)\frac{\partial \Phi[\mathbf{x}(t_f), \dot{\mathbf{x}}(t_f), t_f]}{\partial \dot{\mathbf{x}}(t_f)} + \int_{t_o}^{t_f} \mathbf{h}^T(t_f)\left\{\frac{\partial \Phi}{\partial \mathbf{x}} - \frac{d}{dt}\frac{\partial \Phi}{\partial \dot{\mathbf{x}}}\right\} dt. \tag{4.1-5}$$

In much of our work, it will be convenient to define a quantity, called the Hamiltonian, by

$$H[\mathbf{x}(t), \boldsymbol{\lambda}(t), t] = \Phi - \dot{\mathbf{x}}^T\frac{\partial \Phi}{\partial \dot{\mathbf{x}}} = \Phi + \dot{\mathbf{x}}^T\boldsymbol{\lambda} \tag{4.1-6}$$

where the Hamiltonian is not a function of $\dot{\mathbf{x}}$; $\mathbf{x}(t)$, and $\boldsymbol{\lambda}(t)$ are called the canonical variables. In terms of the Hamiltonian, the first variation of Eq. (4.1-1), which is Eq. (4.1-5) becomes

$$\delta J = -\boldsymbol{\delta}\mathbf{x}^T(t_f)\boldsymbol{\lambda}(t_f) + H[\mathbf{x}(t_f), \boldsymbol{\lambda}(t_f), t_f]\,\delta t_f + \int_{t_0}^{t_f} \mathbf{h}^T(t)\left\{\frac{\partial H}{\partial \mathbf{x}} + \frac{d\boldsymbol{\lambda}}{dt}\right\} dt. \tag{4.1-7}$$

To establish a necessary condition for a minimum, it is necessary that the integrand in Eqs. (4.1-5) and (4.1-7) vanish and also that the transversality condition, as obtained from Eq. (4.1-7),

$$-\boldsymbol{\delta}\mathbf{x}^T(t_f)\boldsymbol{\lambda}(t_f) + H[\mathbf{x}(t_f), \boldsymbol{\lambda}(t_f), t_f]\,\delta t_f = 0 \tag{4.1-8}$$

be satisfied.

4.2 Weierstrass-Erdmann conditions

Thus far in our development, admissible trajectories have been constrained to be continuously differentiable with respect to $\mathbf{x}$ and t. This functional constraint on the class of all admissible trajectories is often unrealistically restrictive, as the following example will show. For this example, an optimal admissible solution does not exist; however, if the functional restriction on an admissible trajectory is sufficiently relaxed, existence of an optimal admissible trajectory is assured. We now examine the consequences of our new definition of an admissible trajectory which are the Weierstrass-Erdmann conditions [1].

Let us consider the problem of minimizing the cost function

$$J = \int_0^1 x^2(2 - \dot{x})^2\,dt$$

subject to

$$x(0) = 0, \qquad x(1) = 1.$$

Physically, it is clear that the absolute minimum for J is 0 and that this is obtained for

$$x(t) = 0, \qquad t \in [0, \tfrac{1}{2}]$$
$$x(t) = 2t - 1, \qquad t \in [\tfrac{1}{2}, 1]$$

which is certainly a solution to the Euler-Lagrange equation for this problem

$$x^2\ddot{x} + x\dot{x}^2 - 4x = 0.$$

There is one disturbing feature about this solution, however, in that the optimum $x(t)$ has a "corner" or discontinuous first derivative which gives rise to formal difficulty since $\ddot{x}$ is contained in the Euler-Lagrange equations. Thus, the solution of the above problem is not admissible. Certainly, this particular function $x(t)$ is continuously differentiable everywhere except at a finite number of points (in this case the single point $t = \frac{1}{2}$). Thus, in relaxing the set of admissible trajectories to allow for functions which are piecewise continuously differentiable, the function $x(t)$ is admissible and the above problem then has an optimal admissible control.

The Weierstrass-Erdmann corner conditions furnish us with necessary conditions for an optimal trajectory to have a discontinuous derivative at a point in the control interval of interest. Specifically, consider the problem of finding a trajectory among the class of all continuously differentiable functions on $[a, b]$ having a corner at $c \in (a, b)$ which satisfies fixed initial and final boundary values such that the functional

$$J(\mathbf{x}) = \int_a^b \Phi[\mathbf{x}(t), \dot{\mathbf{x}}(t), t]\, dt$$

has an extremum. It is of course clear that, for $t \in [a, c]$ and $t \in [c, b]$, the function $\mathbf{x}(t)$ must satisfy the Euler-Lagrange equations for a minimum

$$\frac{d}{dt}\frac{\partial \Phi}{\partial \dot{\mathbf{x}}} - \frac{\partial \Phi}{\partial \mathbf{x}} = \mathbf{0}.$$

We may rewrite the cost function as a sum of two cost functions:

$$\begin{aligned} J(\mathbf{x}) &= \int_a^c \Phi[\mathbf{x}(t), \dot{\mathbf{x}}(t), t]\, dt + \int_c^b \Phi[\mathbf{x}(t), \dot{\mathbf{x}}(t), t]\, dt \\ &= J_1(\mathbf{x}) + J_2(\mathbf{x}). \end{aligned}$$

We may now take the first variation $\delta J_1(\mathbf{x})$ and $\delta J_2(\mathbf{x})$ separately. We assume, for the moment only, that a and b are fixed, and we require that the $\hat{\mathbf{x}}(t)$ calculated from $J_1(\mathbf{x})$ and $J_2(\mathbf{x})$ is the same at $t = c$ which is unknown. Since c is arbitrary, the first variation of $J_1(\mathbf{x})$ is

$$\delta J_1(\mathbf{x}) = \boldsymbol{\delta}\mathbf{x}^T(a)\frac{\partial \Phi[\mathbf{x}(a), \dot{\mathbf{x}}(a), a]}{\partial \dot{\mathbf{x}}(a)} +$$

$$\left\{\Phi[\mathbf{x}(c), \dot{\mathbf{x}}(c), c] - \dot{\mathbf{x}}^T(c)\frac{\partial \Phi[\mathbf{x}(c), \dot{\mathbf{x}}(c), c]}{\partial \dot{\mathbf{x}}(c)}\right\} \delta c + \boldsymbol{\delta}\mathbf{x}^T(c)\frac{\partial \Phi[\mathbf{x}(c), \dot{\mathbf{x}}(c), c]}{\partial \dot{\mathbf{x}}(c)} +$$

$$\int_a^c \mathbf{h}^T(t)\left\{\frac{\partial \Phi}{\partial \mathbf{x}} - \frac{d}{dt}\frac{\partial \Phi}{\partial \dot{\mathbf{x}}}\right\} dt.$$

Since $\mathbf{x}(t)$ satisfies the Euler-Lagrange equations for an extremal and since $\boldsymbol{\delta}\mathbf{x}(a) = \mathbf{0}$, we have

$$\delta J_1(\mathbf{x}) = \boldsymbol{\delta}\mathbf{x}^T(\tau)\frac{\partial \Phi[\mathbf{x}(\tau), \dot{\mathbf{x}}(\tau), \tau]}{\partial \dot{\mathbf{x}}(\tau)} +$$

$$\left\{\Phi[\mathbf{x}(\tau), \dot{\mathbf{x}}(\tau), \tau] - \dot{\mathbf{x}}^T(\tau)\frac{\partial \Phi[\mathbf{x}(\tau), \dot{\mathbf{x}}(\tau), \tau]}{\partial \dot{\mathbf{x}}(\tau)}\right\} \delta\tau \qquad (\text{for } \tau = c - 0).$$

In a similar fashion, we can show that the first variation for the extremal solution of $J_2(\mathbf{x})$ is

$$\delta J_2(\mathbf{x}) = -\boldsymbol{\delta}\mathbf{x}^T(\tau)\frac{\partial\Phi[\mathbf{x}(\tau), \dot{\mathbf{x}}(\tau), \tau]}{\partial\dot{\mathbf{x}}(\tau)} - \left\{\Phi[\mathbf{x}(\tau), \dot{\mathbf{x}}(\tau), \tau] - \dot{\mathbf{x}}^T(\tau)\frac{\partial\Phi[\mathbf{x}(\tau), \dot{\mathbf{x}}(\tau), \tau]}{\partial\dot{\mathbf{x}}(\tau)}\right\}\delta\tau \qquad (\text{for } \tau = c + 0).$$

In order to obtain the extremum, the extremal solution must satisfy

$$\delta J(\mathbf{x}) = \delta J_1(\mathbf{x}) + \delta J_2(\mathbf{x}) = 0$$

Thus

$$\left.\frac{\partial\Phi}{\partial\dot{\mathbf{x}}}\right|_{t=c-0} = \left.\frac{\partial\Phi}{\partial\dot{\mathbf{x}}}\right|_{t=c+0} \tag{4.2-1}$$

$$\left.\Phi - \dot{\mathbf{x}}^T\frac{\partial\Phi}{\partial\dot{\mathbf{x}}}\right|_{t=c-0} = \left.\Phi - \dot{\mathbf{x}}^T\frac{\partial\Phi}{\partial\dot{\mathbf{x}}}\right|_{t=c+0} \tag{4.2-2}$$

since $\boldsymbol{\delta}\mathbf{x}$ and δt_f are arbitrary. These requirements, Eqs. (4.2-1) and (4.2-2), are called the *Weierstrass-Erdmann corner conditions* and must hold at any point c where the extremal has a corner. If we use the Hamiltonian canonical variables

$$H = \Phi - \dot{\mathbf{x}}^T\frac{\partial\Phi}{\partial\dot{\mathbf{x}}} = \Phi + \boldsymbol{\lambda}^T\dot{\mathbf{x}}$$

$$\boldsymbol{\lambda} = -\frac{\partial\Phi}{\partial\dot{\mathbf{x}}},$$

we immediately see that the Weierstrass-Erdmann conditions simply require H and $\boldsymbol{\lambda}$ to be continuous on the optimum trajectory at all points where there are corners.

4.3 The Bolza problem—no inequality constraints

In Sec. 3.7 we considered the solution of Lagrange problems with equality constraints of the form $\mathbf{g}(\mathbf{x}, \dot{\mathbf{x}}, t) = \mathbf{0}$ for all t in the control interval of interest. A special case of this equality constraint which is well-recognized as a model of a large and important class of physical systems is

$$\dot{\mathbf{x}}(t) = \mathbf{f}[\mathbf{x}(t), \mathbf{u}(t), t], \tag{4.3-1}$$

where the m-vector $\mathbf{u}$ represents the control function to be selected and the n-vector $\mathbf{x}$ represents the resulting trajectory. We will assume that $\mathbf{f}$ has continuous partial derivatives with respect to $\mathbf{x}$ and $\mathbf{u}$. Often it is the case that such smoothness assumptions guarantee that for any piecewise continuous function $\mathbf{u}$, there exists a unique, admissible trajectory $\mathbf{x}$ to Eq. (4.3-1). We therefore define the set of admissible control functions to be the class of piecewise continuous functions and assume that, for an admissible $\mathbf{u}$ and a given initial condition $\mathbf{x}(t_o)$, Eq. (4.3-1) defines a unique, admissible solution over the control interval of interest.

Throughout the remainder of this section and the following section, we will consider the development of necessary conditions for Bolza problems subject to the equality constraint in Eq. (4.3-1). Subsections (4.3-1) and (4.3-2) will consider the fixed final time and the unspecified final time cases when no restrictions are imposed on the value that $\mathbf{u}$ can take at each time t during the control interval of interest. Section 4.4 considers two cases of inequality constraints on the control function and its associated trajectory over the control interval.

4.3-1
Continuous optimal control problems—fixed beginning and terminal times—no inequality constraints

We now consider the problem of determining an admissible control function $\mathbf{u}$ in order to minimize the criterion

$$J = \theta[\mathbf{x}(t), t]\Big|_{t=t_o}^{t=t_f} + \int_{t_o}^{t_f} \phi[\mathbf{x}(t), \mathbf{u}(t), t]\, dt, \tag{4.3-2}$$

where θ and ϕ possess continuous partial derivatives in $\mathbf{x}$ and $\mathbf{u}$.

We use the method of Lagrange multipliers discussed in the last chapter to adjoin the system differential equality constraint to the cost function, which gives us

$$J = \theta[\mathbf{x}(t), t]\Big|_{t=t_o}^{t=t_f} + \int_{t_o}^{t_f} \{\phi[\mathbf{x}(t), \mathbf{u}(t), t] + \boldsymbol{\lambda}^T(t)[\mathbf{f}[\mathbf{x}(t), \mathbf{u}(t), t] - \dot{\mathbf{x}}]\}\, dt. \tag{4.3-3}$$

We define a scalar function, the Hamiltonian, as

$$H[\mathbf{x}(t), \mathbf{u}(t), \boldsymbol{\lambda}(t), t] = \phi[\mathbf{x}(t), \mathbf{u}(t), t] + \boldsymbol{\lambda}^T(t)\mathbf{f}[\mathbf{x}(t), \mathbf{u}(t), t]. \tag{4.3-4}$$

Thus the cost function becomes

$$J = \theta[\mathbf{x}(t), t]\Big|_{t=t_o}^{t=t_f} + \int_{t_o}^{t_f} \{H[\mathbf{x}(t), \mathbf{u}(t), \boldsymbol{\lambda}(t), t] - \boldsymbol{\lambda}^T(t)\dot{\mathbf{x}}\}\, dt. \tag{4.3-5}$$

If we integrate the last term in the integrand of Eq. (4.3-5) by parts, we obtain

$$J = \{\theta[\mathbf{x}(t), t] - \boldsymbol{\lambda}^T(t)\mathbf{x}(t)\}\Big|_{t=t_o}^{t=t_f} + \int_{t_o}^{t_f} \{H[\mathbf{x}(t), \mathbf{u}(t), \boldsymbol{\lambda}(t), t] + \dot{\boldsymbol{\lambda}}^T\mathbf{x}(t)\}\, dt. \tag{4.3-6}$$

We now take the first variation of J for variations in the control vector and, consequently, in the state vector about the optimal control and optimal state vector. This gives us

$$\delta J = \left\{ \boldsymbol{\delta}\mathbf{x}^T \left[\frac{\partial \theta}{\partial \mathbf{x}} - \boldsymbol{\lambda} \right] \right\} \Bigg|_{t=t_o}^{t=t_f} + \int_{t_o}^{t_f} \left\{ \boldsymbol{\delta}\mathbf{x}^T \left[\frac{\partial H}{\partial \mathbf{x}} + \dot{\boldsymbol{\lambda}} \right] + \boldsymbol{\delta}\mathbf{u}^T \left[\frac{\partial H}{\partial \mathbf{u}} \right] \right\} dt. \qquad (4.3\text{-}7)$$

A necessary condition for a minimum is that the first variation in J vanish for arbitrary variations $\boldsymbol{\delta}\mathbf{x}$ and $\boldsymbol{\delta}\mathbf{u}$. Thus we have as the necessary condition for a minimum the very important relations

$$\boldsymbol{\delta}\mathbf{x}^T \left[\frac{\partial \theta}{\partial \mathbf{x}} - \boldsymbol{\lambda} \right] = 0, \qquad \text{for} \quad t = t_o, t_f \qquad (4.3\text{-}8)$$

$$\dot{\boldsymbol{\lambda}} = -\frac{\partial H}{\partial \mathbf{x}}, \qquad \dot{\mathbf{x}} = \mathbf{f}(\mathbf{x}, \mathbf{u}, t) = \frac{\partial H}{\partial \boldsymbol{\lambda}} \qquad (4.3\text{-}9)$$

$$\frac{\partial H}{\partial \mathbf{u}} = \mathbf{0}. \qquad (4.3\text{-}10)$$

We now consider in more detail the transversality conditions expressed in Eq. (4.3-8).

For a large class of optimal control problems, the initial state of the system is specified but the terminal state is unspecified. In that case, Eq. (4.3-8) yields the transversality conditions as

$$\mathbf{x}(t_o) = \mathbf{x}_o, \qquad \boldsymbol{\lambda}(t_f) = \frac{\partial \theta[\mathbf{x}(t_f), t_f]}{\partial \mathbf{x}(t_f)} \qquad (4.3\text{-}11)$$

since $\boldsymbol{\delta}\mathbf{x}(t_o) = \mathbf{0}$, $\mathbf{x}(t_o)$ is fixed, and $\boldsymbol{\delta}\mathbf{x}(t_f)$ is completely arbitrary. In another broad class of problems $\mathbf{x}(t_o)$ and $\mathbf{x}(t_f)$ are fixed. In this case $\boldsymbol{\delta}\mathbf{x}(t_o)$ and $\boldsymbol{\delta}\mathbf{x}(t_f)$ must be zero, and $\mathbf{x}(t_o)$ and $\mathbf{x}(t_f)$ are the boundary conditions for the two-point boundary value problem. For many estimation problems, neither $\mathbf{x}(t_o)$ nor $\mathbf{x}(t_f)$ are fixed and $\theta = 0$. In that case, Eq. (4.3-8) yields $\boldsymbol{\lambda}(t_o) = \boldsymbol{\lambda}(t_f) = 0$ as the boundary conditions for the problem since $\boldsymbol{\delta}\mathbf{x}(t_o)$ and $\boldsymbol{\delta}\mathbf{x}(t_f)$ are arbitrary. In still another case, we might have $\mathbf{x}(t_o) = \mathbf{x}_o$, $\theta = 0$, and $||\mathbf{x}(t_f)||^2 = 1$. In this event, it is easy for us to show that the final transversality conditions are obtained if we solve the two scalar equations, each in n variables

$$\boldsymbol{\delta}\mathbf{x}^T(t_f)\mathbf{x}(t_f) = 0, \qquad \boldsymbol{\delta}\mathbf{x}^T(t_f)\boldsymbol{\lambda}(t_f) = 0. \qquad (4.3\text{-}12)$$

We now give a more general and precise interpretation to the transversality conditions. For the general case where the initial manifold is

$$\mathbf{M}[\mathbf{x}(t_o), t_o] = 0 \qquad (4.3\text{-}13)$$

and the terminal manifold is

$$\mathbf{N}[\mathbf{x}(t_f), t_f] = 0, \qquad (4.3\text{-}14)$$

we adjoin these conditions to the θ function by means of Lagrange multipliers, ξ and ν and obtain for the cost function

$$J = \theta[\mathbf{x}(t), t]\Big|_{t=t_o}^{t=t_f} - \boldsymbol{\xi}^T\mathbf{M}[\mathbf{x}(t_o), t_o] + \mathbf{v}^T\mathbf{N}[\mathbf{x}(t_f), t_f] + \int_{t_o}^{t_f} \{H[\mathbf{x}(t), \mathbf{u}(t), \boldsymbol{\lambda}(t), t] - \boldsymbol{\lambda}^T(t)\dot{\mathbf{x}}\}\, dt. \tag{4.3-15}$$

We now apply the usual variational techniques to obtain for the transversality conditions at the initial time:

$$\boldsymbol{\lambda}(t_o) = \frac{\partial\theta}{\partial\mathbf{x}} + \left(\frac{\partial\mathbf{M}^T}{\partial\mathbf{x}}\right)\boldsymbol{\xi}, \qquad \mathbf{M}[\mathbf{x}(t), t] = \mathbf{0}, \qquad t = t_o. \tag{4.3-16}$$

The n initial conditions are obtained from this, with r parameters to be found in Eq. (4.3-16) such that we satisfy the r conditions of Eq. (4.3-13). In a similar fashion, the terminal condition is

$$\boldsymbol{\lambda}(t_f) = \frac{\partial\theta}{\partial\mathbf{x}} + \left(\frac{\partial\mathbf{N}^T}{\partial\mathbf{x}}\right)\mathbf{v}, \qquad \mathbf{N}[\mathbf{x}(t), t] = \mathbf{0}, \qquad t = t_f; \tag{4.3-17}$$

n terminal conditions are obtained from this with q parameters $\mathbf{v}$ found in Eq. (4.3-17) such that the q conditions of Eq. (4.3-14) are satisfied.

The n vector differential equation obtained from Eq. (4.3-9) will be called the adjoint equation. Equation (4.3-10) provides the coupling relation between the original plant dynamics, Eq. (4.3-1), and the adjoint equation, the $\dot{\boldsymbol{\lambda}}$ equation of Eq. (4.3-9). This coupling equation was obtained from

$$\delta J = \ldots + \int_{t_o}^{t_f} \left\{\boldsymbol{\delta}\mathbf{u}^T\frac{\partial H}{\partial\mathbf{u}} + \ldots\right\} dt,$$

and it is important to note that $\boldsymbol{\delta}\mathbf{u}$ must be completely arbitrary in order for us to draw the conclusion that $\partial H/\partial\mathbf{u} = 0$ to obtain the optimal control. For the problem posed here where the admissible control set is infinite, $\boldsymbol{\delta}\mathbf{u}$ can be completely arbitrary. Where the admissible control is bounded, $\boldsymbol{\delta}\mathbf{u}$ cannot be completely arbitrary, and $\partial H/\partial\mathbf{u} = \mathbf{0}$ may not be the correct requirement. We will have more to say about this later. The solution we have obtained for this problem is a special case of the Pontryagin maximum principle.

It is also interesting to note that, since $H = \phi + \boldsymbol{\lambda}^T\mathbf{f}$, we may compute the total derivative with respect to time as

$$\frac{dH}{dt} = \frac{\partial\phi}{\partial t} + \dot{\mathbf{x}}^T\left[\frac{\partial\phi}{\partial\mathbf{x}} + \left(\frac{\partial\mathbf{f}^T}{\partial\mathbf{x}}\right)\boldsymbol{\lambda}\right] + \dot{\mathbf{u}}^T\left[\frac{\partial\phi}{\partial\mathbf{u}} + \left(\frac{\partial\mathbf{f}^T}{\partial\mathbf{u}}\right)\boldsymbol{\lambda}\right] + \dot{\boldsymbol{\lambda}}^T\mathbf{f} + \boldsymbol{\lambda}^T\frac{\partial\mathbf{f}}{\partial t} \tag{4.3-18}$$

but from Eqs. (4.3-9) and (4.3-4) we have

$$\dot{\boldsymbol{\lambda}} = -\frac{\partial H}{\partial\mathbf{x}} = -\frac{\partial\phi}{\partial\mathbf{x}} - \left(\frac{\partial\mathbf{f}^T}{\partial\mathbf{x}}\right)\boldsymbol{\lambda} \tag{4.3-19}$$

and from Eq. (4.3-4)

$$\frac{\partial H}{\partial\mathbf{u}} = \frac{\partial\phi}{\partial\mathbf{u}} + \left(\frac{\partial\mathbf{f}^T}{\partial\mathbf{u}}\right)\boldsymbol{\lambda}. \tag{4.3-20}$$

Thus, since $\dot{\mathbf{x}}^T\boldsymbol{\lambda} = \dot{\boldsymbol{\lambda}}^T\mathbf{f}$, Eq. (4.3-18) becomes

$$\frac{dH}{dt} = \frac{\partial \phi}{\partial t} + \boldsymbol{\lambda}^T \frac{\partial \mathbf{f}}{\partial t} + \dot{\mathbf{u}}^T \frac{\partial H}{\partial \mathbf{u}}. \tag{4.3-21}$$

We see that, if ϕ and $\mathbf{f}$ are not explicit functions of time, the Hamiltonian is constant along an optimal trajectory where $\partial H/\partial \mathbf{u} = \mathbf{0}$. It can be shown that this is always true along an optimal trajectory, even if we cannot require $\partial H/\partial \mathbf{u} = 0$. We will make use of this fact in a later development.

In order that J be a minimum, the second variation of J must be nonnegative along all trajectories such that Eq. (4.3-1) is satisfied. Therefore we need to compute the second variation of J in Eq. (4.3-6) and impose the requirement that the variation of Eq. (4.3-1) is zero, or that

$$\boldsymbol{\delta}\dot{\mathbf{x}} - \left(\frac{\partial \mathbf{f}}{\partial \mathbf{x}}\right)\boldsymbol{\delta}\mathbf{x} - \left(\frac{\partial \mathbf{f}}{\partial \mathbf{u}}\right)\boldsymbol{\delta}\mathbf{u} = \mathbf{0}. \tag{4.3-22}$$

Applying this condition and taking the quadratic part of the Taylor series expansion of $J(\mathbf{x} + \boldsymbol{\delta}\mathbf{x}, \mathbf{u} + \boldsymbol{\delta}\mathbf{u}) - J(\mathbf{x}, \mathbf{u})$, we have for the second variation

$$\delta^2 J = \frac{1}{2}\left[\boldsymbol{\delta}\mathbf{x}^T \frac{\partial^2 \theta}{\partial \mathbf{x}^2}\boldsymbol{\delta}\mathbf{x}\right]\Bigg|_{t=t_0}^{t=t_f} +$$

$$\frac{1}{2}\int_{t_0}^{t_f}[\boldsymbol{\delta}\mathbf{x}^T\ \boldsymbol{\delta}\mathbf{u}^T]\begin{bmatrix} \dfrac{\partial^2 H}{\partial \mathbf{x}^2} & \dfrac{\partial}{\partial \mathbf{u}}\dfrac{\partial H}{\partial \mathbf{x}} \\ \left[\dfrac{\partial}{\partial \mathbf{u}}\dfrac{\partial H}{\partial \mathbf{x}}\right]^T & \dfrac{\partial^2 H}{\partial \mathbf{u}^2}\end{bmatrix}\begin{bmatrix}\boldsymbol{\delta}\mathbf{x} \\ \boldsymbol{\delta}\mathbf{u}\end{bmatrix} dt \tag{4.3-23}$$

and this must be nonnegative for a minimum. This will be the case if the $n + m$ square matrix under the integral sign and $\partial^2\theta/\partial\mathbf{x}^2$ are nonnegative definite.

Example 4.3-1. We are given the differential system consisting of three cascaded integrators

$$\dot{x}_1 = x_2 \qquad x_1(0) = 0$$
$$\dot{x}_2 = x_3 \qquad x_2(0) = 0$$
$$\dot{x}_3 = u \qquad x_3(0) = 0.$$

We wish to drive the system so that we reach the terminal manifold

$$x_1^2(1) + x_2^2(1) = 1$$

such that the cost function

$$J = \tfrac{1}{2}\int_0^1 u^2\,dt$$

is minimized. The solution to the problem proceeds as follows. We compute the Hamiltonian from Eq. (4.3-4) as

$$H = \tfrac{1}{2}u^2 + \lambda_1 x_2 + \lambda_2 x_3 + \lambda_3 u$$

and determine the coupling relation, Eq. (4.3-10),

$$\frac{\partial H}{\partial u} = 0 = u + \lambda_3$$

and the adjoint Eq. (4.3-9),

$$\dot{\lambda}_1 = -\frac{\partial H}{\partial x_1} = 0$$

$$\dot{\lambda}_2 = -\frac{\partial H}{\partial x_2} = -\lambda_1$$

$$\dot{\lambda}_3 = -\frac{\partial H}{\partial x_3} = -\lambda_2.$$

From Eqs. (4.3-14) and (4.3-17) we see that the transversality condition at the terminal time is

$$x_1^2(1) + x_2^2(1) = 1$$

$$\boldsymbol{\lambda}(1) = \frac{\partial \theta}{\partial \mathbf{x}} + \left(\frac{\partial \mathbf{N}^T}{\partial \mathbf{x}}\right)\mathbf{v}, \qquad t = t_f$$

where

$$N[\mathbf{x}(t_f), t_f] = x_1^2(t_f) + x_2^2(t_f) - 1 = 0, \qquad t_f = 1.$$

Thus

$$\boldsymbol{\lambda}(1) = \begin{bmatrix} \lambda_1(1) \\ \lambda_2(1) \\ \lambda_3(1) \end{bmatrix} = \begin{bmatrix} 2x_1(1)\nu \\ 2x_2(1)\nu \\ 0 \end{bmatrix}.$$

Thus the problem of finding the optimal control and associated trajectories for this example is completely resolved when we solve the two-point boundary value problem represented by

$$\begin{aligned}
&\dot{x}_1 = x_2 && x_1(0) = 0 \\
&\dot{x}_2 = x_3 && x_2(0) = 0 \\
&\dot{x}_3 = -\lambda_3 && x_3(0) = 0 \\
&\dot{\lambda}_1 = 0 && \lambda_1(1) = 2x_1(1)\nu \\
&\dot{\lambda}_2 = -\lambda_1 && \lambda_2(1) = 2x_2(1)\nu \\
&\dot{\lambda}_3 = -\lambda_2 && \lambda_3(1) = 0
\end{aligned}$$

$$\left.\begin{matrix} \lambda_1(1) = 2x_1(1)\nu \\ \lambda_2(1) = 2x_2(1)\nu \end{matrix}\right\} x_1^2(1) + x_2^2(1) = 1$$

Although the six first-order differential equations represented above are perfectly linear and time invariant, the solution to this problem is complicated by the nonlinear nature of the terminal conditions. We shall discover various iterative schemes for overcoming this difficulty in Chapter 10.

4.3-2
Continuous optimal control problems— fixed beginning and unspecified terminal times— no inequality constraints

The material of the previous subsection may be easily extended to the case where the terminal manifold equation is a function of the terminal time, and the terminal time is unspecified. For convenience we will assume that the initial time and the initial state vector are specified. Solution may then easily be obtained for the case where the initial time and initial state vector are unspecified. Therefore the problem becomes one of minimizing the cost function

$$J = \theta[\mathbf{x}(t_f), t_f] + \int_{t_o}^{t_f} \phi[\mathbf{x}(t), \mathbf{u}(t), t]\, dt \tag{4.3-24}$$

for the system described by

$$\dot{\mathbf{x}} = \mathbf{f}[\mathbf{x}(t), \mathbf{u}(t), t], \qquad \mathbf{x}(t_o) = \mathbf{x}_o \tag{4.3-25}$$

where t_o is fixed and where, at the unspecified terminal time $t = t_f$, the q vector terminal manifold equation

$$\mathbf{N}[\mathbf{x}(t_f), t_f] = \mathbf{0} \tag{4.3-26}$$

is satisfied. It may be noted here that the terminal manifold line, $x(t_f) = c(t_f)$ of the previous chapter becomes $\mathbf{N}[\mathbf{x}(t_f), t_f] = \mathbf{0}$, which is more general. We adjoin the equality constraints to the cost function via Lagrange multipliers to obtain

$$J = \theta[\mathbf{x}(t_f), t_f] + \mathbf{v}^T\mathbf{N}[\mathbf{x}(t_f), t_f] + \int_{t_o}^{t_f} \{\phi[\mathbf{x}(t), \mathbf{u}(t), t] + \boldsymbol{\lambda}^T(t)[\mathbf{f}[\mathbf{x}(t), \mathbf{u}(t), t] - \dot{\mathbf{x}}]\}\, dt. \tag{4.3-27}$$

As before, we define the Hamiltonian

$$H[\mathbf{x}(t), \mathbf{u}(t), \boldsymbol{\lambda}(t), t] = \phi[\mathbf{x}(t), \mathbf{u}(t), t] + \boldsymbol{\lambda}^T(t)\mathbf{f}[\mathbf{x}(t), \mathbf{u}(t), t]$$

and integrate a portion of the cost function, Eq. (4.3-27), to obtain

$$J = \theta[\mathbf{x}(t_f), t_f] + \mathbf{v}^T\mathbf{N}[\mathbf{x}(t_f), t_f] - \boldsymbol{\lambda}^T(t_f)\mathbf{x}(t_f) + \boldsymbol{\lambda}^T(t_o)\mathbf{x}(t_o) + \int_{t_o}^{t_f} \{H[\mathbf{x}(t), \mathbf{u}(t), \boldsymbol{\lambda}(t), t] + \dot{\boldsymbol{\lambda}}^T\mathbf{x}(t)\}\, dt. \tag{4.3-28}$$

We again form the first variation by letting

$$\mathbf{x}(t) = \hat{\mathbf{x}}(t) + \mathbf{h}(t), \qquad \mathbf{u}(t) = \hat{\mathbf{u}}(t) + \boldsymbol{\delta}\mathbf{u}(t), \qquad t_f = \hat{t}_f + \delta t_f \tag{4.3-29}$$

and then we form the difference $J[\mathbf{x}, \mathbf{u}, t_f] - J[\hat{\mathbf{x}}, \hat{\mathbf{u}}, \hat{t}_f]$ and retain only the linear terms. Thus we have, after dropping the ^ notation for convenience,

$$\delta J = \delta t_f \left\{ H[\mathbf{x}(t_f), \mathbf{u}(t_f), \boldsymbol{\lambda}(t_f), t_f] + \frac{\partial \Theta}{\partial t_f} \right\} + \boldsymbol{\delta}\mathbf{x}^T(t_f) \left\{ \frac{\partial \Theta}{\partial \mathbf{x}} - \boldsymbol{\lambda}(t_f) \right\} + \int_{t_o}^{t_f} \left\{ \mathbf{h}^T(t) \left[\frac{\partial H}{\partial \mathbf{x}} + \dot{\boldsymbol{\lambda}} \right] + \boldsymbol{\delta}\mathbf{u}^T(t) \left[\frac{\partial H}{\partial \mathbf{u}} \right] \right\} dt \tag{4.3-30}$$

where

$$\Theta[\mathbf{x}(t_f), \mathbf{v}, t_f] = \theta[\mathbf{x}(t_f), t_f] + \mathbf{v}^T\mathbf{N}[\mathbf{x}(t_f), t_f]. \tag{4.3-31}$$

We must set this first variation equal to zero to obtain the necessary conditions for a minimum. Therefore, the equations which determine the optimal control and state vector are

$$H = \phi[\mathbf{x}(t), \mathbf{u}(t), t] + \boldsymbol{\lambda}^T(t)\mathbf{f}[\mathbf{x}(t), \mathbf{u}(t), t] \tag{4.3-32}$$

$$\frac{\partial H}{\partial \boldsymbol{\lambda}} = \dot{\mathbf{x}} = \mathbf{f}[\mathbf{x}(t), \mathbf{u}(t), t] \tag{4.3-33}$$

$$\frac{\partial H}{\partial \mathbf{x}} = -\dot{\boldsymbol{\lambda}} = \frac{\partial \mathbf{f}^T[\mathbf{x}(t), \mathbf{u}(t), t]}{\partial \mathbf{x}}\boldsymbol{\lambda}(t) + \frac{\partial \phi[\mathbf{x}(t), \mathbf{u}(t), t]}{\partial \mathbf{x}} \tag{4.3-34}$$

$$\frac{\partial H}{\partial \mathbf{u}} = 0 = \frac{\partial \phi[\mathbf{x}(t), \mathbf{u}(t), t]}{\partial \mathbf{u}} + \frac{\partial \mathbf{f}^T[\mathbf{x}(t), \mathbf{u}(t), t]}{\partial \mathbf{u}}\boldsymbol{\lambda}(t). \tag{4.3-35}$$

These represent the $2n$ differential equations for the two-point boundary value problems. The conditions at the initial time are

$$\mathbf{x}(t_o) = \mathbf{x}_o \tag{4.3-36}$$

whereas those at the final time are

$$\boldsymbol{\lambda}(t_f) = \frac{\partial \Theta}{\partial \mathbf{x}(t_f)} = \frac{\partial \theta}{\partial \mathbf{x}(t_f)} + \left[\frac{\partial \mathbf{N}^T}{\partial \mathbf{x}(t_f)}\right]\mathbf{v} \tag{4.3-37}$$

$$\mathbf{N}[\mathbf{x}(t_f), t_f] = \mathbf{0} \tag{4.3-38}$$

and

$$H[\mathbf{x}(t_f), \mathbf{u}(t_f), \boldsymbol{\lambda}(t_f), t_f] + \frac{\partial \theta}{\partial t_f} + \left(\frac{\partial \mathbf{N}^T}{\partial t_f}\right)\mathbf{v} = 0. \tag{4.3-39}$$

Equation (4.3-37) provides n conditions with q Lagrange multipliers to be determined. Equation (4.3-38) provides q equations to eliminate the Lagrange multipliers, and Eq. (4.3-39) provides the one additional equation which we must have to determine the unspecified terminal time.

Example 4.3-2. For the first-order single integration system

$$\dot{x} = u, \qquad x(0) = 1,$$

we desire to find the control $u(t)$ which makes $x(t_f) = 0$, where t_f is unspecified, such as to make, for specified values of α and β,

$$J = t_f^\alpha + \tfrac{1}{2}\beta \int_0^{t_f} u^2\, dt$$

a minimum. For this problem

$$N[x(t_f), t_f] = x(t_f) = 0, \qquad \phi = \tfrac{1}{2}\beta u^2$$

$$\theta = t_f^\alpha, \qquad H = \tfrac{1}{2}\beta u^2 + \lambda u.$$

The canonic equations are:

$$\dot{x} = u = -\frac{\lambda}{\beta}, \qquad \dot{\lambda} = 0$$

with the boundary conditions $x(0) = 1$, $x(t_f) = 0$, where we determine the final time by solving Eq. (4.3-39) which becomes, for this example,

$$-\frac{\lambda^2(t_f)}{2\beta} + \alpha t_f^{\alpha-1} = 0.$$

The solutions to the canonic equations are

$$x(t) = -\frac{\lambda(t_f)t}{\beta} + 1, \qquad \lambda(t) = \lambda(t_f).$$

But since $x(t_f) = 0$, $t_f = \beta\lambda^{-1}(t_f)$, and in the particular case where $\beta = \alpha = 1$, we can easily show from the foregoing that $\lambda(t_f) = +(2)^{1/2}$, which determines the solution to this example. The optimum control is $u(t) = -\lambda(t) = -2^{1/2}$; the corresponding trajectory is $x(t) = 1 - 2^{1/2}t$, with $t_f = 2^{-1/2}$.

Example 4.3-3. A problem which will be of considerable interest to us later will be the "minimum time" problem. In that case

$$\theta[\mathbf{x}(t_f), t_f] = t_f, \qquad \phi = 0,$$

and we specify the optimal control and corresponding trajectory by solving Eqs. (4.3-32) through (4.3-35), which become

$$H[\mathbf{x}(t), \mathbf{u}(t), \boldsymbol{\lambda}(t), t] = \boldsymbol{\lambda}^T(t)\mathbf{f}[\mathbf{x}(t), \mathbf{u}(t), t]$$

$$\frac{\partial H}{\partial \lambda} = \dot{\mathbf{x}} = \mathbf{f}[\mathbf{x}(t), \mathbf{u}(t), t]$$

$$\frac{\partial H}{\partial \mathbf{x}} = -\dot{\boldsymbol{\lambda}} = \frac{\partial \mathbf{f}^T[\mathbf{x}(t), \mathbf{u}(t), t]}{\partial \mathbf{x}}\lambda(t)$$

$$\frac{\partial H}{\partial \mathbf{u}} = \mathbf{0} = \frac{\partial \mathbf{f}^T[\mathbf{x}(t), \mathbf{u}(t), t]}{\partial \mathbf{u}}\lambda(t)$$

with the boundary conditions specified by Eqs. (4.3-36) through (4.3-39)

$$\mathbf{x}(t_o) = \mathbf{x}_o$$

$$\boldsymbol{\lambda}(t_f) = \frac{\partial \mathbf{N}^T}{\partial \mathbf{x}(t_f)}\nu$$

$$\mathbf{N}[\mathbf{x}(t_f), t_f] = \mathbf{0}$$

$$H[\mathbf{x}(t_f), \mathbf{u}(t_f), t_f] = -1 - \left(\frac{\partial \mathbf{N}^T}{\partial t_f}\right)\nu.$$

In many cases, the system is brought to rest at the unspecified time, and the terminal manifold is the origin, so that

$$\mathbf{N}[\mathbf{x}(t_f), t_f] = \mathbf{x}(t_f) = \mathbf{0}.$$

Then the foregoing expressions reduce to

$$\mathbf{x}(t_o) = \mathbf{x}_o, \qquad \mathbf{x}(t_f) = \mathbf{0}$$

$$H[\mathbf{x}(t_f), \mathbf{u}(t_f), \boldsymbol{\lambda}(t_f), t_f] = -1.$$

If the Hamiltonian is not an explicit function of time, Eq. (4.3-21) which applies here as well, yields $dH/dt = 0$; therefore, for this minimum time problem

$$H[\mathbf{x}(t), \mathbf{u}(t), \boldsymbol{\lambda}(t), t] = H[\mathbf{x}(t), \mathbf{u}(t), \boldsymbol{\lambda}(t)] = -1.$$

It should be emphasized that we are not solving the usual minimum time problem since we have imposed no inequality constraints on the control (or state) variables. An alternate version of this problem would be to consider $\theta = 0$ and $\phi = 1$. This changes the Hamiltonian for this particular problem, but it certainly does not change the optimal control and state vector, as the reader can easily verify.

4.4 The Bolza problem with inequality constraints

The last section treated the Bolza problem with differential equation equality constraint, Eq. (4.3-1). There were no inequality constraints on either the control function or the state trajectory. We now consider the Bolza problem subject to one or more inequality constraints on the control and/or state variable.

4.4-1 The maximum principle with control variable inequality constraints

We now consider the determination of an admissible control function which minimizes the criterion function of Eq. (4.3-2) subject to both the equality constraint, Eq. (4.3-1), and the control variable constraint,

$$\mathbf{u}(t) \in \mho, \; t \in [t_o, t_f], \tag{4.4-1}$$

where $\mho$ is a given subset of R^m. We assume also that the initial state $\mathbf{x}(t_o) = \mathbf{x}_o$ and the initial time t_o are fixed and that the terminal time t_f is defined by the vector terminal manifold equation

$$\mathbf{N}[\mathbf{x}(t_f), t_f] = \mathbf{0}, \tag{4.4-2}$$

where $\mathbf{N}\colon R^{n+1} \longrightarrow R^q$.

Equation (4.4-1) distinguishes the problem posed above from the problems we considered in Sec. 4.3. Such a restricting assumption has important modeling significance since the controls that can be applied to many physical systems must be constrained in magnitude or in the number of feasible control settings. For example, the thrust of a rocket has a maximum and a minimum value; also, there may be a finite set of possible thrust values. We now present necessary conditions for this important problem, leading to the celebrated maximum principle due to McShane [16], Pontryagin [2], and others. Proof of the result for a simplified case will then follow.

We assume $\mathbf{u}$, having corresponding trajectory $\mathbf{x}$, is optimal on $[t_o, t_f]$. Then, recalling the definition of the Hamiltonian from Eq. (4.3-4),

$$H[\mathbf{x}(t), \mathbf{u}(t), \boldsymbol{\lambda}(t), t] \leq H[\mathbf{x}(t), \mathbf{v}, \boldsymbol{\lambda}(t), t], \quad \mathbf{v} \in \mho \tag{4.4-3}$$

$$\frac{\partial H}{\partial \boldsymbol{\lambda}} = \dot{\mathbf{x}} \tag{4.4-4}$$

$$\frac{\partial H}{\partial \mathbf{x}} = -\dot{\boldsymbol{\lambda}} \tag{4.4-5}$$

subject to the two-point boundary conditions

$$\mathbf{x}(t_o) = \mathbf{x}_o$$

$$\mathbf{N}[\mathbf{x}(t_f), t_f] = \mathbf{0}$$

$$\frac{\partial \theta}{\partial t_f} + \left(\frac{\partial \mathbf{N}^T}{\partial t_f}\right)\mathbf{v} + H = 0, \qquad \text{at} \quad t = t_f \tag{4.4-6}$$

$$\frac{\partial \theta}{\partial \mathbf{x}} + \left(\frac{\partial \mathbf{N}^T}{\partial \mathbf{x}}\right)\mathbf{v} - \boldsymbol{\lambda} = \mathbf{0}, \qquad \text{at} \quad t = t_f. \tag{4.4-7}$$

We frequently wish to transfer the system to the origin in minimum time so that we have

$$\mathbf{N}[\mathbf{x}(t_f), t_f] = \mathbf{0} = \mathbf{x}(t_f)$$

$$\theta[\mathbf{x}(t_f), t_f] = t_f$$

$$\phi = 0.$$

In this particular case, the transversality conditions become

$$\mathbf{x}(t_o) = \mathbf{x}_o$$

$$\mathbf{x}(t_f) = \mathbf{0}$$

$$H = -1, \qquad \text{at} \quad t = t_f.$$

Example 4.4-1. Let us consider briefly the time optimal control problem for a linear time-invariant system where the length of the control vector is constrained. We wish to minimize

$$J = t_f$$

for the system

$$\dot{\mathbf{x}} = \mathbf{A}\mathbf{x}(t) + \mathbf{B}\mathbf{u}(t)$$

$$\mathbf{x}(t_o) = \mathbf{x}_o$$

where $\mathbf{u}(t) \in \mathcal{U}$ means $\|\mathbf{u}(t)\| \leq 1$.

The Hamiltonian becomes

$$H[\mathbf{x}(t), \mathbf{u}(t), \boldsymbol{\lambda}(t), t] = \boldsymbol{\lambda}^T(t)[\mathbf{A}\mathbf{x}(t) + \mathbf{B}\mathbf{u}(t)].$$

To make H as small as possible with respect to a choice of $\mathbf{u}(t)$, we must have

$$\mathbf{u}(t) = \frac{-\mathbf{B}^T\boldsymbol{\lambda}(t)}{\|\mathbf{B}^T\boldsymbol{\lambda}(t)\|}.$$

The canonic equations become

$$\frac{\partial H}{\partial \boldsymbol{\lambda}} = \dot{\mathbf{x}} = \mathbf{A}\mathbf{x}(t) + \mathbf{B}\mathbf{u}(t), \qquad \frac{\partial H}{\partial \mathbf{x}} = -\dot{\boldsymbol{\lambda}} = \mathbf{A}^T\boldsymbol{\lambda}(t)$$

with the boundary conditions

$$\mathbf{x}(t_o) = \mathbf{x}_o, \qquad \mathbf{x}(t_f) = \mathbf{0}$$

where we determine t_f by solving

$$H[\mathbf{x}(t_f), \boldsymbol{\lambda}(t_f), \mathbf{u}(t_f)] = -1.$$

But, from Eq. (4.3-21) we see that $dH/dt = 0$ since the Hamiltonian does not depend explicitly on t. Thus the above equation becomes

$$H[\mathbf{x}(t), \mathbf{u}(t), \boldsymbol{\lambda}(t)] = -1 = \boldsymbol{\lambda}^T(t)[\mathbf{A}\mathbf{x}(t) + \mathbf{B}\mathbf{u}(t)]$$

which is the additional relation needed to determine the terminal time.

In a manner similar to [21], we now sketch a proof of the above conditions for optimality under the assumptions that: t_f is fixed, $\mathbf{x}(t_f)$ is unspecified, $\theta = 0$, the problem is time-invariant, and the function $\mathbf{f}$ satisfies the (uniform Lipschitz) condition

$$||\mathbf{f}(\mathbf{x}, \mathbf{u}) - \mathbf{f}(\mathbf{y}, \mathbf{v})|| \leq M(||\mathbf{x} - \mathbf{y}|| + ||\mathbf{u} - \mathbf{v}||), \tag{4.4-8}$$

where M is a known constant and where the above norms are finite-dimensional. The proof requires preliminary establishment of the Lipschitz condition

$$||\mathbf{x} - \mathbf{y}||_X \leq k\,||\mathbf{u} - \mathbf{v}||_U, \tag{4.4-9}$$

where $\mathbf{y}$ is the unique solution of Eq. (4.3-1) for given control function $\mathbf{v}$,

$$||\mathbf{x}||_X = \max_{t_o \leq t \leq t_f} ||\mathbf{x}(t)||, \tag{4.4-10}$$

and

$$||\mathbf{u}||_U = \int_{t_o}^{t_f} ||\mathbf{u}(s)||\, ds. \tag{4.4-11}$$

This is shown as follows. We note that

$$\delta\dot{\mathbf{x}}(t) = \frac{\partial \mathbf{f}}{\partial \mathbf{x}}[\mathbf{x}(t), \mathbf{u}(t)]\, \delta\mathbf{x}(t) + \frac{\partial \mathbf{f}}{\partial \mathbf{u}}[\mathbf{x}(t), \mathbf{u}(t)]\, \delta\mathbf{u}(t) \tag{4.4-12}$$

where $\delta\mathbf{x}(t_o) = \mathbf{0}$ since $\mathbf{x}(t_o)$ is fixed. Thus, there is a unique fundamental matrix $\boldsymbol{\Phi}(t, s)$ such that

$$\delta\mathbf{x}(t) = \int_{t_o}^{t} \boldsymbol{\Phi}(t, s) \frac{\partial \mathbf{f}}{\partial \mathbf{u}} [\mathbf{x}(s), \mathbf{u}(s)]\, \delta\mathbf{u}(s)\, ds \tag{4.4-13}$$

which can easily be shown to imply

$$||\delta\mathbf{x}(t)|| \leq Me^{M(t_f - t_o)} \int_{t_o}^{t_f} ||\delta\mathbf{u}(s)||\, ds = K||\delta\mathbf{u}||_U$$

for all $t \in [t_o, t_f]$, and the preliminary result follows directly.

We now show that for any admissible control function $\mathbf{v}$,

$$J(\mathbf{u}) - J(\mathbf{v}) = \int_{t_o}^{t_f} [H(\mathbf{x}, \mathbf{u}, \boldsymbol{\lambda}) - H(\mathbf{x}, \mathbf{v}, \boldsymbol{\lambda})]\, dt + \text{H.O.T.}(||\mathbf{u} - \mathbf{v}||_U). \tag{4.4-14}$$

To prove this result, assume $\mathbf{y}$ is the trajectory associated with $\mathbf{v}$. Then,

$$\phi(\mathbf{x}, \mathbf{u}) - \phi(\mathbf{y}, \mathbf{v}) = \phi(\mathbf{x}, \mathbf{u}) - \phi(\mathbf{x}, \mathbf{v}) + \phi(\mathbf{x}, \mathbf{v}) - \phi(\mathbf{y}, \mathbf{v}) =$$

$$\phi(\mathbf{x}, \mathbf{u}) - \phi(\mathbf{x}, \mathbf{v}) + \left[\frac{\partial \phi(\mathbf{x}, \mathbf{u})}{\partial \mathbf{x}}\right]^T (\mathbf{x} - \mathbf{y}) + \left[\frac{\partial \phi(\mathbf{y}, \mathbf{v})}{\partial \mathbf{x}} - \frac{\partial \phi(\mathbf{x}, \mathbf{u})}{\partial \mathbf{x}}\right]^T (\mathbf{x} - \mathbf{y}) +$$

$$\text{H.O.T.}(\|\mathbf{x} - \mathbf{y}\|_{\mathbf{X}}) = \phi(\mathbf{x}, \mathbf{u}) - \phi(\mathbf{x}, \mathbf{v}) +$$

$$\left[\frac{\partial \phi(\mathbf{x}, \mathbf{u})}{\partial \mathbf{x}}\right]^T (\mathbf{x} - \mathbf{y}) + \text{H.O.T.}(\|\mathbf{u} - \mathbf{v}\|_{\mathbf{U}}).$$

Similarly,

$$\dot{\mathbf{x}} - \dot{\mathbf{y}} = \mathbf{f}(\mathbf{x}, \mathbf{u}) - \mathbf{f}(\mathbf{x}, \mathbf{v}) + \left[\frac{\partial \mathbf{f}(\mathbf{x}, \mathbf{u})}{\partial \mathbf{x}}\right]^T (\mathbf{x} - \mathbf{y}) + \text{H.O.T.}(\|\mathbf{u} - \mathbf{v}\|_{\mathbf{U}}).$$

Thus,

$$J(\mathbf{u}) - J(\mathbf{v}) = \int_{t_o}^{t_f} [H(\mathbf{x}, \mathbf{u}, \boldsymbol{\lambda}) - H(\mathbf{x}, \mathbf{v}, \boldsymbol{\lambda})]\, dt + \int_{t_o}^{t_f} \left[\frac{\partial H(\mathbf{x}, \mathbf{u})}{\partial \mathbf{x}}\right]^T (\mathbf{x} - \mathbf{y})\, dt -$$

$$\int_{t_o}^{t_f} \boldsymbol{\lambda}^T(\dot{\mathbf{x}}, \dot{\mathbf{y}})\, dt + \text{H.O.T.}(\|\mathbf{u} - \mathbf{v}\|_{\mathbf{U}}). \tag{4.4-15}$$

Integration by parts and use of Eq. (4.4-5) indicate that the last two integrals on the right-hand side of Eq. (4.4-15) equal

$$-\int_{t_o}^{t_f} \dot{\boldsymbol{\lambda}}^T(\mathbf{x} - \mathbf{y})\, dt + \int_{t_o}^{t_f} \dot{\boldsymbol{\lambda}}^T(\mathbf{x} - \mathbf{y})\, dt + \boldsymbol{\lambda}^T(\mathbf{x} - \mathbf{y})\Big|_{t_o}^{t_f} = 0,$$

and Eq. (4.4-14) is validated.

We now proceed to prove Eq. (4.4-3) by contradiction. Suppose there is a $\bar{t} \in [t_o, t_f]$ and a $\mathbf{w} \in \mho$ such that $H[\mathbf{x}(\bar{t}), \mathbf{u}(\bar{t}), \boldsymbol{\lambda}(\bar{t})] > H[\mathbf{x}(\bar{t}), \mathbf{w}, \boldsymbol{\lambda}(\bar{t})]$. The piecewise continuity of $\mathbf{u}$ and the continuity of $\mathbf{x}$, $\boldsymbol{\lambda}$, $\mathbf{f}$, and ϕ imply the existence of an interval $[t_a, t_b] \in [t_o, t_f]$, $\bar{t} \in [t_a, t_b]$, and an $\epsilon > 0$ such that for all $t \in [t_a, t_b]$, $H[\mathbf{x}(t), \mathbf{u}(t), \boldsymbol{\lambda}(t)] - H[\mathbf{x}(t), \mathbf{w}, \boldsymbol{\lambda}(t)] > \epsilon$. We choose $\mathbf{v}$ so that $\mathbf{v}(t) = \mathbf{u}(t)$ for $t \notin [t_a, t_b]$, and $\mathbf{v}(t) = \mathbf{w}$ for $t \in [t_a, t_b]$. Then,

$$J(\mathbf{u}) - J(\mathbf{v}) > \epsilon\,(t_b - t_a) + \text{H.O.T.}(\|\mathbf{u} - \mathbf{v}\|_{\mathbf{U}}).$$

We note, however, that $\|\mathbf{u} - \mathbf{v}\|_{\mathbf{U}} = (t_b - t_a)[\text{H.O.T.}(t_b - t_a)]$. Thus, selection of $t_b - t_a$ small enough implies $J(\mathbf{u}) - J(\mathbf{v})$ can be made positive, which contradicts the optimality of $\mathbf{u}$; hence, Eq. (4.4-3) is proved. Considerably more exotic derivations of the maximum principle may be presented. For proofs of some of the more general forms of the maximum principle, see [17], [2], [6], and the references contained therein.

Often, Eq. (4.4-1) can conveniently be described by a set of inequalities of the form

$$\mathbf{g}[\mathbf{u}(t), t] \geq \mathbf{0}, \tag{4.4-16}$$

where $\mathbf{g}: R^{m+1} \longrightarrow R^r$. We may convert this inequality constraint to an equality constraint by writing for each component of $\mathbf{g}$ either

$$(\dot{z}_i)^2 = g_i[\mathbf{u}(t), t], \qquad z_i(t_o) = 0, \qquad i = 1, 2, \ldots, r \tag{4.4-17}$$

or

$$(y_i)^2 = g_i[\mathbf{u}(t), t], \qquad i = 1, 2, \ldots, r. \tag{4.4-18}$$

It is apparent that either of these two equations force g_i to be greater than or equal to zero since $(\dot{z}_i)^2$ and $(y_i)^2$ must certainly be greater than or equal to zero. This technique was apparently first proposed by Valentine [8] and extended by Berkovitz [6]. It is quite similar to the penalty function technique of Kelly [9] as we shall see in our section concerning the gradient and second variation methods for the computation of optimal controls. The choice between Eqs. (4.4-17) and (4.4-18) will depend largely upon the particular computer [for an analog computer, Eq. (4.4-17) is generally easier to implement than Eq. (4.4-18)] and the particular computational algorithms used (for the quasilinearization method, Eq. (4.4-18) is considerably simpler to use than Eq. (4.4-17) and also results in less computer solution time).

Example 4.4-2. It is quite easy to see that the constraint used here includes, as a special case, that considered in Sec. 3.8. For example, if we require for a scalar control u, $u_{\min} \le u \le u_{\max}$, then we may write

$$g_1[u(t), t] = u_{\max} - u \ge 0, \qquad g_2[u(t), t] = u - u_{\min} \ge 0,$$

and we convert these inequality constraints to equality constraints by writing

$$(y_1)^2 = u_{\max} - u, \qquad (y_2)^2 = u - u_{\min}$$

for which

$$(y_1 y_2)^2 = (u_{\max} - u)(u - u_{\min})$$

which is precisely the constraint used in Sec. 3.8.

For the problem at hand we adjoin, via the Lagrange multiplier, constraints (4.3-1), (4.4-2), and (4.4-16) to Eq. (4.3-2) to obtain

$$J = \theta[\mathbf{x}(t_f), t_f] + \mathbf{v}^T\mathbf{N}[\mathbf{x}(t_f), t_f] + \int_{t_o}^{t_f} \Big\{H[\mathbf{x}(t), \dot{\mathbf{w}}(t), \boldsymbol{\lambda}(t), t] -$$

$$\boldsymbol{\lambda}^T(t)\dot{\mathbf{x}} - \boldsymbol{\Gamma}^T(t)\{\mathbf{g}[\dot{\mathbf{w}}(t), t] - \dot{\mathbf{z}}^2\}\Big\}\, dt$$

where

$$(\mathbf{z}^2)^T = [z_1^2, z_2^2, z_3^2, \ldots, z_r^2]$$

$$H[\mathbf{x}(t), \dot{\mathbf{w}}(t), \boldsymbol{\lambda}(t), t] = \phi[\mathbf{x}(t), \dot{\mathbf{w}}(t), t] + \boldsymbol{\lambda}^T(t)\mathbf{f}[\mathbf{x}(t), \dot{\mathbf{w}}(t), t]$$

$$\dot{\mathbf{w}} = \mathbf{u}(t), \qquad \mathbf{w}(t_o) = \mathbf{0}.$$

We may now apply the Euler-Lagrange equations to the above cost function or take a first variation of it in order to obtain the necessary conditions for a minimum. It is thus convenient to define a scalar function $\boldsymbol{\Phi}$, the Lagrangian, as

$$\Phi[\mathbf{x}(t), \dot{\mathbf{x}}(t), \dot{\mathbf{w}}(t), \boldsymbol{\lambda}(t), \boldsymbol{\Gamma}(t), \dot{\mathbf{z}}(t), t] = H[\mathbf{x}(t), \dot{\mathbf{w}}(t), \boldsymbol{\lambda}(t), t] - \boldsymbol{\lambda}^T(t)\dot{\mathbf{x}} - \boldsymbol{\Gamma}^T(t)\{\mathbf{g}[\dot{\mathbf{w}}(t), t] - \dot{\mathbf{z}}^2\} \tag{4.4-19}$$

We will use the Euler-Lagrange Eqs. (3.5-3). Since there are no $\mathbf{w}(t)$ and $\mathbf{z}(t)$ terms in Eq. (4.4-19) we may write the Euler-Lagrange equations as

$$\frac{d}{dt}\frac{\partial \Phi}{\partial \dot{\mathbf{x}}} - \frac{\partial \Phi}{\partial \mathbf{x}} = 0 \tag{4.4-20}$$

$$\frac{d}{dt}\frac{\partial \Phi}{\partial \dot{\mathbf{w}}} = 0 \tag{4.4-21}$$

$$\frac{d}{dt}\frac{\partial \Phi}{\partial \dot{\mathbf{z}}} = 0. \tag{4.4-22}$$

Each piecewise continuously differentiable solution of the Euler-Lagrange equations (4.4-20), (4.4-21), and (4.4-22) will be called an extremal curve or an extremal trajectory of the associated variational problem. It can be shown that the function Φ need be only piecewise smooth, and thus the Euler-Lagrange equations require that every arc of the extremal trajectory on which the first derivatives of Φ have no discontinuities be a solution of the Euler-Lagrange equations. The corner condition will answer our questions concerning what happens at possible points of discontinuity of some of the derivatives of the state or control variables. This corner condition will ensure continuity of the state and control variables by forcing $\partial\Phi/\partial\dot{\mathbf{z}}$ to be zero everywhere since it is zero at the terminal time.

The transversality conditions for this problem are obtained in the usual fashion as explained in Chapter 3 and the previous three sections. For this problem, they are easily shown to be Eq. (4.4-2), the initial conditions, and

$$\frac{\partial \theta}{\partial t_f} + \left(\frac{\partial \mathbf{N}^T}{\partial t_f}\right)\mathbf{v} + \phi - \dot{\mathbf{x}}^T\frac{\partial \Phi}{\partial \dot{\mathbf{x}}} = 0, \qquad \text{for} \quad t = t_f$$

$$\frac{\partial \theta}{\partial \mathbf{x}} + \left(\frac{\partial \mathbf{N}^T}{\partial \mathbf{x}}\right)\mathbf{v} - \boldsymbol{\lambda} = \mathbf{0}, \qquad \text{for} \quad t = t_f.$$

Also, we have for the final transversality condition

$$\delta\mathbf{z}^T(t_f)\left[\frac{\partial \Phi}{\partial \dot{\mathbf{z}}}\right] = \delta\mathbf{z}^T(t_f)\begin{bmatrix} 2\Gamma_1\dot{z}_1 \\ 2\Gamma_2\dot{z}_2 \\ \cdot \\ \cdot \\ \cdot \\ 2\Gamma_r\dot{z}_r \end{bmatrix} = 0, \qquad \text{for} \quad t = t_f$$

which allows us to write, because of Eq. (4.4-21),

$$\frac{\partial \Phi}{\partial \dot{\mathbf{z}}} = \mathbf{0}, \qquad \forall t \in [t_o, t_f].$$

Since when $\Gamma_i \neq 0$, $\dot{z}_i = 0 = g_i$, and when $z_i \neq 0$, $\Gamma_i = 0$,

$$\Gamma_i z_i = 0, \qquad i = 1, 2, \ldots, r, \qquad \forall t \in [t_o, t_f].$$

Also, with similar reasoning, we have

$$\frac{\partial \Phi}{\partial \dot{\mathbf{w}}} = \mathbf{0}, \qquad \forall t \in [t_o, t_f].$$

4.4-2 *The maximum principle with state (and control) variable inequality constraints*

We now wish to extend the work of Sec. 4.4-1 to include inequality constraints on some or all of the state variables. We will represent this inequality constraint by the s vector equation

$$\mathbf{h}[\mathbf{x}(t), t] \geq \mathbf{0} \tag{4.4-23}$$

where each component of $\mathbf{h}$ is assumed to be continuously differentiable in state space. There are several methods whereby we may convert Eq. (4.4-23) to an equality constraint. We may define a new variable x_{n+1} by

$$\dot{x}_{n+1} = f_{n+1} = [h_1(\mathbf{x}, t)]^2 \mathrm{H}(h_1) + [h_2(\mathbf{x}, t)]^2 \mathrm{H}(h_2) + \cdots + [h_s(\mathbf{x}, t)]^2 \mathrm{H}(h_s) \tag{4.4-24}$$

where $\mathrm{H}[h_s(\mathbf{x}, t)]$ is a modified Heaviside step defined such that

$$\mathrm{H}[h_s(\mathbf{x}, t)] = \begin{cases} 0, & \text{if} \quad h_s(\mathbf{x}, t) \geq 0 \\ K_s, & \text{if} \quad h_s(\mathbf{x}, t) < 0 \end{cases}$$

$$K_s > 0, \qquad s = 1, 2, \ldots, s \tag{4.4-25}$$

and where the initial condition is

$$x_{n+1}(t_o) = 0.$$

Thus we see that $x_{n+1}(t_f)$ is a direct measure of penetration of the state variable inequality constraint

$$x_{n+1}(t_f) = \int_{t_o}^{t_f} \dot{x}_{n+1}(t)\, dt = \int_{t_o}^{t_f} \{[h_1(\mathbf{x}, t)]^2 \mathrm{H}(h_1) + \cdots + [h_s(\mathbf{x}, t)]^2 \mathrm{H}(h_s)\}\, dt.$$

We will require that the final value of $x_{n+1}(t_f)$ is zero,

$$x_{n+1}(t_f) = 0,$$

which will impose the restriction that we do not violate the inequality constraint. This approach is a modification by McGill [10] of a similar procedure by Kelley [9] which converts the s inequality constraints to s equality constraints of the form

$$
\begin{aligned}
\dot{x}_{n+1} &= [h_1(\mathbf{x}, t)]^2 \mathrm{H}(h_1), \qquad & x_{n+1}(t_0) &= 0 \\
\dot{x}_{n+2} &= [h_2(\mathbf{x}, t)]^2 \mathrm{H}(h_2), \qquad & x_{n+2}(t_o) &= 0 \\
&\vdots & &\vdots \\
\dot{x}_{n+s} &= [h_s(\mathbf{x}, t)]^2 \mathrm{H}(h_s), \qquad & x_{n+s}(t_o) &= 0
\end{aligned}
$$

which are then added to the cost function to obtain

$$J_{\text{modified}} = J_{\text{original}} + \sum_{j=1}^{s} x_{n+j}(t_f).$$

The multipliers K_s are thus the penalty functions, and J_{modified} is minimized such that the constraint region is entered only slightly, if at all. If we require $x_{n+j}(t_f) = 0$ for $j = 1, 2, \ldots, s$, the constraint is, of course, not exceeded at all.

A slight modification of the penalty-function approach can be obtained if we define s new state variables

$$
\begin{aligned}
(\dot{x}_{n+1})^2 &= K_1 h_1(\mathbf{x}, t), \qquad & x_{n+1}(t_o) &= 0 \\
(\dot{x}_{n+2})^2 &= K_2 h_2(\mathbf{x}, t), \qquad & x_{n+2}(t_o) &= 0 \\
&\vdots & &\vdots \\
(\dot{x}_{n+s})^2 &= K_s h_s(\mathbf{x}, t), \qquad & x_{n+s}(t_o) &= 0.
\end{aligned}
$$

Berkovitz [7] suggests yet another method for converting the inequality constraint to an equality constraint. For the case of a scalar constraint, a variable

$$\gamma(\mathbf{x}, \eta, t) = \begin{cases} \eta^4 - h(\mathbf{x}, t) & \text{if } \eta > 0 \\ \quad\;\; h(\mathbf{x}, t) & \text{if } \eta < 0 \end{cases}$$

is introduced, and we convert the inequality constraint $h(\mathbf{x}, t) \geq 0$ to an equality constraint by writing

$$\frac{\partial \gamma}{\partial \eta} \frac{d\eta}{dt} = \frac{\partial h}{\partial t} + \frac{\partial h}{\partial \mathbf{x}} \frac{d\mathbf{x}}{dt}$$

which satisfies the constraint if we have the end conditions

$$\gamma[\mathbf{x}(t_o), \eta(t_o), t_o] = \gamma[\mathbf{x}(t_f), \eta(t_f), t_f] = 0.$$

The Euler-Lagrange equations can, of course, be used to determine the differential equations for an extremum, and the associated transversality conditions can be used to specify the two-point boundary values. If inequality constraints on the control variables are present, we must of necessity incorporate these into our problem formulation. The Hamiltonian formulation may also be used. These methods provide us with necessary conditions only.

From Eq. (4.4-19) it follows that the Lagrangian for the problem at hand is

$$\tilde{\Phi} = \Phi + \lambda_{n+1}[f_{n+1} - \dot{x}_{n+1}]$$

$$\tilde{\Phi} = H - \boldsymbol{\lambda}^T\dot{\mathbf{x}} - \boldsymbol{\Gamma}^T[\mathbf{g} - \dot{\mathbf{z}}^2] + \lambda_{n+1}[f_{n+1} - \dot{x}_{n+1}] \qquad (4.4\text{-}26)$$

where Φ is the Lagrangian for no inequality state constraint. We are using the equality constraint method of Eqs. (4.4-24) and (4.4-25). The Euler-Lagrange equations yield,

$$\frac{d}{dt}\frac{\partial\Phi}{\partial\dot{\mathbf{x}}} - \frac{\partial\Phi}{\partial\mathbf{x}} - \frac{\partial f_{n+1}}{\partial\mathbf{x}}\lambda_{n+1} = \mathbf{0} \qquad (4.4\text{-}27)$$

$$\frac{\partial\Phi}{\partial\mathbf{u}} = \frac{d}{dt}\frac{\partial\Phi}{\partial\dot{\mathbf{w}}} = \mathbf{0}$$

$$\frac{d}{dt}\frac{\partial\Phi}{\partial\dot{\mathbf{z}}} = \mathbf{0}$$

which are, except for the f_{n+1} term, exactly the same as Eqs. (4.4-20), (4.4-21), and (4.4-22). Also, we see that

$$\frac{d}{dt}\lambda_{n+1}(t) = 0$$

with the transversality conditions exactly as before and, in addition,

$$x_{n+1}(t_o) = x_{n+1}(t_f) = 0.$$

It is desirable to reinterpret these results in terms of the Hamiltonian, just as we have done for the case of control variable constraints only. We can do this easily by combining Eqs. (4.4-26) and (4.4-27) which yields

$$\dot{\boldsymbol{\lambda}} = \frac{d\boldsymbol{\lambda}(t)}{dt} = -\frac{\partial H}{\partial\mathbf{x}} - \frac{\partial f_{n+1}[\mathbf{x}(t), t]}{\partial\mathbf{x}}\lambda_{n+1}$$

$$\dot{\mathbf{x}} = \frac{d\mathbf{x}(t)}{dt} = \frac{\partial H}{\partial\boldsymbol{\lambda}}$$

$$\dot{x}_{n+1} = \frac{dx_{n+1}(t)}{dt} = f_{n+1} = [h_1(\mathbf{x}, t)]^2\mathrm{H}(h_1) + \cdots + [h_s(\mathbf{x}, t)]^2\mathrm{H}(h_s)$$

$$\dot{\lambda}_{n+1} = \frac{d\lambda_{n+1}(t)}{dt} = 0$$

where

$$H[\mathbf{x}(t), \mathbf{u}(t), \boldsymbol{\lambda}(t), t] = \phi[\mathbf{x}(t), \mathbf{u}(t), t] + \boldsymbol{\lambda}^T(t)\mathbf{f}[\mathbf{x}(t), \mathbf{u}(t), t]$$

$$H[\mathbf{x}(t), \hat{\mathbf{u}}(t), \boldsymbol{\lambda}(t), t] \leq H[\mathbf{x}(t), \mathbf{u}(t), \boldsymbol{\lambda}(t), t], \quad \mathbf{u} \in \mho$$

with the two-point boundary conditions (transversality conditions)

$$\mathbf{x}(t_o) = \mathbf{x}_o$$

$$\mathbf{N}[\mathbf{x}(t_f), t_f] = \mathbf{0}$$

$$\left.\begin{aligned} \frac{\partial \theta}{\partial t_f} + \left(\frac{\partial \mathbf{N}^T}{\partial t_f}\right)\mathbf{v} + H = 0 \\ \frac{\partial \theta}{\partial \mathbf{x}} + \left(\frac{\partial N^T}{\partial \mathbf{x}}\right)\mathbf{v} - \boldsymbol{\lambda} = 0 \end{aligned}\right\} \quad \text{at} \quad t = t_f$$

$$x_{n+1}(t_o) = x_{n+1}(t_f) = 0.$$

Further discussion of the state-space constraint problem can be found in [18].

Example 4.4-3. As an example of optimization with a state variable constraint, we consider the brachistochrone problem previously treated by McGill [10] and Dreyfus [11]. A particle is falling for a specified time, $t_f - t_o$, under the influence of a constant gravitational acceleration g. The particle has initial velocity $x_3(t_o) = x_{3o}$. We wish to find the path that maximizes the final value of the horizontal coordinate $x_2(t_f)$. The final value of the vertical coordinate $x_2(t_f)$ and the velocity $x_3(t_f)$ are unspecified. The path is constrained by a line $h[x_1, x_2] \geq 0$ in the $x_1 x_2$ plane, where it is known that the unconstrained solution intersects the line. The system dynamics are described by

$$\dot{x}_1 = x_3 \cos u, \qquad x_1(t_o) = x_{1o}$$

$$\dot{x}_2 = x_3 \sin u, \qquad x_2(t_o) = x_{2o}$$

$$\dot{x}_3 = g \sin u, \qquad x_3(t_o) = x_{3o}$$

where the control u is the slope of the path. The cost function is

$$J = -x_1(t_f)$$

with no specified endpoint equality constraints, and the state vector inequality constraint,

$$h(x_1 x_2) = a x_1 + b - x_2 \geq 0,$$

which is converted to the equality constraint

$$\dot{x}_4 = f_4 = [h(x_1, x_2)]^2 \mathrm{H}(h).$$

We can easily compute the requisite nonlinear two-point boundary value problem by direct application of the maximum principle given in this section. The equations for this TPBVP are

$$\dot{x}_1 = x_3^2 \lambda_1 [(\lambda_1 x_3)^2 + (\lambda_2 x_3 + \lambda_3 g)^2]^{-1/2}, \qquad x_1(t_0) = x_{1o}$$

$$\dot{x}_2 = x_3(\lambda_2 x_3 + \lambda_3 g)[(\lambda_1 x_3)^2 + (\lambda_2 x_3 + \lambda_3 g)^2]^{-1/2}, \qquad x_2(t_o) = x_{2o}$$

$$\dot{x}_3 = g(\lambda_2 x_3 + \lambda_3 g)[(\lambda_1 x_3)^2 + (\lambda_2 x_3 + \lambda_3 g)^2]^{-1/2}, \qquad x_3(t_o) = x_{3o}$$

$$\dot{x}_4 = h(x_1, x_2)\mathrm{H}(h), \qquad x_4(t_o) = 0$$

$$\dot{\lambda}_1 = -2a\lambda_4 h(x_1, x_2)\mathrm{H}(h), \qquad \lambda_1(t_o) = -1$$

$$\dot{\lambda}_2 = 2\lambda_4 h(x_1, x_2)\mathrm{H}(h), \qquad \lambda_2(t_f) = 0$$
$$\dot{\lambda}_3 = -\lambda_1^2 x_3[(\lambda_1 x_3)^2 + (\lambda_2 x_3 + \lambda_3 g)^2]^{-1/2}$$
$$-\lambda_2 (\lambda_2 x_3 + \lambda_3 g)[(\lambda_1 x_3)^2 + (\lambda_2 x_3 + \lambda_3 g)^2]^{-1/2}, \qquad \lambda_3(t_f) = 0$$
$$\dot{\lambda}_4 = 0, \qquad x_4(t_f) = 0.$$

The solution of this set of nonlinear differential equations with the associated boundary conditions establishes the optimal trajectory and optimal control.

4.5 Hamilton-Jacobi equation and continuous time dynamic programming

We now return to the problem formulated in Subsection 4.4-1 and rederive the necessary condition Eq. (4.4-3) using an argument based on the interchange of the minimization and integration operations. This approach to the development of conditions for optimality is often called the *Principle of Optimality* or *dynamic programming*, the difference of which is elucidated in detail in [19] and [20].

We define

$$V[\mathbf{x}(t), t] = \min_{\mathbf{U}_t} \left\{\theta[\mathbf{x}(t_f), t_f] + \int_t^{t_f} \phi[\mathbf{x}(s), \mathbf{u}(s), s]\, ds\right\} \tag{4.5-1}$$

where $\mathbf{U}_t = \{\mathbf{u}(s), t \leq s \leq t_f\}$, and $\mathbf{x}(s), t \leq s \leq t_f$, is the trajectory associated with an optimal control function over the interval $[t, t_f]$, given $\mathbf{x}(t)$. The function V is therefore the optimal cost to be accrued over the interval $[t, t_f]$, given initial condition $\mathbf{x}(t)$. Clearly, V must satisfy the boundary condition

$$V[\mathbf{x}(t_f), t_f] = \theta[\mathbf{x}(t_f), t_f] \tag{4.5-2}$$

for all pairs $(\mathbf{x}(t_f), t_f)$ satisfying Eq. (4.4-2). We assume that V exists, is continuous, and has continuous first- and second-partial derivatives for all points of interest in R^{n+1}. It follows from Eq. (4.3-1) that $\mathbf{u}(t_1), t \leq t_1$, does not affect $\mathbf{x}(s), s \leq t$; thus, we may interchange the appropriate minimization and integration operations to produce

$$V[\mathbf{x}(t), t] = \min_{\mathbf{u}(\tau)} \left\{\int_t^{t+\Delta t} \phi[\mathbf{x}(\tau), \mathbf{u}(\tau), \tau]\, d\tau + \right.$$
$$\left. \min_{\mathbf{U}_{t+\Delta t}} \left\{\theta[\mathbf{x}(t_f), t_f] + \int_{t+\Delta t}^{t_f} \phi[\mathbf{x}(s), \mathbf{u}(s), s]\, ds\right\}\right\} \simeq$$
$$\min_{\mathbf{u}(\tau)} \{\phi[\mathbf{x}(t), \mathbf{u}(t), t]\, \Delta t + V[\mathbf{x}(t + \Delta t), t + \Delta t]\} =$$
$$\min_{\mathbf{u}(\tau)} \left\{\phi[\mathbf{x}(t), \mathbf{u}(t), t]\, \Delta t + V[\mathbf{x}(t), t] + \frac{\partial V[\mathbf{x}(t), t]}{\partial t} \Delta t + \right.$$
$$\left. \left[\frac{\partial V[\mathbf{x}(t), t]}{\partial \mathbf{x}}\right]^T \mathbf{f}[\mathbf{x}(t), \mathbf{u}(t), t]\, \Delta t + \text{H.O.T.}\,(\Delta t)\right\}, \tag{4.5-3}$$

where for the last equality, $V[\mathbf{x}(t+\Delta t), t+\Delta t]$ is expanded in a Taylor series and the minimization is taken over $\tau \in [t, t+\Delta t]$. The smoothness assumptions and the interchange of the minimization and limit operations as $\Delta t \to 0$ (justification of which can be found in [6] and [11]) imply that V must necessarily satisfy the partial differential equation

$$\frac{-\partial V[\mathbf{x}(t), t]}{\partial t} = \min_{\mathbf{v} \in \mathcal{U}} \phi[\mathbf{x}(t), \mathbf{v}, t)] + \left[\frac{\partial V[\mathbf{x}(t), t]}{\partial \mathbf{x}}\right]^T \mathbf{f}[\mathbf{x}(t), \mathbf{v}, t] \tag{4.5-4}$$

called the Hamilton-Jacobi or Hamilton-Jacobi-Bellman equation [11–14]. A repetition of the above argument for given optimal control function $\mathbf{u}$ produces

$$\frac{-\partial V[\mathbf{x}(t), t]}{\partial t} = \phi[\mathbf{x}(t), \mathbf{u}(t), t] + \left[\frac{\partial V[\mathbf{x}(t), t]}{\partial \mathbf{x}}\right]^T \mathbf{f}[\mathbf{x}(t), \mathbf{u}(t), t]; \tag{4.5-5}$$

hence, $\mathbf{u}$ must necessarily satisfy Eq. (4.4-3) for all $t \in [t_0, t_f]$, where we assume

$$\boldsymbol{\lambda}(t) = \frac{\partial V[\mathbf{x}(t), t]}{\partial \mathbf{x}}. \tag{4.5-6}$$

Although the Hamilton-Jacobi equation is quite difficult to solve in general, when it can be solved, a candidate for an optimal control function is found as a function of the state trajectory—a highly desirable feedback form.

Example 4.5-1. Let us consider the linear constant differential system described by

$$\dot{\mathbf{x}} = \mathbf{A}\mathbf{x}(t) + \mathbf{b}u(t), \qquad \mathbf{x}(0) = \mathbf{x}_o$$

where $\mathbf{A}$ is an $n \times n$ matrix and $\mathbf{b}$ is an n vector. Any $u(t)$ is assumed to be admissible. We wish to find $u(t)$ as a function of $\mathbf{x}(t)$ such that

$$J = \tfrac{1}{2}\int_0^\infty [\mathbf{x}^T\mathbf{Q}\mathbf{x} + ru^2]\, dt$$

is a minimum. $\mathbf{Q}$ is a positive constant semidefinite matrix, and r is positive. The Hamiltonian for the problem is

$$H(\mathbf{x}, u, \boldsymbol{\lambda}, t) = \tfrac{1}{2}\mathbf{x}^T\mathbf{Q}\mathbf{x} + \tfrac{1}{2}ru^2 + \boldsymbol{\lambda}^T\mathbf{A}\mathbf{x} + \boldsymbol{\lambda}^T\mathbf{b}u.$$

We need to find the control u which minimizes the Hamiltonian. This is

$$\frac{\partial H}{\partial u} = 0 = ru + \mathbf{b}^T\boldsymbol{\lambda}$$

so

$$u = -\mathbf{b}^T\boldsymbol{\lambda}r^{-1}$$

and the Hamiltonian becomes

$$H(\mathbf{x}, \boldsymbol{\lambda}, t) = \tfrac{1}{2}\mathbf{x}^T\mathbf{Q}\mathbf{x} + \boldsymbol{\lambda}^T\mathbf{A}\mathbf{x} - \tfrac{1}{2}\boldsymbol{\lambda}^T\mathbf{b}\mathbf{b}^T\boldsymbol{\lambda}r^{-1}.$$

Since the system and the $\mathbf{Q}$ and r terms are time invariant and since the optimization is for a process of infinite duration, it follows that $V(\mathbf{x}, t)$ will depend only upon the initial state $\mathbf{x}$. This implies that

$$\frac{\partial V(\mathbf{x}, t)}{\partial t} = 0.$$

Therefore, since $\boldsymbol{\lambda} = \partial V/\partial \mathbf{x}$, the Hamilton-Jacobi equation becomes

$$\frac{1}{2}\mathbf{x}^T\mathbf{Q}\mathbf{x} + \left(\frac{\partial V}{\partial \mathbf{x}}\right)^T \mathbf{A}\mathbf{x} - \frac{1}{2}\left[\left(\frac{\partial V}{\partial \mathbf{x}}\right)^T \mathbf{b}\right]^2 r^{-1} = 0.$$

If we assume a solution

$$V(\mathbf{x}, t) = \tfrac{1}{2}\mathbf{x}^T\mathbf{P}\mathbf{x},$$

we see that

$$\frac{\partial V}{\partial \mathbf{x}} = \mathbf{P}\mathbf{x}$$

and the Hamilton-Jacobi equation may be written as

$$\mathbf{x}^T[\tfrac{1}{2}\mathbf{Q} + \tfrac{1}{2}\mathbf{P}\mathbf{A} + \tfrac{1}{2}\mathbf{A}^T\mathbf{P} - \tfrac{1}{2}\mathbf{P}\mathbf{b}\mathbf{b}^T\mathbf{P}r^{-1}]\mathbf{x} = 0$$

which says that, for any nonzero $\mathbf{x}(t)$, the matrix $\mathbf{P}$ must satisfy the $n(n + 1)/2$ algebraic equations (the $\mathbf{P}$ matrix is symmetric)

$$\mathbf{Q} + \mathbf{P}\mathbf{A} + \mathbf{A}^T\mathbf{P} - \mathbf{P}\mathbf{b}\mathbf{b}^T\mathbf{P}r^{-1} = 0.$$

This equation is solved for $\mathbf{P}$, and then the control is computed from

$$u = -\mathbf{b}^T\boldsymbol{\lambda}r^{-1} = -\mathbf{b}^T r^{-1}\left(\frac{\partial V}{\partial \mathbf{x}}\right) = -\mathbf{b}^T\mathbf{P}\mathbf{x}r^{-1}.$$

If we further consider the system

$$\dot{x}_1 = x_2, \qquad x_1(0) = x_{10}$$

$$\dot{x}_2 = u, \qquad x_2(0) = x_{20},$$

and the cost function

$$J = \tfrac{1}{2}\int_0^\infty (4x_1^2 + u^2)\, dt,$$

it is easy for us to show that the optimum control is given by

$$u = -2x_1 - 2x_2.$$

Example 4.5-2. Consider the system

$$\dot{x} = -x^3 + u, \qquad x(0) = x_0$$

with cost function

$$J = \tfrac{1}{2}\int_0^{t_f} (x^2 + u^2)\, dt$$

where it is desired to determine the optimal feedback control. We accomplish this by forming the Hamiltonian

$$H(x, u, \lambda, t) = \tfrac{1}{2}x^2 + \tfrac{1}{2}u^2 + \lambda u - \lambda x^3.$$

We then set $\partial H/\partial u = 0$ and note that $\lambda = \partial V/\partial x$ to obtain $u = -\lambda$; then

$$H\left(x, \frac{\partial V}{\partial x}\right) = \frac{1}{2}x^2 - \frac{1}{2}\left[\frac{\partial V(x, t)}{\partial x}\right]^2 - \left[\frac{\partial V(x, t)}{\partial x}\right]x^3.$$

The Hamilton-Jacobi equation is

$$\frac{\partial V(x, t)}{\partial t} - \frac{1}{2}\left[\frac{\partial V(x, t)}{\partial x}\right]^2 - \left[\frac{\partial V(x, t)}{\partial x}\right]x^3 + \frac{1}{2}x^2 = 0$$

with $V[x(t_f), t_f] = 0$.

If the optimization interval is infinite, then $\partial V/\partial t = 0$, and we need to solve the differential equation

$$\left[\frac{dV(x)}{dx}\right]^2 + 2\left[\frac{dV(x)}{dx}\right]x^3 - x^2 = 0$$

with $V(0) = 0$ as the initial condition. We may approximate the solution to this ordinary differential equation by a series expansion

$$V(x) = p_0 + p_1 x + \frac{1}{2}p_2 x^2 + \frac{1}{3!}p_3 x^3 + \frac{1}{4!}p_4 x^4 + \dots.$$

If we terminate the series after the fourth-order term, substitute the assumed solution into the differential equation, and equate like powers of x (up to x^4), we obtain $p_0 = p_1 = p_3 = 0, p_2 = 1, p_4 = -6$. Thus the approximate closed-loop control is

$$u = -\lambda = -\frac{dV}{dx} = -x + x^3.$$

We naturally may question the stability of the approximate control. However, with u as obtained, the system differential equation becomes

$$\dot{x} = -x^3 + u = -x$$

which is certainly stable.

A similar procedure to this could have been used to obtain an approximate solution to the nonlinear partial differential equation that is the Hamilton-Jacobi equation for this example. In this case, the p's would be functions of time, and we would obtain matrix Riccati-type equations [15]. This approach has many attractive features. In particular, only initial condition problems need be solved. However, there are two disadvantages: There is no assurance of system stability; the number of matrix Riccati differential equations which must be solved increases greatly with the order of the differential system and the order of the polynomial in $\mathbf{x}$ for the approximate solution to $V(\mathbf{x}, t)$. If an expansion in $\mathbf{x}$ to the N order is used for an n vector differential system, the number of distinct Riccati-type differential equations is

$$E = \sum_{j=1}^{N} \frac{(n - 1 + j)!}{(n - 1)!j!}$$

for an assumed solution of the form

$$V(\mathbf{x}, t) = \sum_{j=1}^{n} p_j x_j + \tfrac{1}{2}\sum_{j=1}^{n}\sum_{k=1}^{n} p_{jk}x_j x_k + \tfrac{1}{6}\sum_{j=1}^{n}\sum_{k=1}^{n}\sum_{l=1}^{n} p_{jkl}x_j x_k x_l + \dots.$$

If, for example, the solution to a four-vector differential system is approximated by terms up to and including the fourth power in $\mathbf{x}$, we need to solve 69 differential equations to obtain the closed-loop control.

Our discussion of the second variation technique, the invariant imbedding procedure, and specific optimal control using the quasilinearization approach will point out many interesting interconnections with the approach used to obtain the solution to this example.

In our development thus far, we have assumed that the terminal time, t_f, is fixed. It is possible to remove this restriction with the result that the

Hamilton-Jacobi equation is still applicable. The initial condition for the Hamilton-Jacobi equation is still Eq. (4.5-2) and, in addition, the terminal time is determined by the relation

$$H\left(\mathbf{x}, \frac{\partial V}{\partial \mathbf{x}}, t\right) + 1 = 0, \qquad \text{at} \quad t = t_f$$

which holds if the problem is a minimum time problem such that

$$V(\mathbf{x}, t) = t_f - t.$$

If, further, the differential system is time invariant, the Hamiltonian is equal to -1 at all times along the optimal trajectory.

We may formally obtain the Pontryagin maximum principle by taking appropriate partial derivatives of the Hamilton-Jacobi equations (Problem 9). However, the resulting maximum principle is not applicable to as broad a class of problems as is possible. The reason for this is that it is necessary that $V(\mathbf{x}, t)$ be smooth or, in other words, twice continuously differentiable with respect to $\mathbf{x}$ in order to obtain the Hamilton canonic equations of the maximum principle. We shall illustrate these difficulties with a simple example.

Example 4.5-3. A second-order example will now be discussed to illustrate that the assumption of the differentiability of $V(\mathbf{x}, t)$ does not hold in some of the simplest cases. We will consider the problem of transferring the system represented by the differential equations

$$\dot{x}_1 = x_2, \qquad \dot{x}_2 = u$$

from an initial state $\mathbf{x}_o$ to the origin in minimum time. We assume that the admissible set for the scalar control is described by $|u(t)| \leq 1$.

This problem can be solved by the Pontryagin maximum principle. In the time optimal problem

$$J = \int_{t_o}^{t_f} (1)\, dt.$$

Therefore, the Hamiltonian is

$$H[\mathbf{x}, u, \boldsymbol{\lambda}, t] = 1 + \lambda_1 x_2 + \lambda_2 u.$$

The adjoint equations are

$$\dot{\lambda}_1 = 0, \qquad \dot{\lambda}_2 = -\lambda_1.$$

The solutions to these equations are

$$\lambda_1 = C_1, \qquad \lambda_2 = C_2 - C_1 t$$

where C_j is the initial condition on λ_j. The control which minimizes the Hamiltonian subject to $|u| \leq 1$ is

$$u = -\text{sign}\, \lambda_2 = -\text{sign}\, (C_2 - C_1 t).$$

The initial conditions C_1 and C_2 are not arbitrary but must be such that $\mathbf{x}(t_f) = 0$ since it is desired to transfer the system $\mathbf{x}_o$ to the origin in minimum time.

When $u = +1$, the solution to the differential system equation is

$$x_2 = t + x_2(0)$$

$$x_1 = \frac{t^2}{2} + x_2(0)t + x_1(0).$$

If t is eliminated from the foregoing, we obtain

$$x_1 = \frac{x_2^2}{2} + x_1(0) - \frac{x_2^2(0)}{2}.$$

When $u = -1$, the solution to the differential system equations is

$$x_2 = -t + x_2'(0)$$

$$x_1 = \frac{-t^2}{2} + x_2'(0)t + x_1'(0)$$

and if t is eliminated in the foregoing, we obtain

$$x_1 = \frac{-x_2^2}{2} + x_1'(0) + \frac{x_2'^2(0)}{2}.$$

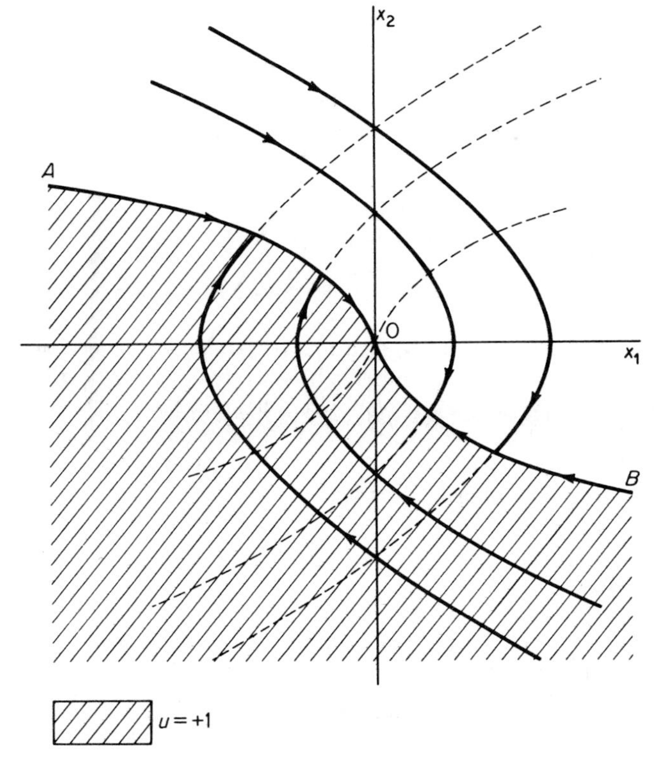

Fig. 4.5-1 Switching curve and trajectories for minimum time, Example (4.5-3).

By determining the constants C_1 and C_2 in terms of x_1 and x_2, it is a straightforward task for us to show that the control law is

$$u = -\text{sign}\,[x_1(t) + \tfrac{1}{2}x_2(t)\,|x_2(t)|].$$

These equations represent the optimal control and trajectories for $u = -1$ and $u = +1$, respectively, and they indicate that these trajectories are segments of parabolas. Figure 4.5-1 is a plot of some of these parabolas.

The segment of the parabola which is not an optimal trajectory has been represented by a broken line. The optimal control can be determined from Fig. 4.5-1 and a knowledge of the state of the system. The curve AOB represents the switching curve. When $\mathbf{x}$ lies below AOB, $u = +1$ until the system state reaches the curve AO, at which time the control switches to -1. If $\mathbf{x}$ lies above AOB, $u = -1$ until it reaches BO, where it switches to $+1$.

The optimal transition time $T(\mathbf{x})$, which is the cost function J or $V(\mathbf{x}, t)$, can be determined from the solutions for x_1 and x_2. Figure 4.5-2 is a plot of $T(\mathbf{x})$, the minimum time to transfer to the origin for the case in which the initial x_2 is held constant ($x_{2o} = -2$), and x_{1o} is varied about the switching line.

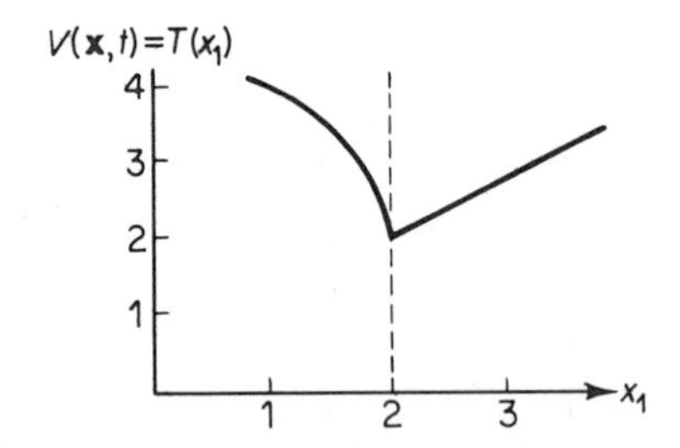

Fig. 4.5-2 Minimum time to origin for fixed x_{2o}, Example (4.5-3).

From the graph it can be seen that $\partial T(\mathbf{x})/\partial x_1$ has a discontinuity at the switching curve. It can be shown analytically that $\partial T(\mathbf{x})/\partial x_1$ "blows up" as x_1 approaches $+2$ from the left. Hence the Hamilton-Jacobi equation would not be applicable in examples of this type. This example indicates the loss of generality which results from deriving the maximum principle from the Hamilton-Jacobi-Bellman equations.

References

1. GELFAND, I.M., and FOMIN, S.V., *Calculus of Variations.* Prentice-Hall, Inc., Englewood Cliffs, New Jersey, 1963.
2. PONTRYAGIN, L.S., et al., *The Mathematical Theory of Optimal Processes.* Wiley, New York, 1962.
3. ROZONOÉR, L.I., "L.S. Pontryagin's Maximum Principle in the Theory of Optimum Systems I." *Automatica i Telemekhanika*, 20, no. 10, (October 1959), 1320–34.

4. ROZONOÉR, L.I., "L.S. Pontryagin's Maximum Principle in the Theory of Optimum Systems II." *Automatica i Telemekhanika*, 20, no. 11, (November 1959), 1441–58.

5. ROZONOÉR, L.I., "L.S. Pontryagin's Maximum Principle in the Theory of Optimum Systems III." *Automatica i Telemekhanika*, 20, no. 12, (December 1959), 1561–78.

6. BERKOVITZ, L.D., "Variational Methods in Problems of Control and Programming." *J. Math. Anal. Appl.*, 3, (1961), 145–69.

7. BERKOVITZ, L.D., "On Control Problems with Bounded State Variables." *J. Math. Anal. Appl.*, 5, (1962), 488–98.

8. VALENTINE, F.A., "The Problem of Lagrange with Differential Inequalities as Added Side Conditions." *Contributions to the Calculus of Variations 1933–1937*, University of Chicago Press, Chicago, 1937.

9. KELLEY, H.J., "Methods of Gradients," G. Leitman, ed., *Optimization Techniques*. Academic Press, New York, 1962, chapter 6.

10. MCGILL, R., "Optimal Control, Inequality State Constraints, and the Generalized Newton-Raphson Algorithm." *SIAM J. Control*, 3, (1965), 291–98.

11. DREYFUS, S.E., *Dynamic Programming and the Calculus of Variations*. Academic Press, New York, 1965.

12. KALMAN, R.E., "Contributions to the Theory of Optimal Control." *Bol. Soc. Mat. Mex.*, 5, (1960), 102–19.

13. BELLMAN, R., ed., *Mathematical Optimization Techniques*. University of California Press, Berkeley, 1963.

14. BRYSON, A.E., Jr., "Optimal Programming and Control." *Proceedings IBM Scientific Computing Symposium on Control Theory and Applications*, IBM Publication 320-1939, 1966.

15. MERRIAM, C.W. III, *Optimization Theory and the Design of Feedback Control Systems*. McGraw-Hill Book Co., New York, 1964.

16. MCSHANE, E.J., "On Multipliers for Lagrange Problems." *American J. Math.* 61, (1939), 809–19.

17. BAUM, R.F., and L. CESARI, "On a Recent Proof of Pontryagin's Necessary Conditions," *SIAM J. Control*, 10, (1972), 56–75.

18. FUNK, J.E., and E.G. GILBERT, "Some Sufficient Conditions for Optimality in Control Problems with State Space Constraints," *SIAM J. Control*, 8, (1970), 498–504.

19. HINDERER, K., *Foundations of Non-Stationary Dynamic Programming with Discrete Time Parameter*. Springer-Verlag, New York, 1970.

20. PORTEUS, E., "An Informal look at the Principle of Optimality." *Management Science*, 21, (1975), 1346–48.

21. LUENBERGER, D.G., *Optimization by Vector Space Methods*. Wiley, New York, 1969.

Problems

1. Find the TPBVP which, when solved, yields the control, $u(t)$, and trajectory, $x(t)$, which minimize

$$J = \frac{1}{2}\int_0^1 (x^2 + u^2)\,dt$$

for the system

$$\dot{x} = -x^3 + u, \qquad x(0) = 1.$$

2. Find the control and trajectory which transfers the system

$$\dot{x}_1 = x_2, \qquad x_1(0) = 0$$
$$\dot{x}_2 = u, \qquad x_2(0) = 0$$

to the line

$$x_1(1) + x_2(1) = 1$$

such that

$$J = \frac{1}{2}\int_0^1 u^2(t)\,dt$$

is minimized.

3. Find the control and trajectory which transfers the system

$$\dot{x} = -x + u$$

from $x(0) = 10$ to $x(1) = 0$ such that

$$J = \frac{1}{2}\int_0^1 (\dot{u})^2\,dt$$

is minimized.

4. Find the control and trajectory which minimizes

$$J = \frac{1}{2}\int_0^4 x^2(t)\,dt$$

subject to the inequality constraint $|u(t)| \leq 1$ for the system $\dot{x} = u$ such that $x(0) = 1$, $x(4) = 1$.

5. Determine the Weierstrass-Erdmann corner conditions for the minimization of the cost function

$$J = \int_0^1 x^2(2 - \dot{x})^2\,dt.$$

6. What inequalities must the constraints a and b satisfy so that the function

$$j = \int_0^1 (\dot{x} + a)^2(\dot{x} + b)^2\,dt, \qquad x(0) = 0, \qquad x(1) = 1,$$

has just one corner?

7. For the system

$$\dot{x}_1 = x_2, \qquad x_1(0) = 10$$
$$\dot{x}_2 = u, \qquad x_2(0) = 0$$

find the control and trajectory which minimizes

$$J = t_f^2 + \tfrac{1}{2}\int_0^{t_f} u^2\,dt$$

if the desired final state is:
(a) $x_1(t_f) = x_2(t_f) = 0$.
(b) $x_1(t_f) = 0, \qquad x_2(t_f) = \text{unspecified}$.

8. Develop a second- and fourth-order approximation to the solution of the Hamilton-Jacobi equation to find the closed-loop control which minimizes

$$J = \tfrac{1}{2}\int_{t_0}^{t_f} (x_1^2 + u^2)\,dt$$

for the system

$$\dot{x}_1 = x_2 + x_2^3$$
$$\dot{x}_2 = x_1 - x_2 + u.$$

Compute and compare the actual numerical results when t_f is infinite.

9. Derive the Pontryagin maximum principle from the Hamilton-Jacobi equation by calculating $(d/dt)(\partial V/\partial \mathbf{x})$ and $\partial V/\partial \boldsymbol{\lambda}$ as outlined in Sec. 4.5. Observe the differentiability requirement on $V(\mathbf{x}, t)$.

10. Find the control vector which minimizes

$$J = \tfrac{1}{2}\int_0^1 (x^2 + u_1^2 + u_2^2)\,dt$$

for the system described by

$$\dot{x} = u_1 + u_2, \qquad x(0) = 1.$$

Use the maximum principle and the Hamilton-Jacobi equations to find the optimum control vector.

11. Set up the differential equations and boundary conditions to minimize for t_f unspecified

$$J = \int_0^{t_f} u^2\,dt + t_f x_2(t_f)$$

subject to the constraints
a) $\dot{x}_1 = x_2, \dot{x}_2 = x_3, \dot{x}_3 = u$
b) $\mathbf{x}(0) = \mathbf{0}$
c) $|u| \leq 1; |x_3| \leq 10$
d) $x_1(t_f) = t_f^2, x_2(t_f) = x_3^2(t_f)$.

12. Set up the equations and boundary conditions to optimize the system

$$\dot{x}_1 = x_2, \qquad \dot{x}_2 = x_3, \qquad \dot{x}_3 = u$$

for the performance index with t_f unspecified

$$J = \int_0^{t_f} x_1^2\,dt + t_f^2 x_2(t_f)$$

subject to all of the following constraints
a) $\mathbf{x}^T(0) = [1, 0, 0]$
b) $x_1(t_f) = x_2(t_f)$
c) $x_3(t_f) = 0$

d) $|u| \leq 1$

e) $\int_0^{t_f} u^2 \, dt = 1.$

13. Find the Hamilton-Jacobi equation for the system

$$\dot{x}_1 = x_2, \qquad \dot{x}_2 = -x_2 - x_1^2 + u$$

if the performance index is

$$J = \int_0^{t_f} (x_1^2 + u^2) \, dt.$$

14. Show that the solution of the Hamilton-Jacobi equation for the system

$$\dot{\mathbf{x}} = \mathbf{A}\mathbf{x} + \mathbf{u}, \qquad \mathbf{A}^T + \mathbf{A} = \mathbf{0}, \qquad \|\mathbf{u}\| \leq 1$$

and the cost function

$$J = \int_0^{t_f} dt = t_f$$

is

$$V(\mathbf{x}) = \|\mathbf{x}\|.$$

What is the optimal control?

15. Find the optimal control to minimize

$$J = \int_0^{t_f} dt$$

for the system

$$\dot{x} = -x + u,$$

when

$$x(0) = 1, \qquad x(t_f) = 0$$

$$|u| \leq 1 + |x|.$$

Optimum systems control examples 5

In this chapter, we will illustrate some, but certainly not all or even most, of the optimal control problems for which closed-form analytic solutions have been obtained. The problems we will solve in this chapter are very important in their own right and illustrate the use of the maximum principle for problems in which closed-form analytic solutions may be obtained. Specifically, we will discuss the linear regulator problem, the first solution of which was due to Kalman [1, 2, 3, 4]. We then discuss the minimum time problem which has been considered by Pontryagin [5], Bellman [6], LaSalle [7], and many others [8 through 13].

A characteristic of some minimum time problems is the possibility of a singular solution. The possibility of singular solutions is well-recognized in the variational calculus literature and has been extensively discussed for control problems by Johnson [14, 15, 16] and others. Minimum fuel problems for linear differential systems are then discussed. A variety of authors, but notably Athans, have discussed various aspects of minimum fuel problems including the possibility of singular solutions [17 through 20]. Although we will not consider the minimum time–fuel–energy control of self-adjoint systems [21] due to its limited practical usefulness, we do note that such systems admit a particularly thorough analysis. For a survey of many other problems plus a lengthy bibliography, we refer to the survey papers of Paiewonsky [22] and Athans [23].

5.1
The linear regulator

We will now study a particular control problem which has as its solution a linear feedback control law. It occurs where we have a linear differential system

$$\dot{\mathbf{x}} = \mathbf{A}(t)\mathbf{x} + \mathbf{B}(t)\mathbf{u}, \qquad \mathbf{x}(t_o) = \mathbf{x}_o \tag{5.1-1}$$

and wish to find the control which minimizes the cost function (for t_f fixed)

$$J = \tfrac{1}{2}\mathbf{x}^T(t_f)\mathbf{S}\mathbf{x}(t_f) + \tfrac{1}{2}\int_{t_o}^{t_f} [\mathbf{x}^T(t)\mathbf{Q}(t)\mathbf{x}(t) + \mathbf{u}^T(t)\mathbf{R}(t)\mathbf{u}(t)]\, dt. \tag{5.1-2}$$

Clearly, there is no loss of generality in assuming $\mathbf{Q}$, $\mathbf{R}$, and $\mathbf{S}$ to be symmetric. We may obtain the solution to this problem via the maximum principle or the Hamilton-Jacobi equation. Here, we will use the former method. The Hamiltonian is

$$H[\mathbf{x}(t), \mathbf{u}(t), \boldsymbol{\lambda}(t), t] = \tfrac{1}{2}\mathbf{x}^T\mathbf{Q}\mathbf{x} + \tfrac{1}{2}\mathbf{u}^T\mathbf{R}\mathbf{u} + \boldsymbol{\lambda}^T\mathbf{A}\mathbf{x} + \boldsymbol{\lambda}^T\mathbf{B}\mathbf{u}. \tag{5.1-3}$$

Application of the maximum principle requires that, for an optimum control,

$$\frac{\partial H}{\partial \mathbf{u}} = \mathbf{0} = \mathbf{R}(t)\mathbf{u}(t) + \mathbf{B}^T(t)\boldsymbol{\lambda}(t) \tag{5.1-4}$$

and

$$\frac{\partial H}{\partial \mathbf{x}} = -\dot{\boldsymbol{\lambda}} = \mathbf{Q}(t)\mathbf{x}(t) + \mathbf{A}^T(t)\boldsymbol{\lambda}(t) \tag{5.1-5}$$

with the terminal condition

$$\boldsymbol{\lambda}(t_f) = \frac{\partial \theta}{\partial \mathbf{x}(t_f)} = \mathbf{S}\mathbf{x}(t_f). \tag{5.1-6}$$

Thus we require that

$$\mathbf{u}(t) = -\mathbf{R}^{-1}(t)\mathbf{B}^T(t)\boldsymbol{\lambda}(t), \tag{5.1-7}$$

and we shall inquire whether we may convert this to a closed-loop control by assuming that the solution for the adjoint is similar to Eq. (5.1-6)

$$\boldsymbol{\lambda}(t) = \mathbf{P}(t)\mathbf{x}(t). \tag{5.1-8}$$

If we substitute this relation into Eqs. (5.1-1) and (5.1-7), we see that we must require

$$\dot{\mathbf{x}} = \mathbf{A}(t)\mathbf{x}(t) - \mathbf{B}(t)\mathbf{R}^{-1}(t)\mathbf{B}^T(t)\mathbf{P}(t)\mathbf{x}(t). \tag{5.1-9}$$

Also, from Eqs. (5.1-8) and (5.1-5) we require

$$\dot{\boldsymbol{\lambda}} = \dot{\mathbf{P}}\mathbf{x}(t) + \mathbf{P}(t)\dot{\mathbf{x}} = -\mathbf{Q}(t)\mathbf{x}(t) - \mathbf{A}^T(t)\mathbf{P}(t)\mathbf{x}(t). \tag{5.1-10}$$

By combining Eqs. (5.1-9) and (5.1-10) we have

$$[\dot{\mathbf{P}} + \mathbf{P}(t)\mathbf{A}(t) + \mathbf{A}^T(t)\mathbf{P}(t) - \mathbf{P}(t)\mathbf{B}(t)\mathbf{R}^{-1}(t)\mathbf{B}^T(t)\mathbf{P}(t) + \mathbf{Q}(t)]\mathbf{x}(t) = \mathbf{0}. \tag{5.1-11}$$

Since this must hold for all nonzero $\mathbf{x}(t)$, the term premultiplying $\mathbf{x}(t)$ must be zero. Thus the $\mathbf{P}$ matrix, which we see is an $n \times n$ symmetric matrix and which has $n(n+1)/2$ different terms, must satisfy the matrix Riccati equation—which, as we shall see later, must be positive definite—

$$\dot{\mathbf{P}} = -\mathbf{P}(t)\mathbf{A}(t) - \mathbf{A}^T(t)\mathbf{P}(t) + \mathbf{P}(t)\mathbf{B}(t)\mathbf{R}^{-1}(t)\mathbf{B}^T(t)\mathbf{P}(t) - \mathbf{Q}(t) \quad (5.1\text{-}12)$$

with a terminal condition given by Eqs. (5.1-6) and (5.1-8)

$$\mathbf{P}(t_f) = \mathbf{S}. \quad (5.1\text{-}13)$$

Thus we may solve the matrix Riccati equation backward in time from t_f to t_o, store the matrix

$$\mathbf{K}(t) = -\mathbf{R}^{-1}(t)\mathbf{B}^T(t)\mathbf{P}(t), \quad (5.1\text{-}14)$$

and then obtain a closed-loop control from

$$\mathbf{u}(t) = +\mathbf{K}(t)\mathbf{x}(t). \quad (5.1\text{-}15)$$

It is important to note that all components of the state vector must be accessible. We will remove this restriction in Chapter 8 when we discuss the ideal observer. A block diagram for accomplishing this solution to the regulator problem is shown in Fig. 5.1-1. If we compute the second variation, we find that

$$\delta^2 J = \tfrac{1}{2}\,\boldsymbol{\delta}\mathbf{x}^T(t_f)\mathbf{S}\,\boldsymbol{\delta}\mathbf{x}(t_f) + \tfrac{1}{2}\int_{t_o}^{t_f} [\boldsymbol{\delta}\mathbf{x}^T(t)\mathbf{Q}(t)\,\boldsymbol{\delta}\mathbf{x}(t) + \boldsymbol{\delta}\mathbf{u}^T(t)\mathbf{R}(t)\,\boldsymbol{\delta}\mathbf{u}(t)]\,dt. \quad (5.1\text{-}16)$$

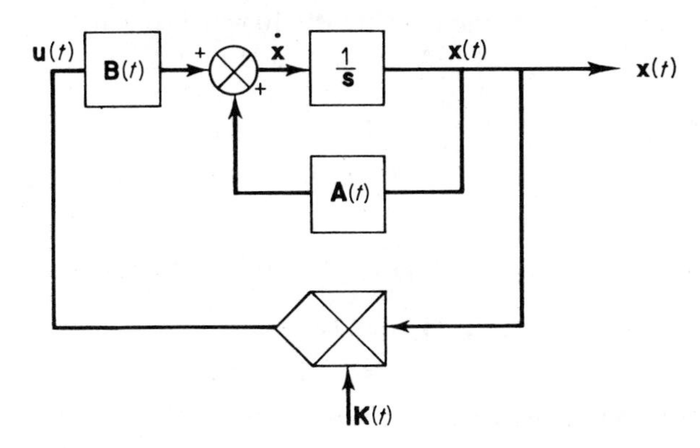

Fig. 5.1-1 Optimum linear closed-loop regulator.

Thus, $\mathbf{Q}$, $\mathbf{R}$, and $\mathbf{S}$ must be at least positive semidefinite in order to establish the sufficient condition for a minimum. In addition, we know from Eq. (5.1-7) that $\mathbf{R}$ must have an inverse;† therefore, it is sufficient that $\mathbf{R}$ be positive definite and the $\mathbf{Q}$ and $\mathbf{S}$ be at least positive semidefinite.

†Approaches that allow this assumption to be relaxed can be found in [24] and [25].

In some cases it may turn out that certain elements of the S matrix are large enough to give computational difficulties. In this case, it is possible and perhaps desirable to obtain an inverse Riccati differential equation; we let

$$\mathbf{P}(t)\mathbf{P}^{-1}(t) = \mathbf{I}, \tag{5.1-17}$$

and, by differentiating, we obtain

$$\dot{\mathbf{P}}\mathbf{P}^{-1}(t) + \mathbf{P}(t)\dot{\mathbf{P}}^{-1} = \mathbf{0} \tag{5.1-18}$$

such that we obtain an "inverse" matrix Riccati equation

$$\dot{\mathbf{P}}^{-1} = \mathbf{A}(t)\mathbf{P}^{-1}(t) + \mathbf{P}^{-1}(t)\mathbf{A}^T(t) - \mathbf{B}(t)\mathbf{R}^{-1}(t)\mathbf{B}^T(t) + \mathbf{P}^{-1}(t)\mathbf{Q}(t)\mathbf{P}^{-1}(t) \tag{5.1-19}$$

with

$$\mathbf{P}^{-1}(t_f) = \mathbf{S}^{-1}. \tag{5.1-20}$$

In this way, for example, it is possible to solve the Riccati equation such that $\mathbf{S}^{-1} = [\mathbf{0}]$, the null matrix, which will require that each and every component of the state vector approach the origin as the time approaches the terminal time. The "gains" $\mathbf{K}(t)$, or at least some components of them, become infinite at the terminal time in this case. It is also necessary to assume certain controllability requirements here, as we shall see in Chapter 7.

It is possible to write the nonlinear $n \times n$ matrix Riccati equation with a terminal condition as a $2n$ vector linear differential equation with two-point boundary conditions. We will use this approach, in part, to solve a Riccati equation associated with a filtering problem in Chapter 8. Our discussion of the second variation method in Chapter 10 will also make use of a Riccati transformation.

Example 5.1-1. Consider the scalar system

$$\dot{x} = -\tfrac{1}{2}x(t) + u(t), \qquad x(t_o) = x_o$$

with the cost function

$$J = \tfrac{1}{2}sx^2(t_f) + \tfrac{1}{2}\int_{t_o}^{t_f} [2x^2(t) + u^2(t)]\, dt.$$

The Riccati equation, Eq. (5.1-12), becomes

$$p = p + p^2 - 2, \qquad p(t_f) = s$$

which has a solution we may write as either

$$p(t) = -0.5 + 1.5 \tanh(-1.5t + \xi_1)$$

or

$$p(t) = -0.5 + 1.5 \coth(-1.5t + \xi_2)$$

where ξ_1 and ξ_2 are adjusted such that $p(t_f) = s$.

For example, if

(a) $s = 0$, $t_f = 1$, then $\xi_1 = 1.845$ radians, which gives

$$K(t) = -R^{-1}B^TP = 0.5 - 1.5 \tanh(-1.5t + 1.845).$$

Since $s = 0$, we are not particularly weighting the state at the final time, and the "gain" (and control) goes to zero at the final time.

(b) $s = 10$, $t_f = 10$, then $\xi_2 = 15.1425$ radians. In this case we are applying a great weight to the error at $t = t_f$, and the gain becomes large (-10) at the terminal time.

(c) $s = \infty$, the Riccati equation cannot be solved directly since it has an infinite initial condition. The inverse Riccati equation can be solved with zero terminal condition to give

$$K^{-1}(t) = 0.25 + 0.75 \tanh(-1.5t + 1.5t_f - 0.346).$$

As t_f becomes infinite, it is easy to show that $K(t)$ becomes unity and, as is expected, the feedback gain becomes constant. Figure 5.1-2 illustrates $K(t)$, the "Kalman gains" as they are sometimes called, for these three cases for this particular problem.

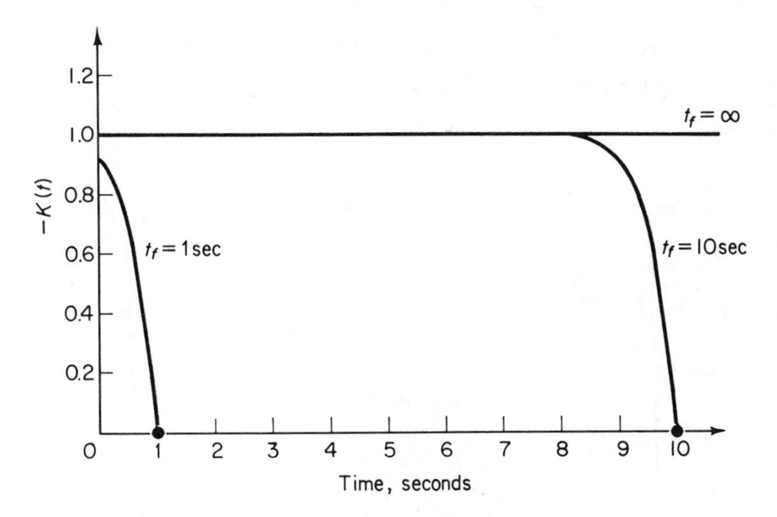

Fig. 5.1-2a (-1) times Kalman gain for controller, $s = 0$.

Example 5.1-2. Let us consider the optimum closed-loop control for a nuclear reactor system. Specifically, we wish to consider a very simple reactor model with zero temperature feedback. Only one group of delayed neutrons will be used.

The reactor kinetics are described by the equations

$$\dot{n} = \frac{(\rho - \beta)n}{\Lambda} + \lambda c, \qquad \dot{c} = \frac{\beta n}{\Lambda} - \lambda c$$

where the neutron density, n, and the precursor concentration, c, are the state variables, and the reactivity ρ is the control variable. The system has the initial conditions $n(0) = n_o$ and $c(0) = c_o$. β, Λ and λ are constants, the average fraction of precursors formed, effective neutron lifetime, and precursor decay constant.

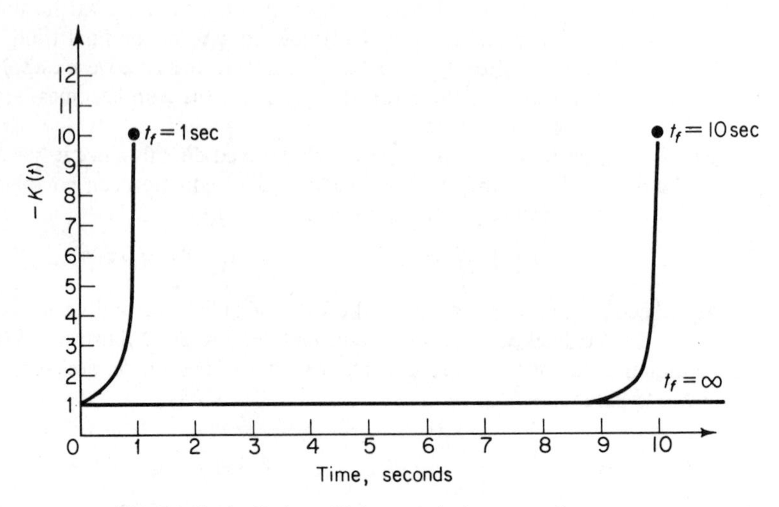

Fig. 5.1-2b (-1) times Kalman gain for controller, $s = 10$.

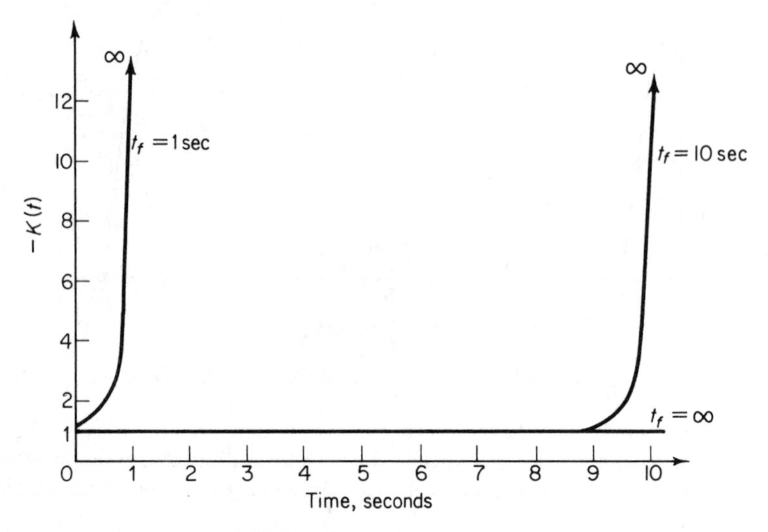

Fig. 5.1-2c (-1) times Kalman gain for controller, $s = \infty$.

The problem is to increase the power from the initial state n_o to a terminal state dn_o, where d is some constant greater than 1.0. The performance index for the system is

$$J_1 = \frac{1}{2}\int_o^{t_f} \dot{\rho}^2\, dt.$$

The control variable therefore becomes $\dot{\rho}$, and ρ, in effect, thus becomes a state variable. The kinetics equations may then be rewritten as

$$\dot{n} = \frac{(\rho - \beta)n}{\Lambda} + \lambda c$$

$$\dot{c} = \frac{\beta n}{\Lambda} - \lambda c$$

$$\dot{\rho} = u$$

where u is the control variable. Chapter 10 on quasilinearization indicates how the nonlinear two-point boundary value problem resulting from the use of optimal control theory may be used to obtain the optimum control and trajectory, which are shown in Fig. 5.1-3, for the following system parameters

$$\lambda = 0.1 \text{ sec}^{-1} \qquad n_o = 10 \text{ kW}$$

$$d = 5$$

$$\Lambda = 10^{-3} \text{ sec} \qquad \beta = 0.0064$$

$$t_f = 0.5 \text{ sec.}$$

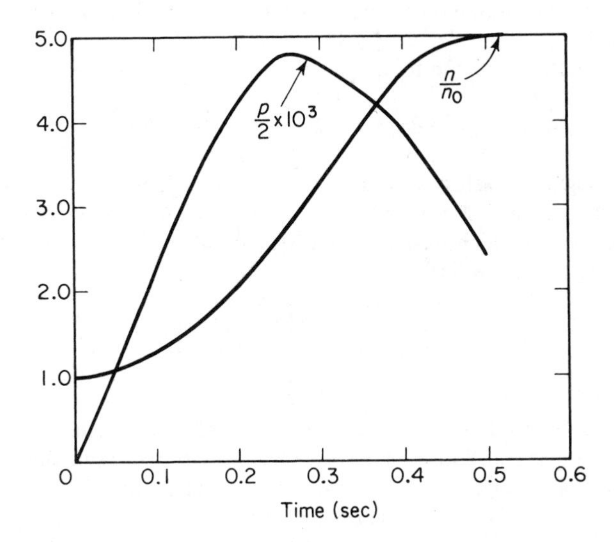

Fig. 5.1-3 Optimal control (reactivity) and trajectory (flux density) for Example (5.1-2).

We will now develop a method of feedback control about the optimal trajectory which minimizes a cost function J_2; it will be quadratic in deviation from the nominal (optimal for J_1) trajectory and control.

Having formulated a model for the nuclear reactor system and determined the optimal trajectories, we now desire to determine the linearized system coefficient matrix about the optimal trajectory. The deviations of the state and control variables about the optimal or nominal trajectories are expressed by

$$n = n_n(t) + \Delta n(t), \qquad c = c_n(t) + \Delta c(t)$$

$$\rho = \rho_n(t) + \Delta \rho(t), \qquad u = u_n(t) + \Delta u(t).$$

The state vector is

$$\Delta \mathbf{x}^T(t) = [\Delta n(t), \Delta c(t), \Delta \rho(t)].$$

The linearized model becomes

$$\Delta \dot{\mathbf{x}} = \begin{bmatrix} a_{11}(t) & \lambda & a_{13}(t) \\ \dfrac{\beta}{\Lambda} & -\lambda & 0 \\ 0 & 0 & 0 \end{bmatrix} \Delta \mathbf{x}(t) + \begin{bmatrix} 0 \\ 0 \\ 1 \end{bmatrix} \Delta u$$

$$= \mathbf{A}(t)\, \Delta \mathbf{x}(t) + \mathbf{b}(t)\, \Delta u(t)$$

where

$$a_{11}(t) = \frac{\rho_n(t) - \beta}{\Lambda}, \qquad a_{13}(t) = \frac{n_n(t)}{\Lambda}.$$

To complete our design of the closed-loop controller, we must evaluate $\mathbf{A}(t)$ and $\mathbf{b}(t)$ about the optimum or nominal trajectories, select the **R**, **Q**, and **S** matrices, and solve the associated Riccati equation. The nominal trajectory, control, and time-varying gains are then stored and used to complete the closed-loop controller design.

The choice of the **R**, **Q**, and **S** matrices to minimize

$$J_2 = \tfrac{1}{2}\, \Delta \mathbf{x}^T(t_f)\mathbf{S}\Delta \mathbf{x}(t_f) + \tfrac{1}{2} \int_{t_0}^{t_f} [\Delta \mathbf{x}^T(t)\mathbf{Q}(t)\Delta \mathbf{x}(t) + r(t)\Delta u^2(t)]\, dt$$

is somewhat arbitrary and can perhaps best be done here by experimentation. We can accomplish this only after we have obtained a knowledge of possible disturbances which may drive the system off the nominal trajectory. Let us assume that we will use

$$\mathbf{Q} = \begin{bmatrix} 1 & 0 & 0 \\ 0 & 0 & 0 \\ 0 & 0 & 10^4 \end{bmatrix}, \qquad \mathbf{S} = \mathbf{0}, \qquad r = 1.$$

In Chapter 10 the second variation and neighboring optimal methods of control-law computation will lead us to a method for choosing the proper weighting matrices for a variety of cases, in particular, for relating J_1 and J_2.

The control, $\Delta u(t)$, is computed from

$$\begin{aligned} \Delta u(t) &= -\mathbf{R}^{-1}(t)\mathbf{B}^T(t)\mathbf{P}(t)\Delta \mathbf{x}(t) \\ &= -[p_{31}(t)\, \Delta n(t) + p_{32}(t)\, \Delta c(t) + p_{33}(t)\, \Delta \rho(t)] \end{aligned}$$

where it is necessary to solve the 3×3 matrix Riccati equation, having six different first-order differential equations, to obtain $\mathbf{P}(t)$. Figure 5.1-4 illustrates the Kalman gains, $-\mathbf{K}^T(t) = [p_{31}(t), p_{32}(t), p_{33}(t)]$, for this example. Figure 5.1-5 indicates how the complete closed-loop controller is obtained. It is interesting to note that, in an actual physical problem, the precursor concentration is not measurable, and therefore we need to add an "observer" of this particular state variable. We also need to discuss many more aspects of this problem such as disturbances and parameter variations. We will postpone further consideration of these important questions until we establish some foundation in state and parameter estimation and optimal adaptive

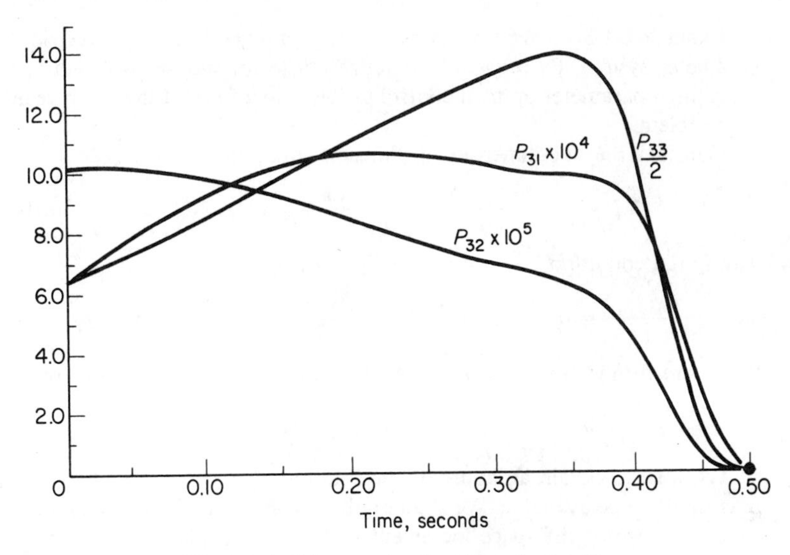

Fig. 5.1-4 Kalman gains for Example (5.1-2).

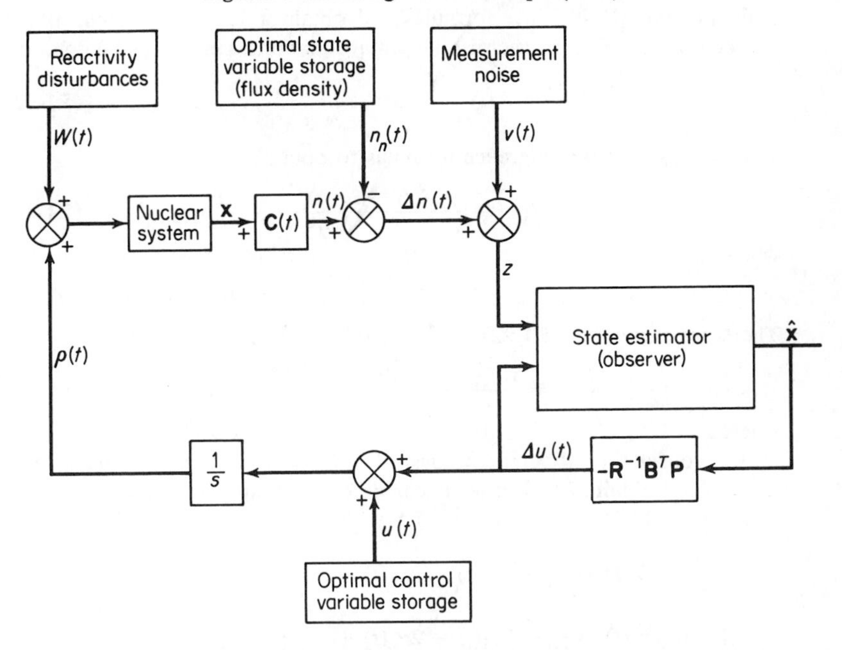

Fig. 5.1-5 Structure of controller for Example (5.1-2).

control. We have, in this example, illustrated how a basically nonlinear problem may be linearized, and a linear time-varying closed-loop controller obtained, if a nominal trajectory is known. Since this can be accomplished for a variety of problems, we see that the linear regulator problem is indeed an important one.

Example 5.1-3. We now consider the optimal control of a distributed parameter system. By a spatial discretization technique, we will reduce the distributed parameter optimal control problem to a form of the linear regulator problem.

Consider the one-dimensional diffusion equation

$$\frac{\partial x(y, t)}{\partial t} = \frac{\partial^2 x(y, t)}{\partial y^2} + u(y, t) \tag{5.1-21}$$

with initial condition $x(y, t_o = 0) = x_o(y)$,

and $\frac{\partial x(y, t)}{\partial y} = 0$ at $y = 0$, $\frac{\partial x(y, t)}{\partial y} = 0$ at $y = y_f$.

We desire to find the control $u(y, t)$ which minimizes the cost function

$$J = \frac{1}{2} \int_o^{t_f} \int_o^{y_f} [Q' x^2(y, t) + R' u^2(y, t)] \, dy \, dt.$$

We wish to obtain an approximate solution of Eq. (5.1-21) where $u(y, t)$ is assumed to be available. We shall establish a spatially discretized model in which the size of the space increment is $\Delta y = y_f/n$, where n is an integer. Physically, this corresponds to cutting a slab of length y_f into n slices. We shall use central difference formulas and obtain a spatially discrete model that can be described by vector differential equations. Let us use the notation

$$\frac{\partial x(y, t)}{\partial t} = \dot{x}_i(t) \qquad \text{where} \quad i = 1, 2, \ldots, n \tag{5.1-22}$$

and then use central difference formulas to obtain

$$\frac{\partial^2 x(y, t)}{\partial y^2} \cong \frac{x_{i+1}(t) - 2x_i(t) + x_{i-1}(t)}{(\Delta y)^2} \tag{5.1-23}$$

where

$$x_{i+1}(t) = x(y + \Delta y, t), \qquad x_i(t) = x(y, t), \qquad x_{i-1}(t) = x(y - \Delta y, t).$$

Therefore, using Eqs. (5.1-22) and (5.1-23) in Eq. (5.1-21), we obtain

$$\dot{x}_i(t) = \frac{x_{i+1}(t) - 2x_i(t) + x_{i-1}(t)}{(\Delta y)^2} + u_i(t) \tag{5.1-24}$$

where $i = 1, 2, \ldots, n$.

By considering different values of i, $i = 1, 2, \ldots, n$, and using Eq. (5.1-24), we obtain n first-order linear differential equations which approximate Eq. (5.1-21). These are

$$\begin{aligned}
\dot{x}_1(t) &= \frac{1}{(\Delta y)^2}[x_2(t) - 2x_1(t) + x_0(t)] + u_1(t) \\
\dot{x}_2(t) &= \frac{1}{(\Delta y)^2}[x_3(t) - 2x_2(t) + x_1(t)] + u_2(t) \\
&\;\;\vdots \\
\dot{x}_{n-1}(t) &= \frac{1}{(\Delta y)^2}[x_n(t) - 2x_{n-1}(t) + x_{n-2}(t)] + u_{n-1}(t) \\
\dot{x}_n(t) &= \frac{1}{(\Delta y)^2}[x_{n+1}(t) - 2x_n(t) + x_{n-1}(t)] + u_n(t).
\end{aligned} \tag{5.1-25}$$

We may use the boundary conditions to obtain $x_0(t)$ and $x_{n+1}(t)$. Then, by using a first difference approximation to the initial boundary condition, we obtain

$$\frac{x_{i+1}(t) - x_i(t)}{(\Delta y)} = 0 \qquad \text{for} \quad y = 0 \quad \text{or, equivalently,} \quad i = 0. \tag{5.1-26}$$

We have therefore established the boundary condition

$$x_0(t) = x_1(t). \tag{5.1-27}$$

In a similar fashion, we may easily show that

$$x_n(t) = x_{n+1}(t). \tag{5.1-28}$$

If use is made of Eqs. (5.1-27) and (5.1-28) in the set of ordinary differential equations given in Eq. (4.3-28), we obtain

$$\begin{aligned}
\dot{x}_1(t) &= \frac{1}{(\Delta y)^2}[x_2(t) - x_1(t)] + u_1(t) \\
\dot{x}_2(t) &= \frac{1}{(\Delta y)^2}[x_3(t) - 2x_2(t) + x_1(t)] + u_2(t) \\
\dot{x}_3(t) &= \frac{1}{(\Delta y)^2}[x_4(t) - 2x_3(t) + x_2(t)] + u_3(t) \\
&\ \ \vdots \\
\dot{x}_{n-1}(t) &= \frac{1}{(\Delta y)^2}[x_n(t) - 2x_{n-1}(t) + x_{n-2}(t)] + u_{n-1}(t) \\
\dot{x}_n(t) &= \frac{1}{(\Delta y)^2}[-x_n(t) + x_{n-1}(t)] + u_n(t).
\end{aligned} \tag{5.1-29}$$

We will now represent this set of ordinary linear differential equations by the vector differential equation

$$\dot{\mathbf{x}}(t) = \mathbf{A}\mathbf{x}(t) + \mathbf{B}\mathbf{u}(t), \qquad \mathbf{x}(0) = \mathbf{x}_o \tag{5.1-30}$$

where: $\mathbf{x}$ is an n-dimensional state vector; $\mathbf{u}$ is an n-dimensional control vector; $\mathbf{A}$ is the $n \times n$ tridiagonal matrix,

$$\mathbf{A} = \frac{1}{(\Delta y)^2}\begin{bmatrix}
-1 & 1 & 0 & 0 & 0 & \cdots & 0 & 0 & 0 & 0 \\
1 & -2 & 1 & 0 & 0 & \cdots & 0 & 0 & 0 & 0 \\
0 & 1 & -2 & 1 & 0 & \cdots & 0 & 0 & 0 & 0 \\
0 & 0 & 1 & -2 & 1 & \cdots & 0 & 0 & 0 & 0 \\
\vdots & \vdots & \vdots & \vdots & \vdots & & \vdots & \vdots & \vdots & \vdots \\
0 & 0 & 0 & 0 & 0 & & 0 & 1 & -2 & 1 \\
0 & 0 & 0 & 0 & 0 & & 0 & 0 & 1 & -1
\end{bmatrix}; \tag{5.1-31}$$

and where $\mathbf{B}$ is the identity matrix of order n, $\mathbf{B} = \mathbf{I}$. It is an easy task to verify that this linear system is always stable.

A discrete approximate form of the performance function is

$$J = \tfrac{1}{2}\Delta y \int_0^{t_f} \{\sum_{i=1}^{n-1} [Q'x_i^2(t) + R'u_i^2(t)] + \tfrac{1}{2}[Q'x_o(t) + Q'x_n(t) + R'u_o(t) + R'u_n(t)]\}\, dt$$

where n is the last discretized spatial stage. We may rewrite this as

$$J = \tfrac{1}{2}\Delta y \int_0^{t_f} [\mathbf{x}^T(t)\mathbf{Q}\mathbf{x}(t) + \mathbf{u}^T(t)\mathbf{R}\mathbf{u}(t)]\, dt.$$

For this problem, the Hamiltonian is

$$H(\mathbf{x}, \mathbf{u}, \boldsymbol{\lambda}, t) = \tfrac{1}{2}\, \Delta y \mathbf{x}^T\mathbf{Q}\mathbf{x} + \tfrac{1}{2}\, \Delta y \mathbf{u}^T\mathbf{R}\mathbf{u} + \boldsymbol{\lambda}^T\mathbf{A}\mathbf{x} + \boldsymbol{\lambda}^T\mathbf{B}\mathbf{u}.$$

Application of the maximum principle to this problem immediately yields the two-point boundary value problem

$$\dot{\mathbf{x}} = \mathbf{A}\mathbf{x} + \mathbf{B}\mathbf{u}, \qquad x_i(0) = 1 + \alpha i y_f/n, \qquad i = 1, 2, \dots, n$$

$$-\dot{\boldsymbol{\lambda}} = \Delta y \mathbf{Q}\mathbf{x}(t) + \mathbf{A}^T\boldsymbol{\lambda}, \qquad \boldsymbol{\lambda}(t_f) = 0$$

$$\mathbf{u} = -\frac{1}{\Delta y}\mathbf{R}^{-1}\mathbf{B}^T\boldsymbol{\lambda}.$$

We shall solve this problem by generating the Riccati equation where, as before in Sec. 5.1, we assume $\boldsymbol{\lambda}(t) = \mathbf{P}(t)\mathbf{x}(t)$. Thus, the optimal control is a linear feedback control determined by solving

$$\dot{\mathbf{P}}(t) + \mathbf{P}(t)\mathbf{A} - \frac{1}{\Delta y}\mathbf{P}(t)\mathbf{B}\mathbf{R}^{-1}\mathbf{B}^T\mathbf{P}(t) + \Delta y\mathbf{Q} + \mathbf{A}^T\mathbf{P}(t) = \mathbf{0}, \qquad \mathbf{P}(t_f) = \mathbf{0}$$

$$\dot{\mathbf{x}} = \mathbf{A}\mathbf{x} - \frac{1}{\Delta y}\mathbf{B}\mathbf{R}^{-1}\mathbf{B}^T\mathbf{P}\mathbf{x}(t), \qquad x_i(0) = 1 + \alpha i y_f/n, \qquad i = 1, 2, \dots, n$$

$$\mathbf{u}(t) = -\frac{1}{\Delta y}\mathbf{R}^{-1}\mathbf{B}^T\mathbf{P}(t)\mathbf{x}(t).$$

Let us consider the following two cases:

Case A $\quad t_f = 1.0, \quad y_f = 4.0, \quad \mathbf{B} = \mathbf{I}, \quad Q' = R' = 1$

$\quad \Delta t = 0.01, \quad \Delta y = 1.0, \quad \alpha = 1$

$$\mathbf{A} = \frac{1}{1}\begin{bmatrix} -1 & 1 & 0 & 0 & 0 \\ 1 & -2 & 1 & 0 & 0 \\ 0 & 1 & -2 & 1 & 0 \\ 0 & 0 & 1 & -2 & 1 \\ 0 & 0 & 0 & 1 & -1 \end{bmatrix}, \qquad \mathbf{Q} = \mathbf{R} = \begin{bmatrix} \frac{1}{2} & 0 & 0 & 0 & 0 \\ 0 & 1 & 0 & 0 & 0 \\ 0 & 0 & 1 & 0 & 0 \\ 0 & 0 & 0 & 1 & 0 \\ 0 & 0 & 0 & 0 & \frac{1}{2} \end{bmatrix}.$$

Case B $\quad t_f = 1.0, \quad \Delta t = 0.01, \quad \mathbf{B} = \mathbf{I}, \quad Q' = R' = 1$

$\quad y_f = 4.0, \quad \Delta y = 0.5, \quad \alpha = 1$

$$\mathbf{A} = 4\begin{bmatrix} -1 & 1 & 0 & 0 & 0 & 0 & 0 & 0 & 0 \\ 1 & -2 & 1 & 0 & 0 & 0 & 0 & 0 & 0 \\ 0 & 1 & -2 & 1 & 0 & 0 & 0 & 0 & 0 \\ 0 & 0 & 1 & -2 & 1 & 0 & 0 & 0 & 0 \\ 0 & 0 & 0 & 1 & -2 & 1 & 0 & 0 & 0 \\ 0 & 0 & 0 & 0 & 1 & -2 & 1 & 0 & 0 \\ 0 & 0 & 0 & 0 & 0 & 1 & -2 & 1 & 0 \\ 0 & 0 & 0 & 0 & 0 & 0 & 1 & -2 & 1 \\ 0 & 0 & 0 & 0 & 0 & 0 & 0 & 1 & -1 \end{bmatrix},$$

$$\mathbf{Q} = \mathbf{R} = \begin{bmatrix} \frac{1}{2} & 0 & 0 & 0 & 0 & 0 & 0 & 0 & 0 \\ 0 & 1 & 0 & 0 & 0 & 0 & 0 & 0 & 0 \\ 0 & 0 & 1 & 0 & 0 & 0 & 0 & 0 & 0 \\ 0 & 0 & 0 & 1 & 0 & 0 & 0 & 0 & 0 \\ 0 & 0 & 0 & 0 & 1 & 0 & 0 & 0 & 0 \\ 0 & 0 & 0 & 0 & 0 & 1 & 0 & 0 & 0 \\ 0 & 0 & 0 & 0 & 0 & 0 & 1 & 0 & 0 \\ 0 & 0 & 0 & 0 & 0 & 0 & 0 & 1 & 0 \\ 0 & 0 & 0 & 0 & 0 & 0 & 0 & 0 & \frac{1}{2} \end{bmatrix}.$$

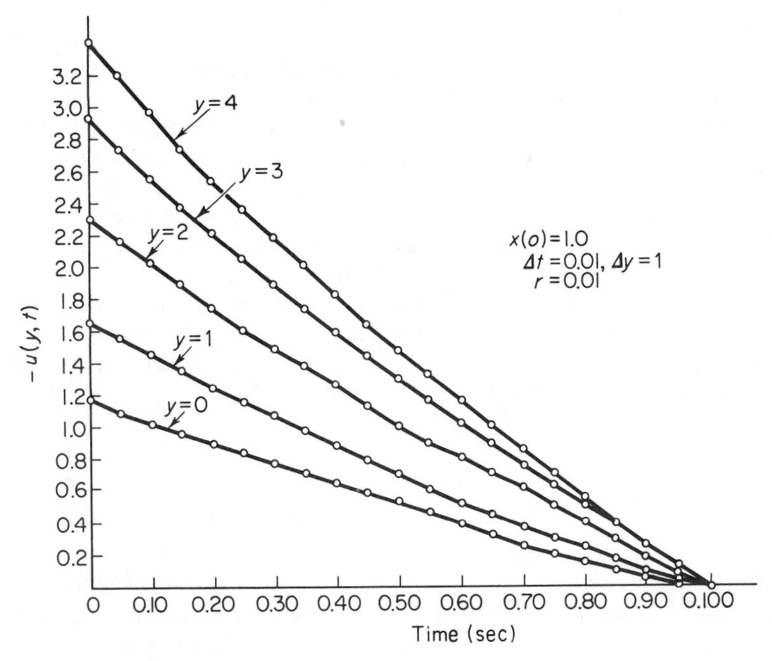

Fig. 5.1-6 Optimal control versus spatial coordinate and time, Example 5.1-3.

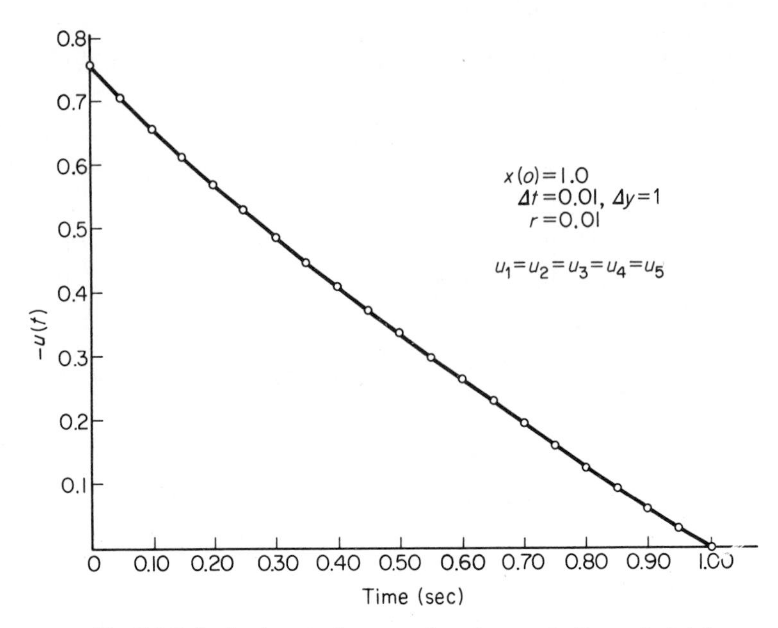

Fig. 5.1-7 Optimal control versus time for $\alpha = 0$, Example 5.1-3.

Solution of the two cases considered yields essentially the same result. This indicates that, for the particular initial condition $\mathbf{x}(y, 0) = 1 + y$, a model with five coordinates yields as good a solution as a model with ten coordinates. Thus we may safely assume that, for this particular initial condition, lumping the distributed system into five states is a satisfactory thing to do. As we change the initial distribution, $x(y, 0) = 1 + \alpha y$, by changing α, the number of necessary states to provide a good lumped model changes. For $\alpha = 0$, where the initial condition is uniform throughout y, a single state suffices for an exact model since $\partial^2 x(y, t)/\partial y^2$ is then always zero, and the distributed system degenerates to a lumped system for this particular case.

Figure 5.1-6 illustrates a plot of the optimal control versus time and distance y for $\alpha = 1$. A plot of the optimal control versus time is shown for the spatially independent case when $\alpha = 0$ in Fig. 5.1-7.

5.2 The linear servomechanism

The linear regulator problem considered in the preceding section can be generalized in several ways. We can assume that we desire to find the control in such a way as to cause the output to track or follow a desired output state, $\boldsymbol{\eta}(t)$. We may also assume that there is a forcing function (not the control) for the system differential equations. Therefore, we will consider the minimization of

$$J = \tfrac{1}{2}\|\boldsymbol{\eta}(t_f) - \mathbf{z}(t_f)\|^2_{\mathbf{S}} + \tfrac{1}{2}\int_{t_o}^{t_f} [\|\boldsymbol{\eta}(t) - \mathbf{z}(t)\|^2_{\mathbf{Q}(t)} + \|\mathbf{u}(t)\|^2_{\mathbf{R}(t)}]\, dt \quad (5.2\text{-}1)$$

for the system which contains a deterministic input or plant "noise" vector $\mathbf{w}(t)$

$$\dot{\mathbf{x}} = \mathbf{A}(t)\mathbf{x}(t) + \mathbf{B}(t)\mathbf{u}(t) + \mathbf{w}(t), \qquad \mathbf{x}(t_o) = \mathbf{x}_o \quad (5.2\text{-}2)$$

$$\mathbf{z}(t) = \mathbf{C}(t)\mathbf{x}(t). \quad (5.2\text{-}3)$$

The requirements on the various matrices are the same as in the preceding section. We proceed in exactly the same fashion as for the regulator problem. The Hamiltonian is, from Eq. (4.3-34),

$$H(\mathbf{x}, \mathbf{u}, \boldsymbol{\lambda}, t) = \tfrac{1}{2}\|\boldsymbol{\eta}(t) - \mathbf{C}(t)\mathbf{x}(t)\|^2_{\mathbf{Q}(t)} + \tfrac{1}{2}\|\mathbf{u}(t)\|^2_{\mathbf{R}(t)} + \boldsymbol{\lambda}^T(t)[\mathbf{A}(t)\mathbf{x}(t) + \mathbf{B}(t)\mathbf{u}(t) + \mathbf{w}(t)]. \quad (5.2\text{-}4)$$

We employ the maximum principle and set $\partial H/\partial \mathbf{u} = \mathbf{0}$ to obtain

$$\mathbf{u}(t) = -\mathbf{R}^{-1}(t)\mathbf{B}^T(t)\boldsymbol{\lambda}(t) \quad (5.2\text{-}5)$$

and

$$\frac{\partial H}{\partial \mathbf{x}} = -\dot{\lambda} = \mathbf{C}^T(t)\mathbf{Q}(t)[\mathbf{C}(t)\mathbf{x}(t) - \boldsymbol{\eta}(t)] + \mathbf{A}^T(t)\boldsymbol{\lambda}(t) \quad (5.2\text{-}6)$$

with the terminal condition

$$\boldsymbol{\lambda}(t_f) = \mathbf{C}^T(t_f)\mathbf{S}[\mathbf{C}(t_f)\mathbf{x}(t_f) - \boldsymbol{\eta}(t_f)]. \quad (5.2\text{-}7)$$

In order to attempt to determine a closed-loop control, we assume

$$\boldsymbol{\lambda}(t) = \mathbf{P}(t)\mathbf{x}(t) - \boldsymbol{\xi}(t). \quad (5.2\text{-}8)$$

We substitute this relation into the canonic equations and determine the requirements for a solution. By a procedure analogous to that of the preceding section, we easily obtain the following requirements:

$$\dot{\mathbf{P}} = -\mathbf{P}(t)\mathbf{A}(t) - \mathbf{A}^T(t)\mathbf{P}(t) + \mathbf{P}(t)\mathbf{B}(t)\mathbf{R}^{1-}(t)\mathbf{B}^T(t)\mathbf{P}(t) - \mathbf{C}^T(t)\mathbf{Q}(t)\mathbf{C}(t) \quad (5.2\text{-}9)$$

$$\mathbf{P}(t_f) = \mathbf{C}^T(t_f)\mathbf{S}\mathbf{C}(t_f), \quad (5.2\text{-}10)$$

and

$$\dot{\boldsymbol{\xi}} = -[\mathbf{A}(t) - \mathbf{B}(t)\mathbf{R}^{-1}(t)\mathbf{B}^T(t)\mathbf{P}(t)]^T\boldsymbol{\xi} + \mathbf{P}(t)\mathbf{w}(t) - \mathbf{C}^T(t)\mathbf{Q}(t)\boldsymbol{\eta}(t) \quad (5.2\text{-}11)$$

$$\boldsymbol{\xi}(t_f) = \mathbf{C}^T(t_f)\mathbf{S}\boldsymbol{\eta}(t_f). \quad (5.2\text{-}12)$$

Thus we see that the linear servomechanism problem is composed of two parts: a linear regulator part, plus a prefilter to determine the optimal driving function from the desired value, $\boldsymbol{\eta}(t)$, of the system output. The optimum control law is linear and is obtained from Eq. (5.2-5) as

$$\mathbf{u}(t) = -\mathbf{R}^{-1}(t)\mathbf{B}^T(t)[\mathbf{P}(t)\mathbf{x}(t) - \boldsymbol{\xi}(t)]. \quad (5.2\text{-}13)$$

Unfortunately, the optimal control is, in practice, often computationally unrealizable because it involves $\boldsymbol{\xi}(t)$ which must be solved backward from t_f to t_o and, therefore, requires a knowledge of $\boldsymbol{\eta}(t)$ and $\mathbf{w}(t)$ for all time $t \in [t_o, t_f]$. This is quite often not known at the initial time t_o.

Example 5.2-1. Let us consider the minimization of the cost function

$$J = \frac{1}{2}\int_0^{t_f} [(x_1 - \eta_1)^2 + u^2]\, dt$$

for the system described by

$$\dot{x}_1 = x_2, \qquad x_1(0) = x_{10}$$
$$\dot{x}_2 = u, \qquad x_2(0) = x_{20}.$$

We first use Eqs. (5.2-9) and (5.2-10) to obtain the Riccati equation for this example

$$\dot{p}_{11} = p_{12}^2 - 1, \qquad p_{11}(t_f) = 0$$
$$\dot{p}_{12} = -p_{11} + p_{12}p_{22}, \qquad p_{12}(t_f) = 0$$
$$\dot{p}_{22} = -2p_{12} + p_{22}^2, \qquad p_{22}(t_f) = 0.$$

If we allow t_f to become infinite, we obtain the solution $p_{11} = p_{22} = \sqrt{2}$, $p_{12} = 1$. Thus we have for the closed-loop control

$$u = -\mathbf{R}^{-1}\mathbf{B}^T[\mathbf{P}\mathbf{x} - \boldsymbol{\xi}] = -x_1 - \sqrt{2}x_2 + \xi_2$$

where we must determine $\boldsymbol{\xi}$ by solving Eqs. (5.2-11) and (5.2-12) which become for this example

$$\dot{\xi}_1 = \xi_1 - \eta_1, \qquad \xi_1(t_f) = 0$$
$$\dot{\xi}_2 = -\xi_1 + \sqrt{2}\xi_2, \qquad \xi_2(t_f) = 0.$$

If $\eta_1 = \alpha$, a constant, for t greater than zero, we are justified in obtaining the equilibrium solution for the $\boldsymbol{\xi}$ equation if $t_f = \infty$ by setting $\dot{\boldsymbol{\xi}} = \mathbf{0}$ to obtain $\xi_2 = 0.707\xi_1 = \eta_1 = \alpha$. If $\eta_1 = 1 - e^{-t}$, we will then find by a simple limiting process that for $t_f = \infty$,

$$\xi_2(t) = 1 + \frac{1}{2 + \sqrt{2}}e^{-t}, \qquad t \geq 0.$$

We may realize this solution as shown in Fig. 5.2-1.

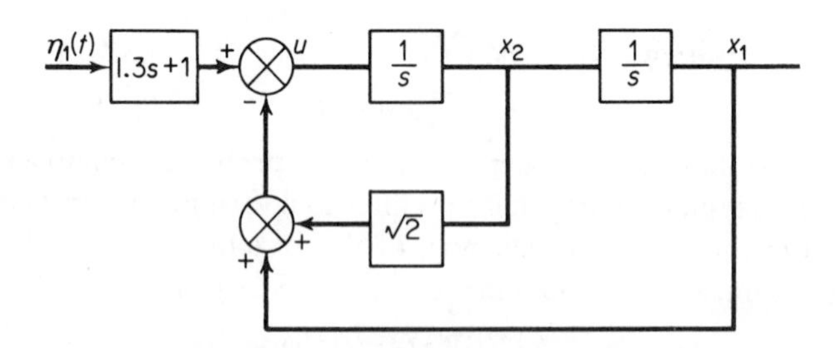

Fig. 5.2-1 Block diagram of optimum servomechanism for Example 5.2-1.

We note that if $\mathbf{w}(t) = \boldsymbol{\eta}(t) = \mathbf{0}$, or for that matter, any vector constant in time, the servomechanism problem reduces to a regulator problem except that it is an "output" regulator problem rather than a "state" regulator problem because of the presence of the output matrix $\mathbf{C}(t)$. It is not necessary for the system to be controllable in order to find a solution to the regulator problem. The only exception to this is in the limiting cases where $\mathbf{S}$ becomes infinite or where t_f becomes infinite. It is, however, necessary that the system be observable in order for a solution to the output regulator problem to exist. We will expand considerably on these ideas when we consider controllability, observability, and the reachable zone problem in Chapter 7.

5.3 Bang bang control and minimum time problems

Maximum effort control problems have become increasingly important in a variety of applications. It is natural that we ask under what circumstances optimal controls will always be maximum effort, or *bang bang*. To do this, we will restrict each component of the control vector, $\mathbf{u}(t)$, to some bounded interval. Let us consider the nonlinear differential system where the control enters in a linear fashion

$$\dot{\mathbf{x}} = \mathbf{f}[\mathbf{x}(t), t] + \mathbf{G}[\mathbf{x}(t), t]\mathbf{u}(t), \qquad \mathbf{x}(t_o) = \mathbf{x}_o \tag{5.3-1}$$

$$a_i \leq u_i \leq b_i, \qquad \forall\, i \tag{5.3-2}$$

and assume a performance index which likewise contains only linear terms in the control variable, such that the Hamiltonian will also be linear in $\mathbf{u}(t)$.

$$J = \theta[\mathbf{x}(t_f), t_f] + \int_{t_0}^{t_f} \{\phi[\mathbf{x}(t), t] + \mathbf{h}^T[\mathbf{x}(t), t]\mathbf{u}(t)\}\, dt \tag{5.3-3}$$

$$H[\mathbf{x}(t), \mathbf{u}(t), \boldsymbol{\lambda}(t), t] = \phi[\mathbf{x}(t), t] + \mathbf{h}^T[\mathbf{x}(t), t]\mathbf{u}(t) + \boldsymbol{\lambda}^T(t)\{\mathbf{f}[\mathbf{x}(t), t] + \mathbf{G}[\mathbf{x}(t), t]\mathbf{u}(t)\}. \tag{5.3-4}$$

Since the Hamiltonian is linear in the control vector, $\mathbf{u}(t)$, minimization of the Hamiltonian with respect to $\mathbf{u}(t)$ requires that

$$u_i = \begin{cases} a_i & \text{if } \{\mathbf{h}^T[\mathbf{x}(t), t] + \boldsymbol{\lambda}^T(t)\mathbf{G}[\mathbf{x}(t), t]\}_i > 0 \\ b_i & \text{if } \{\mathbf{h}^T[\mathbf{x}(t), t] + \boldsymbol{\lambda}^T(t)\mathbf{G}[\mathbf{x}(t), t]\}_i < 0. \end{cases} \tag{5.3-5}$$

Thus we see that when the control vector appears linearly in both the equation of motion of the differential system and the performance index, and if, in addition, each component of the control vector is bounded, the optimal control is bang bang. The only exception to this occurs in cases where

$$\mathbf{h}^T[\mathbf{x}(t), t] + \boldsymbol{\lambda}^T(t)\mathbf{G}[\mathbf{x}(t), t] = \mathbf{0}, \tag{5.3-6}$$

for then the Hamiltonian is not a function of $\mathbf{u}(t)$ and cannot be minimized with respect to $\mathbf{u}(t)$. When Eq. (5.3-6) holds for more than isolated points in time, the optimization problem is said to possess a singular solution, a problem which we will discuss in detail in the next section. A singular solution is possible with respect to a particular control component, u_i, if the ith component of Eq. (5.3-6) is zero.

For this problem, the canonic equations are obtained as

$$\dot{\mathbf{x}} = \frac{\partial H}{\partial \boldsymbol{\lambda}} = \mathbf{f}[\mathbf{x}(t), t] + \mathbf{G}[\mathbf{x}(t), t]\mathbf{u}(t) \tag{5.3-7}$$

$$-\dot{\boldsymbol{\lambda}} = \frac{\partial H}{\partial \mathbf{x}} = \frac{\partial \phi[\mathbf{x}(t), t]}{\partial \mathbf{x}} + \frac{\partial \mathbf{h}^T[\mathbf{x}(t), t]}{\partial \mathbf{x}}\mathbf{u}(t) + \frac{\partial \mathbf{f}^T[\mathbf{x}(t), t]}{\partial \mathbf{x}}\boldsymbol{\lambda}(t) + \frac{\partial \{\mathbf{G}[\mathbf{x}(t), t]\mathbf{u}(t)\}^T}{\partial \mathbf{x}(t)}\boldsymbol{\lambda}(t) \tag{5.3-8}$$

where $\mathbf{u}(t)$ is determined via Eq. (5.3-5). Since we have not specifically stated the end conditions, we have carried the general problem about as far as possible. When we specify information concerning the desired states at the terminal time and the initial condition vector, we have, as before, a two-point boundary value problem with half of the conditions specified at the initial time and half at the terminal time. A possible method of solution of the canonic equations for this formulation consists of reversing time in the canonic equations. Starting at the determined or specified terminal vector, which often is the origin of the state vector, we integrate back from this point with a constant control until a switching point is obtained from Eq. (5.3-5). Since no terminal conditions are present for half of the state variables, the method is, of necessity, cut and try. Chapter 10 provides more systematic methods for solving this type of two-point boundary value problem.

We shall now illustrate various solutions to a particular case which results in bang bang control—the minimum time problem for constant linear systems with a scalar input. In this problem, we desire to transfer an n vector constant differential system

$$\dot{\mathbf{x}} = \mathbf{A}\mathbf{x}(t) + \mathbf{b}u(t), \qquad \mathbf{x}(t_o) = \mathbf{x}_o \tag{5.3-9}$$

to the origin, $\mathbf{x}(t_f) = \mathbf{0}$, in minimum time, such that we have for the cost function

$$J = \int_{t_o}^{t_f} (1)\, dt = t_f - t_o \tag{5.3-10}$$

with the restriction that

$$-1 \le u(t) \le +1. \tag{5.3-11}$$

The Hamiltonian for our problem is

$$H[\mathbf{x}(t), u(t), \boldsymbol{\lambda}(t)] = 1 + \boldsymbol{\lambda}^T(t)\mathbf{A}\mathbf{x}(t) + \boldsymbol{\lambda}^T(t)\mathbf{b}u(t). \tag{5.3-12}$$

We must minimize the Hamiltonian with respect to a choice of $u(t)$, so we require

$$u(t) = -\text{sign}\,[\boldsymbol{\lambda}^T(t)\mathbf{b}]. \tag{5.3-13}$$

Thus the Hamiltonian with the control optimum is

$$H[\mathbf{x}(t), \boldsymbol{\lambda}(t)] = 1 + \boldsymbol{\lambda}^T(t)\mathbf{A}\mathbf{x}(t) - |\boldsymbol{\lambda}^T(t)\mathbf{b}|. \tag{5.3-14}$$

Since the terminal time is free, and since H does not depend explicitly on t, we know from Eqs. (4.3-21) and (4.3-39) that

$$H[\mathbf{x}(t), \boldsymbol{\lambda}(t)] = 0, \qquad \forall t \in [t_o, t_f] \tag{5.3-15}$$

on the optimal trajectory. The canonic equations are

$$\dot{\mathbf{x}} = \frac{\partial H}{\partial \boldsymbol{\lambda}} = \mathbf{A}\mathbf{x}(t) + \mathbf{b}u(t) = \mathbf{A}(t)\mathbf{x}(t) - \mathbf{b}\,\text{sign}\,[\boldsymbol{\lambda}^T(t)\mathbf{b}] \tag{5.3-16}$$

$$\dot{\boldsymbol{\lambda}} = -\frac{\partial H}{\partial \mathbf{x}} = -\mathbf{A}^T\boldsymbol{\lambda}(t). \tag{5.3-17}$$

To avoid a singular solution, we must ensure that $\boldsymbol{\lambda}^T(t)\mathbf{b}$ cannot be zero over a time interval of nonzero length. From Eq. (5.3-17) we see that this is almost certainly the case unless $\boldsymbol{\lambda}(t_o)$ were identically $\mathbf{0}$, which is not possible. The only other requirement is that the system be controllable as shall be defined in Chapter 7. If the system were not controllable, we could not, in general, transfer its motion to the origin. For this problem we shall discover stronger requirements than controllability which we must place on the system if we are to transfer it to the origin. These requirements come about because of the restriction $|u| \leq 1$ which always results in initial states in an unstable plant which cannot be transferred to the origin. The solution to Eq. (5.3-17) is

$$\boldsymbol{\lambda}(t) = e^{-\mathbf{A}^T(t-t_f)}\boldsymbol{\lambda}(t_f). \tag{5.3-18}$$

It is convenient for us to rewrite Eq. (5.3-16) in terms of the time to go by letting $t_o = 0$ and

$$\tau = t_f - t \tag{5.3-19}$$

$$\boldsymbol{\xi}(\tau) = \mathbf{x}(t) = \mathbf{x}(t_f - \tau). \tag{5.3-20}$$

This gives us

$$\frac{d\boldsymbol{\xi}}{d\tau} = -\mathbf{A}\boldsymbol{\xi}(\tau) + \mathbf{b}\,\text{sign}\,[\boldsymbol{\lambda}^T(t_f)e^{\mathbf{A}\tau}\mathbf{b}] \tag{5.3-21}$$

which has its solution, since $\boldsymbol{\xi}(0) = \mathbf{x}(t_f) = \mathbf{0}$,

$$\boldsymbol{\xi}(\tau) = \int_0^\tau e^{-\mathbf{A}(\tau-p)}\mathbf{b}\,\text{sign}\,[\boldsymbol{\lambda}^T(t_f)e^{\mathbf{A}p}\mathbf{b}]\,dp. \tag{5.3-22}$$

A state $\mathbf{x}(t_o) = \mathbf{x}_o$ from which the origin can be reached in a specified minimum time t_f may now be obtained if we substitute a value of $\boldsymbol{\lambda}(t_f)$ in Eq. (5.3-22) and then calculate $\mathbf{x}(t_o) = \mathbf{x}(0) = \boldsymbol{\xi}(t_f)$. This alludes to a relation between $\mathbf{x}(t_o)$ and $\boldsymbol{\lambda}(t_f)$ and suggests that an alternate name for $\boldsymbol{\lambda}$ might well be "influence function." Chapter 10 will develop this very point in considerably more detail.

Since it is clear that the direction and not the magnitude of the $\boldsymbol{\lambda}(t_f)$ vector determines sign $[\boldsymbol{\lambda}^T(t_f)e^{\mathbf{A}p}\mathbf{b}]$, all states which can be reached in a given minimum time may be determined if we allow $\boldsymbol{\lambda}(t_f)$ to assume all values over a unit sphere. At points where $\boldsymbol{\lambda}^T(t_f)e^{\mathbf{A}\tau}\mathbf{b}$ is $\mathbf{0}$, we have a switching point. It is possible to show [7] that if the eigenvalues of $\mathbf{A}$ are real, there are, at most, $n - 1$ switchings or changes of sign of the control. We will now give several simple examples of calculations of the minimum time control.

Example 5.3-1. Let us now consider the simplest nontrivial time optimal control problem, the time optimal control of a pure inertia or double integration system. The system dynamics are described by

$$\begin{bmatrix}\dot{x}_1(t)\\ \dot{x}_2(t)\end{bmatrix} = \begin{bmatrix}0 & 1\\ 0 & 0\end{bmatrix}\begin{bmatrix}x_1(t)\\ x_2(t)\end{bmatrix} + \begin{bmatrix}0\\ 1\end{bmatrix}u(t), \qquad \begin{matrix}x_1(0) = x_{10}\\ x_2(0) = x_{20}.\end{matrix}$$

We need to compute the transition matrix which is

$$e^{\mathbf{A}t} = e^{\begin{bmatrix}0 & 1\\ 0 & 0\end{bmatrix}t} = \mathbf{I} + \mathbf{A}t + \frac{\mathbf{A}^2t^2}{2!} + \cdots = \begin{bmatrix}1 & t\\ 0 & 1\end{bmatrix}.$$

Thus, from Eq. (5.3-22) we have

$$\boldsymbol{\xi}(\tau) = \begin{bmatrix}1 & -\tau\\ 0 & 1\end{bmatrix}\int_0^\tau \begin{bmatrix}p\\ 1\end{bmatrix}\text{sign}\,[\lambda_1(t_f)p + \lambda_2(t_f)]\,dp.$$

For a fixed $\lambda(t_f)$, a switch may occur only at $\tau_s = -\lambda_2(t_f)/\lambda_1(t_f)$, verifying, in part, the statement made earlier that, at most, $n - 1$ switchings occur in a plant with n real poles. If we attempt to compute the switching line, we have at the switching point

$$\mathbf{x}(t_s) = \xi(\tau_s) = \begin{bmatrix}1 & -\tau_s\\ 0 & 1\end{bmatrix}\int_0^{\tau_s}\begin{bmatrix}p\\ 1\end{bmatrix}\text{sign}\,\{\lambda_1(t_f)[p - \tau_s]\}\,dp$$

$$= -\begin{bmatrix}1 & -\tau_s\\ 0 & 1\end{bmatrix}\int_0^{\tau_s}\begin{bmatrix}q\\ 1\end{bmatrix}\text{sign}\,[\lambda_1(t_f)]\,dq = \{\text{sign}\,\lambda_1(t_f)\}\begin{bmatrix}\frac{\tau_s^2}{2}\\ -\tau_s\end{bmatrix}.$$

Thus we see that, if $\lambda_1(t_f)$ is greater than zero, the switch points are $x_1 = \tau_s^2/2$, $x_2 = -\tau_s$; whereas if $\lambda_1(t_f)$ is less than zero, the switch points are $x_1 = -\tau_s^2/2$, $x_2 = \tau_s$. Therefore, the equation for the switching boundary is

$$x_1 + \tfrac{1}{2}x_2\,|x_2| = 0$$

which has the parabolic shape previously obtained in Fig. 4.5-1. It is now evident to us that the control law for this minimum time problem is

$$u = -\text{sign}\,[x_1(t) + \tfrac{1}{2}x_2(t)\,|x_2(t)|].$$

Example 5.3-2. Let us consider the slightly more complicated minimum time problem for the plant with real eigenvalues

$$\dot{x}_1 = x_2, \qquad x_1(0) = x_{10}$$

$$\dot{x}_2 = -\alpha x_2 + u, \qquad x_2(0) = x_{20}$$

or

$$\begin{bmatrix} \dot{x}_1 \\ \dot{x}_2 \end{bmatrix} = \begin{bmatrix} 0 & 1 \\ 0 & -\alpha \end{bmatrix} \begin{bmatrix} x_1(t) \\ x_2(t) \end{bmatrix} + \begin{bmatrix} 0 \\ 1 \end{bmatrix} u(t), \qquad \begin{bmatrix} x_1(0) \\ x_2(0) \end{bmatrix} = x_o.$$

For this problem, it is a simple matter to compute the transition matrix as

$$e^{\mathbf{A}t} = \begin{bmatrix} 1 & \frac{1}{\alpha}(1 - e^{-\alpha t}) \\ 0 & e^{-\alpha t} \end{bmatrix}$$

and the equation for the switching boundary as

$$\xi(\tau_s) = \begin{bmatrix} 1 & \frac{1}{\alpha}(1 - e^{\alpha \tau_s}) \\ 0 & e^{\alpha \tau_s} \end{bmatrix} \int_0^{\tau_s} \begin{bmatrix} \frac{1}{\alpha}(1 - e^{-\alpha p}) \\ e^{-\alpha p} \end{bmatrix} \times$$

$$\text{sign}\left\{\lambda_1(t_f)\left[\frac{1}{\alpha}(1 - e^{\alpha p})\right] + \lambda_2(t_f)[e^{-\alpha p}]\right\} dp.$$

Upon evaluation of this expression, we see that the switching boundary and control are given by the equations

$$x_1(t) + \frac{1}{\alpha} x_2(t) - \left\{\frac{\text{sign}\,[x_2(t)]}{\alpha^2}\right\}\{\ln\,[1 + \alpha\,|x_2(t)|]\} = 0$$

$$u(t) = -\text{sign}\left[x_1(t) + \frac{1}{\alpha} x_2(t) - \left\{\frac{\text{sign}\,[x_2(t)]}{\alpha^2}\right\}\{\ln\,[1 + \alpha\,|x_2(t)|]\}\right].$$

These results are, of course, quite similar to those in the previous examples. There is, at most, one switching point, and the equation for the switching boundary is relatively simple. The switching boundary is much more complicated if the **A** matrix has complex or complex-conjugate eigenvalues as we shall now see.

Example 5.3-3. We will now consider the minimum time control of a sinusoidal oscillator. The equations of motion are

$$\begin{bmatrix} \dot{x}_1(t) \\ \dot{x}_2(t) \end{bmatrix} = \begin{bmatrix} 0 & 1 \\ -0 & 0 \end{bmatrix} \begin{bmatrix} x_1(t) \\ x_2(t) \end{bmatrix} + \begin{bmatrix} 0 \\ 1 \end{bmatrix} u(t), \qquad \begin{matrix} x_1(0) = x_{10} \\ x_2(0) = x_{20} \end{matrix}$$

and the transition matrix is

$$e^{\mathbf{A}t} = \begin{bmatrix} \cos t & \sin t \\ -\sin t & \cos t \end{bmatrix}$$

which yields for the switching boundary

$$\xi(\tau_s) = \begin{bmatrix} \cos \tau_s & -\sin \tau_s \\ \sin \tau_s & \cos \tau_s \end{bmatrix} \int_0^{\tau} \begin{bmatrix} \sin p \\ \cos p \end{bmatrix} \text{sign}\,[\lambda_1(t_f) \sin p + \lambda_2(t_f) \cos p]\, dp.$$

Switching occurs when

$$\lambda_1(t_f) \sin p + \lambda_2(t_f) \cos p = 0$$

which indicates that there may be an infinite number of switchings for this problem. It is considerably more difficult to evaluate $\xi(\tau_s)$ in this case than in the previous two examples. It is not difficult, however, to verify that the proper switching line and control are as shown in Fig. 5.3-1. A typical trajectory with four switchings is also shown in the figure.

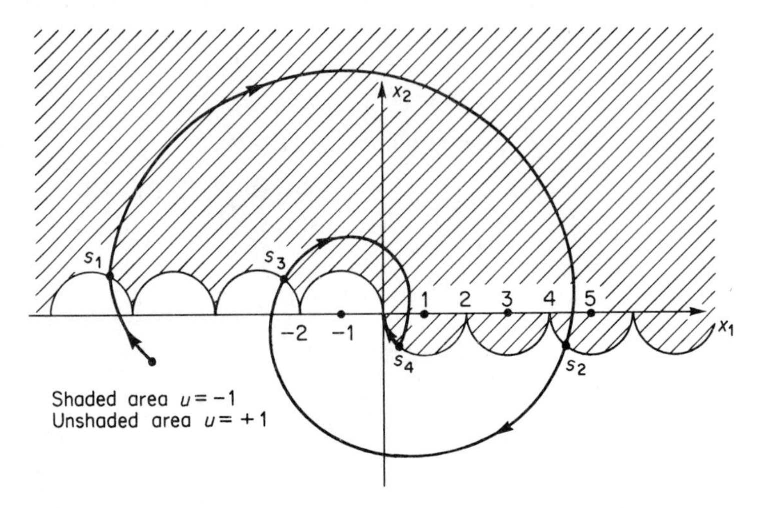

Fig. 5.3-1 Switching line for time optimum control of oscillator and typical trajectory, Example 5.3-3.

Example 5.3-4. As a final example of minimum time control of linear systems with a scalar input control, consider the unstable system with two positive real eigenvalues of $+1$:

$$\dot{x}_1 = x_2, \qquad x_1(0) = x_{10}$$

$$\dot{x}_2 = -x_1 + 2x_2 + u, \qquad x_2(0) = x_{20}.$$

The transition matrix for this unstable system is

$$e^{At} = \begin{bmatrix} 1-t & t \\ -t & 1+t \end{bmatrix} e^t$$

which gives for the switching boundary

$$\xi(\tau) = \begin{bmatrix} 1+\tau & -\tau \\ \tau & 1-\tau \end{bmatrix} e^{-\tau} \int_0^{\tau} \begin{bmatrix} pe^p \\ (1+p)e^p \end{bmatrix} \operatorname{sign}\{p\lambda_1(t_f) + [p+1]\lambda_2(t_f)\}\, dp.$$

We see that there is, at most, one switching which occurs at

$$p = \tau_s = -\frac{\lambda_2(t_f)}{[\lambda_1(t_f) + \lambda_2(t_f)]}.$$

Rather than use the $\xi(t)$ equation to determine the switching boundary and reachable zone, an ancillary approach will be used which gives insight into an additional method of determining solutions to time optimal control problems.

If we run in reverse time from the origin with a (+1) control (the sign { } term in the expression for $\xi(\tau)$ is +1), we obtain the switching curve

$$x_1(\tau) = +(\tau + 1)e^{-\tau} - 1$$
$$x_2(\tau) = +\tau e^{-\tau}.$$

A similar expression results for applying a (−1) control. The switching line is shown in Fig. 5.3-2. By solving the original equations of motion in the forward direction, we see that many initial values for x_{10} and x_{20} lead to trajectories which cannot hit the switching line. Therefore we have a region of state space inside which we can get to the origin. If the initial conditions are outside of this region, we cannot hit the origin no matter what control strategy we apply as long as $|u(t)| \leq 1$.

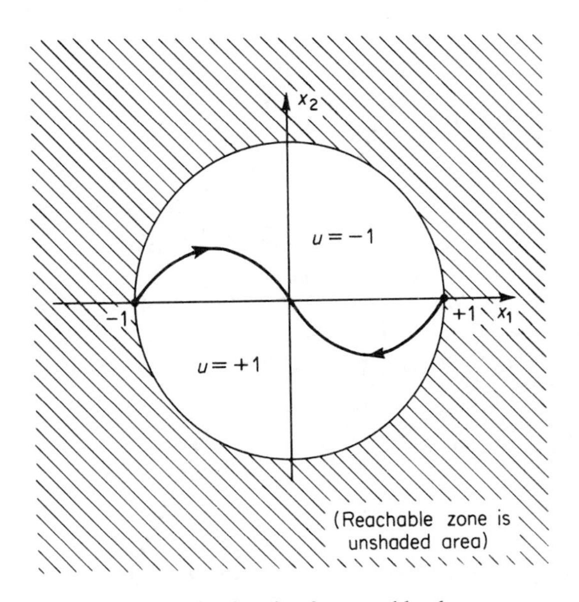

Fig. 5.3-2 Switching line for unstable plant.

Considerable insight can be gained into the analysis of optimal control problems, particularly the minimum time problem, by investigating such problems geometrically. We now examine the geometry of the linear minimum time problem, following the approach of [26], [27], and [28]. Let $K(t)$, called the *set of attainability*, denote the set of all trajectories at time t associated with admissible control functions having values at each time t constrained to the set $\mho \subset R^m$. Thus,

$$K(t) = \{\mathbf{x}(t, \mathbf{u}): \ \mathbf{u} \text{ is admissible}, \mathbf{u}(s) \in \mho, s \leq t\},$$

where $\mathbf{x}(\cdot, \mathbf{u})$ designates the unique trajectory associated with control function $\mathbf{u}$. It can be shown that for each t, if $\mho$ is compact (closed and bounded), then $K(t)$ is convex and compact [28]. Additionally, $K(t)$ inherits a type of continuity (in the sense of the Hausdorff metric (see Appendix B)) from the smoothness characteristics of the admissible trajectories. It is easy to visualize $K(t)$ expanding (which it does under reasonable assumptions [28]) and touching a given target set for the first time at time t^*, where: t^* is then the minimum time; a point of $K(t^*)$ which touches the target set is associated with an optimal trajectory; and a control function taking the initial condition to a point of intersection is an optimal control. It is easily argued that an optimal control $\mathbf{u}^*$ must take the initial condition to the boundary of $K(t^*)$. Otherwise, due to the continuity of $K(t)$, some point in $K(t)$ would have intersected the target set at some time $t < t^*$, contradicting the optimality of $\mathbf{u}^*$. Thus, an optimal control is necessarily an extremal control at time t^*, where we define an extremal control at time t_1 as a control taking the initial point to the boundary of the set of attainability at time t_1. The analysis of extremal controls forms the basis for the geometric development of necessary conditions. Two key results are as follows: Control $\mathbf{u}$ is extremal at time t_1 if and only if there exists a nontrivial solution of

$$\dot{\boldsymbol{\lambda}}(t) = -\mathbf{A}^T(t)\boldsymbol{\lambda}(t) \tag{5.3-23}$$

such that

$$\boldsymbol{\lambda}^T(t)\mathbf{B}(t)\mathbf{u}(t) = \min_{\mathbf{v} \in \mho} \boldsymbol{\lambda}^T(t)\mathbf{B}(t)\mathbf{v}, \tag{5.3-24}$$

for all $t \in [t_0, t_1]$; corollary to this fact is the second result: If $\mathbf{u}$ is extremal at time t_1, then $\mathbf{u}$ is extremal for all time $t \in [t_0, t_1]$. These facts imply that the trajectory $\mathbf{x}^*(\cdot, \mathbf{u}^*)$ associated with an optimal control $\mathbf{u}^*$ lies on the boundary of $K(t)$ for all $t \in [t_0, t^*]$. Necessary conditions for a control to be optimal then follow directly from the first key result: If $\mathbf{u}^*$ is an optimal control, then there exists a nontrivial solution of

$$\dot{\boldsymbol{\lambda}}(t) = -\mathbf{A}^T(t)\boldsymbol{\lambda}(t) \tag{5.3-25}$$

such that

$$\boldsymbol{\lambda}^T(t)\mathbf{B}(t)\mathbf{u}^*(t) = \min_{\mathbf{v} \in \mho} \boldsymbol{\lambda}^T(t)\mathbf{B}(t)\mathbf{v}, \tag{5.3-26}$$

for all $t \in [t_0, t^*]$. We now sketch a proof of the first result, using an argument following [28] which gives a geometric interpretation to the adjoint solution $\boldsymbol{\lambda}(t)$. Proof of the corollary is straightforward and is presented in [28].

Let $\mathbf{u}$ be an extremal control at time t_1 with associated trajectory $\mathbf{x}(\cdot, \mathbf{u})$; thus, $\mathbf{x}(t_1, \mathbf{u})$ is on the boundary of $K(t_1)$. The convexity of $K(t_1)$ then implies that $\mathbf{x}(t_1, \mathbf{u})$ necessarily and sufficiently satisfies the inequality

$$\boldsymbol{\lambda}^T(t_1)[\mathbf{x}(t_1, \mathbf{v}) - \mathbf{x}(t_1, \mathbf{u})] \geq 0, \tag{5.3-27}$$

where $\boldsymbol{\lambda}(t_1)$ is the unit inward normal to the hyperplane tangent to $K(t_1)$ at $\mathbf{x}(t_1, \mathbf{u})$, and $\mathbf{v}$ is any admissible control such that $\mathbf{v}(\tau) \in \mho$ for all $\tau \leq t_1$. Define the adjoint response

$$\dot{\boldsymbol{\lambda}}(t) = -\mathbf{A}^T(t)\boldsymbol{\lambda}(t)$$

having solution

$$\boldsymbol{\lambda}^T(t) = \boldsymbol{\lambda}^T(t_0)\boldsymbol{\Phi}^{-1}(t, t_0),$$

where $\boldsymbol{\Phi}$ is the fundamental matrix associated with the solution of the state trajectory; i.e.,

$$\mathbf{x}(t, \mathbf{u}) = \boldsymbol{\Phi}(t, t_0)\mathbf{x}(t_0) + \int_{t_0}^{t} \boldsymbol{\Phi}(t, s)\mathbf{B}(s)\mathbf{u}(s)\, ds.$$

It then follows that

$$\boldsymbol{\lambda}^T(t)\mathbf{x}(t, \mathbf{u}) = \boldsymbol{\lambda}^T(t_0)\mathbf{x}(t_0) + \int_{t_0}^{t} \boldsymbol{\lambda}^T(s)\mathbf{B}(s)\mathbf{u}(s)\, ds. \tag{5.3-28}$$

Suppose that Eq. (5.3-24) is not true over some interval $(t_a, t_b) \subset [t_0, t_1]$, $t_b > t_a$, and let $\tilde{\mathbf{u}}$ be such that it satisfies Eq. (5.3-26) for all $t \in [t_0, t_1]$. It then follows from Eq. (5.3-28) that

$$\boldsymbol{\lambda}^T(t_1)\mathbf{x}(t_1, \tilde{\mathbf{u}}) < \boldsymbol{\lambda}^T(t_1)\mathbf{x}(t_1, \mathbf{u}),$$

which violates Eq. (5.3-27) and contradicts the assumption that $\mathbf{u}$ is an extremal at t_1. Thus, if $\mathbf{u}$ is an extremal at t_1, then Eq. (5.3-24) holds for all $t \in [t_0, t_1]$.

Conversely, assume an admissible control $\mathbf{u}$, $\mathbf{u}(\tau) \in \mathfrak{U}$ for $\tau \leq t_1$, satisfies Eq. (5.3-24) for all $t \in [t_0, t_1]$ but that $\mathbf{x}(t_1, \mathbf{u})$ is an interior point of $K(t_1)$. There then exists an admissible control $\tilde{\mathbf{u}}$, $\tilde{\mathbf{u}}(\tau) \in \mathfrak{U}$ for $\tau \leq t_1$, such that $\boldsymbol{\lambda}^T(t_1)\mathbf{x}(t_1, \mathbf{u}) > \boldsymbol{\lambda}^T(t_1)\mathbf{x}(t_1, \tilde{\mathbf{u}})$. By hypothesis,

$$\boldsymbol{\lambda}^T(t)\mathbf{B}(t)\tilde{\mathbf{u}}(t) \leq \boldsymbol{\lambda}^T(t)\mathbf{B}(t)\mathbf{u}(t)$$

for all $t \in [t_0, t_1]$, which implies from Eq. (5.3-28) that $\boldsymbol{\lambda}^T(t_1)\mathbf{x}(t_1, \mathbf{u}) \leq \boldsymbol{\lambda}^T(t_1)\mathbf{x}(t_1, \tilde{\mathbf{u}})$; thus, $\mathbf{u}$ is an extremal control at t_1, and the result is proved.

Of course, the necessary conditions of Eqs. (5.3-25) and (5.3-26) are only restatements of a slight generalization of Eqs. (5.3-12) and (5.3-17), where we note that all that is subject to minimization with respect to $\mathbf{u}$ in Eq. (5.3-12) is the term examined in Eq. (5.3-26).

We noted from Eqs. (5.3-5) and (5.3-13) that when $\mathfrak{U}$ is a rectangle or cube, as defined in Eqs. (5.3-2) and (5.3-11), the set of all allowable control values could be substantially restricted without affecting minimum cost. This bang bang result can be generalized to include compact sets $\mathfrak{U} \subset R^m$ as follows. Define the convex hull of $\mathfrak{U}$, denoted by $\mathcal{H}(\mathfrak{U})$, as the smallest convex set in R^m containing $\mathfrak{U}$. Let $\mathfrak{U}'$ be the smallest subset of $\mathfrak{U}$ such that $\mathcal{H}(\mathfrak{U}') = \mathcal{H}(\mathfrak{U})$. The bang bang principle then states that it is sufficient to consider only control values in $\mathfrak{U}'$ without affecting minimum cost [27, 28]. Since the convex hull of the vertices of a rectangle equals (the convex hull of) the rectangle, the results stated in Eqs. (5.3-5) and (5.3-12) are special cases of the bang bang principle.

5.4 Singular solutions

Let us consider the problem of obtaining the optimal control which minimizes the performance index

$$J = \frac{1}{2}\int_0^2 x^2\,dt \tag{5.4-1}$$

for the linear differential system

$$\dot{x} = u(t), \qquad x(0) = 1 \tag{5.4-2}$$

where the magnitude of the control effort is restricted by $|u(t)| \leq 1$.

The Hamiltonian for this simple problem is

$$H[x(t), u(t), \lambda(t)] = \tfrac{1}{2}x^2(t) + \lambda(t)u(t), \tag{5.4-3}$$

and it is clear that, for nonzero $\lambda(t)$, we should make

$$u(t) = -\text{sign}\,\lambda(t) = \begin{cases} +1 & \lambda(t) < 0 \\ -1 & \lambda(t) > 0. \end{cases} \tag{5.4-4}$$

This indicates, as expected, that the optimal control operates on the boundary. It is possible, however, for the control to be within the ± 1 boundary, in which case we should set

$$\frac{\partial H}{\partial u} = 0 = \lambda(t). \tag{5.4-5}$$

Thus we see that $\lambda(t) = 0$ is a possible solution, in which case the Hamiltonian does not depend on $u(t)$ at all, and therefore cannot be minimized with respect to a choice of $u(t)$. Problems such as this are called *singular problems*. More generally, an extremal arc of the optimal control problem is said to be singular if the determinant of the matrix $H_{\mathbf{uu}}$, $|H_{\mathbf{uu}}|$, vanishes at any point along the arc [29].

We may solve the above singular solution problem as follows. The canonic equations are

$$\dot{x} = \frac{\partial H}{\partial \lambda} = u(t), \qquad \dot{\lambda} = -\frac{\partial H}{\partial x} = -x(t). \tag{5.4-6}$$

We assume that initially $\lambda(t)$ is positive, such that $u(t) = -\,\text{sign}\,\lambda(t) = -1$. The equations of motion are then

$$x(t) = 1 - t, \qquad \lambda(t) = \lambda(0) - t + \frac{t^2}{2}\cdot \tag{5.4-7}$$

On the singular arc, we have found that $\lambda(t) = 0$, and this requires that $\dot{\lambda}$ be zero, as well as $x(t)$. But from the foregoing, we see that $x(t)$ is zero at $t = 1$, and all the requirements for a singular solution for $t \in [1, 2]$ are established if the initial condition on the adjoint $\lambda(t_o)$ is equal to $\frac{1}{2}$. The

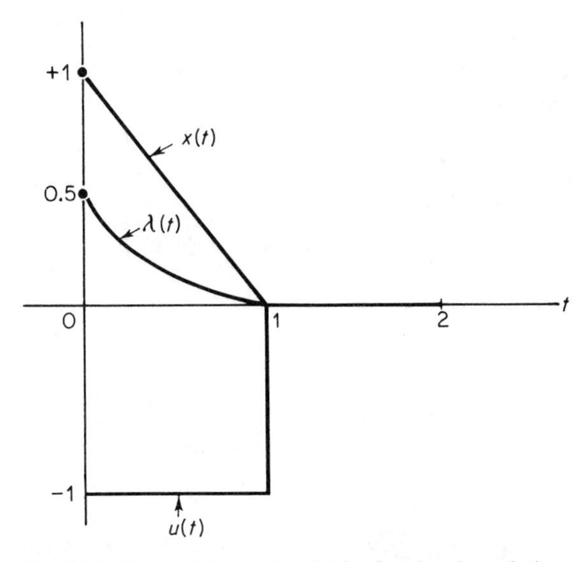

Fig. 5.4-1 State and control variables for singular solution.

optimal control and trajectory are shown in Fig. 5.4-1. Clearly, this satisfies the transversality condition that $\lambda(2) = 0$.

We may obtain an alternate approach to this problem by converting the control inequality constraint to an equality constraint as outlined in Chapters 3 and 4. We therefore use $1 - u^2(t) = \alpha^2(t)$ and write the Hamiltonian as

$$H[x(t), u(t), \lambda(t), \alpha(t)] = \tfrac{1}{2}x^2(t) + \lambda_1(t)u(t) + \lambda_2(t)[1 - u^2(t) - \alpha^2(t)]. \tag{5.4-8}$$

We then find

$$\frac{\partial H}{\partial u} = 0 = \lambda_1(t) - 2\lambda_2(t)u(t) \tag{5.4-9}$$

$$\frac{\partial H}{\partial x} = -\dot{\lambda}_1 = x(t) \tag{5.4-10}$$

$$\frac{\partial H}{\partial \alpha} = 0 = -2\lambda_2(t)\alpha(t). \tag{5.4-11}$$

Thus we need to solve the following equations

$$\begin{aligned} &\dot{x} = u, \qquad x(0) = 1 \\ &\dot{\lambda}_1 = -x(t), \qquad \lambda_1(2) = 0 \\ &1 - u^2(t) = \alpha^2(t) \\ &\lambda_2(t)\alpha(t) = 0 \\ &\lambda_1(t) = 2\lambda_2(t)u(t). \end{aligned} \tag{5.4-12}$$

From these equations, we get precisely the same results as when we incorporate the inequality constraint into the maximum principle. The only difference is that, in a computational solution of the problem, the singular control does not arise except as the solution becomes exact (after a large number of iterations in practice) when $\lambda_2(t) = \lambda_1(t) = \alpha$, $\forall t \in [1, 2]$. An alternate method of eliminating the singular solution is to use the penalty-function technique and write the cost function as

$$J = \tfrac{1}{2}x_2^2(t_f)s + \tfrac{1}{2}\int_0^2 x_1^2(t)\,dt \tag{5.4-13}$$

for the system (where there is no inequality control constraint and where H denotes the Heaviside step) described by

$$\dot{x}_1 = u, \qquad x_1(0) = 1 \tag{5.4-14}$$

$$\dot{x}_2 = K\mathrm{H}(1 - u^2), \qquad x_2(0) = 0 \tag{5.4-15}$$

$$K > 0, \qquad \mathrm{H}(a) = \begin{cases} 1 & \text{if } a > 0 \\ 0 & \text{if } a < 0. \end{cases}$$

We shall now extend these ideas to more complex systems and performance indices. In the previous section, we considered the general nonlinear problem wherein the control entered the Hamiltonian in a linear fashion, Eq. (5.3-4). The possibility of a singular solution exists if

$$\frac{\partial H}{\partial \mathbf{u}} = \mathbf{0} = \mathbf{h}[\mathbf{x}(t), t] + \mathbf{G}^T[\mathbf{x}(t), t]\boldsymbol{\lambda}(t) = \mathbf{0} \tag{5.4-16}$$

such that the Hamiltonian does not depend on $\mathbf{u}(t)$

$$H[\mathbf{x}(t), \boldsymbol{\lambda}(t), t] = \phi[\mathbf{x}(t), t] + \boldsymbol{\lambda}^T(t)\mathbf{f}[\mathbf{x}(t), t]. \tag{5.4-17}$$

If the Hamiltonian is insensitive to the control for any finite range of time, then a singular solution does, in fact, exist. It is generally easy to verify whether a singular solution possibility does or does not exist. It is normally considerably more difficult to actually determine an optimum singular solution.

Let us now generalize the work of this section to include the time-varying linear system with a quadratic performance index in the state variables only. We will consider the determination of the control vector $\mathbf{u}(t)$ which minimizes

$$J = \tfrac{1}{2}\mathbf{x}^T(t_f)\mathbf{S}\mathbf{x}(t_f) + \tfrac{1}{2}\int_{t_o}^{t_f} \mathbf{x}^T(t)\mathbf{Q}(t)\mathbf{x}(t)\,dt \tag{5.4-18}$$

for the system

$$\dot{\mathbf{x}} = \mathbf{A}(t)\mathbf{x}(t) + \mathbf{B}(t)\mathbf{u}(t), \qquad \mathbf{x}(t_o) = \mathbf{x}_o, \tag{5.4-19}$$

where t_f is fixed and where the control vector is bounded. The Hamiltonian

$$H[\mathbf{x}(t), \mathbf{u}(t), \boldsymbol{\lambda}(t), t] = \tfrac{1}{2}\mathbf{x}^T(t)\mathbf{Q}(t)\mathbf{x}(t) + \boldsymbol{\lambda}^T(t)\mathbf{A}(t)\mathbf{x}(t) + \boldsymbol{\lambda}^T(t)\mathbf{B}(t)\mathbf{u}(t) \tag{5.4-20}$$

is again linear in the control variable, and for nonzero $\boldsymbol{\lambda}^T(t)\mathbf{B}(t)$, we know that $\mathbf{u}(t)$ operates on the boundary of the admissible input set. It is still possible for

$$\frac{\partial H}{\partial \mathbf{u}} = 0 = \mathbf{B}^T(t)\boldsymbol{\lambda}(t) \tag{5.4-21}$$

to be a solution which minimizes the Hamiltonian. In this case, we have

$$H[\mathbf{x}(t), \boldsymbol{\lambda}(t), t] = \tfrac{1}{2}\mathbf{x}^T(t)\mathbf{Q}(t)\mathbf{x}(t) + \boldsymbol{\lambda}^T(t)\mathbf{A}(t)\mathbf{x}(t), \tag{5.4-22}$$

and if the Hamiltonian is not a function of $\mathbf{u}(t)$ for some finite range of t, we have a singular solution.

Recent effort has been concerned with the development of necessary, sufficient, and necessary and sufficient conditions for singular solutions from the second variation of the cost function. Of particular interest are results in [30, 31, 32]. A recent survey of singular problems can be found in [33].

Example 5.4-1. Let us consider the following particular example [14]. The cost function is

$$J = \tfrac{1}{2}\int_0^{t_f} x_1^2 \, dt$$

for the system

$$\dot{x}_1 = x_2(t) + u(t), \qquad x_1(0) = x_{10}$$

$$\dot{x}_2 = -u(t), \qquad x_2(0) = x_{20},$$

where it is desired to drive the system to the origin at a fixed time, t_f, $x_1(t_f) = x_2(t_f) = 0$ with a bounded control $|u(t)| \leq K$.

The Hamiltonian is

$$H[\mathbf{x}(t), \mathbf{u}(t), \boldsymbol{\lambda}(t)] = \lambda_1(t)[x_2(t) + u(t)] - \lambda_2(t)u(t) + \tfrac{1}{2}x_1^2(t).$$

Singular arcs are such that

$$\frac{\partial H}{\partial u} = 0 = \lambda_1(t) - \lambda_2(t)$$

for a finite time interval with the canonic equations

$$\dot{x}_1 = x_2(t) + u(t), \qquad x_1(0) = x_{10}$$

$$\dot{x}_2 = -u(t), \qquad x_2(0) = x_{20}$$

$$\dot{\lambda}_1 = -\frac{\partial H}{\partial x_1} = -x_1(t), \qquad x_1(t_f) = 0$$

$$\dot{\lambda}_2 = -\frac{\partial H}{\partial x_2} = -\lambda_1(t), \qquad x_2(t_f) = 0.$$

For a singular solution, $\lambda_1(t) = \lambda_2(t)$; therefore, $\dot{\lambda}_1 = \dot{\lambda}_2$, which yields from the canonic equations for the adjoint variables $x_1(t) = \lambda_1(t)$. Also, on the singular arc

$$\frac{d^2}{dt^2}\frac{\partial H}{\partial u} = 0 = -\dot{x}_1 + \dot{\lambda}_1 = -x_2(t) - u(t) - x_1(t).$$

Since the Hamiltonian depends only implicitly on time, it must be constant about an optimal control and trajectory and, therefore, on a singular arc

$$H[\mathbf{x}(t), \boldsymbol{\lambda}(t)] = x_1(t)x_2(t) + \tfrac{1}{2}x_1^2(t) = \text{constant}.$$

Thus we have discovered that, on the singular arc, the closed-loop control is

$$u(t) = -x_1(t) - x_2(t),$$

and the singular arcs are described by

$$x_1(t)x_2(t) + \tfrac{1}{2}x_1^2(t) = \text{constant}.$$

On a singular arc, the optimum closed-loop control yields the system equations:

$$\dot{x}_1 = x_2 - (x_1 + x_2) = x_1(t)$$

$$\dot{x}_2 = x_1(t) + x_2(t)$$

so

$$x_1(t) = e^{-(t-t_1)}x_1(t_1)$$

$$x_2(t) = \frac{e^{(t-t_1)} - e^{-(t-t_1)}}{2}\bigg| x_1(t_1) + e^{(t-t_1)}x_2(t_1).$$

A typical trajectory involves three parts. An extreme $\pm K$ value of the control is used initially to transfer the system to the singular arc. The singular control $u(t) = -x_1(t) - x_2(t)$ is then applied until another application of the extreme value of the control transfers the state of the system to the origin. We may now consider two cases:

(a) $K \longrightarrow \infty$ such that the control is unbounded. In this case, impulse controls are used initially to transfer to the singular arc which is followed until the time $x_1(t) + x_2(t) = 0$, where application of another impulse transfers the system to the origin. Since the system must arrive at $x_1(t) + x_2(t) = 0$ at $t = t_f$, and since the impulse transfers the state to the origin in zero time, the particular value of the constant Hamiltonian which determines the particular singular arc is determined. Figure 5.4-2 illustrates a possible trajectory in this case and also shows the straight line directions of motion due to application of impulse control inputs.

(b) $t_f \longrightarrow \infty$ such that $H[\mathbf{x}(t), \boldsymbol{\lambda}(t)] = 0$. In this case, the singular arcs are the two lines $x_1(t) = 0$ and $x_1(t) + 2x_2(t) = 0$, as we easily obtain by setting the Hamiltonian equal to zero for the singular solution. On the line $x_1(t) = 0$, the closed-loop control becomes $u(t) = -x_2(t)$ and, consequently,

$$x_1(t) = 0, \qquad x_2(t) = e^{(t-t_1)}x_2(t_1).$$

On the line $x_1(t) + 2x_2(t) = 0$, the closed-loop control is $u(t) = x_2(t)$ and

$$x_1(t) = x_1(t) + [1 - e^{-(t-t_1)}]x_2(t_1), \qquad x_2(t) = e^{-(t-t_1)}x_2(t_1).$$

If there is an inequality constraint on u, $|u(t)| \leq K$, then the largest values S_1 and S_2 that the singular arc can have are also limited. The

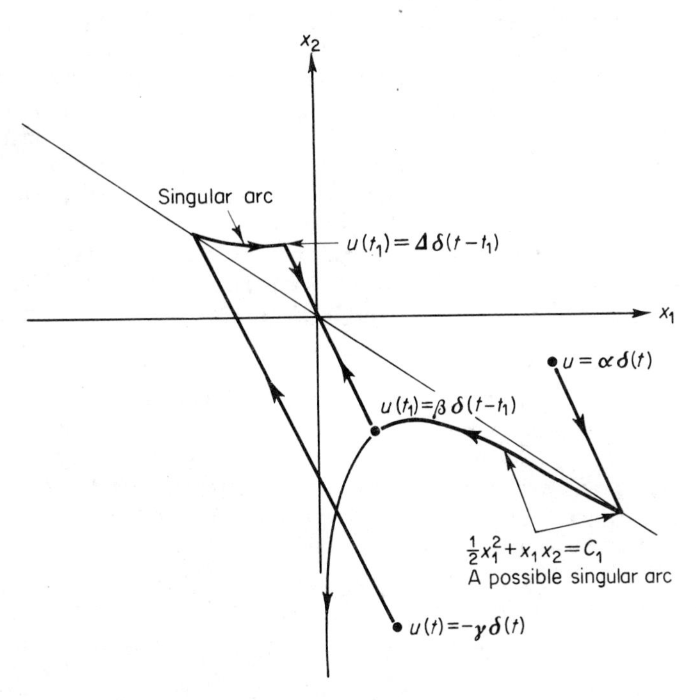

Fig. 5.4-2 Possible system trajectories for Example 5.4-1(a).

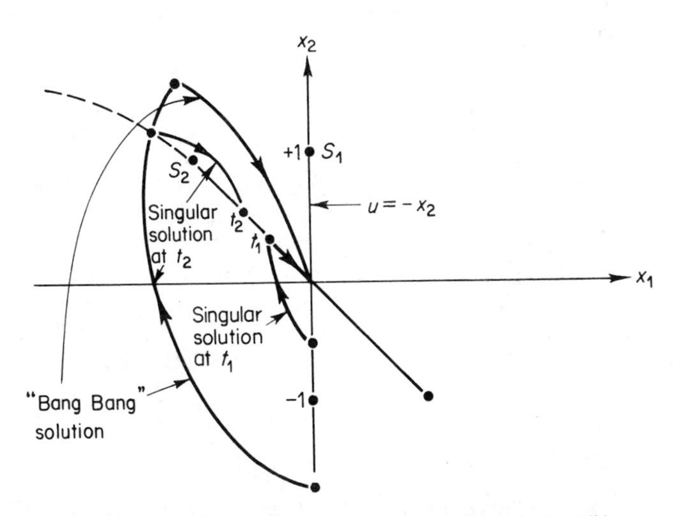

Fig. 5.4-3 Possible system trajectories for Example 5.4-1(b).

motion on the singular arc where $u(t) = -x_2(t)$ is unstable in that it is directed away from the origin. In this case, the optimal control is at the extreme value until the singular arc is reached, at which time the singular arc is followed until the origin is reached. As shown in Fig. 5.4-3, which illustrates the singular arc and several optimal trajectories, there is no guarantee that the singular control is minimizing or that it will always enter into the optimal solution even if the possibility of a singular solution does exist. In this example too large an initial condition produces a bang bang control to drive the system to the singular arc. The resulting cost function turns out to be less than the cost function with pure bang bang control.

Example 5.4-2. [33, 34] Historically, singular solution problems initially arose in aerospace applications. Consider the following formulation of the fundamental problem of space navigation. Let (x_1, x_2, x_2) and (v_1, v_2, v_3) designate the Cartesian spatial and velocity coordinates for a space vehicle at time t, and assume its rocket thrust acts in a direction having direction cosines (l_1, l_2, l_3). Let the control $m(t)$ be the mass rate of propellant consumption. The equations of motion are

$$\dot{v}_i = (cml_i)/M + g_i(x_1, x_2, x_3, t)$$
$$\dot{x}_i = v_i, \qquad i = 1, 2, 3$$
$$\dot{M} = -m, \quad 0 \leq m \leq \bar{m},$$

where g_i, $i = 1, 2, 3$, are the components of the gravitational field, c is the rocket exhaust velocity, and M is the rocket mass. The direction cosines can be expressed as

$$l_1 = \sin\theta \cos\phi$$
$$l_2 = \sin\theta \sin\phi$$
$$l_3 = \cos\theta,$$

where the controls θ and ϕ are the spherical polar coordinates. The control problem is to determine θ, ϕ, and m so as to minimize fuel consumption,

$$J = -M(t_f),$$

subject to the following boundary conditions: The initial and final states $x_i(t_o)$ and $x_i(t_f)$, $i = 1, 2, 3$, are specified; the initial and final velocities $v_i(t_o)$ and $v_i(t_f)$, $i = 1, 2, 3$, are specified; the initial mass of the rocket $M(t_o)$ is specified; and the initial and final times can be free or fixed. It is easily shown that the Hamiltonian for this problem is linear in the fuel consumption rate $m(t)$; therefore, the control function will be singular. Recent applications of singular solution theory to aerospace problems can be found in [35, 36].

References

1. KALMAN, R.E., "Contributions to the Theory of Optimal Control." *Bol. Soc. Mat. Mex.*, 5, (1960), 102–19.

2. KALMAN, R.E., et al., "Fundamental Study of Adaptive Control Systems." *Wright-Patterson Air Force Base Tech. Rept.*, ASD-TR-61-27, Vol. 1, (April 1962).
3. KALMAN, R.E., "The Theory of Optimal Control and the Calculus of Variations." R. Bellman, ed., *Mathematical Optimization Techniques*. University of California Press, Berkeley, 1963.
4. KALMAN, R.E., "Mathematical Description of Linear Dynamical Systems." *SIAM J. Control*, Ser. A, 1, (1963), 152–92.
5. PONTRYAGIN, L.S., BOLTYANSKII, V., GAMKRELIDZE, R., and MISHEHENKO, E., *The Mathematical Theory of Optimal Processes*. Interscience Publishers, New York, 1962.
6. BELLMAN, R., GLICKSBERG, I., and GROSS, O., "On the Bang-Bang Control Problem." *Quart. Appl. Math.*, 14, (1956), 11–18.
7. LASALLE, J.P., "The Time-Optimal Control Problem." *Contributions to Differential Equations*, Vol. V, Princeton University Press, Princeton, New Jersey, 1960, 1–24.
8. DESOER, C.A., "The Bang Bang Servo Problem Treated by Variational Techniques." *Information and Control*, 2, (1959), 333–48.
9. FRIEDLAND, B., "A Minimum Response-Time Controller for Amplitude and Energy Constraints." *IRE Trans. Autom. Control*, AC-7, (1962), 73–74.
10. ATHANASSIADES, M., "Optimal Control for Linear Time Invariant Plants with Time-, Fuel-, and Energy Constraints." *IEEE Trans. Appl. Ind.*, 81, (1963), 321–25.
11. LASALLE, J.P., "The Bang-Bang Principle." *Proceedings First IFAC Congress*, 1960, 493–97.
12. LEE, E.B., "Mathematical Aspects of the Synthesis of Linear Minimum Response Time Controllers." *IRE Trans. Autom. Control*, AC-5, (1960), 283–89.
13. LEE, E.B., "On the Time-Optimal Regulation of Plants with Numerator Dynamics." *IRE Trans. Autom. Control*, AC-6, (1961), 351–52.
14. JOHNSON, C.D., and GIBSON, J.E., "Singular Solutions in Problems of Optimal Control." *IEEE Trans. Autom. Control*, AC-8, (1963), 4–14.
15. JOHNSON, C.D., and GIBSON, J.E., "Optimal Control of a Linear Regulator with Quadratic Index of Performance and Fixed Terminal Time." *IEEE Trans. Autom. Control*, AC-9, (1964), 355–60.
16. JOHNSON, C.D., "Singular Solutions in Problems of Optimal Control." C.T. Leondes, ed., *Advances in Control Systems: Theory and Applications*, Vol. 11. Academic Press, New York, 1965.
17. ATHANS, M., "Minimum Fuel Feedback Control Systems: Second Order Case." *IEEE Trans. Appl. Ind.*, 82, (1963), 8–17.
18. ATHANS, M., "Minimum Fuel Control of Second Order Systems with Real Poles." *IEEE Trans. Appl. Ind.*, 83, (1964), 148–53.

19. ATHANS, M., and CANON, M.D., "Fuel-Optimal Singular Control of a Nonlinear Second Order System." *Preprints 1964 Joint Autom. Control Conf.*, June 1964, 245–55.

20. FOY, W.H., "Fuel Minimization in Flight Vehicle Attitude Control." *IEEE Trans. Autom. Control*, AC-8, (1963), 84–88.

21. ATHANS, M., FALB, P.L., and LACOSS, R.T., "On Optimal Control of Self Adjoint Systems." *IEEE Trans. Appl. Ind.*, 83, (1964), 161–66.

22. PAIEWONSKY, B., "Optimal Control: a Review of Theory and Practice." *AIAA J.*, 3, (1965), 1985–2006.

23. ATHANS, M., "Status of Optimal Control Theory and Applications for Deterministic Systems." *IEEE Trans. Autom. Control*, AC-11, No. 3, (1966), 580–96.

24. RAO, C.R., and MITRA, S.K., *Generalized Inverse Matrices and Its Applications*. Wiley, New York, 1971.

25. BOULLION, T.L., and O'DELL, P.L., *Generalized Inverse Matrices*. Wiley, New York, 1971.

26. LEITMANN, G., *An Introduction to Optimal Control*. McGraw-Hill Book Co., New York, 1966.

27. HERMES, H., and LASALLE, J., *Functional Analysis and Time Optimal Control*. Academic Press, New York, 1969.

28. LEE, E.B., and MARKUS, L., *Foundations of Optimal Control Theory*. Wiley, New York, 1967.

29. BLISS, G.A., *Lectures on the Calculus of Variations*. University of Chicago Press, Chicago, 1946.

30. JACOBSON, D.H., "On Singular Arcs and Surfaces in a Class of Quadratic Minimization Problems." *J. Math. Anal. Appl.*, 37, (1972), 185–201.

31. SPEYER, J.L., and JACOBSON, D.H., "Necessary and Sufficient Conditions for Optimality for Singular Control Problems: A Transformation Approach." *J. Math. Anal. Appl.*, 33, (1971), 163–87.

32. MCDANELL, J.P., and POWERS, W.F., "Necessary Conditions for Joining Optimal Singular and Nonsingular Subarcs." *SIAM J. Control*, 9, (1971), 161–73.

33. BELL, D.J., "Singular Problems in Optimal Control—A Survey." *Int. J. Control*, 21, (1975), 319–31.

34. LAWDEN, D.F., *Optimal Trajectories for Space Navigation*. Butterworth, Washington, D.C., 1963.

35. POWERS, W.F., and MCDANELL, J.P., "Switching Conditions and a Synthesis Technique for the Singular Saturn Guidance Problem." *J. Spacecraft Rockets*, 8, (1971), 1027–31.

36. SPEYER, J.L., "On the Fuel Optimality of Cruise." *J. Aircraft*, 10, (1973), 763–64.

Problems

1. Obtain the closed-loop control for the regulator problem where the dynamics are

$$\dot{x}_1 = x_2, \qquad x_1(0) = x_{10}$$

$$\dot{x}_2 = u, \qquad x_2(0) = x_{20}$$

$$J = \tfrac{1}{2}\mathbf{x}^T(t_f)\begin{bmatrix} s_{11} & 0 \\ 0 & s_{22} \end{bmatrix}\mathbf{x}(t_f) + \tfrac{1}{2}\int_0^{t_f}\left\{u^2(t) + \mathbf{x}^T(t)\begin{bmatrix} q_{11} & 0 \\ 0 & q_{22} \end{bmatrix}\mathbf{x}(t)\right\} dt.$$

For $t_f = \infty$ we may solve the Riccati equation $\dot{\mathbf{P}} = \mathbf{0}$ and obtain two values for $\mathbf{P}$. How do we determine which one to use?

2. In Problem 1, let $q_{11} = q_{22} = 0$ and $s_{11} = s_{22} = \infty$. Determine the closed-loop control for this case as a function of t_f and $\mathbf{x}(t)$.

3. In Problem 1, let $z(t) = x_1(t) + x_2(t)$. Determine the closed-loop control which minimizes

$$J = \tfrac{1}{2}sz^2(t_f) + \tfrac{1}{2}\int_{t_0}^{t_f} [u^2(t) + qz^2(t)]\, dt.$$

4. In Problem 3, let $q = 0$, $s = \infty$. Determine the closed-loop control for this case as a function of t_f and $\mathbf{x}(t)$.

5. Determine the switching surfaces for the minimum time control which transfers the system

$$\dot{\mathbf{x}} = \begin{bmatrix} -2 & 2 \\ 0 & -1 \end{bmatrix}\mathbf{x}(t) + \begin{bmatrix} 0 \\ 1 \end{bmatrix}u(t), \qquad \mathbf{x}(t_o) = \mathbf{x}_o, \qquad |u| \le 1$$

to the origin.

6. Determine the switching surfaces for the minimum time control which transfers

$$\dot{\mathbf{y}} = \begin{bmatrix} 0 & 1 \\ -2 & -3 \end{bmatrix}\mathbf{y} + \begin{bmatrix} 0 \\ 2 \end{bmatrix}u(t), \qquad \mathbf{y}(t_o) = \mathbf{y}_o, \qquad |u| \le 1$$

to the origin.

7. Show that Problem 5 is related to Problem 6 by the linear transformation

$$\mathbf{y}(t) = \begin{bmatrix} 1 & 0 \\ -2 & 2 \end{bmatrix}\mathbf{x}(t).$$

How does this linear transformation affect the switching surfaces of the two problems?

8. Determine the switching surfaces for the minimum time control which transfers to the origin the system

$$\dot{x}_1 = x_2, \qquad \dot{x}_2 = x_3, \qquad \dot{x}_3 = u, \qquad \mathbf{x}(t_o) = \mathbf{x}_o, \qquad |u| \le 1.$$

9. Discuss in detail the nature of the optimum control which minimizes

$$J = \tfrac{1}{2}\int_0^{t_f} x^2(t)\, dt, \qquad |u| \le 1$$

for the system

$$\dot{x} = -ax(t) + u(t), \qquad x(0) = x_0, \qquad -\infty < a < \infty.$$

10. Discuss the singular solution possibility for the system where t_f is unspecified:

$$\dot{x}_1 = x_2(t), \qquad x_1(0) = x_{10}$$

$$\dot{x}_2 = -x_2(t) - x_1(t)u(t), \qquad x_2(0) = x_{20}$$

$$J = \tfrac{1}{2}\int_0^{t_f} [q_{11}x_1^2(t) + x_2^2(t)]\,dt, \qquad |u(t)| \leq 1.$$

The required terminal manifold is $x_1^2(t_f) + x_2^2(t_f) = 1$. Show that the singular solution trajectory is $\sqrt{q_{11}}x_1(t) = x_2(t)$, $\sqrt{q_{11}}x_1(t) = -x_2(t)$, and that the optimal control on the singular trajectory is

$$u = \frac{-(q_{11}x_1 + x_2)}{x_1}.$$

11. Find the control $u(t)$ which minimizes

$$J[u] = 0\int_0^{t_f} [x_1^2(t) + x_2^2(t)]\,dt, \qquad |u(t)| \leq 1$$

for the system where t_f is (a) specified, and (b) unspecified:

$$\dot{x}_1 = x_2, \qquad x_1(0) = x_{10}, \qquad x_1(t_f) = 0$$

$$\dot{x}_2 = u, \qquad x_2(0) = x_{20}, \qquad x_2(t_f) = 0.$$

12. Determine and justify the switching boundary as illustrated in Fig. 5.4-3 for Example 5.4-1(b).

13. Determine the equations which specify the switching lines of Example 5.5-1 as a function of q_1 and q_2. Investigate the optimality of the possible singular solution as a function of q_1 and q_2 [20].

14. Discuss the nature of the time-optimal control for the system

$$\dot{x}_1 = x_2 + u_1, \qquad \dot{x}_2 = -x_1 + u_2$$

with $|u_1| \leq 1$ and $|u_2| \leq 1$. Sketch the general nature of the switching curves.

15. Discuss the nature of the solution for the optimization problem

$$\dot{x}_1 = -x_1 + x_2, \qquad \dot{x}_2 = -2x_2 + u, \qquad |u| \leq 1$$

$$J = \int_0^{t_f} (x_1^2 + |u|)\,dt$$

for t_f fixed as well as t_f free. Include a treatment of singular solutions, conditions for which a singular solution is possible and a possible trajectory including a singular subarc.

16. You are discussing the soft lunar-landing problem described by the system of equations

$$\dot{x}_1 = x_2, \qquad \dot{x}_2 = -ku/x_3 - g, \qquad \dot{x}_3 = u$$

where: x_1 is the vertical distance from surface, x_2 is the vertical velocity, x_3 is the vehicle mass, u is the mass flow rate, and k and g are positive constants.

The control u is constrained by $-1 \leq u \leq 0$. It is desired to take the vehicle from some initial state $\mathbf{x}(0)$ to the terminal state $x_1(t_f) = x_2(t_f) = 0$ while minimizing the fuel consumption

$$J = \int_0^{t_f} -u(t)\, dt = \int_0^{t_f} -\dot{x}_3(t)\, dt = x_3(0) - x_3(t_f)$$

where t_f is the unspecified touch-down time.

By immediately recognizing that

$$\dot{x}_3/x_3 = d(\ln x_3)/dt$$

and hence that

$$\dot{x}_2 = -kd(\ln x_3)/dt - g,$$

you state that this minimum fuel problem is equivalent to a minimum time problem with which you are very familiar. You immediately suggest that, since the problem is time-optimal, u can only have values equal to 0 and -1. A cohort suggests the possibility of singular solutions with $u \neq 0$ or 1. However, you quickly argue that, if $\lambda_1(t_f) \neq 0$, no singular solutions can exist since $\lambda_3(t_f) = 0$. Discuss the complete solution including: a rough sketch of the switching curve, a demonstration that no singular solutions exist if $\lambda_1(t_f) \neq 0$, and an argument that this soft lunar-landing formulation is equivalent to a time optimal problem.

17. Assume $\ddot{x} = u$ and $x(0) = \dot{x}(0) = 0$. Graph $K(1)$ for the following $\mho$:
(a) $\mho = [-1, 1]$
(b) $\mho = \{-1, 1\}$
(c) $\mho = [0, 1]$.

18. In the development of Eqs. (5.3-23) and (5.3-24), $\boldsymbol{\lambda}(t)$ was assumed to be the inward normal of the hyperplane tangent to $\mathbf{x}(t, \mathbf{u})$, where $\mathbf{u}$ is an extremal control. How does the assumption that $\boldsymbol{\lambda}(t)$ is an outward normal alter the forms of Eqs. (5.3-23) and (5.3-24)?

19. Consider the linear system

$$\frac{\partial^2 x}{\partial t^2} = \frac{a\partial^2 x}{\partial y^2} + u(y, t)$$

having performance index

$$J = \tfrac{1}{2}\int_{t_0}^{t_f}\int_{y_0}^{y_f} \{Ru^2(y, t) + \|x(y, t)\|_Q^2\}\, dt,$$

which we wish to minimize. Determine a spatially discrete approximation to this distributed optimal control problem. What are the appropriate initial and boundary conditions?

Discrete variational calculus and the discrete maximum principle 6

Perhaps the most useful single technique in modern control system theory is that branch of mathematics known as the calculus of variations. Variational principles have been applied to physical problems, such as wave propagation, since the time of Huygens. The Hamiltonian formulation of the variational problem has existed since the early nineteenth century in the works of Hamilton, Jacobi, and others. The most significant contribution in recent times was made by L. S. Pontryagin. The work of Pontryagin [1, 2] extended the variational method to include problems wherein the available control and state vector is bounded, as we have seen in the previous two chapters.

Recently, the maximum principle has been applied to problems involving discrete-data systems [3, 4]. In reality, the maximum principle is not universally valid for the case of discrete systems [1]. Due to restrictions on possible variations of the control signal, the maximum principle must be modified for the general discrete case. Jordan and Polak [5] discuss the limitations and derive a modified form of the maximum principle, which is applicable to the general discrete problem. Pearson and Sridhar [6] investigate the discrete maximum principle using the framework of nonlinear programming [7]. Further discussion can be found in [8]. The results in [9] represent a particularly general development of necessary conditions for the discrete time case.

We begin this chapter by determining a discrete version of the Euler-Lagrange equations and transversality conditions. The discrete maximum principle is then stated and compared with the continuous maximum principle. The final section explores the relationship between discrete optimal control and mathematical programming.

6.1 Derivation of the discrete Euler-Lagrange equations

In our previous work we minimized cost functions which were integrals of scalar functions. Here we are interested in minimization of cost functions which are summations of scalar functions. Thus we are concerned with minimizing (or maximizing) functions such as

$$J = \sum_{k=k_o}^{k_f-1} \Phi(\mathbf{x}_k, \mathbf{x}_{k+1}, k) = \sum_{k=k_o}^{k_f-1} \Phi_k \tag{6.1-1}$$

where $\mathbf{x}_k = \mathbf{x}(t_k)$. For the case of synchronous sampling, or sampling with an equal time interval between samples, $\mathbf{x}_k = \mathbf{x}(kT)$, where T is the sampling period.†

It should also be noted that the function Φ_k represents the incremental cost for one stage of the discrete process. For a cost function equivalent to that of an integral for a continuous process, Φ_k will contain, as a multiplying factor, the sampling period T. We let $\mathbf{x}_k$ and $\mathbf{x}_{k+1}$ take on variations

$$\mathbf{x}_k = \hat{\mathbf{x}}_k + \epsilon \boldsymbol{\eta}_{x_k}, \qquad \mathbf{x}_{k+1} = \hat{\mathbf{x}}_{k+1} + \epsilon \boldsymbol{\eta}_{x_{k+1}} \tag{6.1-2}$$

where $\hat{\mathbf{x}}$ denotes the solution to the optimization problem. We use the same procedure as used previously for continuous systems. We substitute the foregoing values assumed for $\mathbf{x}(t_k)$ and $\mathbf{x}(t_{k+1})$ into the given cost function J. We compute $\partial J/\partial \epsilon$ and set it equal to zero at $\epsilon = 0$ independent of the variation $\boldsymbol{\eta}_{x_k}$ and $\boldsymbol{\eta}_{x_{k+1}}$. Thus we obtain

$$\sum_{k=k_o}^{k_f-1} \left\{\left\langle \frac{\partial \Phi_k}{\partial \hat{\mathbf{x}}_k}, \boldsymbol{\eta}_{x_k} \right\rangle + \left\langle \frac{\partial \Phi_k}{\partial \hat{\mathbf{x}}_{k+1}}, \boldsymbol{\eta}_{x_{k+1}} \right\rangle \right\} = 0 \tag{6.1-3}$$

where the notation $\langle \mathbf{x}, \mathbf{y} \rangle$ is used to indicate the inner product, or $\mathbf{x}^T\mathbf{y}$.

If we use variational notation, Eq. (6.1-3) can be written in a simpler form as follows. (We remember that the first variation δJ is set equal to 0 to extremize J.)

$$\delta J = \sum_{k=k_o}^{k_f-1} \left\{ \delta \mathbf{x}_k^T \frac{\partial \Phi_k}{\partial \hat{\mathbf{x}}_k} + \delta \mathbf{x}_{k+1}^T \frac{\partial \Phi_k}{\partial \hat{\mathbf{x}}_{k+1}} \right\} = 0. \tag{6.1-4}$$

When we manipulate the last term of Eq. (6.1-4) into a more convenient form by exchanging summation indices (replacing k by $m - 1$), and when we

---

†We will also use T to represent the transpose of a vector or matrix. Clearly, no confusion should result from this.

drop the ^ superscript, we obtain

$$\sum_{k=k_o}^{k_f-1} \delta \mathbf{x}_{k+1}^T \frac{\partial \Phi_k}{\partial \mathbf{x}_{k+1}} = \sum_{m=k_o+1}^{k_f} \delta \mathbf{x}_m^T \frac{\partial \Phi[\mathbf{x}_{m-1}, \mathbf{x}_m, m-1]}{\partial \mathbf{x}_m}. \tag{6.1-5}$$

If we rewrite this equation letting $k = m$ and starting the summation at $k = k_o$ and ending at $k_f - 1$, we get the result

$$\sum_{k=k_o}^{k_f-1} \delta \mathbf{x}_{k+1}^T \frac{\partial \Phi_k}{\partial \mathbf{x}_{k+1}} = \sum_{k=k_o}^{k_f-1} \delta \mathbf{x}_k^T \frac{\partial \Phi[\mathbf{x}_{k-1}, \mathbf{x}_k, k-1]}{\partial \mathbf{x}_k} + \delta \mathbf{x}_k^T \frac{\partial \Phi[\mathbf{x}_{k-1}, \mathbf{x}_k, k-1]}{\partial \mathbf{x}_k}\bigg|_{k=k_o}^{k=k_f}. \tag{6.1-6}$$

Therefore, Eq. (6.1-4) becomes

$$\sum_{k=k_o}^{k_f-1} \delta \mathbf{x}_k^T \left\{\frac{\partial \Phi[\mathbf{x}_k, \mathbf{x}_{k+1}, k]}{\partial \mathbf{x}_k} + \frac{\partial \Phi[\mathbf{x}_{k-1}, \mathbf{x}_k, k-1]}{\partial \mathbf{x}_k}\right\} + \delta \mathbf{x}_k^T \frac{\partial \Phi[\mathbf{x}_{k-1}, \mathbf{x}_k, k-1]}{\partial \mathbf{x}_k}\bigg|_{k=k_o}^{k=k_f} = 0. \tag{6.1.7}$$

For Eq. (6.1-7) to be equal to zero for arbitrary variations, the following vector difference equation, which is necessary for an extremum of the cost function, Eq. (6.1-1), must hold:

$$\frac{\partial \Phi[\mathbf{x}_k, \mathbf{x}_{k+1}, k]}{\partial \mathbf{x}_k} + \frac{\partial \Phi[\mathbf{x}_{k-1}, \mathbf{x}_k, k-1]}{\partial \mathbf{x}_k} = \mathbf{0}. \tag{6.1-8}$$

This may be spoken of as the *discrete Euler-Lagrange equation*. The transversality condition is obtained when we set the last term in Eq. (6.1-7) equal to zero:

$$\delta \mathbf{x}_k^T \frac{\partial \Phi[\mathbf{x}_{k-1}, \mathbf{x}_k, k-1]}{\partial \mathbf{x}_k} = 0 \qquad \text{for} \quad k = k_o, k_f. \tag{6.1-9}$$

The discussions in Chapters 3 and 4 regarding application of the transversality conditions apply well here. Also, the discussion of the Lagrange multiplier method to treat equality constraints applies almost without modification. This will be illustrated by an example.

Example 6.1-1. In this example, we will consider a simple scalar problem and solve it by very elementary techniques. The cost function to be minimized is

$$J = \tfrac{1}{2} \sum_{k=0}^{9} u^2(k).$$

The cost is minimized subject to the equality constraints

$$x(k+1) = x(k) + \alpha u(k), \qquad x(0) = 1, \qquad x(10) = 0.$$

We adjoin to the original cost function the given constraint via a Lagrange multiplier. This yields

$$J' = \sum_{k=0}^{9} [\tfrac{1}{2}u^2(k) + \lambda(k+1)\{-x(k+1) + x(k) + \alpha u(k)\}].$$

The reader may well question why the stage $k + 1$ is associated with the Lagrange multiplier. The reason is simplicity of the final result, as will be apparent in the next section. For this example, we have

$$\Phi_k = \tfrac{1}{2}u^2(k) + \lambda(k+1)\{-x(k+1) + x(k) + \alpha u(k)\}$$

$$\frac{\partial\Phi[x_k, x_{k+1}, k]}{\partial x_k} = +\lambda(k+1), \qquad \frac{\partial\Phi[x_{k-1}, x_k, k-1]}{\partial x_k} = -\lambda(k)$$

$$\frac{\partial\Phi[x_k, x_{k+1}, k]}{\partial u_k} = \alpha\lambda(k+1) + u(k), \qquad \frac{\partial\Phi[x_{k-1}, x_k, k-1]}{\partial u_k} = 0.$$

Thus the discrete Euler-Lagrange equation (6.1-8) yields

$$\lambda(k) - \lambda(k+1) = 0, \qquad u(k) + \alpha\lambda(k+1) = 0.$$

Also, the original equation must hold, subject to the stated boundary conditions

$$x(k+1) = x(k) + \alpha u(k), \qquad x(0) = 1, \qquad x(10) = 0.$$

Solving the last two equations, we obtain

$$\lambda(k) = \text{constant} = c, \qquad u(k) = -\alpha c, \qquad x(k+1) = x(k) - \alpha^2 c.$$

This is the final difference equation to be solved. By solving it stage by stage, we obtain

$$\begin{aligned} x(1) &= x(0) - \alpha^2 c \\ x(2) &= x(1) - \alpha^2 c = x(0) - 2\alpha^2 c \\ x(3) &= x(2) - \alpha^2 c = x(0) - 3\alpha^2 c \\ &\vdots \qquad \vdots \\ x(k) &= x(0) - k\alpha^2 c. \end{aligned}$$

Therefore, to satisfy the boundary conditions, we must have

$$x(10) = 0 = x(0) - 10\alpha^2 c, \quad c = \frac{+x(0)}{10\alpha^2} = \frac{+1}{10\alpha^2}.$$

Hence the control to be applied to this discrete system is $u(k) = -1/10\alpha$. The resulting trajectory is $x(k) = 1 - k/10$.

Example 6.1-2. We now return to the distributed parameter system discussed in Example 5.1-3 where we now discretize both in time and space. We wish to reduce the problem to the form

$$\mathbf{x}(k+1) = \mathbf{A}\mathbf{x}(k) + \mathbf{B}\mathbf{u}(k)$$

where

$$\mathbf{x}(k) = \begin{bmatrix} x_{o,k} \\ x_{1,k} \\ \cdot \\ \cdot \\ \cdot \\ x_{m,k} \end{bmatrix}, \qquad \mathbf{u}(k) = \begin{bmatrix} u_{o,k} \\ u_{1,k} \\ \cdot \\ \cdot \\ \cdot \\ u_{m,k} \end{bmatrix}$$

$$\mathbf{A} = \begin{bmatrix} 1-r & r & 0 & 0 & \dots & 0 & 0 \\ r & 1-2r & r & 0 & \dots & 0 & 0 \\ 0 & r & 1-2r & r & \dots & 0 & 0 \\ \cdot & & & & & & \\ \cdot & & & & & & \\ \cdot & & & & & & \\ 0 & 0 & 0 & 0 & \dots & r & 1-r \end{bmatrix}$$

$$\mathbf{B} = \Delta t \begin{bmatrix} 1 & 0 & 0 & \dots & 0 \\ 0 & 1 & 0 & \dots & 0 \\ \cdot & & & & \\ \cdot & & & & \\ \cdot & & & & \\ 0 & 0 & 0 & \dots & 1 \end{bmatrix}, \qquad r = \frac{\Delta t}{(\Delta y)^2}.$$

The discrete version of the performance index is approximated as

$$J = \tfrac{1}{2} \int_0^{t_f} \int_0^{y_f} \{Q' x^2(y, t) + R' u^2(y, t)\}\, dy\, dt$$

$$\approx \tfrac{1}{2}\, \Delta y\, \Delta t \sum_{k=0}^{K-1} \{\mathbf{x}^T(k)\mathbf{Q}\mathbf{x}(k) + \mathbf{u}^T(k)\mathbf{R}\mathbf{u}(k)\}.$$

We shall now use the discrete maximum principle to obtain the optimal control $\mathbf{u}(k)$. The Hamiltonian is

$$H[\mathbf{x}(k), \mathbf{u}(k), \boldsymbol{\lambda}(k+1), k] = \tfrac{1}{2}\, \Delta y\, \Delta t\, \mathbf{x}^T(k)\mathbf{Q}\mathbf{x}(k) + \tfrac{1}{2}\, \Delta y\, \Delta t\, \mathbf{u}^T(k)\mathbf{R}\mathbf{u}(k) + \boldsymbol{\lambda}^T(k+1)[\mathbf{A}\mathbf{x}(k) + \mathbf{B}\mathbf{u}(k)].$$

If we apply the discrete maximum principle, we need to solve

$$\mathbf{x}(k+1) = \mathbf{A}\mathbf{x}(k) + \mathbf{B}\mathbf{u}(k), \qquad x_i(0) = 1 + \alpha i y_f/n, \qquad i = 1, 2, \dots, n$$

$$\boldsymbol{\lambda}(k) = \mathbf{A}^T\boldsymbol{\lambda}(k+1) + \Delta y\, \Delta t\, \mathbf{Q}\mathbf{x}(k), \qquad \boldsymbol{\lambda}(k) = \mathbf{0}$$

$$\mathbf{u}(k) = -\frac{1}{\Delta y\, \Delta t}\mathbf{R}^{-1}\mathbf{B}^T\boldsymbol{\lambda}(k+1).$$

We shall solve this problem by generating the closed-loop control and the discrete matrix Riccati equation, where we assume $\boldsymbol{\lambda}(k) = \mathbf{P}(k)\mathbf{x}(k)$. The optimal control is discerned by solution of the difference equations

$$\mathbf{P}(k) = \Delta y\, \Delta t\, \mathbf{Q} + \mathbf{A}^T\mathbf{P}(k+1)\left[\mathbf{A}^{-1} + \frac{1}{\Delta y\, \Delta t}\mathbf{A}^{-1}\mathbf{B}\mathbf{R}^{-1}\mathbf{B}^T\mathbf{P}(k+1)\right]^{-1},$$

$$\mathbf{P}(K) = \mathbf{0}$$

$$\mathbf{x}(k) = \left[\mathbf{I} + \frac{1}{\Delta y\, \Delta t}\mathbf{B}\mathbf{R}^{-1}\mathbf{B}^T\mathbf{P}(k)\right]^{-1}\mathbf{A}\mathbf{x}(k-1), \qquad x_i(0) = 1 + \alpha i y_f/n$$

$$\mathbf{u}(k) = -\frac{1}{\Delta y\, \Delta t}\mathbf{R}^{-1}\mathbf{B}^T\mathbf{P}(k+1)\mathbf{x}(k+1)$$

$$= -\frac{1}{\Delta y\, \Delta t}\mathbf{R}^{-1}\mathbf{B}^T\mathbf{A}^{-T}[\mathbf{P}(k) - \mathbf{Q}]\mathbf{x}(k).$$

Let us consider the following two cases.

Case A $t_f = 1.0, \quad \Delta t = 0.01$
$y_f = 4.0, \quad \Delta y = 1.0$
$\alpha = 1, \quad Q' = 1, \quad R' = 1$
$r = \Delta t/(\Delta y)^2 = 0.01, \quad \mathbf{B} = 0.01\mathbf{I}.$

$$\mathbf{A} = \begin{bmatrix} 0.99 & 0.01 & 0 & 0 & 0 \\ 0.01 & 0.98 & 0.01 & 0 & 0 \\ 0 & 0.01 & 0.98 & 0.01 & 0 \\ 0 & 0 & 0.01 & 0.98 & 0.01 \\ 0 & 0 & 0 & 0.01 & 0.99 \end{bmatrix}, \qquad \mathbf{R} = \mathbf{Q} = \begin{bmatrix} \frac{1}{2} & 0 & 0 & 0 & 0 \\ 0 & 1 & 0 & 0 & 0 \\ 0 & 0 & 1 & 0 & 0 \\ 0 & 0 & 0 & 1 & 0 \\ 0 & 0 & 0 & 0 & \frac{1}{2} \end{bmatrix}.$$

Case B $t_f = 1.0, \quad \Delta t = 0.01$
$y_f = 4.0, \quad \Delta y = 0.5$
$\alpha = 1.0, \quad Q' = R' = 1$
$r = 0.04, \quad \mathbf{B} = 0.01\mathbf{I}.$

$$\mathbf{A} = \begin{bmatrix} 0.96 & 0.04 & 0 & 0 & 0 & 0 & 0 & 0 & 0 \\ 0.04 & 0.92 & 0.04 & 0 & 0 & 0 & 0 & 0 & 0 \\ 0 & 0.04 & 0.92 & 0.04 & 0 & 0 & 0 & 0 & 0 \\ 0 & 0 & 0.04 & 0.92 & 0.04 & 0 & 0 & 0 & 0 \\ 0 & 0 & 0 & 0.04 & 0.92 & 0.04 & 0 & 0 & 0 \\ 0 & 0 & 0 & 0 & 0.04 & 0.92 & 0.04 & 0 & 0 \\ 0 & 0 & 0 & 0 & 0 & 0.04 & 0.92 & 0.04 & 0 \\ 0 & 0 & 0 & 0 & 0 & 0 & 0.04 & 0.92 & 0.04 \\ 0 & 0 & 0 & 0 & 0 & 0 & 0 & 0.04 & 0.96 \end{bmatrix},$$

$$\mathbf{Q} = \mathbf{R} = \begin{bmatrix} \frac{1}{2} & 0 & 0 & 0 & 0 & 0 & 0 & 0 & 0 \\ 0 & 1 & 0 & 0 & 0 & 0 & 0 & 0 & 0 \\ 0 & 0 & 1 & 0 & 0 & 0 & 0 & 0 & 0 \\ 0 & 0 & 0 & 1 & 0 & 0 & 0 & 0 & 0 \\ 0 & 0 & 0 & 0 & 1 & 0 & 0 & 0 & 0 \\ 0 & 0 & 0 & 0 & 0 & 1 & 0 & 0 & 0 \\ 0 & 0 & 0 & 0 & 0 & 0 & 1 & 0 & 0 \\ 0 & 0 & 0 & 0 & 0 & 0 & 0 & 1 & 0 \\ 0 & 0 & 0 & 0 & 0 & 0 & 0 & 0 & \frac{1}{2} \end{bmatrix}.$$

As indicated in Fig. 6.1-1, there is a slight difference in the computed controls for these two cases. A third trial with $\Delta t = 0.01$, $\Delta y = 0.25$, and $r = 0.16$ indicates that the result for $r = 0.04$ is acceptable in that there is no noticeable change in the controls computed for the two values of r. Again the number of spatial coordinates required for an accurate model is a function of α with only a single coordinate required for $\alpha = 0$. Figures 6.1-2 and 6.1-3 illustrate optimum system behavior for $\alpha = 20$.

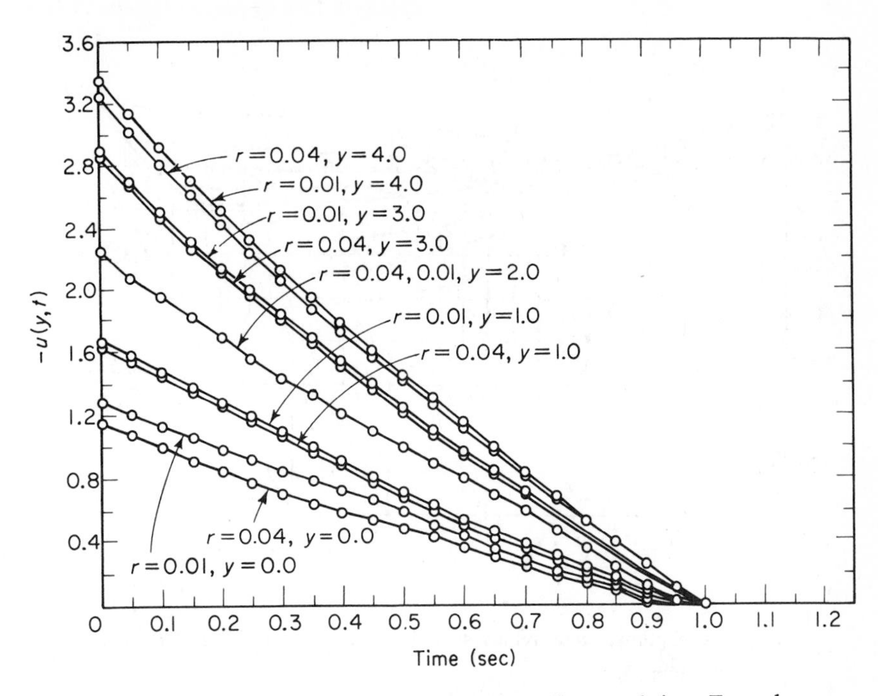

Fig. 6.1-1 Optimal control versus spatial coordinate and time, Example 6.1-2.

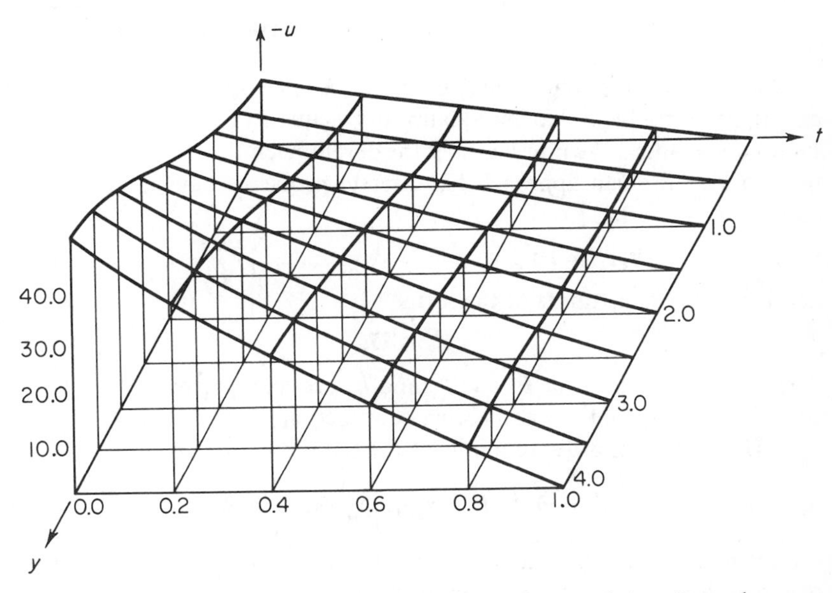

Fig. 6.1-2 Optimal control versus spatial coordinate and time, Example 6.1-2, $\alpha = 20$.

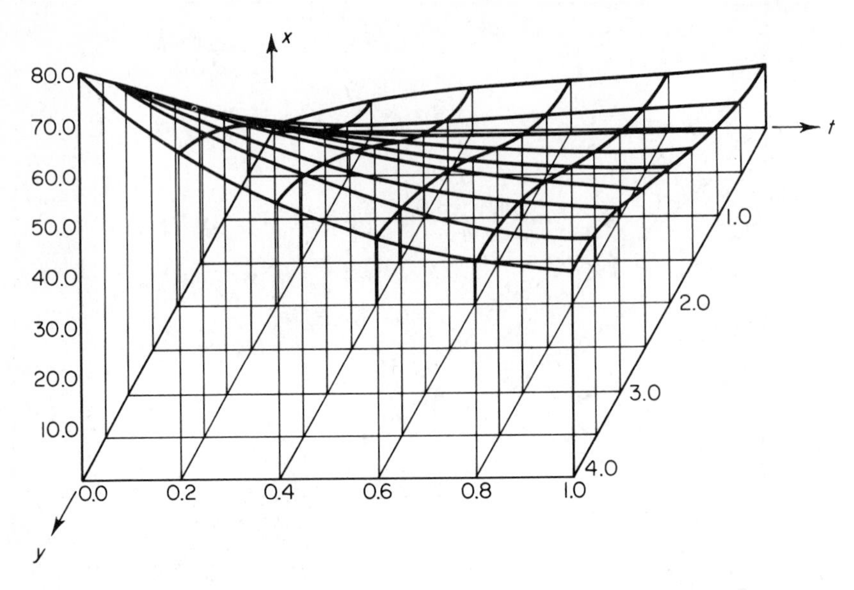

Fig. 6.1-3 Optimal state versus spatial coordinate and time, Example 6.1-2, $\alpha = 20$.

6.2 The discrete maximum principle

Analogous to the continuous time case, use of the discrete Euler-Lagrange equations for problems having equality and inequality constraints can become quite cumbersome. We now consider the development of necessary conditions using a Hamiltonian approach for discrete problems having equality constraints of the form

$$\mathbf{x}_{k+1} = \mathbf{f}(\mathbf{x}_k, \mathbf{u}_k, k), \qquad k = k_o, \ldots, k_f - 1 \tag{6.2-1}$$

and inequality constraints of the form

$$\mathbf{u}_k \in \mathbf{\mho}, \tag{6.2-2}$$

where $\mathbf{\mho}$ is a given set in R^r and k_o and k_f are fixed integers. The problem considered is to find an admissible sequence $\mathbf{u}_k$, $k = k_o, \ldots, k_f - 1$, i.e., $\mathbf{u}_k \in \mathbf{\mho}$ for all k, in order to minimize the criterion

$$J = \Theta(\mathbf{x}_k, k)\big|_{k=k_o}^{k=k_f} + \sum_{k=k_o}^{k_f-1} \phi(\mathbf{x}_k, \mathbf{u}_k, k), \tag{6.2-3}$$

subject to Eq. (6.2-1), with respect to the set of all admissible control sequences. We now state a maximum principle for this problem.

Let $\hat{\mathbf{u}}_k$, $k = k_o, \ldots, k_f - 1$, be an optimal sequence, and let $\hat{\mathbf{x}}_k$, $k = k_o, \ldots, k_f$, be the state response of $\hat{\mathbf{u}}$ uniquely defined by Eq. (6.2-1).

Then, under reasonable assumptions†, there exists a nontrivial function $\hat{\boldsymbol{\lambda}}$ satisfying

$$\hat{\boldsymbol{\lambda}}_k = \frac{\partial H(\hat{\mathbf{x}}_k, \hat{\mathbf{u}}_k, \hat{\boldsymbol{\lambda}}_{k+1}, k)}{\partial \hat{\mathbf{x}}_k}, \tag{6.2-4}$$

$$\boldsymbol{\eta}_{k_o}^T = \left[\hat{\boldsymbol{\lambda}}_{k_o} - \left(\frac{\partial \theta_{k_o}}{\partial \hat{\mathbf{x}}_{k_o}}\right)\right] = \mathbf{0}, \tag{6.2-5}$$

and

$$\boldsymbol{\eta}_{k_f}^T = \left[\hat{\boldsymbol{\lambda}}_{k_f} - \left(\frac{\partial \theta_{k_f}}{\partial \hat{\mathbf{x}}_{k_f}}\right)\right] = \mathbf{0}, \tag{6.2-6}$$

where

$$H(\mathbf{x}_k, \mathbf{u}_k, \boldsymbol{\lambda}_{k+1}, k) = \phi(\mathbf{x}_k, \mathbf{u}_k, k) + \boldsymbol{\lambda}_{k+1}^T \mathbf{f}(\mathbf{x}_k, \mathbf{u}_k, k) \tag{6.2-7}$$

such that for all $k = k_o, \ldots, k_f - 1$,

$$H(\hat{\mathbf{x}}_k, \hat{\mathbf{u}}_k, \hat{\boldsymbol{\lambda}}_{k+1}, k) = \min_{\mathbf{v} \in \mathcal{U}} H(\hat{\mathbf{x}}_k, \hat{\mathbf{v}}, \hat{\boldsymbol{\lambda}}_{k+1}, k). \tag{6.2-8}$$

A proof of these necessary conditions can follow analogously to the proof of the continuous time maximum principle presented in Chapter 4. A more general discrete problem statement, involving state-space constraints, and proof of the discrete necessary conditions can be found in [9]. Sufficient conditions for a related problem are presented in [10].

We note that for the unconstrained case where $\mathcal{U} = R^r$, Eq. (6.2-8) implies the necessary condition

$$\frac{\partial H(\hat{\mathbf{x}}_k, \hat{\mathbf{u}}_k, \hat{\boldsymbol{\lambda}}_{k+1}, k)}{\partial \mathbf{u}_k} = \mathbf{0}, \qquad k = k_o, \ldots, k_f - 1. \tag{6.2-9}$$

A perturbation method can be used to develop necessary conditions for this case as follows. By means of the Lagrange multiplier $\boldsymbol{\lambda}_k$, an equivalent cost function can be written as

$$J' = \theta(\mathbf{x}_k, k)\Big|_{k=k_o}^{k=k_f} + \sum_{k=k_o}^{k_f-1} \{\phi(\mathbf{x}_k, \mathbf{u}_k, k) - \boldsymbol{\lambda}_{k+1}^T[\mathbf{x}_{k+1} - \mathbf{f}(\mathbf{x}_k, \mathbf{u}_k, k)]\} \tag{6.2-10}$$

which becomes, upon introduction of the Hamiltonian Eq. (6.2-7),

$$J' = \theta(\mathbf{x}_k, k)\Big|_{k=k_o}^{k=k_f} + \sum_{k=k_o}^{k_f-1} [H_k - \boldsymbol{\lambda}_{k+1}^T \mathbf{x}_{k+1}]. \tag{6.2-11}$$

Let

$$\mathbf{x}_k = \hat{\mathbf{x}}_k + \epsilon \boldsymbol{\eta}_k$$
$$\mathbf{x}_{k+1} = \hat{\mathbf{x}}_{k+1} + \epsilon \boldsymbol{\eta}_{k+1}$$
$$\mathbf{u}_k = \hat{\mathbf{u}}_k + \epsilon \mathbf{v}_k.$$

---

†These conditions are for each $k = k_o, \ldots, k_f - 1$, $\phi(\cdot, \cdot, k)\colon R^n x R^m \longrightarrow R$ is continuously differentiable (*c.d.*) in both arguments, $\theta(\cdot, k)\colon R^n \longrightarrow R$ is *c.d.* for $k = k_o$, k_f, $f(\cdot, \mathbf{v}, k)\colon R^n \longrightarrow R^n$ is *c.d.* for $k = k_o, \ldots, k_f - 1$ and $\mathbf{v} \in \mathcal{U}$, and the set $\{f(\mathbf{x}, \mathbf{v}, k)\colon \mathbf{v} \in \mathcal{U}\}$ is convex for $k = k_o, \ldots, k_f - 1$ and $\mathbf{x} \in R^n$. These assumptions are presented in a substantially weakened form in [9].

We note that the perturbations at different stages are independent; hence, $\boldsymbol{\eta}_k$, $\boldsymbol{\eta}_{k+1}$, and $\mathbf{v}_k$ are all mutually independent.

Introducing the perturbations into Eq. (6.2-11), we obtain

$$J' = \theta(\hat{\mathbf{x}}_{k_f} + \epsilon\boldsymbol{\eta}_{k_f}, k_f) - \theta(\hat{\mathbf{x}}_{k_o} + \epsilon\boldsymbol{\eta}_{k_o}, k_o) + \sum_{k=k_o}^{k_f-1} [H(\hat{\mathbf{x}}_k + \epsilon\boldsymbol{\eta}_k, \hat{\mathbf{u}}_k + \epsilon\mathbf{v}_k, \boldsymbol{\lambda}_{k+1}, k) - \boldsymbol{\lambda}_{k+1}^T[\hat{\mathbf{x}}_{k+1} + \epsilon\boldsymbol{\eta}_{k+1}]].$$

From our previous work, we know that a minimum of J' requires

$$\frac{\partial J'}{\partial \epsilon} = 0, \qquad \frac{\partial^2 J'}{\partial \epsilon^2} > 0 \tag{6.2-12}$$

for $\epsilon = 0$, independent of the variations. In this development, we will assume that the second-derivative requirement is satisfied for all cost functions and systems of interest. Equating to zero the first derivative in Eq. (6.2-12) requires that

$$\left(\frac{\partial \theta_{k_f}}{\partial \hat{\mathbf{x}}_{k_f}}\right)^T \boldsymbol{\eta}_{k_f} - \left(\frac{\partial \theta_{k_o}}{\partial \hat{\mathbf{x}}_{k_o}}\right)^T \boldsymbol{\eta}_{k_o} + \sum_{k=k_o}^{k_f-1} \left(\frac{\partial H_k}{\partial \hat{\mathbf{x}}_k}\right)^T \boldsymbol{\eta}_k - \sum_{k=k_o}^{k_f-1} \boldsymbol{\lambda}_{k+1}^T \boldsymbol{\eta}_{k+1} + \sum_{k=k_o}^{k_f-1} \left(\frac{\partial H_k}{\partial \hat{\mathbf{u}}_k}\right)^T \mathbf{v}_k = 0. \tag{6.2-13}$$

Employing the discrete version of integration by parts, we can write the fourth term of Eq. (6.2-13) as:

$$-\sum_{k=k_o}^{k_f-1} \boldsymbol{\lambda}_{k+1}^T \boldsymbol{\eta}_{k+1} = -\sum_{k=k_o+1}^{k_f} \boldsymbol{\lambda}_k^T \boldsymbol{\eta}_k = -\sum_{k=k_o}^{k_f-1} [\boldsymbol{\lambda}_k^T \boldsymbol{\eta}_k] - \boldsymbol{\lambda}_{k_f}^T \boldsymbol{\eta}_{k_f} + \boldsymbol{\lambda}_{k_o}^T \boldsymbol{\eta}_{k_o}. \tag{6.2-14}$$

Using Eq. (6.2-14) in Eq. (6.2-13), combining terms, and dropping the ^ notation, we obtain

$$\left[\left(\frac{\partial \theta_{k_f}}{\partial \mathbf{x}_{k_f}}\right)^T - \boldsymbol{\lambda}_{k_f}^T\right] \boldsymbol{\eta}_{k_f} - \left[\left(\frac{\partial \theta_{k_o}}{\partial \mathbf{x}_{k_o}}\right)^T - \boldsymbol{\lambda}_{k_o}^T\right] \boldsymbol{\eta}_{k_o} + \sum_{k=k_o}^{k_f-1} \left[\left(\frac{\partial H_k}{\partial \mathbf{x}_k}\right)^T - \boldsymbol{\lambda}_k^T\right] \boldsymbol{\eta}_k + \sum_{k=k_o}^{k_f-1} \left(\frac{\partial H_k}{\partial \mathbf{u}_k}\right)^T \mathbf{v}_k = 0. \tag{6.2-15}$$

The necessary conditions Eqs. (6.2-4), (6.2-5), (6.2-6), and (6.2-9) hold due to the mutual independence of the appropriate variations in Eq. (6.2-15). This development implies a discrete, two-point boundary value problem of the form

$$\mathbf{x}_{k+1} = \mathbf{f}(\mathbf{x}_k, \boldsymbol{\lambda}_{k+1}, k), \qquad \boldsymbol{\lambda}_{k+1} = \mathbf{g}(\mathbf{x}_k, \boldsymbol{\lambda}_k, k). \tag{6.2-16}$$

If desired, Eq. (6.2-4) can be used to eliminate $\boldsymbol{\lambda}_{k+1}$ from the first part of Eq. (6.2-10). The equations which result from the foregoing are the canonical equations of the required optimal system. The nonlinear, two-point boundary value problem represented by these equations must be solved, in general, by reiterative techniques. Since these reiterative techniques will be used on a digital computer, the discrete maximum principle provides an optimization method which is the natural one to use for many continuous problems after an accurate discrete model has been chosen. We will present several methods for the solution of the resulting two-point boundary value problems in later chapters. The discrete linear regulator will now be considered, since (with a quadratic cost function) the two-point boundary value problem can be easily overcome.

Example 6.2-1. Discrete Linear Regulator We consider a general discrete system represented by

$$\mathbf{x}_{k+1} = \mathbf{A}\mathbf{x}_k + \mathbf{B}\mathbf{u}_k, \qquad \mathbf{x}(0) = x_0, \qquad k = 0, 1, 2, \ldots, k_f$$

where $\mathbf{A}$ and $\mathbf{B}$ may be functions of k. The cost function is

$$J = \tfrac{1}{2}\,||\mathbf{x}_{k_f}||^2_{\mathbf{S}} + \tfrac{1}{2}\sum_{k=0}^{k_f-1}\{||\mathbf{x}_k||^2_{\mathbf{Q}} + ||\mathbf{u}_k||^2_{\mathbf{R}}\}$$

where the weighting matrices $\mathbf{Q}$ and $\mathbf{R}$ may be functions of the stage, k. We thus form the Hamiltonian given by

$$H_k = \tfrac{1}{2}\mathbf{x}_k^T\mathbf{Q}\mathbf{x}_k + \tfrac{1}{2}\mathbf{u}_k^T\mathbf{R}\mathbf{u}_k + \boldsymbol{\lambda}_{k+1}^T[\mathbf{A}\mathbf{x}_k + \mathbf{B}\mathbf{u}_k].$$

From Eq. (6.2-4) the adjoint vector equation is given by

$$\boldsymbol{\lambda}_k = \mathbf{Q}\mathbf{x}_k + \mathbf{A}^T\boldsymbol{\lambda}_{k+1}.$$

We see that this equation cannot be solved for $\boldsymbol{\lambda}_{k+1}$ in terms of $\boldsymbol{\lambda}_k$ unless $\mathbf{A}^{-1}$ exists. Since $\mathbf{A}$ is a state transition matrix,† it will always have an inverse. And since the terminal state is unspecified, the boundary condition is obtained from Eq. (6.2-6) as

$$\boldsymbol{\lambda}(k_f) = \mathbf{S}\mathbf{x}(k_f).$$

From Eq. (6.2-9) we have

$$\frac{\partial H}{\partial \mathbf{u}_k} = \mathbf{0} = \mathbf{R}\mathbf{u}_k + \mathbf{B}^T\boldsymbol{\lambda}_{k+1}.$$

Therefore we have linear difference equations to solve, the solution of which will yield an optimum open-loop control. These equations are

$$\mathbf{x}_{k+1} = \mathbf{A}\mathbf{x}_k - \mathbf{B}\mathbf{R}^{-1}\mathbf{B}^T\boldsymbol{\lambda}_{k+1}, \qquad \mathbf{x}(k_0) = \mathbf{x}_0$$

$$\boldsymbol{\lambda}_k = \mathbf{Q}\mathbf{x}_k + \mathbf{A}^T\boldsymbol{\lambda}_{k+1}, \qquad \boldsymbol{\lambda}(k_f) = \mathbf{S}\mathbf{x}(k_f).$$

We now guess a solution for these equations of the form

$$\boldsymbol{\lambda}_k = \mathbf{P}_k\mathbf{x}_k$$

and substitute in order to eliminate $\boldsymbol{\lambda}$. This yields

$$\mathbf{x}_{k+1} = \mathbf{A}\mathbf{x}_k - \mathbf{B}\mathbf{R}^{-1}\mathbf{B}^T\mathbf{P}_{k+1}\mathbf{x}_{k+1}$$

$$\mathbf{P}_k\mathbf{x}_k = \mathbf{Q}\mathbf{x}_k + \mathbf{A}^T\mathbf{P}_{k+1}\mathbf{x}_{k+1}.$$

†We have a linear difference equation, the homogeneous part of which is

$$\mathbf{x}(t_{k+1}) = \mathbf{A}(t_{k+1}, t_k)\mathbf{x}(t_k).$$

Thus $\mathbf{A}$ is clearly a state transition matrix.

By solving for $\mathbf{x}_{k+1}$ and eliminating it, we obtain

$$\mathbf{P}_k\mathbf{x}_k = \mathbf{Q}\mathbf{x}_k + \mathbf{A}^T\mathbf{P}_{k+1}[\mathbf{I} + \mathbf{B}\mathbf{R}^{-1}\mathbf{B}^T\mathbf{P}_{k+1}]^{-1}\mathbf{A}\mathbf{x}_k$$

which will hold for arbitrary $\mathbf{x}_k$ only if

$$\mathbf{P}_k = \mathbf{Q} + \mathbf{A}^T\mathbf{P}_{k+1}[\mathbf{I} + \mathbf{B}\mathbf{R}^{-1}\mathbf{B}^T\mathbf{P}_{k+1}]^{-1}\mathbf{A} = \mathbf{Q} + \mathbf{A}^T[\mathbf{P}_{k+1}^{-1} + \mathbf{B}\mathbf{R}^{-1}\mathbf{B}^T]^{-1}\mathbf{A}$$

with the condition at the final stage obtained as

$$\mathbf{P}_{k_f} = \mathbf{S}.$$

If the matrix Riccati difference equation is solved backward in time from $k = k_f$ to $k = 0$, certain "gain" functions are obtained which are stored after they are precomputed and applied to the physical system as it runs forward in real time. Thus we have designed a closed-loop optimal discrete system. Most of the remarks in the last chapters on the continuous linear regulator apply here. It is necessary that $\mathbf{Q}$, $\mathbf{R}$, and $\mathbf{S}$ be nonnegative definite in order for the second variation to be positive. Also, $\mathbf{R}$ must be positive definite since its inverse must exist to compute $\mathbf{u}_k$. The "gains" precomputed by this method are called "Kalman gains" and are instrumented as shown in Fig. 6.2-1. The closed-loop control is obtained from the prestored memory as

$$\mathbf{u}_k = -\mathbf{R}^{-1}\mathbf{B}^T\mathbf{A}^{-T}[\mathbf{P}_k - \mathbf{Q}]\mathbf{x}_k = -\mathbf{R}^{-1}\mathbf{B}^T[\mathbf{P}_{k+1}^{-1} + \mathbf{B}\mathbf{R}^{-1}\mathbf{B}^T]^{-1}\mathbf{A}\mathbf{x}_k$$

which is, of course, very similar to the way in which the closed-loop control is obtained for the continuous linear regulator.

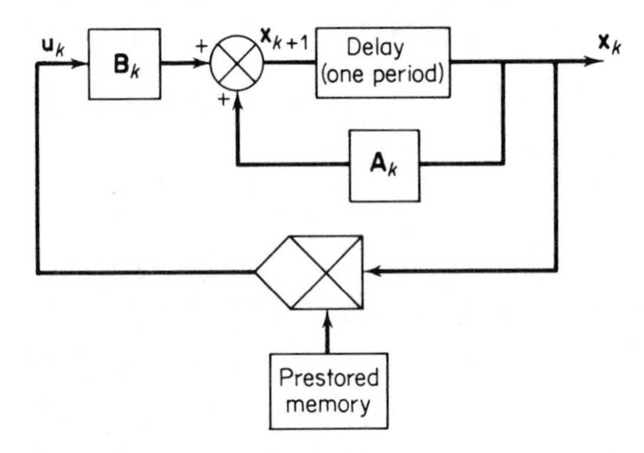

Fig. 6.2-1 Block diagram of closed–loop controller for discrete linear regulator, Example 6.2-1.

6.3 Comparison between the discrete and continuous maximum principle

Having discussed both the continuous and discrete maximum principle, we shall now inquire about the comparison and interconnections between the two. It is only natural to expect that both the continuous and discrete maximum principle will yield very similar (and perhaps the same) solutions to a given problem. We shall see in this section that the two-point discrete boundary value problems which we solve are different in each case. For reasonable sample periods, however, the computational solutions for the two approaches will be essentially the same. Consider the Lagrange problem of the variational calculus. We desire to minimize

$$J = \int_{t_o}^{t_f} \phi(\mathbf{x}, \mathbf{u}, t)\, dt \tag{6.3-1}$$

subject to the equality (vector) constraint

$$\dot{\mathbf{x}} = \mathbf{f}(\mathbf{x}, \mathbf{u}, t) \tag{6.3-2}$$

$$\mathbf{x}(t_o) = \mathbf{x}_o. \tag{6.3-3}$$

The TPBVP is obtained from the maximum principle as follows. We define the Hamiltonian

$$H(\mathbf{x}, \mathbf{u}, \boldsymbol{\lambda}, t) = \phi(\mathbf{x}, \mathbf{u}, t) + \boldsymbol{\lambda}^T(t)\mathbf{f}(\mathbf{x}, \mathbf{u}, t). \tag{6.3-4}$$

The optimum control is determined by

$$\frac{\partial H}{\partial \mathbf{u}} = \mathbf{0} = \frac{\partial \phi(\mathbf{x}, \mathbf{u}, t)}{\partial \mathbf{u}} + \left[\frac{\partial \mathbf{f}^T(\mathbf{x}, \mathbf{u}, t)}{\partial \mathbf{u}}\right]\boldsymbol{\lambda}(t). \tag{6.3-5}$$

The adjoint equations and associated boundary conditions are

$$-\dot{\boldsymbol{\lambda}} = \frac{\partial H}{\partial \mathbf{x}} = \frac{\partial \phi(\mathbf{x}, \mathbf{u}, t)}{\partial \mathbf{x}} + \left[\frac{\partial \mathbf{f}^T(\mathbf{x}, \mathbf{u}, t)}{\partial \mathbf{x}}\right]\boldsymbol{\lambda}(t) \tag{6.3-6}$$

$$\boldsymbol{\lambda}(t_f) = \mathbf{0}. \tag{6.3-7}$$

Thus the continuous TPBVP to be solved is given by Eqs. (6.3-2) and (6.3-6), with the boundary conditions of Eqs. (6.3-3) and (6.3-7), and the coupling equation, Eq. (6.3-5). If a digital computer is used to solve this nonlinear TPBVP, with the first difference expression being used for $\dot{\mathbf{x}}$ and $\dot{\boldsymbol{\lambda}}$, we use the first difference approximations

$$\dot{\mathbf{x}}\big|_{t=kT} = \frac{\mathbf{x}_{k+1} - \mathbf{x}_k}{T} = \frac{\mathbf{x}(\overline{k+1}T) - \mathbf{x}(kT)}{T} \tag{6.3-8}$$

$$\dot{\boldsymbol{\lambda}}\big|_{t=kT} = \frac{\boldsymbol{\lambda}_{k+1} - \boldsymbol{\lambda}_k}{T} = \frac{\boldsymbol{\lambda}(\overline{k+1}T) - \boldsymbol{\lambda}(kT)}{T}. \tag{6.3-9}$$

The resulting discrete TPBVP becomes

$$\mathbf{x}_{k+1} = \mathbf{x}_k + T\mathbf{f}(\mathbf{x}_k, \mathbf{u}_k, k) \tag{6.3-10}$$

$$\boldsymbol{\lambda}_{k+1} = \boldsymbol{\lambda}_k - T\frac{\partial \phi(\mathbf{x}_k, \mathbf{u}_k, k)}{\partial \mathbf{x}_k} - T\left[\frac{\partial \mathbf{f}^T(\mathbf{x}_k, \mathbf{u}_k, k)}{\partial \mathbf{x}_k}\right]\boldsymbol{\lambda}_k \tag{6.3-11}$$

$$\mathbf{x}_{k_o} = \mathbf{x}_o \tag{6.3-12}$$

$$\boldsymbol{\lambda}_{k_f} = \mathbf{0} \tag{6.3-13}$$

$$\frac{\partial \phi(\mathbf{x}_k, \mathbf{u}_k, k)}{\partial \mathbf{u}_k} + \left[\frac{\partial \mathbf{f}^T(\mathbf{x}_k, \mathbf{u}_k, k)}{\partial \mathbf{u}_k}\right]\boldsymbol{\lambda}_k = \mathbf{0}. \tag{6.3-14}$$

The alternate approach using the discrete maximum principle will now be given. The first difference approximation to the equality constraint yields

$$\mathbf{x}_{k+1} = \mathbf{x}_k + T\mathbf{f}(\mathbf{x}_k, \mathbf{u}_k, k) \tag{6.3-15}$$

$$\mathbf{x}(k_o) = \mathbf{x}_o. \tag{6.3-16}$$

Discretization of the cost function yields

$$J = T\sum_{k=k_o}^{k_f-1} \phi(\mathbf{x}_k, \mathbf{u}_k, k). \tag{6.3-17}$$

Thus the Hamiltonian for the discrete maximum principle is

$$H_k = T\phi(\mathbf{x}_k, \mathbf{u}_k, k) + \boldsymbol{\lambda}_{k+1}^T[\mathbf{x}_k + T\mathbf{f}(\mathbf{x}_k, \mathbf{u}_k, k)]. \tag{6.3-18}$$

The control vector is determined from

$$\frac{\partial H_k}{\partial \mathbf{u}_k} = \mathbf{0} = T\frac{\partial \phi(\mathbf{x}_k, \mathbf{u}_k, k)}{\partial \mathbf{u}_k} + T\left[\frac{\partial \mathbf{f}^T(\mathbf{x}_k, \mathbf{u}_k, k)}{\partial \mathbf{u}_k}\right]\boldsymbol{\lambda}_{k+1}. \tag{6.3-19}$$

The adjoint equation and associated transversality condition are:

$$\boldsymbol{\lambda}_k = \frac{\partial H_k}{\partial \mathbf{x}_k} = T\frac{\partial \phi(\mathbf{x}_k, \mathbf{u}_k, k)}{\partial \mathbf{x}_k} + \left[\mathbf{I} + T\frac{\partial \mathbf{f}^T(\mathbf{x}_k, \mathbf{u}_k, k)}{\partial \mathbf{x}_k}\right]\boldsymbol{\lambda}_{k+1} \tag{6.3-20}$$

$$\boldsymbol{\lambda}_{k_f} = \mathbf{0}. \tag{6.3-21}$$

Equations (6.3-15) and (6.3-20) constitute the discrete two-point boundary value nonlinear difference equation to be solved. The boundary conditions are Eqs. (6.3-16) and (6.3-21) with the coupling equation, Eq. (6.3-19). It is immediately apparent that these equations are not the same as those obtained when we discretize the continuous nonlinear TPBVP [Eqs. (6.3-10) through (6.3-14)]. The equality constraint equations (6.3-10) and (6.3-15) and their initial conditions, Eqs. (6.3-11) and (6.3-16), are the same, as are terminal conditions. However, although similar, the coupling equation and the adjoint equation are different. If the $\boldsymbol{\lambda}_{k+1}$ term in the coupling equation, Eq. (6.3-19), is replaced by $\boldsymbol{\lambda}_k$, the two coupling equations are the same. For a "small enough" sampling period, this is not an unreasonable approximation since $\boldsymbol{\lambda}_k$ will not change much from stage to stage. If the adjoint equation, Eqs. (6.3-20), is solved for $\boldsymbol{\lambda}_{k+1}$, there results

$$\boldsymbol{\lambda}_{k+1} = \left[\mathbf{I} + T\frac{\partial \mathbf{f}^T(\mathbf{x}_k, \mathbf{u}_k, k)}{\partial \mathbf{x}_k}\right]^{-1}\left[\boldsymbol{\lambda}_k - T\frac{\partial \phi(\mathbf{x}_k, \mathbf{u}_k, k)}{\partial \mathbf{x}_k}\right]. \tag{6.3-22}$$

Now if the inverse transpose expression is expanded in a Taylor series about T (the sampling period) equals zero, and if, in multiplying the result by the term on the right-hand side of Eq. (6.3-22), powers of T (sampling period) higher than the first are neglected, we obtain

$$\boldsymbol{\lambda}_{k+1} = \boldsymbol{\lambda}_k - T\left[\frac{\partial \mathbf{f}^T(\mathbf{x}_k, \mathbf{u}_k, k)}{\partial \mathbf{x}_k}\right]\boldsymbol{\lambda}_k - T\frac{\partial \phi(\mathbf{x}_k, \mathbf{u}_k, k)}{\partial \mathbf{x}_k}. \tag{6.3-23}$$

This is identical to Eq. (6.3-11), the discretized adjoint equation for the continuous maximum principle. Therefore, as previously stated, computational results for the two approaches will be essentially the same for "small enough" sampling periods. Further discussion of this point can be found in [11]. Such a result does not necessarily hold true for more general problem formulations, and care must be used in determining a sequence of discrete optimal control problems which (in some sense) converge to the associated continuous optimal control problem. Reference [12] elaborates on this point, while [13] presents a constructive technique for proper determination of the sequence of discrete optimal control problems.

6.4 Discrete optimal control and mathematical programming

Section 6.3 explored the relationship between discrete and continuous time optimal control problems. We now investigate the relationship between optimal control and mathematical programming problems, which we recognize both as being multivariate extremization problems subject to various equality and inequality constraints. We begin our investigation by noting the equivalence of the fixed final time discrete optimal control problem and the mathematical problem posed in Sec. 2.3. The various discretization techniques employed in the last section then allow for direct examination of continuous time optimal control and mathematical programming. Further discussion of reducing discrete time optimal control problems to mathematical programming problems by partitioning the time base can be found in [14, 15]. A discretization scheme due to Canon, Cullum, and Polak [16], involving a restrictive definition of an admissible control but allowing an exact solution via mathematical programming, will also be presented.

In order to analytically show the relationship, we define the vector $\mathbf{z}$ as follows:

$$\begin{aligned}
\mathbf{z}_1 &= \mathbf{u}_{k_o} \\
&\vdots \\
\mathbf{z}_K &= \mathbf{u}_{k_f-1} \\
\mathbf{z}_{K+1} &= \mathbf{x}_{k_o} \\
&\vdots \\
\mathbf{z}_{2K+1} &= \mathbf{x}_{k_f},
\end{aligned}$$

where $K = k_f - k_o - 1$. Note that $\mathbf{z}_i \in \mathbf{\mho}$, $i = 1, \ldots, K$, $\mathbf{z}_{K+1}$, and $\mathbf{z}_{2K+1}$ may be restricted to lie in given initial and final target sets, and the equality constraints

$$\mathbf{z}_{K+i+1} = \mathbf{f}(\mathbf{z}_{K+i}, \mathbf{z}_i, i)$$

must be satisfied for all $i = 1, \ldots, K$. These constraints are easily described in the form $\mathbf{\Lambda}(\mathbf{z}) \leq \mathbf{0}$. We define

$$\hat{\theta}(\mathbf{z}) = \theta(\mathbf{z}_{2K+1}, k_f) - \theta(\mathbf{z}_{K+1}, k_o) + \sum_{k=1}^{k-1} \phi(\mathbf{z}_{k+k}, \mathbf{z}_k, k).$$

The discrete optimal control problem is now in the form of the mathematical programming problem:

$$\min \hat{\theta}(\mathbf{z})$$

subject to

$$\mathbf{\Lambda}(\mathbf{z}) \leq \mathbf{0}.$$

Reduction of discrete optimal control problems having state–space constraints to mathematical programming problems (and conversely) is also straightforward [16].

If the discrete optimal control problem was generated from a continuous optimal control problem, clearly the solution to the associated mathematical programming problem will, in general, only approximate the solution to the original continuous problem. We now examine a class of continuous time optimal control problems, having admissible controls of a special structure, which can be converted into mathematical programming problems without resorting to approximation. Consider the continuous time optimal control problem formulated in Sec. 4.4; however, assume a control function is admissible if and only if it is of the following piecewise constant form:

$$\mathbf{u}(t) = \mathbf{u}_k \in \mathbf{\mho}$$

where $t_k \leq t < t_{k+1}$, $t_o = t_{k_o} < \ldots < t_{k_f} = t_f$. Following [16], define

$$\mathbf{x}_k^o(t) = \mathbf{x}_k^o + \int_{t_k}^{t} \phi[\mathbf{x}_k(s), \mathbf{u}_k, s]\, ds$$

for $t_k < t < t_{k+1}$, $\mathbf{x}_{k+1} = \mathbf{x}_k(t_{k+1})$,

$$\phi_k(\mathbf{x}_k, \mathbf{u}_k) = \int_{t_k}^{t_{k+1}} \phi[\mathbf{x}_k(s), \mathbf{u}_k, s]\, ds,$$

and

$$\mathbf{f}_k(\mathbf{x}_k, \mathbf{u}_k) = \mathbf{x}_k + \int_{t_k}^{t_{k+1}} \mathbf{f}[\mathbf{x}_k(s), \mathbf{u}_k, s]\, ds.$$

The continuous time optimal control problem with piecewise constant admissible controls has thus been reduced without approximate discretization to a discrete control problem of the form defined in Sec. 6.2. Reduction to the mathematical programming problem then proceeds as before. As mentioned in [16], other useful restrictions on the admissible controls which imply discretization without approximation are controls of the form

$$\mathbf{u}(t) = \sum_{\alpha=o}^{A} \mathbf{u}_{k\alpha} t^{\alpha}, \qquad t_k < t < t_{k+1}$$

and

$$\mathbf{u}(t) = \sum_{\alpha=o}^{A} u_{\alpha} \mathbf{h}_{\alpha}(t),$$

for functions $h_\alpha : R \longrightarrow R^m$. We note that the last definition of an admissible control does not require that the interval $[t_o, t_f]$ be partitioned. Hence, in reformulating this problem as a mathematical programming problem, there is no discrete time optimal control problem representing a middle step.

Chapter 10 will study the utilization of some of the computational algorithms of mathematical programming for the numerical solution of optimal control problems. Reference [17] presents a comprehensive example of an application of mathematical programming to an optimal control problem.

References

1. ROZONOER, L.I., "L.S. Pontryagin's Maximum Principle in Optimal Control Theory." *Automation and Remote Control*, 1, No. 20, October, November, and December, 1959.
2. PONTRYAGIN, L.S., et al., *The Mathematical Theory of Optimal Processes*. Interscience Publishers, New York, 1962.
3. TOU, J., *Modern Control Theory*. McGraw-Hill Book Co., New York, 1964.
4. FAN, L.T., and WANG, C.S., *The Discrete Maximum Principle*. Wiley, New York, 1964.
5. JORDAN, B.W., and POLAK, E., "Theory of a Class of Discrete Optimal Control Systems." *J. Electronics Control*, 17, December 1964.
6. PEARSON, J.B., and SRIDHAR, R., "A Discrete Optimal Control Problem." *IEEE Trans. Autom. Control*, AC-11, No. 2, April 1966.
7. KUHN, H.W., and TUCKER, A.W., "Nonlinear Programming." *Proceedings 2nd Berkeley Symposium on Mathematical Statistics and Probability*. University of California Press, Berkeley, 1951.
8. HOLTZMAN, J.M., "Convexity and the Maximum Principle for Discrete Systems." *IEEE Trans. Autom. Control*, AC-11, January 1966.
9. HAUTUS, M.L.J., "Necessary Conditions for Multiple Constraint Optimization Problems." *SIAM J. Control*, 11, No. 4, (1973), 653–69.
10. MCCLAMROCH, N.H., "A Sufficiency Condition for Discrete Optimal Processes." *Int. J. Control*, 12, (1970), 157–61.
11. BUDAK, B.M., "Difference Approximations in Optimal Control Problems." *SIAM J. Control*, 7, (1969), 18–31.
12. CULLUM, J., "Discrete Approximations to Continuous Optimal Control Problems." *SIAM J. Control*, 7, (1969), 32–49. (Also presented at *J.A.C.C.*, 1970).
13. CULLUM, J., "An Explicit Method for Discretizing Continuous Optimal Control Problems." *J.O.T.A.*, 5, 1970.
14. TABAK, D., and KUO, B.C., "Application of Mathematical Programming in the Design of Optimal Control Systems." *Int. J. Control*, 10, (1969), 545–52.
15. TABAK, D., and KUO, B.C., *Optimal Control and Mathematical Programming*. Prentice-Hall, Inc., Englewood Cliffs, New Jersey, 1971.
16. CANON, M.D., CULLUM, C.D., JR., and POLAK, E., *Theory of Optimal Control and Mathematical Programming*. McGraw-Hill Book Co., New York, 1970.

17. FEGLEY, K.A., et. al., "Stochastic and Deterministic Design and Control via Linear and Quadratic Programming." *IEEE Trans. Autom. Control*, AC-16, (1971), 759–66.

Problems

1. For the differential system

$$\dot{x}_1 = x_2, \qquad x_1(0) = 1$$
$$\dot{x}_2 = u, \qquad x_2(0) = 0$$

with cost function

$$J = \tfrac{1}{2}\int_0^1 u^2\,dt$$

and equality constraints

$$x_1(1) = x_2(1) = 0$$

an "equivalent" discrete problem is

$$(x_1)_{k+1} = (x_1)_k + 0.1(x_2)_k, \qquad (x_1)_0 = 1$$
$$(x_2)_{k+1} = (x_2)_k + 0.1u_k, \qquad (x_2)_0 = 0$$
$$J = 0.05\sum_{k=0}^{9} u_k^2$$
$$(x_1)_{10} = (x_2)_{10} = 0.$$

Find the open-loop control and trajectory which minimize the discrete cost function subject to the given equality constraints.

2. Find a discrete closed-loop system which solves the problem posed in Problem 1.

3. Compare the solution of Problem 1 with that obtained by discretizing the TPBVP resulting from application of the continuous maximum principle to Problem 1.

4. Find the discrete canonic equations whose solutions minimize

$$J = \tfrac{1}{2}\sum_{k=k_o}^{k_f-1} x_k^2 + u_k^2$$

subject to the equality constraints

$$x_{k+1} = x_k + \alpha x_k^3 + \beta u_k, \qquad x_{k_o} = x_o, \qquad x_{k_f} = 0.$$

5. Rederive the discrete maximum principle for the case where there is the terminal manifold constraint $\mathbf{N}[\mathbf{x}(k_f), k_f] = \mathbf{0}$.

6. By use of the matrix inversion lemma (Appendix D),

$$\begin{aligned}\mathbf{P}_{n+1} &= [\mathbf{P}_n^{-1} + \mathbf{H}_{n+1}^T\mathbf{R}_{n+1}^{-1}\mathbf{H}_{n+1}]^{-1}\\ &= \mathbf{P}_n - \mathbf{P}_n\mathbf{H}_{n+1}^T(\mathbf{H}_{n+1}\mathbf{P}_n\mathbf{H}_{n+1}^T + \mathbf{R}_{n+1})^{-1}\mathbf{H}_{n+1}\mathbf{P}_n,\end{aligned}$$

show that the discrete matrix Riccati equation for the regulator problem of Example (6.2-1) may be rewritten as

$$\mathbf{P}_k = \mathbf{A}^T\mathbf{P}_{k+1}\mathbf{A} - \mathbf{A}^T\mathbf{P}_{k+1}\mathbf{B}[\mathbf{B}^T\mathbf{P}_{k+1}\mathbf{B} + \mathbf{R}]^{-1}\mathbf{B}^T\mathbf{P}_{k+1}\mathbf{A} + \mathbf{Q}$$

which will simplify the computational burden of matrix inversion considerably if $\mathbf{u}$ is of lower dimension than $\mathbf{x}$. What is a similar computationally simple form for the control $\mathbf{u}(t) = \mathbf{K}(t)\mathbf{x}(t)$?

7. From [8], prepare a discussion of the convexity requirement and its application to the discrete maximum principle, with inequality constraints on either the control or the state variables or both.

8. Find the optimal control sequence $u(0)$, $u(1)$, $u(2)$ for the system

$$\mathbf{x}(k+1) = \begin{bmatrix} 0 & 1 \\ -1 & 1 \end{bmatrix}\mathbf{x}(k) + \begin{bmatrix} 0 \\ 1 \end{bmatrix}u(k), \qquad \mathbf{x}(0) = \begin{bmatrix} 1 \\ 1 \end{bmatrix}$$

with

$$J = \sum_{k=0}^{2} [x_1^2(k+1) + u^2(k)]$$

and $x_1(3)$ unspecified and $x_2(3) = 0$.

9. Find the closed-loop controller which minimizes

$$J = \tfrac{1}{2} \sum_{k=0}^{k_f-1} \{\|\mathbf{x}_{k+1}\|^2_{\mathbf{Q}_{k+1}} + \|\mathbf{u}_k\|^2_{\mathbf{R}_k}\}$$

for the system

$$\mathbf{x}_{k+1} = \mathbf{A}_k\mathbf{x}_k + \mathbf{B}_k\mathbf{u}_k, \qquad \mathbf{x}(0) = \mathbf{x}_o.$$

How does this controller differ from the controller of Example (6.2-1)?

10. Develop the discrete maximum principle such as to minimize the cost function

$$J = \sum_{k=k_o}^{k_f-1} \phi[\mathbf{x}(k+1), \mathbf{x}(k), \mathbf{u}(k+1), \mathbf{u}(k), k]$$

for the system

$$\mathbf{x}(k+1) = \mathbf{f}[\mathbf{x}(k), \mathbf{u}(k), k]$$

with the equality constraints

$$\mathbf{x}(k_o) = \mathbf{x}_o, \qquad \mathbf{N}[\mathbf{x}(k_f), k_f] = \mathbf{0}.$$

Consider the case where k_f is free as well as the case for k_f fixed. How does this maximum principle differ from the version derived in this chapter? What does the maximum principle become if the system equation is

(a) $\mathbf{x}(k+1) = \mathbf{f}[\mathbf{x}(k), \mathbf{u}(k+1), k]$
(b) $\mathbf{x}(k+1) = \mathbf{x}(k) + \mathbf{f}[\mathbf{x}(k), \mathbf{u}(k), k]$?

In particular, does the result of part (b) resemble the continuous maximum principle? Will the change in the cost function from that of Eq. (6.2-3) to the one in this example lead to any significant change in $\mathbf{u}(k)$ for the two approaches for a reasonably large number of stages?

11. Using the techniques described in Sec. 4.5, derive the discrete version of the Hamilton-Jacobi equation for the problem formulated in Sec. 6.2 for the case where $\boldsymbol{\theta}(\mathbf{x}_{k_o}, k_o) = 0$.

12. Determine a time and spatially discrete approximation to the distributed system optimal control problem posed in Problem 19, Chapter 5.

Systems concepts

7

This chapter considers the fundamental systems concepts of *controllability* and *observability*, *sensitivity*, and *stability*. The first two sections of this chapter are concerned with the notions of controllability and observability. The observability property states that the value of the state of the system in the past can be determined by observing the systems output. The closely related concept of controllability is concerned with whether or not a control exists that can cause the state or output of a system to follow some desired path. Original efforts in this area are due to Kalman, Ho, and Narendra [1, 2, 3, 4], Kreindler and Sarachik [5], Lee [6], and Gilbert [7]. The issue of system sensitivity to deviation of system components and parameters is considered in the third section. The last section is concerned with system (Lyapunov) stability for linear and nonlinear systems.

7.1 Observability in linear dynamic systems

For a system to be observable, it must be possible to determine the state of the unforced system from the knowledge of the output of the system over some time interval. Specifically, a system is observable over the interval $[t_o, t_f]$ if and only if the state of the system at time t_o can be uniquely determined from the output of the system over the interval $[t_o, t_f]$. We shall first discuss observability for linear discrete systems and then proceed to a discussion of linear continuous systems.

7.1-1
Observability in time-varying discrete systems

Let us suppose that we have a system whose state is described by the homogeneous n-vector difference equation

$$\mathbf{x}(k+1) = \mathbf{A}(k)\mathbf{x}(k), \tag{7.1-1}$$

and suppose that we observe an m-vector $z(k)$ described by

$$\mathbf{z}(k) = \mathbf{C}(k)\mathbf{x}(k). \tag{7.1-2}$$

It is clear that

$$\mathbf{x}(k_o + k) = \boldsymbol{\varphi}(k_o + k, k_o)\mathbf{x}(k_o) \tag{7.1-3}$$

where

$$\boldsymbol{\varphi}(k_o + k, k_o) = \mathbf{A}(k_o + k - 1) \ldots \mathbf{A}(k_o), \tag{7.1-4}$$

and

$$\boldsymbol{\varphi}(k_o, k_o) = \mathbf{I}. \tag{7.1-5}$$

Thus,

$$\mathbf{z}(k) = \mathbf{C}(k)\boldsymbol{\varphi}(k, k_o)\mathbf{x}(k_o). \tag{7.1-6}$$

The unique representation of $\mathbf{x}(k_o)$ in terms of $\mathbf{z}(k)$, $k_o \leq k \leq k_f$, is intimately related to the rank of the matrix

$$\mathbf{M}(k_o, k_f) = \sum_{k=k_o}^{k_f} \boldsymbol{\varphi}^T(k, k_o)\mathbf{C}^T(k)\mathbf{C}(k)\boldsymbol{\varphi}(k, k_o), \tag{7.1-7}$$

which is often referred to as the *observability Gramian*. In fact, $\mathbf{x}(k_o)$ is observable if and only if $\mathbf{M}(k_o, k_f)$ is nonsingular. To prove this result, note that premultiplication of both sides of Eq. (7.1-6) by $\boldsymbol{\varphi}^T(k, k_o)\mathbf{C}^T(k)$ yields

$$\boldsymbol{\varphi}^T(k, k_o)\mathbf{C}^T(k)\mathbf{z}(k) = \boldsymbol{\varphi}^T(k, k_o)\mathbf{C}^T(k)\mathbf{C}(k)\boldsymbol{\varphi}(k, k_o)\mathbf{x}(k_o).$$

Upon summing and using Eq. (7.1-7), we obtain

$$\sum_{k=k_o}^{k_f} \boldsymbol{\varphi}^T(k, k_o)\mathbf{C}^T(k)\mathbf{z}(k) = \mathbf{M}(k_o, k_f)\mathbf{x}(k_o). \tag{7.1-8}$$

We recall from elementary studies in linear algebra that Eq. (7.1-8) has a unique solution if and only if the number of independent columns of $\mathbf{M}(k_o, k_f)$ equals the dimension of the vector on the left-hand side of Eq. (7.1-8). The vector on the left-hand side of Eq. (7.1-8) has dimension n; therefore, a necessary and sufficient condition for Eq. (7.1-8) to have a unique solution is that $\mathbf{M}(k_o, k_f)$ have full rank; i.e., be nonsingular.

Example 7.1-1. Suppose we have two integrators in cascade as in Fig. 7.1-1(a). We ask: Can we estimate $\mathbf{x}^T = [x_1, x_2]$ by observing $\mathbf{z}$? Obviously not, because we do not know the initial condition on the second integrator. In this case we would find $\mathbf{M}$ to be singular and thus not positive definite.

Now suppose that we add a switch to the system as shown in Fig. 7-1.1(b). We begin by observing $\mathbf{z}^T = [z_1, z_2]$ at some time $t_o < t_1$. Can we estimate $\mathbf{x}$?

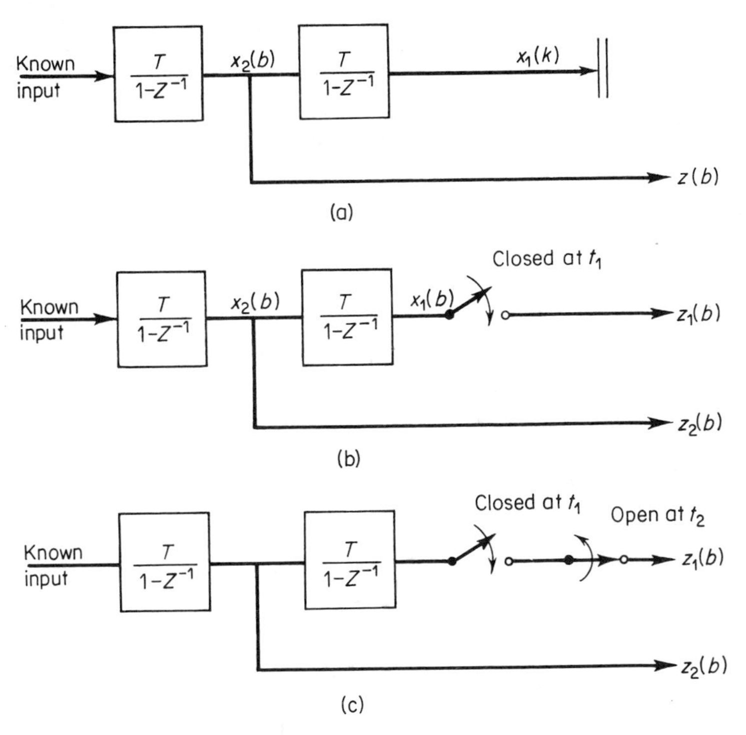

Fig. 7.1-1 A simple system which is (a) unobservable, (b) and (c) observable for $t > t_1$.

We would find that **M** is singular for $t < t_1$ and nonsingular thereafter, indicating that the system is observable for $t > t_1$, and nonobservable for $t < t_1$. This is what we could expect intuitively. Lastly, we add another switch, which we open at time t_2 as shown in Fig. 7.1-1(c). In this case, the system would be nonobservable for $t < t_1$, but observable thereafter, even for $t > t_2$. This is because once we know the value of x_1 for some time t_o, we know x_1 for all time, provided x_2 is known, and we are always observing x_2. Thus, **M** will be singular for $t < t_1$ and nonsingular thereafter. There is a general theorem we could have applied to the third part of this example [2] which states that the rank of $\mathbf{M}(k_f, k_o)$ is nondecreasing with increasing time or, here, increasing k_f.

An alternative necessary and sufficient condition for the observability of the system described by Eqs. (7.1-1) and (7.1-2) is

$$\text{rank}\ \mathbf{\Delta}(k_o, k_f) = n \tag{7.1-9}$$

where

$$\mathbf{\Delta}(k_o, k_f) = [\mathbf{C}^T(k_o), \boldsymbol{\varphi}^T(k_o + 1, k_o)\mathbf{C}^T(k_o + 1), \ldots, \boldsymbol{\varphi}^T(k_f, k_o)\mathbf{C}^T(k_f)] \tag{7.1-10}$$

and where we note from Eq. (7.1-6) that

$$\begin{bmatrix} \mathbf{z}(k_o) \\ \cdot \\ \cdot \\ \cdot \\ \mathbf{z}(k_f) \end{bmatrix} = \mathbf{\Delta}^T(k_o, k_f)\mathbf{x}(k_o). \tag{7.1-11}$$

It is easily shown that

$$\mathbf{M}(k_o, k_f) = \mathbf{\Delta}(k_o, k_f)\mathbf{\Delta}^T(k_o, k_f). \tag{7.1-12}$$

The alternative necessary and sufficient condition for observability then follows from the previously determined result that: $\mathbf{x}(k_o)$ is observable if and only if $\mathbf{M}(k_o, k_f)$ is nonsingular, and $\mathbf{M}(k_o, k_f)$ is nonsingular if and only if $\mathbf{\Delta}(k_o, k_f)$ has rank n.

The alternative equivalent condition for observability, Eq. (7.1-9), reduces to a particularly useful form in the time-invariant case:

$$\operatorname{rank}[\mathbf{C}^T, \mathbf{A}^T\mathbf{C}^T, \ldots, (\mathbf{A}^T)^{k_f-k_o}\mathbf{C}^T] = n. \tag{7.1-13}$$

The dimensions of the matrices imply that in the time-invariant case, rank $\mathbf{\Delta}(0, k) = \operatorname{rank} \mathbf{\Delta}(0, n)$ for all $k \geq n$. Since, in general, $k \geq n$ implies rank $\mathbf{\Delta}(0, k) \leq \operatorname{rank} \mathbf{\Delta}(0, n)$, we need only examine Eq. (7.1-13) for the $k_f - k_o = n$ case; hence,

$$\operatorname{rank}[\mathbf{C}^T, \mathbf{A}^T\mathbf{C}^T, \ldots, (\mathbf{A}^T)^{n-1}\mathbf{C}^T] = n \tag{7.1-14}$$

is an equivalent condition for observability for the time-invariant case.

7.1-2
Observability in continuous time systems

We now consider a continuous time dynamic system represented by the n-vector equation

$$\dot{\mathbf{x}}(t) = \mathbf{A}(t)\mathbf{x}(t) \tag{7.1-15}$$

where we measure the output

$$\mathbf{z}(t) = \mathbf{C}(t)\mathbf{x}(t). \tag{7.1-16}$$

The solution of Eq. (7.1-15) is

$$\mathbf{x}(t) = \boldsymbol{\varphi}(t, \tau)\mathbf{x}(\tau), \tag{7.1-17}$$

where $\boldsymbol{\varphi}(t, \tau)$ satisfies

$$\frac{\partial \boldsymbol{\varphi}(t, \tau)}{\partial t} = \mathbf{A}(t)\boldsymbol{\varphi}(t, \tau), \qquad \boldsymbol{\varphi}(t, t) = \mathbf{I}. \tag{7.1-18}$$

Thus,

$$\mathbf{z}(t) = \mathbf{C}(t)\boldsymbol{\varphi}(t, t_o)\mathbf{x}(t_o). \tag{7.1-19}$$

The continuous-time observability Gramian, in analogy to Eq. (7.1-7), is

$$\mathbf{M}(t_o, t_f) = \int_{t_o}^{t_f} \boldsymbol{\varphi}^T(t, t_o)\mathbf{C}^T(t)\mathbf{C}(t)\boldsymbol{\varphi}(t, t_o)\, dt. \tag{7.1-20}$$

A necessary and sufficient condition for the system described by Eqs. (7.1-15) and (7.1-16) to be observable is that $\mathbf{M}(t_o, t_f)$ be nonsingular. Proof of this follows the analogous proof in Sec. 7.1-1. Again, Eq. (7.1-9) gives a necessary and sufficient condition for the observability of a time-invariant system; i.e., $\mathbf{A}(t) = \mathbf{A}$, and $\mathbf{C}(t) = \mathbf{C}$. Proof of this result for continuous systems can be found in [8, 9, 10, 11]. Also, we may obtain the continuous time observability conditions by means of a least-squares curve fit.

We may now distinguish between several types of observability. A system is said to be observable on the interval $[t_o, t_f]$ if, for a specified t_o and specified t_f, every state $\mathbf{x}(t_o)$ may be determined from knowledge of $\mathbf{z}(t)\ \forall\ t \in [t_o, t_f]$. In other words, the $\mathbf{M}$ matrix is positive definite or the rank test is satisfied for the fixed t_o and fixed t_f. If this is true for all t_o and some $t_f > t_o$, we say that the system is *completely observable*. If this is true for every t_o and every $t_f > t_o$, the system is said to be *totally observable*. The only modification of this statement needed to treat discrete systems is that there are a finite number of states, as discussed in Sec. 7.1-1, before a discrete system will become observable.

7.2 Controllability in linear systems

We consider the continuous time system described by

$$\dot{\mathbf{x}}(t) = \mathbf{A}(t)\mathbf{x}(t) + \mathbf{B}(t)\mathbf{u}(t) \tag{7.2-1}$$

$$\mathbf{z}(t) = \mathbf{C}(t)\mathbf{x}(t). \tag{7.2-2}$$

The system described above is said to be state (output) controllable on the interval $[t_o, t_f]$ if and only if there exists a control function defined on $[t_o, t_f]$ which transfers any initial state vector $\mathbf{x}(t_o)$ (output vector $\mathbf{z}(t_o)$) to any final state vector $\mathbf{x}(t_f)$ (output vector $\mathbf{z}(t_f)$).

We recall that the solution of Eq. (7.2-1) is given by

$$\mathbf{x}(t) = \boldsymbol{\varphi}(t, t_o)\mathbf{x}(t_o) + \int_{t_o}^{t} \boldsymbol{\varphi}(t, s)\mathbf{B}(s)\mathbf{u}(s)\, ds; \tag{7.2-3}$$

thus, $\mathbf{x}_o$ at time t_o can be transferred to $\mathbf{x}_f$ at time t_f if and only if there exists some input $\mathbf{u}(t)$ on $[t_o, t_f]$ such that

$$\mathbf{x}_f - \boldsymbol{\varphi}(t_f, t_o)\mathbf{x}(t_o) = \int_{t_o}^{t_f} \boldsymbol{\varphi}(t_f, s)\mathbf{B}(s)\mathbf{u}(s)\, ds. \tag{7.2-4}$$

We define the $n \times n$ controllability Gramian as

$$\mathbf{W}(t_o, t_f) = \int_{t_o}^{t_f} \boldsymbol{\varphi}(t_f, s)\mathbf{B}(s)\mathbf{B}^T(s)\boldsymbol{\varphi}^T(t_f, s)\, ds. \tag{7.2-5}$$

We now show that $(t_o, \mathbf{x}_o)$ can be transferred to $(t_f, \mathbf{x}_f)$ if and only if there is some n-vector $\boldsymbol{\eta}$ such that

$$\mathbf{x}_f - \boldsymbol{\varphi}(t_f, t_o)\mathbf{x}_o = \mathbf{W}(t_o, t_f)\boldsymbol{\eta}. \tag{7.2-6}$$

If there exists an $\boldsymbol{\eta}$ such that Eq. (7.2-6) holds, then the control $\mathbf{u}(t) = \mathbf{B}^T(t)\boldsymbol{\varphi}^T(t_f, t)\boldsymbol{\eta}$, when submitted into Eq. (7.2-4), implies that $(t_o, \mathbf{x}_o)$ is transferred to $(t_f, \mathbf{x}_f)$. To prove the converse, assume there exists no $\boldsymbol{\eta}$ such that Eq. (7.2-6) is true. It then follows from [12, p. 357] that there exists a $\mathbf{y} \in R^n$ such that $\mathbf{W}(t_o, t_f)\mathbf{y} = \mathbf{0}$ and $\mathbf{y}^T[\mathbf{x}_f - \boldsymbol{\varphi}(t_f, t_o)\mathbf{x}_o] \neq \mathbf{0}$. Assume that $\mathbf{u}(t)$ is a control on $[t_o, t_f]$ which transfers $(t_o, \mathbf{x}_o)$ to $(t_f, \mathbf{x}_f)$. Then, from Eq. (7.2-4),

$$\mathbf{y}^T[\mathbf{x}_f - \boldsymbol{\varphi}(t_f, t_o)\mathbf{x}_o] = \int_{t_o}^{t_f} \mathbf{y}^T\boldsymbol{\varphi}(t_f, s)\mathbf{B}(s)\mathbf{u}(s)\, ds \neq \mathbf{0}. \tag{7.2-7}$$

From the properties of $\mathbf{y}$,

$$\begin{aligned}\mathbf{y}^T\mathbf{W}(t_o, t_f)\mathbf{y} &= \int_{t_o}^{t_f} \mathbf{y}^T\, \boldsymbol{\varphi}(t_f, s)\mathbf{B}(s)\mathbf{B}^T(s)\boldsymbol{\varphi}^T(t_f, s)\mathbf{y}\, ds \\ &= \int_{t_o}^{t_f} \|\mathbf{B}^T(s)\boldsymbol{\varphi}^T(t_f, s)\mathbf{y}\|^2\, ds = 0\end{aligned}$$

which is true (by the continuity of $\mathbf{B}$, $\boldsymbol{\varphi}$, and the norm operator) if and only if $\mathbf{B}^T(s)\boldsymbol{\varphi}^T(t_f, s)\mathbf{y} = \mathbf{0}$ for $s \in [t_o, t_f]$. But $\mathbf{B}^T(s)\boldsymbol{\varphi}^T(t_f, s)\mathbf{y} = \mathbf{0}$ on $[t_o, t_f]$ implies that

$$\int_{t_o}^{t_f} \mathbf{y}^T\boldsymbol{\varphi}(t_f, s)\mathbf{B}(s)\mathbf{u}(s)\, ds = \mathbf{0},$$

which contradicts Eq. (7.2-7). Thus, no control exists which transfers $(t_o, \mathbf{x}_o)$ to $(t_f, \mathbf{x}_f)$.

The above result implies that the system described by Eq. (7.2-1) is controllable on $[t_o, t_f]$ if and only if for any $\mathbf{y} \in R^n$ there is an $\boldsymbol{\eta} \in R^n$ such that $\mathbf{y} = \mathbf{W}(t_o, t_f)\boldsymbol{\eta}$. In general the matrix equation $\mathbf{M}\boldsymbol{\eta} = \mathbf{y}$ has a solution for any $\mathbf{y}$ if and only if the number of independent columns of $\mathbf{M}$ equals the dimension of $\mathbf{y}$. Thus, a necessary and sufficient condition for state controllability is that $\mathbf{W}(t_o, t_f)$ be nonsingular. It then follows directly from Eq. (7.2-4) that, if the system described by Eq. (7.2-1) is controllable, the control

$$\mathbf{u}(t) = \mathbf{B}^T(t)\boldsymbol{\varphi}(t_f, t)\mathbf{W}^{-1}(t_o, t_f)[\mathbf{x}_f - \boldsymbol{\varphi}(t_f, t_o)\mathbf{x}_o] \tag{7.2-8}$$

transfers $(t_o, \mathbf{x}_o)$ to $(t_f, \mathbf{x}_f)$. These results may also be obtained by solving an optimization problem.

We can offer an alternative approach to this problem. We shall do this now for the output controllability problem which reduces to the state controllability problem when $\mathbf{C}(t) = \mathbf{I}$. The solution to Eqs. (7.2-1) and (7.2-2) is the m-vector output due to the r-vector control

$$\mathbf{z}(t) - \mathbf{C}(t)\boldsymbol{\varphi}(t, t_o)\mathbf{x}(t_o) = \mathbf{C}(t) \int_{t_o}^{t} \boldsymbol{\varphi}(t, \tau)\mathbf{B}(\tau)\mathbf{u}(\tau)\, d\tau. \tag{7.2-9}$$

At time t_f, the left-hand side of this equation is simply equal to some specified value $\mathbf{z}_d(t_f)$ such that we may write

$$\mathbf{z}_d(t_f) = \mathbf{z}(t_f) - \mathbf{C}(t_f)\boldsymbol{\varphi}(t_f, t_o)\mathbf{x}(t_o) = \int_{t_o}^{t_f} \mathbf{C}(t_f)\boldsymbol{\varphi}(t_f, \tau)\mathbf{B}(\tau)\mathbf{u}(\tau)\, d\tau. \quad (7.2\text{-}10)$$

A necessary and sufficient condition for output controllability on $[t_o, t_f]$ is that the columns of $\mathbf{C}(t_f)\boldsymbol{\varphi}(t_f, \tau)\mathbf{B}(\tau)$ be linearly independent, which means that, for arbitrary m-vector η, we have the r-vector equation [15, 16]

$$\boldsymbol{\eta}^T\mathbf{C}(t_f)\boldsymbol{\varphi}(t_f, \tau)\mathbf{B}(\tau) \neq \mathbf{0}^T, \qquad t_o \leq \tau \leq t_f. \quad (7.2\text{-}11)$$

We may develop another output controllability condition from this condition. This proof will proceed by the method of contradiction. Suppose that there exists at least one nonzero vector $\boldsymbol{\eta}$, such that Eq. (7.2-11) is, in fact, true. Repeated differentiation of Eq. (7.2-11) with respect to τ yields

$$\boldsymbol{\eta}^T\mathbf{C}(t_f)\boldsymbol{\varphi}(t_f, \tau)\boldsymbol{\Gamma}_j(\tau) = \mathbf{0}^T, \qquad j = 1, 2, \ldots, n \quad (7.2\text{-}12)$$

where, since $\partial\boldsymbol{\varphi}(t_f, \tau)/\partial\tau = -\boldsymbol{\varphi}(t_f, \tau)\mathbf{A}(\tau)$,

$$\begin{aligned} \boldsymbol{\Gamma}_1(\tau) &= \mathbf{B}(\tau) \\ \boldsymbol{\Gamma}_k(\tau) &= \frac{\partial\boldsymbol{\Gamma}_{k-1}(\tau)}{\partial\tau} - \mathbf{A}(\tau)\boldsymbol{\Gamma}_{k-1}(\tau). \end{aligned} \quad (7.2\text{-}13)$$

Then, if we define the n by nm matrix $\boldsymbol{\Gamma}$

$$\boldsymbol{\Gamma} = [\boldsymbol{\Gamma}_1, \boldsymbol{\Gamma}_2, \ldots, \boldsymbol{\Gamma}_n], \quad (7.2\text{-}14)$$

the condition of Eq. (7.2-12) becomes, for the $n\boldsymbol{\eta}$ vectors $\mathfrak{N}$,

$$\mathfrak{N}^T\mathbf{C}(t_f)\boldsymbol{\varphi}(t_f, \tau)\boldsymbol{\Gamma} = \mathbf{0}^T \quad (7.2\text{-}15)$$

which would tell us that $\boldsymbol{\Gamma}$ could not be of rank n since $\boldsymbol{\varphi}$ is nonsingular (excluding for the moment the possibility of $\mathbf{C}$ being singular). But Eq. (7.2-15) cannot be zero by Eq. (7.2-11), and so $\boldsymbol{\Gamma}$ must then be of rank n, and Eq. (7.2-15) will not, in fact, be zero. Although this requirement holds for time-varying systems, it is particularly easy to apply in the case of constant systems, for then, as is easily verified, for $\boldsymbol{\Gamma}' = [\boldsymbol{\Gamma}_1, -\boldsymbol{\Gamma}_2, \ldots, (-1)^{n+1}\boldsymbol{\Gamma}_n]$,

$$\boldsymbol{\Gamma}' = [\mathbf{B} \,\vdots\, \mathbf{AB} \,\vdots\, \mathbf{A}^2\mathbf{B} \,\vdots\, \ldots \,\vdots\, \mathbf{A}^{n-1}\mathbf{B}], \quad (7.2\text{-}16)$$

and this must be of rank n. This is only the requirement for state controllability since, if a constant system is controllable at all, it is controllable at $t_f = t_o$ (impulse control required). Therefore, from Eq. (7.2-15), the output controllability requirement is that

$$[\mathbf{CB} \,\vdots\, \mathbf{CAB} \,\vdots\, \mathbf{CA}^2\mathbf{B} \,\vdots\, \ldots \,\vdots\, \mathbf{CA}^{n-1}\mathbf{B}] \quad (7.2\text{-}17)$$

be of rank m. For the general time-varying case, the $\mathbf{C}(t_f)\boldsymbol{\Gamma}$ term of Eq. (7.2-15) must be of rank m since we know that $\boldsymbol{\varphi}$ must be nonsingular.

If, in Eq. (7.2-10), we let

$$\mathbf{u}(t) = \mathbf{B}^T(t)\boldsymbol{\varphi}^T(t_f, t)\mathbf{C}^T(t_f)\boldsymbol{\lambda}(t_f), \quad (7.2\text{-}18)$$

we have

$$\boldsymbol{\lambda}(t_f) = -\mathbf{V}^{-1}(t_o, t_f)\mathbf{z}_d(t_f), \quad (7.2\text{-}19)$$

where

$$\mathbf{V}(t_o, t_f) = \int_{t_o}^{t_f} \mathbf{C}(t_f)\boldsymbol{\varphi}(t_f, \tau)\mathbf{B}(\tau)\mathbf{B}^T(\tau)\boldsymbol{\varphi}^T(t_f, t)\mathbf{C}^T(t_f)\, d\tau \tag{7.2-20}$$

and must be positive definite for a controllable system. For state controllability, we may treat $\mathbf{C} = \mathbf{I}$; then we can easily show that

$$\mathbf{V}(t_o, t_f) = \boldsymbol{\varphi}(t_f, t_o)\mathbf{W}(t_o, t_f)\boldsymbol{\varphi}^T(t_f, t_o) \tag{7.2-21}$$

where $\mathbf{W}(t_o, t_f)$ is defined by Eq. (7.2-5).

It is quite easy for us to show that all of these results carry over exactly to the discrete system described by

$$\mathbf{x}(k + 1) = \mathbf{A}(k)\mathbf{x}(k) + \mathbf{B}(k)\mathbf{u}(k) \tag{7.2-22}$$

$$\mathbf{z}(k) = \mathbf{C}(k)\mathbf{x}(k) \tag{7.2-23}$$

except that discrete transition matrices and summations are used rather than continuous transition matrices and integrations. The time interval $[t_o, t_f]$ is then replaced by the sequence $k_o, k_o + 1, \ldots, k_f$. Thus, for instance, the discrete equivalent of Eq. (7.2-15) is

$$\mathbf{W}(k_o, k_f) = \sum_{k=k_o}^{k=k_f} \boldsymbol{\varphi}(k_o, k)\mathbf{B}(k)\mathbf{B}^T(k)\boldsymbol{\varphi}^T(k_o, k). \tag{7.2-24}$$

Analogous to the discrete observability requirement, a controllable discrete system can be transferred to the origin in at most n stages, where $\mathbf{x}$ is an n-vector.

Example 7.2-1. Let us consider the linear system described by

$$\dot{x}_1 = x_2(t) + u(t), \qquad z_1(t) = x_1(t)$$

$$\dot{x}_2 = -x_1(t) - 2x_2(t) - u(t), \qquad z_2(t) = x_1(t) + x_2(t).$$

The system dynamics can also be written as

$$\dot{\mathbf{x}} = \mathbf{A}\mathbf{x}(t) + \mathbf{b}u(t), \qquad \mathbf{z}(t) = \mathbf{C}\mathbf{x}(t)$$

where

$$\mathbf{A} = \begin{bmatrix} 0 & 1 \\ -1 & -2 \end{bmatrix}, \qquad \mathbf{b} = \begin{bmatrix} 1 \\ -1 \end{bmatrix}, \qquad \mathbf{C} = \begin{bmatrix} 1 & 0 \\ 1 & 1 \end{bmatrix}.$$

We wish to determine the observability and controllability of the system. From the preceding section we know that the system is observable if the $n \times nm$ matrix

$$[\mathbf{C}^T \,\vdots\, \mathbf{A}^T\mathbf{C}^T \,\vdots\, \ldots \,\vdots\, \mathbf{A}^{T^{n-1}}\mathbf{C}^T] = \begin{bmatrix} 1 & 1 & 0 & -1 \\ 0 & 1 & 1 & -1 \end{bmatrix}$$

is of rank 2. This is the case, and so the system is observable. To discern state controllability, we must examine the matrix

$$[\mathbf{B} \,\vdots\, \mathbf{AB} \,\vdots\, \mathbf{A}^2\mathbf{B} \,\vdots\, \ldots \,\vdots\, \mathbf{A}^{n-1}\mathbf{B}] = \begin{bmatrix} 1 & -1 \\ -1 & 1 \end{bmatrix}$$

to see if it is of rank 2. Clearly it is not, and so this system is not state controllable. Neither is the system output controllable, because the matrix

$$[\mathbf{CB} \mid \mathbf{CAB} \mid \mathbf{CA}^2\mathbf{B} \mid \ldots \mid \mathbf{CA}^{n-1}\mathbf{B}] = \begin{bmatrix} 1 & -1 \\ 0 & 0 \end{bmatrix}$$

is not of rank 2.

Let us now examine the reasons for this uncontrollability. Figure 7.2-1 illustrates a possible block diagram for this system. Appropriate transfer functions for the system are

$$\frac{x_1(s)}{u(s)} = \frac{1}{s+1}, \qquad \frac{x_2(s)}{u(s)} = \frac{-1}{s+1},$$

and we observe that the physical reason the system is not state controllable is that the state vector $\mathbf{x}(t)$ can be controlled only along or parallel to a straight line $x_1(t) + x_2(t) = 0$. This is certainly not in two dimensions; therefore, the system is not state controllable. Appropriate transfer functions for the output state are

$$\frac{z_1(s)}{u(s)} = \frac{x_1(s)}{u(s)} = \frac{1}{s+1}, \qquad \frac{z_2(s)}{u(s)} = \frac{x_1(s) + x_2(s)}{u(s)} = 0.$$

Since the output $z_2(t)$ cannot be controlled by the input, the entire system is not output controllable. If the output were just $z_1(t)$, a scalar, then the system would not be state controllable but output controllable. This means that we could determine an input which could drive $z_1(t)$ to any given value but could not drive $x_1(t)$ and $x_2(t)$ to any value which lies off the line $x_1(t) + x_2(t) = 0$. We note that we were given a second-order system but found a first order transfer function from control input to state and output state variables. This implies that the given system is "reducible" in order. Choate and Sage [13] have shown that systems which are not totally controllable must be reducible.

The important duality concept relating observability and controllability is presented in the problems. The relationship between the realization of a

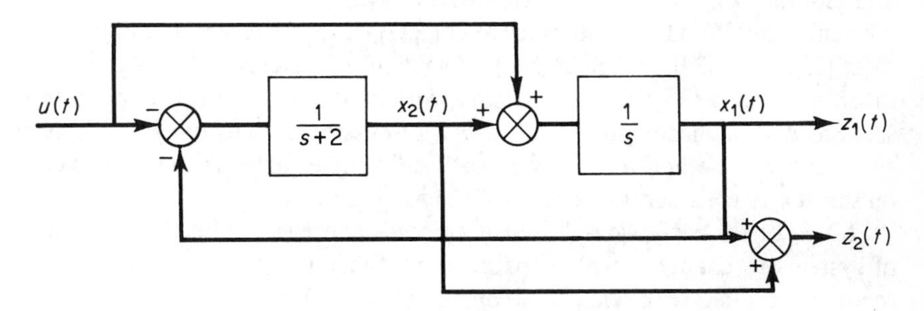

Fig. 7.2-1 Block diagram of uncontrollable system in Example 7.2-1.

linear system and the notion of controllability and observability will not be explored here; the interested reader is referred to [9] and [10].

7.3
Sensitivity in optimum systems control

To accomplish successful system design, it is necessary for us to consider the deviation of a system from its nominal behavior caused by deviation of system components and parameters from their nominal performance characteristics. This is the essence of *sensitivity analysis* which we shall study in this section.

Since it is not possible, for a variety of reasons, for us to duplicate precisely the nominal performance characteristics of system components, we see that sensitivity questions arise whenever we attempt to construct a physical system from a set of mathematical specifications. In this section, we will explore various techniques which provide approaches to the problem of component or parameter tolerance specifications. One of the goals of a sensitivity study is to assign accuracy requirements for system parameters consistent with sensitivity significance in a system model.

Sensitivity seems to have been first introduced in a book by Bode [14]. The utilization of sensitivity in system design was extended in the late 50s and early 60s by many authors who contributed during this period by relating stability and other system characteristics to the sensitivity function. Thus, sensitivity became a more important part of control system design and synthesis. The application of sensitivity to control systems has continued with work by Lindorff [15], Tomovic [16, 17], Horowitz [18], VanNess et al. [19], and Siljak and Stojic [20]. More recent papers are presented in [37].

In 1963, Dorato [21] showed that sensitivity could be useful in the design of optimum control systems. Since that time, Cruz and Perkins [22], Rohrer and Sobral [23], Macko and Mesarovic [24], Pagurek [25], Dorato and Kestenbaum [26], Holtzman and Horing [27], Witsenhausen [28], Siljak and Dorf [29], Dorf [30], and Belanger [30] have investigated sensitivity problems in optimal control systems. It is this work that we will emphasize in this section. Although not presented here, a discussion of sensitivity in discrete systems and a method for the systematic adjustment of sampling rates based on sensitivity considerations can be found in [32 through 36].

Much of the early work in sensitivity was concerned with the sensitivity of systems to changes in plant parameters. Since this work is fundamental to an understanding of sensitivity considerations in optimal systems, we shall give an introduction of some of the work that has been accomplished in this area.

7.3-1
Parameter sensitivity

PARAMETER SENSITIVITY IN CONTINUOUS SYSTEMS: Parameter sensitivity of control systems was perhaps first mentioned by Bode in his book published in 1945. This definition of the sensitivity of the system gain, to variations in a parameter, was

$$S^{-1} = \frac{dp}{dT}\frac{T}{p}, \tag{7.3-1}$$

where T is the system gain and p is a parameter of the system. For almost ten years after its introduction, there was very little written on sensitivity. However, some related work in differential equations and error analysis was carried on during this period. Beginning in 1955, work began to appear in which Bode's definition of sensitivity was related to other system characteristics. During the years 1957 through 1962, sensitivity received increased attention. The pole-zero sensitivity was studied, and sensitivity was related in root-locus properties. The use of sensitivity analysis in linear system theory was presented and extended to sampled-data systems; also presented was the use of sensitivity coefficients for system identification and adaptive control.

One of the first unified presentations of parameter sensitivity is due to Tomovic, who considers a dynamic system with the mathematical model

$$F(\ddot{x}, \dot{x}, x, p, t) = 0, \tag{7.3-2}$$

where x is the output state of the system and p is a single system parameter. We define the *dynamic sensitivity coefficient* as the change in the output state x due to variations in the parameter. We express this as

$$v(p, t) = \frac{\partial x(p, t)}{\partial p}. \tag{7.3-3}$$

We then obtain the sensitivity equation by first taking the partial derivative of Eq. (7.3-2) with respect to p, which gives us

$$\frac{\partial F}{\partial \ddot{x}}\frac{\partial \ddot{x}}{\partial p} + \frac{\partial F}{\partial \dot{x}}\frac{\partial \dot{x}}{\partial p} + \frac{\partial F}{\partial x}\frac{\partial x}{\partial p} + \frac{\partial F}{\partial p} = 0. \tag{7.3-4}$$

We then substitute the relations

$$\frac{\partial \dot{x}}{\partial p} = \dot{v}, \qquad \frac{\partial \ddot{x}}{\partial p} = \ddot{v} \tag{7.3-5}$$

to obtain the sensitivity equation

$$\frac{\partial F}{\partial \ddot{x}}\ddot{v} + \frac{\partial F}{\partial \dot{x}}\dot{v} + \frac{\partial F}{\partial x}v = -\frac{\partial F}{\partial p}. \tag{7.3-6}$$

This sensitivity equation is a linear differential equation which we can solve to determine the sensitivity $v(p, t)$. However, machine solution methods are certainly to be preferred. One such method is the simultaneous solution of the system, Eq. (7.3-2), and the sensitivity, Eq. (7.3-6), on an analog computer. This makes use of the connections between the two equations. Also, the structural similarity of the two equations facilitates solution by a digital computer.

The sensitivity coefficient, as a function of time, about a particular parameter p_o is useful, but a knowledge of its values in parameter space is even more useful. We may study such problems as structural sensitivity and the effect of the variation of a number of parameters by means of the sensitivity coefficients in parameter space. Similarly, we may also discuss the problem of inverse sensitivity. In this particular formulation, we assume that we know the variations in x and desire to determine the corresponding variations in the parameter p.

Equation (7.3-2) is not in a form directly related to most of the system equations we have considered thus far. So, let us consider the system

$$\dot{\mathbf{x}} = \mathbf{f}[\mathbf{x}(t), \mathbf{p}, t], \qquad \mathbf{x}(t_o) = \mathbf{x}_o, \tag{7.3-7}$$

where $\mathbf{x}(t)$ is an n-dimensional state vector and $\mathbf{p}$ is an r-dimensional parameter vector. We will now derive a sensitivity equation and sensitivity coefficients for parameters other than those which change system order and initial conditions. For small changes in $\mathbf{p}$, a first-order approximation for the corresponding change in the state vector $\mathbf{x}$ is

$$\Delta\mathbf{x} = \sum_{j=1}^{r} \mathbf{v}_j \Delta p_j + \ldots \tag{7.3-8}$$

where $\mathbf{v}_j$ is the sensitivity vector

$$\mathbf{v}_j = \left[\frac{\partial x_1}{\partial p_j} \cdots \frac{\partial x_n}{\partial p_j}\right]^T. \tag{7.3-9}$$

The ith component of $\mathbf{v}_j$ is the sensitivity coefficient

$$v_{ij} = \frac{\partial x_i}{\partial p_j} \tag{7.3-10}$$

which represents the variation in the ith component of the state vector $\mathbf{x}$ due to a change in the jth component of the parameter vector $\mathbf{p}$. Thus, we see that the sensitivity equation is

$$\dot{v}_{kj} = \sum_{i=1}^{n} \left[\frac{\partial f_k}{\partial x_i}\right] v_{ij} + \frac{\partial f_k}{\partial p_j}, \tag{7.3-11}$$

where $k = 1, \ldots, n; j = 1, \ldots, r$; and the initial conditions are $v_{kj}(t_o) = 0$. We obtain this equation in a straightforward way by time differentiation of Eq. (7.3-10).

PARAMETER SENSITIVITY IN DISCRETE SYSTEMS: We now desire to develop, for discrete systems, the sensitivity of state variables to changes in system parameters. In the following, we will develop discrete sensitivity by two methods. The change in state will first be expressed in terms of a perturbation matrix and then developed by means of a sensitivity-vector function.

The most general representation of a lumped discrete system is given by the equation†

$$\mathbf{x}(k) = \mathbf{f}[\mathbf{x}(k-1), \mathbf{u}(k-1), \mathbf{p}, t_{k-1}, t_k], \tag{7.3-12}$$

where $\mathbf{x}(k)$ is an n-dimensional state vector; $\mathbf{u}(k)$ is an m-dimensional control vector; and $\mathbf{p}$ is a constant r-dimensional parameter vector. The sampling interval T_k is the time between two sampling instants

$$T_k = t_{k+1} - t_k. \tag{7.3-13}$$

For convenience, the arguments of $\mathbf{x}$ and $\mathbf{u}$ will not include the time t explicitly. Instead, time will be designated by the particular sampling interval, as for example, $\mathbf{x}(t_k) = \mathbf{x}(k)$. If the discrete system is linear, we may represent it by the equation

$$\mathbf{x}(k) = \mathbf{A}(k-1)\mathbf{x}(k-1) + \mathbf{B}(k-1)\mathbf{u}(k-1), \tag{7.3-14}$$

where $\mathbf{A}(k-1)$‡ and $\mathbf{B}(k-1)$ are $\mathbf{A}(t_{k-1}, t_k)$ and $\mathbf{B}(t_{k-1}, t_k)$ during the sampling interval $T_{k-1} = (t_k - t_{k-1})$. If the system represented by Eq. (7.3-14) is also stationary, the matrices $\mathbf{A}$ and $\mathbf{B}$ are constant, $\mathbf{A}(j) = \mathbf{A}(i) = \mathbf{A}$, and $\mathbf{B}(j) = \mathbf{B}(i) = \mathbf{B}$. In the following development, some of the methods are applicable only for linear systems, and Eq. (7.3-14) will be used.

One method for determining the variation in the state $\mathbf{x}(k)$ for changes in system parameters makes use of a perturbation matrix. This method is applicable only to linear systems, and thus we will develop it from Eq. (7.3-14) and its solution (see Appendix A).

$$\mathbf{x}(k) = \boldsymbol{\varphi}(k, 0)\mathbf{x}(0) + \sum_{j=0}^{k-1} \boldsymbol{\varphi}(k, j+1)\mathbf{B}(j)\mathbf{u}(j) \tag{7.3-15}$$

where the transition matrix is given by

$$\boldsymbol{\varphi}(k, j) = \prod_{i=j}^{k-1} \mathbf{A}(i) = \mathbf{A}(k-1)\mathbf{A}(k-2)\ldots\mathbf{A}(j+1)\mathbf{A}(j) \tag{7.3-16}$$

$$\boldsymbol{\varphi}(k, k) = \mathbf{I}.$$

If the system is stationary, the transition matrix is

$$\boldsymbol{\varphi}(k, j) = \boldsymbol{\varphi}(k-j) = \mathbf{A}^{k-j}, \tag{7.3-17}$$

and we may write the solution vector as

$$\mathbf{x}(k) = \boldsymbol{\varphi}(k)\mathbf{x}(0) + \sum_{j=0}^{k-1} \boldsymbol{\varphi}(j)\mathbf{B}\mathbf{u}(k-j-1). \tag{7.3-18}$$

---

†Previously, we have used the equivalent equation $\mathbf{x}(k+1) = \mathbf{f}[\mathbf{x}(k), \mathbf{u}(k), \mathbf{p}, t_k, t_{k+1}]$. The current usage is preferable with respect to the aims of this chapter.

‡We use this notation to provide for the possibility of nonequally spaced samples.

Let us suppose for the moment that we have a stationary system. To determine the effects of a change in parameters, we let the matrix $\mathbf{A}$ of a stationary system change by an amount $\Delta\mathbf{A}$. Equation (7.3-14) can be written for the perturbed system as

$$\tilde{\mathbf{x}}(k) = [\mathbf{A} + \Delta\mathbf{A}]\tilde{\mathbf{x}}(k-1) + \mathbf{B}\mathbf{u}(k-1). \tag{7.3-19}$$

If we write Eq. (7.3-19) for successive values of k, we will derive the system transition equation for $\tilde{\mathbf{x}}(k)$ in terms of the control, the initial value $\mathbf{x}(0)$, and the system parameters. The resulting equation is easily shown to be

$$\tilde{\mathbf{x}}(k) = \boldsymbol{\varphi}(k)\mathbf{x}(0) + \sum_{j=0}^{k-1} \boldsymbol{\varphi}(j)\mathbf{B}\mathbf{u}(k-j-1) + \sum_{j=0}^{k-1} \boldsymbol{\varphi}(k-j-1)\Delta\mathbf{A}\boldsymbol{\varphi}(j)\mathbf{x}(0) + \sum_{i=0}^{k-2}\sum_{j=0}^{k-i-2} \boldsymbol{\varphi}(k-j-i-2)\Delta\mathbf{A}\boldsymbol{\varphi}(j)\mathbf{B}\mathbf{u}(i) + 0(\epsilon^2). \tag{7.3-20}$$

We may determine the change in $\mathbf{x}(k)$, $\Delta\mathbf{x}(k) = \tilde{\mathbf{x}}(k) - \mathbf{x}(k)$, by subtracting Eq. (7.3-18) from Eq. (7.3-20). The first-order change in $\Delta\mathbf{x}(k)$ is obtained if we neglect the higher-order terms in the resulting expression. For a stationary linear system, we obtain the first-order change as

$$\Delta\mathbf{x}(k) \cong \sum_{j=0}^{k-1} \boldsymbol{\varphi}(k-j-1)\Delta\mathbf{A}\boldsymbol{\varphi}(j)\mathbf{x}(0) + \sum_{i=0}^{k-2}\sum_{j=0}^{k-i-2} \boldsymbol{\varphi}(k-j-i-2)\Delta\mathbf{A}\boldsymbol{\varphi}(j)\mathbf{B}\mathbf{u}(i). \tag{7.3-21}$$

A similar expression gives the first-order change in $\mathbf{x}(k)$ for a time-varying linear system

$$\Delta\mathbf{x}(k) \cong \sum_{j=0}^{k-1} \boldsymbol{\varphi}(k, j+1)\Delta\mathbf{A}(j)\boldsymbol{\varphi}(j, 0)\mathbf{x}(0) + \sum_{i=0}^{k-2}\sum_{j=0}^{k-i-2} \boldsymbol{\varphi}(k, j+i+2)\Delta\mathbf{A}(j+i+1)\boldsymbol{\varphi}(j+i+1, i+1)\mathbf{B}(i)\mathbf{u}(i). \tag{7.3-22}$$

The first-order effects of $\Delta\mathbf{A}$ can also be obtained in the form of a difference equation as

$$\Delta\mathbf{x}(k) \cong \mathbf{A}(k-1)\Delta\mathbf{x}(k-1) + \Delta\mathbf{A}(k-1)\mathbf{x}(k-1) \tag{7.3-23}$$

with $\Delta\mathbf{x}(0) = \mathbf{0}$. Thus, the first-order change at each sampling instant can be calculated from the state and the first-order error at the preceding sampling instant.

An exact difference equation for the change in a state, $\mathbf{x}(k)$, due to the perturbations, $\Delta\mathbf{A}(k)$, is obtained by subtracting Eq. (7.3-14) from the stage varying version of Eq. (7.3-19). This is

$$\Delta\mathbf{x}(k) = [\mathbf{A}(k-1) + \Delta\mathbf{A}(k-1)]\Delta\mathbf{x}(k-1) + \Delta\mathbf{A}(k-1)\mathbf{x}(k-1) \tag{7.3-24}$$

where $\Delta\mathbf{x}(0) = \mathbf{0}$. We note that the exact change given by Eq. (7.3-24) differs only by one term from the first-order change of Eq. (7.3-23). The equations obtained in this section give the change in the state due to known perturbations of system prarameters. By careful selection of the elements of the matrix $\Delta\mathbf{A}$, we may vary any number of the system parameters of $\mathbf{A}$.

Another approach to the sensitivity of discrete systems is very similar to the sensitivity vectors and sensitivity functions of continuous systems. In this method, we do not require the system to be stationary or linear. We define a sensitivity vector as

$$\mathbf{v}_i(k) = \frac{\partial \mathbf{x}(k)}{\partial p_i} = \left[\frac{\partial x_1(k)}{\partial p_i} \frac{\partial x_2(k)}{\partial p_i}, \dots, \frac{\partial x_n(k)}{\partial p_i}\right]^T \tag{7.3-25}$$

where the sensitivity functions are the components,

$$v_{ij}(k) = \frac{\partial x_j(k)}{\partial p_i}. \tag{7.3-26}$$

We will have a sensitivity vector for each of the r parameters of the parameter vector $\mathbf{p}$. By using the sensitivity vectors, we obtain a first-order approximation of the change in $\mathbf{x}(k)$ as

$$\Delta\mathbf{x}(k) \cong \sum_{j=1}^{r} \mathbf{v}_j(k)\Delta p_j. \tag{7.3-27}$$

The method used to calculate the sensitivity coefficients is of primary importance. In continuous analysis, a differential equation is formulated, and its solution yields the sensitivity functions. For the discrete approach, a difference equation is desirable. To formulate a sensitivity difference equation, we take the partial derivative of Eq. (7.3-12) with respect to p_i, which gives us the equation

$$\frac{\partial \mathbf{x}(k)}{\partial p_i} = \left[\frac{\partial \mathbf{f}(k-1)}{\partial \mathbf{x}(k-1)}\right]\left[\frac{\partial \mathbf{x}(k-1)}{\partial p_i}\right] + \frac{\partial \mathbf{f}(k-1)}{\partial p_i}. \tag{7.3-28}$$

Terms involving $\mathbf{u}$ are not present, since we assume, for this development, that the control is not dependent upon the changing parameters. By combining Eqs. (7.3-25) and (7.3-28), we obtain a difference equation for the sensitivity vectors of the discrete system

$$\mathbf{v}_i(k) = \frac{\partial \mathbf{f}(k-1)}{\partial \mathbf{x}(k-1)}\mathbf{v}_i(k-1) + \frac{\partial \mathbf{f}(k-1)}{\partial p_i} \tag{7.3-29}$$

where $\mathbf{v}_i(0) = \mathbf{0}$. The sensitivity functions are then determined from an equation for the various components of Eq. (7.3-29)

$$v_{ij}(k) = \sum_{s=1}^{n} \frac{\partial f_j(k-1)}{\partial x_s(k-1)} v_{is}(k-1) + \frac{\partial j_j(k-1)}{\partial p_i} \tag{7.3-30}$$

where $v_{ij}(0) = 0$.

The sensitivity vector function approach of Eqs. (7.3-29) and (7.3-30) is, in general, more useful than the transition matrix approach which we discussed previously. The primary advantage is that the varying parameters are not restricted to the elements of $\mathbf{A}$, and that the system need not be linear.

We must still know the control vector however. We shall remove this restriction in the next section where we will consider closed- and open-loop optimal control and sensitivity determination for optimal systems.

Example 7.3-1. Let us consider the system of Fig. 7.3-1. The equation for the continuous system is

$$\dot{x}(t) = -ax(t) + u(t).$$

A possible discrete version of the system is

$$x(k) = Ax(k-1) + Bu(k-1)$$

where

$$A = \exp\left[-a(t_k - t_{k-1})\right], \qquad B = \int_{t_{k-1}}^{t_k} \exp\left[-a(t_k - \eta)\right] d\eta.$$

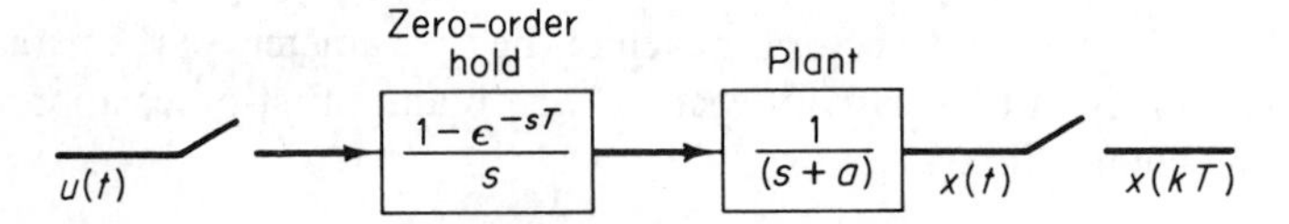

Fig. 7.3-1 First-order system of Example 7.3-1.

For a sampling interval of 0.1 second and $a = 1.0$, these coefficients are $A = 0.90484$ and $B = 0.09516$. The system response for a unit step input is as shown in Fig. 7.3-2.

To use the perturbation matrix approach for parameter variations, we must solve Eq. (7.3-23) for the generalized first-order change. The solution of this equation for a unit step input is also illustrated in Fig. 7.3-2. Of course, we know that the response is not continuous, as shown in the figure, but is actually a series of discrete data points.

For variations in A, this first-order system will have only one sensitivity vector with one component. The solution to Eq. (7.3-29) is identical to that for Eq. (7.3-23). Thus we see that, for this linear system, the sensitivity vector and the perturbation matrix methods have the same solution. Figure 7.3-2

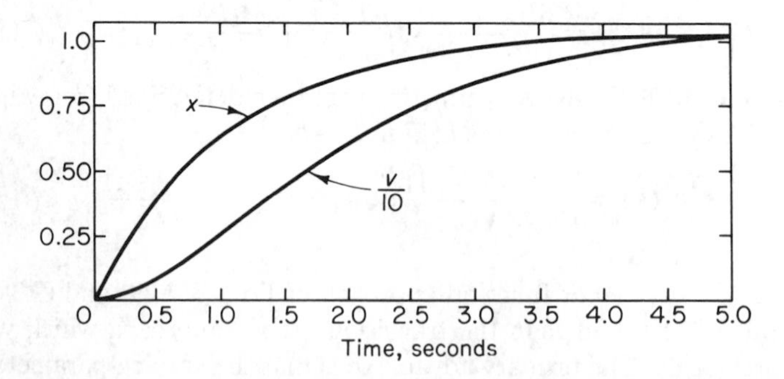

Fig. 7.3-2 Unit step response and parameter sensitivity of Example 7.3-1.

thus represents the sensitivity of state x to variations in the parameter A when determined by either method.

To check the accuracy of the methods, we may let $A = 0.89584$, $B = 0.9516$, and $T = 0.1$ second. This is equivalent to perturbing A by -0.009. Figure 7.3-3 illustrates the difference between the original response x and perturbed response x. For comparison, we also include a curve for the first-order change calculated from Eq. (7.3-23) with $\Delta A = -0.009$, and with the exact change given by Eq. (7.3-24).

The accuracy of the sensitivity vector approach was checked by means of Eq. (7.3-27) with $p = -0.009$. The calculated first-order change in x is identical to that of the perturbation matrix approach. Thus, the first-order error curve of Fig. 7.3-3 represents both methods.

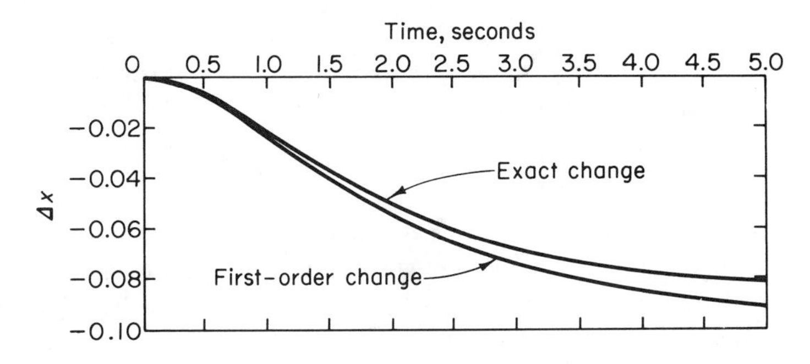

Fig. 7.3-3 Exact and first-order change of Example 7.3-1.

7.3-2 *Sensitivity in optimal control*

In the previous subsection, we presented some concepts concerning classical sensitivity in continuous and discrete systems. In this subsection, we present a discussion of the sensitivity of quantities such as the performance index, the state vector, and the terminal state, for variations in the plant parameters, the state vector, and the control vector.

PERFORMANCE INDEX SENSITIVITY: In one of the first works on optimal control sensitivity, Dorato [21] used sensitivity to determine the variation in the performance index

$$J = \int_{t_o}^{t_f} \phi[\mathbf{x}(t), \mathbf{u}(t), t]\, dt \tag{7.3-31}$$

due to changes in the plant parameters. We assume that the plant state vector $\mathbf{x}(t)$ is related to the plant control vector $\mathbf{u}(t)$ by the vector differential equation

$$\dot{\mathbf{x}} = \mathbf{f}[\mathbf{x}(t), \mathbf{u}(t), \mathbf{p}, t], \qquad \mathbf{x}(t_o) = \mathbf{x}_o \tag{7.3-32}$$

where $\mathbf{p}$ represents the set of plant parameters. We assume that the optimal closed-loop control law is of the form

$$\mathbf{u}(t) = \mathbf{g}[\mathbf{x}(t), \mathbf{p}, t]. \tag{7.3-33}$$

For changes in $\mathbf{p}$ from the nominal $\tilde{\mathbf{p}}$, the change in the performance index will be written as $\Delta J = J(\mathbf{p}) - J(\tilde{\mathbf{p}})$.

We now consider small variations in $\mathbf{p}$ and write for the change in J,

$$\Delta J = dJ = \frac{\partial J}{\partial \tilde{p}_1}\Delta p_1 + \cdots + \frac{\partial J}{\partial \tilde{p}_m}\Delta p_m = \left[\frac{\partial J}{\partial \tilde{\mathbf{p}}}\right]^T \Delta \mathbf{p}, \tag{7.3-34}$$

where $\partial J/\partial \tilde{\mathbf{p}}$ is the performance index sensitivity vector and is evaluated at $\tilde{\mathbf{p}} = \mathbf{p}$. This can be written as

$$\frac{\partial J}{\partial \mathbf{p}} = \int_{t_o}^{t_f} \left[\frac{\partial \mathbf{x}}{\partial \mathbf{p}}\right]^T \frac{\partial \phi}{\partial \mathbf{x}}\, dt, \tag{7.3-35}$$

where $\partial \mathbf{x}/\partial \mathbf{p}$ is the matrix which is the solution to the sensitivity equation (7.3-11). Using the present notation, we can write Eq. (7.3-11) as

$$\frac{d}{dt}\left[\frac{\partial \mathbf{x}}{\partial \mathbf{p}}\right] = \left[\frac{\partial \mathbf{f}}{\partial \mathbf{x}}\right]\left[\frac{\partial \mathbf{x}}{\partial \mathbf{p}}\right] + \left[\frac{\partial \mathbf{f}}{\partial \mathbf{p}}\right] = \dot{\mathbf{V}} = \frac{\partial \mathbf{f}}{\partial \mathbf{x}}\mathbf{V} + \frac{\partial \mathbf{f}}{\partial \mathbf{p}}, \tag{7.3-36}$$

where $\partial \mathbf{f}/\partial \mathbf{x}$ and $\partial \mathbf{f}/\partial \mathbf{p}$ are evaluated about the nominal optimal trajectory. This result suggests that sensitivity might be a useful criterion to use in comparing open- and closed-loop control. In particular, since Eq. (7.3-36) is linear, we may write the solution as

$$\mathbf{V} = \left[\frac{\partial \mathbf{x}}{\partial \mathbf{p}}\right] = \int_{t_o}^{t_f} \boldsymbol{\varphi}(t, \tau)\left[\frac{\partial \mathbf{f}\{\hat{\mathbf{x}}(\tau), \mathbf{g}[\hat{\mathbf{x}}(\tau), \mathbf{p}, \tau], \mathbf{p}, \tau\}}{\partial \mathbf{p}}\right] d\tau \tag{7.3-37}$$

where

$$\frac{\partial \boldsymbol{\varphi}(t, \tau)}{\partial t} = \left[\frac{\partial\{\mathbf{f}[\hat{\mathbf{x}}(t), \mathbf{g}[\hat{\mathbf{x}}(t), \mathbf{p}, t], \mathbf{p}, t\}}{\partial \hat{\mathbf{x}}}\right]\boldsymbol{\varphi}(t, \tau) \tag{7.3-38}$$

and where $\boldsymbol{\varphi}(t, t) = \mathbf{I}$. The initial condition matrix $\partial \mathbf{x}/\partial \mathbf{p}$ is zero at the initial time since $\mathbf{x}(t_o)$ is known. Thus there is no $\boldsymbol{\varphi}(t, t_o)[\partial \mathbf{x}(t_o)/\partial \mathbf{p}]$ term appearing in Eq. (7.3-37). We can therefore evaluate the performance index sensitivity vector, Eq. (7.3-35). Clearly, for any nontrivial system, this evaluation will have to be accomplished on a digital computer.

Cruz and Perkins have investigated this problem and the more general problem of the sensitivity of multivariable systems [22]. To illustrate this procedure, we introduce a sensitivity matrix $\mathbf{S}(s)$ which relates $\Delta\mathbf{X}_o(s)$ to $\Delta\mathbf{X}_c(s)$ by the relationship $\Delta\mathbf{X}_c(s) = \mathbf{S}(s)\Delta\mathbf{X}_o(s)$, where $\Delta\mathbf{X}_o(s)$ is the output error due to the plant parameter variation of an open-loop realization of the system, and where $\Delta\mathbf{X}_c(s)$ is the output error due to plant parameter variations of a closed-loop realization of the system. To have a meaningful comparison, the output of both our open- and closed-loop systems must be equal if there are no plant parameter variations. The open- and closed-loop systems are illustrated in Fig. 7.3-4.

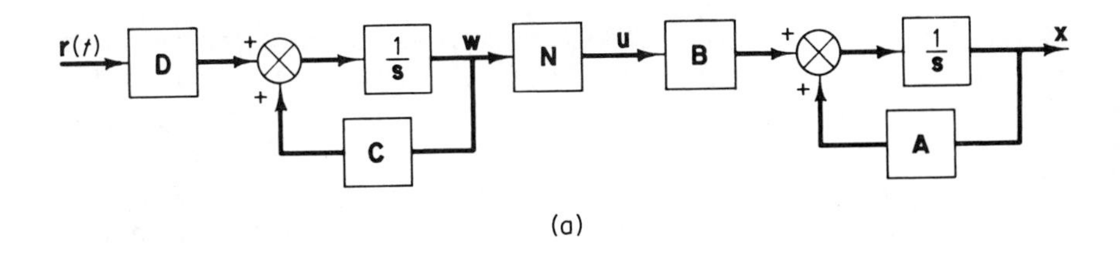

(a)

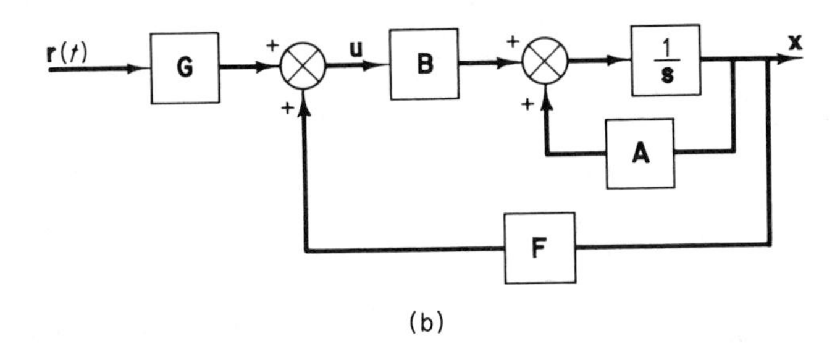

(b)

Fig. 7.3-4 (a) Open- and (b) closed-loop systems for a Cruz and Perkins sensitivity study.

Let us consider an open-loop linear constant system represented by

$$\dot{\mathbf{x}} = \mathbf{Ax}(t) + \mathbf{Bu}(t), \tag{7.3-39}$$

where the control is obtained from a reference input $\mathbf{r}(t)$ by

$$\dot{\mathbf{w}} = \mathbf{Cw}(t) + \mathbf{D}(t)\mathbf{r}(t), \qquad \mathbf{u}(t) = \mathbf{Nw}(t). \tag{7.3-40}$$

We now perturb parameter matrices $\mathbf{A}$ and $\mathbf{B}$ to the value $\tilde{\mathbf{A}} = \mathbf{A} + \Delta\mathbf{A}$, $\tilde{\mathbf{B}} = \mathbf{B} + \Delta\mathbf{B}$, and also $\tilde{\mathbf{x}} = \mathbf{x} + \Delta\mathbf{x}_o$. Then we have for Eq. (7.3-39)

$$\dot{\tilde{\mathbf{x}}} = \tilde{\mathbf{A}}\tilde{\mathbf{x}}(t) + \tilde{\mathbf{B}}\mathbf{u}(t) \tag{7.3-41}$$

since the control $\mathbf{u}(t)$ is not changed for this, the open-loop case. If we take the Laplace transform of Eq. (7.3-41), we obtain

$$(s\mathbf{I} - \tilde{\mathbf{A}})\tilde{\mathbf{X}}(s) = \tilde{\mathbf{x}}(0) + \tilde{\mathbf{B}}\mathbf{U}(s). \tag{7.3-42}$$

From the Laplace transform of Eq. (7.3-39), we have for the open-loop error

$$\Delta\mathbf{X}_0(s) = [s\mathbf{I} - \tilde{\mathbf{A}}]^{-1}[\Delta\mathbf{A}\mathbf{X}(s) + \Delta\mathbf{B}\mathbf{U}(s)]. \tag{7.3-43}$$

For the closed-loop system, we assume dynamics of the form

$$\dot{\mathbf{x}} = \mathbf{Ax}(t) + \mathbf{Bu}(t), \qquad \mathbf{u}(t) = \mathbf{Fx}(t) + \mathbf{Gr}(t). \tag{7.3-44}$$

In a fashion similar to that for the open-loop case, we may easily show that

$$\Delta\mathbf{X}_c(s) = [s\mathbf{I} - \tilde{\mathbf{A}} - \tilde{\mathbf{B}}\mathbf{F}]^{-1}[\Delta\mathbf{A}\mathbf{X}(s) + \Delta\mathbf{B}\mathbf{U}(s)]. \tag{7.3-45}$$

If we compare these two relations for the errors due to varying parameters, Eqs. (7.3-43) and (7.3-45), we obtain

$$\Delta\mathbf{X}_c(s) = \mathbf{S}(s)\Delta\mathbf{X}_o(s), \qquad \mathbf{S}(s) = [s\mathbf{I} - \tilde{\mathbf{A}} - \tilde{\mathbf{B}}\mathbf{F}]^{-1}[s\mathbf{I} - \tilde{\mathbf{A}}]. \tag{7.3-46}$$

It is possible to relate the sensitivity matrix $\mathbf{S}(s)$ to a matrix generalization of return difference for multivariable, linear, time-invariant feedback systems. Also, for single-input, single-output systems, the sensitivity matrix is compatible with the classical definition of sensitivity. However, in the application of sensitivity to optimal control systems, one of the most useful aspects of this work is the idea of a "comparative" sensitivity. For example, if we wish to develop a scalar performance index involving the sensitivity matrix, we may choose the weighted sum of integrated square errors as a performance index and use

$$\int_0^\infty ||\Delta\mathbf{x}_c(t)||_{\mathbf{Q}}^2\, dt < \int_0^\infty ||\Delta\mathbf{x}_o(t)||_{\mathbf{Q}}^2\, dt, \tag{7.3-47}$$

where $\mathbf{Q}$ is a positive definite matrix to guarantee that the closed-loop design is better than an open-loop design. By use of Parseval's theorem (Appendix D), the previous expression may be written as

$$\int_{-j\infty}^{j\infty} \Delta\mathbf{X}_c^T(-s)\mathbf{Q}\Delta\mathbf{X}_c(s)\, ds < \int_{-j\infty}^{j\infty} \Delta\mathbf{X}_o^T(-s)\mathbf{Q}\Delta\mathbf{X}(s)\, ds. \tag{7.3-48}$$

But $\Delta\mathbf{X}_c(s) = \mathbf{S}(s)\Delta\mathbf{X}_o(s)$, which yields from the foregoing

$$\int_{-j\infty}^{j\infty} \Delta\mathbf{X}_o^T(-s)[\mathbf{S}^T(-s)\mathbf{Q}\mathbf{S}(s) - \mathbf{Q}]\Delta\mathbf{X}_o(s)\, ds < 0. \tag{7.3-49}$$

Thus, if the matrix $[\mathbf{S}^T(-s)\mathbf{Q}\mathbf{S}(s) - \mathbf{Q}]$ is negative definite for all ω, we are guaranteed that the closed-loop system is better than the open-loop system according to the stated cost criteria, Eq. (7.3-47). It is possible to show [22] that the closed-loop, optimal, linear, single-input, constant-coefficient regulator with infinite time interval of optimization satisfies this requirement. It is also possible to give a discrete time interpretation to these requirements. We have partially accomplished this in the preceding subsection.

COMPARATIVE SENSITIVITY: Another definition of sensitivity that is of a comparative nature is that introduced by Rohrer and Sobral [23]. To avoid having to completely specify the conditions associated with the normal or "absolute" definitions of sensitivity, we define a relative sensitivity [23] to control variations as

$$S^R[\mathbf{u}(t), \mathbf{p}] = \frac{J[\mathbf{u}(t), \mathbf{p}] - J[\hat{\mathbf{u}}(t), \mathbf{p}]}{|J[\hat{\mathbf{u}}(t), \mathbf{p}]|}, \tag{7.3-50}$$

where $J[\hat{\mathbf{u}}(t), \mathbf{p}]$ represents the performance index associated with the system when driven by the optimum control $\mathbf{u}(t)$ for the given set of plant parameters $\mathbf{p}$. $J[\mathbf{u}(t), \mathbf{p}]$ is the performance index when the control $\mathbf{u}(t)$ is not the optimum for the given set of parameters $\mathbf{p}$. For small departures†, $\delta\mathbf{u} = \hat{\mathbf{u}}(t) - \mathbf{u}(t)$

†It is necessary to introduce three "error vectors" in our discussion, the difference between actual and nominal control $\Delta\mathbf{u} = \tilde{\mathbf{u}} - \mathbf{u}$, the difference between optimum and nominal $\delta\mathbf{u} = \hat{\mathbf{u}} - \mathbf{u}$, and the difference between optimum and actual $\delta\hat{\mathbf{u}} = \hat{\mathbf{u}} - \tilde{\mathbf{u}}$. Similar relations hold for $\mathbf{x}$ and $\mathbf{p}$.

from optimal, the relative sensitivity may be approximated as

$$S^R[\mathbf{u}(t), \mathbf{p}] \approx \frac{\delta^2 J[\hat{\mathbf{u}}(t), \boldsymbol{\delta} u(t), \mathbf{p}]}{|J[\hat{\mathbf{u}}(t), \mathbf{p}]|} \tag{7.3-51}$$

for $\mathbf{u}(t)$ interior to its allowable set $\mathfrak{U}$ since

$$J[\mathbf{u}(t), \mathbf{p}] - J[\hat{\mathbf{u}}(t), \mathbf{p}] = \delta J[\hat{\mathbf{u}}(t), \boldsymbol{\delta}\mathbf{u}(t), \mathbf{p}] + \delta^2 J[\hat{\mathbf{u}}(t), \boldsymbol{\delta}\mathbf{u}(t), \mathbf{p}] + \cdots, \tag{7.3-52}$$

and since the first variation of the cost function must, by definition, be zero, $\delta J[\hat{\mathbf{u}}(t), \boldsymbol{\delta}\mathbf{u}(t), \mathbf{p}] = 0$, about the optimal trajectory. Certainly, if the cost function is actually minimum along the optimal control $\hat{\mathbf{u}}(t)$, then the second variation $\delta^2 J[\hat{\mathbf{u}}(t), \boldsymbol{\delta}\mathbf{u}(t), \mathbf{p}(t)]$ must be positive. However, when the control $\hat{\mathbf{u}}(t)$ is the on boundary of $\mathfrak{U}$, the first variation of the performance index will not vanish, and the first necessary condition for (at least) a local minimum is $\delta^2 J[\hat{\mathbf{u}}(t), \boldsymbol{\delta}\mathbf{u}(t), \mathbf{p}] \geq 0$. In this case, the relative sensitivity becomes approximately

$$S^R[\mathbf{u}(t), \mathbf{p}] \approx \frac{\delta J[\hat{\mathbf{u}}(t), \boldsymbol{\delta}\mathbf{u}(t), \mathbf{p}]}{|J[\hat{\mathbf{u}}(t), \mathbf{p}]|}. \tag{7.3-53}$$

We note that relative sensitivity approaches zero as a system approaches its optimal performance. Also, we see that the relative sensitivity is a function of both the control $\mathbf{u}(t)$ and the parameters $\mathbf{p}$. To develop a design criterion, it is convenient for us to use relative sensitivity to define a plant sensitivity

$$S^M[\mathbf{u}(t)] = \max_{\mathbf{p} \in P} \{S^R[\mathbf{u}(t), \mathbf{p}]\} \tag{7.3-54}$$

where P is the admissible set of plant parameter variations. This expression represents the maximum value of relative sensitivity for all parameters of the allowable set P. This definition of plant sensitivity could be used as a design criterion, and the optimization would seek the control $\hat{\mathbf{u}}^*(t)$ which minimizes the plant sensitivity $S^M[\mathbf{u}(t)]$. Therefore, if the plant parameters were known to vary over a certain range, the control $\hat{\mathbf{u}}^*(t)$ would minimize the maximum deviation from optimal performance for parameter variations over the specified range.

A plant sensitivity that is useful in systems in which the plant parameters are given random variables may also be defined. This definition is based on the expected value of the relative sensitivity which we may state as

$$S^{\mathcal{E}}[\mathbf{u}(t)] = \underset{p \in P}{\mathcal{E}} \{S^R[\mathbf{u}(t), \mathbf{p}]\} \tag{7.3-55}$$

where $\mathcal{E}$, as before, indicates the expected value. Here we seek the control $\hat{\mathbf{u}}^*(t)$ which minimizes $S^{\mathcal{E}}[\mathbf{u}(t)]$. This would result in a control which will minimize the average or expected deviation from the optimal performance.

A similar optimization procedure using a game theory approach has been formulated. This method is useful since both large and small plant parameter variations can be considered, and the controller structure need not be fixed.

Specifically, Dorato and Kestenbaum [26] consider a fixed controller structure with a controller parameter $\mathbf{p}_c$, where the plant dynamics are given by

$$\dot{\mathbf{x}}(t) = \mathbf{f}[\mathbf{x}(t), \mathbf{u}(t), \mathbf{p}_p, t], \tag{7.3-56}$$

where $\mathbf{p}_p$ is the plant parameter. If $\mathbf{p}_c = \mathbf{p}_p$, then the controller generates the control

$$\mathbf{u}(t) = \mathbf{g}[\mathbf{x}(t), \mathbf{p}_c, t] \tag{7.3-57}$$

which is optimal, and consequently, our performance index

$$J = \int_{t_0}^{t_f} \phi[\mathbf{x}(t), \mathbf{u}(t), t]\, dt \tag{7.3-58}$$

is minimized. In this problem formulation, all we know about $\mathbf{p}_p$ is that it lies somewhere in the range $\mathbf{p}_1 \leq \mathbf{p}_p \leq \mathbf{p}_2$, and therefore we see that the performance index is a function of $\mathbf{p}_p$ and $\mathbf{p}_c$. The object of the optimization is to determine the "best" value of $\mathbf{p}_c$. Since $\mathbf{p}_p$ is known to range only from $\mathbf{p}_1$ to $\mathbf{p}_2$, the desirable controller parameter $\hat{\mathbf{p}}_c$ should keep the performance index equal to or less than some value $\hat{J}$ for all values of $\mathbf{p}_p$ in its expected range. This can be expressed as

$$J(\hat{\mathbf{p}}_c, \mathbf{p}_p) \leq \hat{J}, \qquad \text{for} \quad \mathbf{p}_1 \leq \mathbf{p}_p \leq \mathbf{p}_2. \tag{7.3-59}$$

Also, the inequality

$$\hat{J} \leq J(\mathbf{p}_c, \hat{\mathbf{p}}_p), \qquad \text{for} \quad \mathbf{p}_1 \leq \mathbf{p}_c \leq \mathbf{p}_2 \tag{7.3-60}$$

must also hold if $\hat{J}$ is to be as low as possible.

In the game theory interpretation of this problem, $J(\mathbf{p}_c, \mathbf{p}_p)$ is the value of the game or the "payoff function," and the players or "antagonists" are $\mathbf{p}_p$ and $\mathbf{p}_c$. The pair $(\mathbf{p}_c, \mathbf{p}_p)$ is an optimal or pure strategy, and the type of game is infinite or continuous. The conditions for optimal strategies to be pure are given [13] as

$$\operatorname*{Min}_{\mathbf{p}_c} \operatorname*{Max}_{\mathbf{p}_p} J(\mathbf{p}_c, \mathbf{p}_p) = \operatorname*{Max}_{\mathbf{p}_p} \operatorname*{Min}_{\mathbf{p}_c} J(\mathbf{p}_c, \mathbf{p}_p). \tag{7.3-61}$$

Also, we must require the existence of numbers $\hat{\mathbf{p}}_c$, $\hat{\mathbf{p}}_p$, and $\hat{J}$ such that

$$J(\hat{\mathbf{p}}_c, \mathbf{p}_p) \leq \hat{J} \leq J(\mathbf{p}_c, \hat{\mathbf{p}}_p). \tag{7.3-62}$$

THE HAMILTON-JACOBI-BELLMAN-EQUATION FORMULATION OF PERFORMANCE INDEX SENSITIVITY: Pagurek [25] also considers open- and closed-loop sensitivity of optimal linear systems. The sensitivity of the performance index is formulated into the structure of the Hamilton-Jacobi equation, and it is shown that the open- and closed-loop performance index sensitivity functions are the same. This approach is useful in that sensitivity analysis can be carried out by the same technique used to obtain the optimal control law. Witsenhausen [28] extends this work to the nonlinear case. Perhaps the most interesting part of this work is: Our intuitive feeling that a closed-loop system is better than an open-loop system from the standpoint of this definition of sensitivity is shown to be false.

We formulate this sensitivity problem as follows. The system and control of Eqs. (7.3-32) and (7.3-33) are again considered:

$$\dot{\mathbf{x}} = \mathbf{f}[\mathbf{x}(t), \mathbf{u}(t), \tilde{\mathbf{p}}, t], \qquad \mathbf{x}(t_o) = \mathbf{x}_o, \qquad \mathbf{u}(t) = \mathbf{g}[\mathbf{x}(t), \tilde{\mathbf{p}}, t]. \quad (7.3\text{-}63)$$

The cost function is slightly different from that of Eq. (7.3-31) in that we allow for the inclusion of the variable parameter vector $\tilde{\mathbf{p}}$. We may write this cost function about an optimal control and trajectory as a function of the initial state $\mathbf{x}(t_o)$, actual parameter vector $\tilde{\mathbf{p}}$, nominal parameter vector $\mathbf{p}$, and initial time t_o:

$$V[\mathbf{x}(t_o), \tilde{\mathbf{p}}, \mathbf{p}, t_o] = J = \theta[\mathbf{x}(t_f), t_f] + \int_{t_o}^{t_f} \phi[\mathbf{x}(t), \hat{\mathbf{u}}(t), \mathbf{p}, t]\, dt \quad (7.3\text{-}64)$$

where the optimum closed-loop control and trajectory, $\hat{\mathbf{x}}(t)$ and $\hat{\mathbf{u}}(t)$, are determined with the actual system parameter $\tilde{\mathbf{p}}$. As in Eq. (7.3-34), we may write a first-order approximation for small parameter variations as

$$\Delta V = J(\mathbf{p}) - J(\tilde{\mathbf{p}}) = \left\{\frac{\partial V[\mathbf{x}(t_o), \tilde{\mathbf{p}}, \mathbf{p}, t_o]}{\partial \tilde{\mathbf{p}}}\right\}^T \Delta\tilde{\mathbf{p}}. \quad (7.3\text{-}65)$$

We are interested in determining the performance index sensitivity vector, Eq. (7.3-35), which becomes

$$\mathbf{W}[\mathbf{x}(t_o), \mathbf{p}, t_o] = \left.\frac{\partial V[\mathbf{x}(t_o), \tilde{\mathbf{p}}, \mathbf{p}, t_o]}{\partial \tilde{\mathbf{p}}}\right|_{\tilde{p}=p}. \quad (7.3\text{-}66)$$

For the closed-loop case, we can demonstrate, as in Chapter 4, that the function V satisfies the Hamilton-Jacobi-Bellman equation

$$\frac{\partial V[\mathbf{x}(t_o), \tilde{\mathbf{p}}, \mathbf{p}, t_o]}{\partial t_o} + H\left[\mathbf{x}(t_o), \frac{\partial V}{\partial \mathbf{x}}, \tilde{\mathbf{p}}, \mathbf{p}, t_o\right] = 0 \quad (7.3\text{-}67)$$

where the Hamiltonian is

$$H\left[\mathbf{x}(t), \frac{\partial V}{\partial \mathbf{x}}, \tilde{\mathbf{p}}, \mathbf{p}, t\right] = \phi[\mathbf{x}(t), \hat{\mathbf{u}}(t), \mathbf{p}, t] + \left[\frac{\partial V[\mathbf{x}(t), \tilde{\mathbf{p}}, \mathbf{p}, t]}{\partial \mathbf{x}(t)}\right]^T \mathbf{f}[\mathbf{x}(t), \hat{\mathbf{u}}(t), \tilde{\mathbf{p}}, t] \quad (7.3\text{-}68)$$

and where the boundary condition is $V[\mathbf{x}(t_f), \tilde{\mathbf{p}}, \mathbf{p}, t_f] = \theta[\mathbf{x}(t_f), t_f]$. By taking the partial derivative of Eq. (7.3-67) with respect to $\tilde{\mathbf{p}}$, then letting $\tilde{\mathbf{p}} = \mathbf{p}$, and interchanging the order of the derivatives in the result, we obtain, by virtue of the performance index sensitivity vector, Eq. (7.3-66),

$$\frac{\partial W^i[\mathbf{x}(t_o), \mathbf{p}, t_o]}{\partial t_o} + H'\left[\mathbf{x}(t_o), \frac{\partial W^i}{\partial \mathbf{x}}, \tilde{\mathbf{p}}, \mathbf{p}, t_o\right] = 0, \qquad W^i[\mathbf{x}(t_f), \mathbf{p}, t_f] = 0. \quad (7.3\text{-}69)$$

This is just another Hamilton–Jacobi–Bellman equation, where the Hamiltonian H' is now

$$H'\left[\mathbf{x}(t), \frac{\partial W^i}{\partial \mathbf{x}}, \tilde{\mathbf{p}}, \mathbf{p}, t_o\right] = \left\{\frac{\partial V[\mathbf{x}(t), \mathbf{p}, t]}{\partial \mathbf{x}(t)}\right\}^T \left\{\frac{\partial \mathbf{f}[\mathbf{x}(t), \hat{\mathbf{u}}(t), \tilde{\mathbf{p}}, t]}{\partial \tilde{p}_i}\right\}\Bigg|_{\tilde{\mathbf{p}}=\mathbf{p}} +$$
$$\frac{\partial \phi[\mathbf{x}(t), \hat{\mathbf{u}}(t), \mathbf{p}, t]}{\partial \tilde{p}_i}\Bigg|_{\tilde{\mathbf{p}}=\mathbf{p}} + \left\{\frac{\partial W^i[\mathbf{x}(t), \mathbf{p}, t]}{\partial \mathbf{x}}\right\}^T \mathbf{f}[\mathbf{x}(t), \hat{\mathbf{u}}(t), \mathbf{p}, t], \quad (7.3\text{-}70)$$

where we recall that $\hat{\mathbf{u}}(t)$, the closed-loop control, is a function of $\tilde{\mathbf{p}}$ because of Eq. (7.3-63) and therefore realize that terms such as $\partial\phi[\mathbf{x}(t), \hat{\mathbf{u}}(t), \mathbf{p}, t]/\partial\tilde{p}_i$ are meaningful. We have r equations in Eq. (7.3-69), one for each component of $\mathbf{W}$, $W^i, i = 1, 2, \ldots, r$. These correspond to the r components of $\mathbf{p}, p_i$, $i = 1, 2, \ldots, r$.

For the open-loop case, $\hat{\mathbf{u}}(t)$ is a function only of the initial state and time and $\mathbf{p}$, the nominal parameter vector, since the open-loop control can be precomputed. Thus it is not possible, in this case, to write a version of the Hamilton-Jacobi equation. The cost function and performance index sensitivity vector can be determined for the open-loop case by straightforward computational means as is evident from the preceding discussion and that of performance index sensitivity.

A principal disadvantage to this formulation is that the Hamilton-Jacobi partial differential equations are most difficult to solve except for a few special cases. One of these is the linear regulator problem (Chapter 5, Sec. 5.1), where the solution to the minimization of

$$J = \tfrac{1}{2}\|\mathbf{x}(t_f)\|^2_S + \tfrac{1}{2}\int_{t_o}^{t_f} [\|\mathbf{x}(t)\|^2_{\mathbf{Q}(t)} + \|\mathbf{u}(t)\|^2_{\mathbf{R}(t)}]\, dt \quad (7.3\text{-}71)$$

for the system,

$$\dot{\mathbf{x}} = \mathbf{A}(t)\mathbf{x}(t) + \mathbf{B}(t)\mathbf{u}(t), \qquad \mathbf{x}(t_o) = \mathbf{x}_o, \quad (7.3\text{-}72)$$

is determined by solving

$$\hat{\mathbf{u}}(t) = -\mathbf{R}^{-1}(t)\mathbf{B}^T(t)\mathbf{P}(t)\mathbf{x}(t) \quad (7.3\text{-}73)$$

$$\dot{\mathbf{P}} = -\mathbf{A}^T\mathbf{P} - \mathbf{P}\mathbf{A} + \mathbf{P}\mathbf{B}\mathbf{R}^{-1}\mathbf{B}^T\mathbf{P} - \mathbf{Q}, \qquad \mathbf{P}(t_f) = \mathbf{S}. \quad (7.3\text{-}74)$$

Now we consider that the feedback is based on the nominal system $\hat{\mathbf{u}}(t) = -\mathbf{R}^{-1}(t)\mathbf{B}^T(t)\mathbf{P}(t)\mathbf{x}(t)$ but that the actual system parameters are $\tilde{\mathbf{A}}$ and $\tilde{\mathbf{B}}$ with weighting matrices $\tilde{\mathbf{Q}}$ and $\tilde{\mathbf{R}}$. The actual parameter vector is $\tilde{\mathbf{p}}$. If the control law, Eq. (7.3-73), is used with the actual plant, we have

$$\dot{\mathbf{x}} = [\tilde{\mathbf{A}}(t) - \tilde{\mathbf{B}}(t)\mathbf{R}^{-1}(t)\mathbf{B}^T(t)\mathbf{P}(t)]\mathbf{x}(t) = \tilde{\mathbf{F}}(t)\mathbf{x}(t). \quad (7.3\text{-}75)$$

The ϕ function for the actual problems is, from Eq. (7.3-71) with the control of Eq. (7.3-73),

$$\tfrac{1}{2}\|\mathbf{x}(t)\|^2_{\tilde{\mathbf{Q}}(t)} + \tfrac{1}{2}\|\mathbf{u}(t)\|^2_{\tilde{\mathbf{R}}(t)} = \tfrac{1}{2}\|\mathbf{x}(t)\|^2_{\mathbf{Q}'(t)} \quad (7.3\text{-}76)$$

where

$$\mathbf{Q}'(t) = \mathbf{P}(t)\mathbf{B}(t)\mathbf{R}^{-1}(t)\tilde{\mathbf{R}}(t)\mathbf{R}^{-1}(t)\mathbf{B}^T(t)\mathbf{P}(t) + \tilde{\mathbf{Q}}(t). \quad (7.3\text{-}77)$$

By solving the Hamilton–Jacobi–Bellman equation, Eq. (7.3-67), for this problem, we obtain

$$V[\mathbf{x}(t), \tilde{\mathbf{p}}, \mathbf{p}, t] = \tfrac{1}{2}\mathbf{x}^T(t)\mathbf{M}(t)\mathbf{x}(t),$$

$$\dot{\mathbf{M}} = -\tilde{\mathbf{F}}^T(t)\mathbf{M}(t) - \mathbf{M}(t)\tilde{\mathbf{F}}(t) - \mathbf{Q}'(t), \qquad \mathbf{M}(t_f) = \mathbf{S}. \tag{7.3-78}$$

In exactly the same way, we find the Hamilton-Jacobi equation for the performance index sensitivity function from Eqs. (7.3-69) and (7.3-70), and with use of Eq. (7.3-78):

$$\frac{\partial W^i[\mathbf{x}(t), \mathbf{p}, t]}{\partial t} + \left\{\frac{\partial W^i[\mathbf{x}(t), \mathbf{p}, t]}{\partial \mathbf{x}(t)}\right\}^T \mathbf{F}(t) +$$

$$\frac{1}{2}\mathbf{x}^T(t)\left[\frac{\partial \mathbf{Q}'(t)}{\partial \tilde{p}_i} + \mathbf{P}(t)\frac{\partial \tilde{\mathbf{F}}}{\partial \tilde{p}_i} + \frac{\partial \tilde{\mathbf{F}}^T}{\partial \tilde{p}_i}\mathbf{P}(t)\right]\Bigg|_{\tilde{\mathbf{p}}=\mathbf{p}} \mathbf{x}(t) = 0 \tag{7.3-79}$$

where $\mathbf{Q}'(t)$ is defined in Eq. (7.3-77), $\tilde{\mathbf{F}}(t)$ in Eq. (7.3-75), and where

$$\mathbf{F}(t) = \mathbf{A}(t) - \mathbf{B}(t)\mathbf{R}^{-1}(t)\mathbf{B}^T(t)\mathbf{P}(t). \tag{7.3-80}$$

We assume a solution $W^i[\mathbf{x}(t), \mathbf{p}, t] = \frac{1}{2}\mathbf{x}^T(t)\mathbf{\Gamma}_i(t)\mathbf{x}(t)$ and obtain

$$\dot{\mathbf{\Gamma}}_i = -\mathbf{F}^T(t)\mathbf{\Gamma}_i(t) - \mathbf{\Gamma}_i(t)\mathbf{F}(t) - \mathbf{Q}_i''(t), \qquad \mathbf{\Gamma}_i(t_f) = \mathbf{0}, \tag{7.3-81}$$

where

$$\mathbf{Q}_i''(t) = \left[\frac{\partial \mathbf{Q}'(t)}{\partial \tilde{p}_i} + \mathbf{P}(t)\frac{\partial \tilde{\mathbf{F}}}{\partial \tilde{p}_i} + \frac{\partial \tilde{\mathbf{F}}^T}{\partial \tilde{p}_i}\mathbf{P}(t)\right]\Bigg|_{\tilde{\mathbf{p}}=\mathbf{p}}. \tag{7.3-82}$$

Solution of this equation for $\mathbf{\Gamma}_i(t)$ then gives the performance index sensitivity vector since we assumed $W^i[\mathbf{x}(t), \mathbf{p}, t] = \frac{1}{2}\|\mathbf{x}(t)\|^2_{\mathbf{\Gamma}_i(t)}$, $i = 1, 2, \ldots, r$. Example 7.3-2 will illustrate again the steps in this procedure.

As previously mentioned, it has been shown [25] that the open-loop linear regulator gives exactly the same sensitivity, Eq. (7.3-35), at the initial time and initial state as that obtained for the closed-loop case in Eq. (7.3-81). This should not, however, lead us into the conclusion that open-loop systems are as good as closed loop systems. After all, only one definition of sensitivity was used for the demonstration. The inclusion by Siljac and Dorf [29], and Holtzman and Horing [27], of the sensitivity function as part of the cost function, such as to give a cost function which includes sensitivity as an integral part, will alter this result, as will the use of a different sensitivity function. Finally, we must recall that the derivation of the performance index sensitivity was based on infinitesimal parameter vector changes. Performance index sensitivity will normally not be the same for open- and closed-loop systems for large parameter variations.

SENSITIVITY IN THE PERFORMANCE INDEX: Siljak and Dorf [29] note that most applications of sensitivity to optimal control do not use sensitivity as a criterion for determining the optimal control, but determine sensitivity after the optimal control has been synthesized. To avoid this, they suggest the use of a time-domain sensitivity technique and introduce a general index of optimality which includes both sensitivity and performance characteristics.

Thus, the optimal control synthesized satisfies sensitivity and optimality requirements simultaneously.

To include sensitivity in a general index, the usual index of performance $J = \int_{t_o}^{t_f} \phi[\mathbf{x}(t), \mathbf{u}(t), t]\, dt$ is altered to also include the sensitivity functions. The resulting generalized index of optimality is

$$J = \int_{t_o}^{t_f} \phi[\mathbf{x}(t), \mathbf{u}(t), \mathbf{v}_1(t), \mathbf{v}_2(t), \ldots, \mathbf{v}_r(t)]\, dt \tag{7.3-83}$$

where the sensitivity functions are $\mathbf{v}_i = \partial\mathbf{x}/\partial p_i$, and p_i is the ith variable parameter. We should ensure that the sensitivity functions $\mathbf{v}_i$ should appear in the index as squares or magnitudes to avoid the canceling effects of a change in sign. Also, the motivation of a particular physical problem will indicate the usefulness of a weighting function which allows certain sensitivity functions to receive different emphasis. The resulting control law will then yield the $\hat{\mathbf{u}}(t)$ that optimizes both sensitivity and performance. Determination of the nonlinear two-point boundary value problem whose solution minimizes the cost function, which in part includes the sensitivity function, is relatively straightforward using any of the optimization procedures of Chapter 3 or 4. Resolution of the resulting nonlinear two-point boundary value problem may be accomplished by one of the numerical methods in our last chapter.

In addition to means for discerning the sensitivity of the performance index to changes in system parameters, it is also of value to be able to compute the sensitivity of the terminal state of an optimal control system to variation in plant parameters or the control applied to the plant.

Holtzman and Horing [27] use variational techniques to study the sensitivity of the terminal conditions of both open- and closed-loop optimal systems. An important part of their work is the inclusion of sensitivity prior to optimization for the open-loop system. A degradation of the cost function due to the sensitivity inclusion is naturally to be expected. However, the inclusion of sensitivity in the cost function allows the sensitivity of terminal conditions to be prespecified or constrained. As a result of this, it is possible to show that, normally, a closed-loop configuration has sensitivity characteristics superior to those of an equivalent open-loop system.

To illustrate the sensitivity of terminal conditions to parameter variations in optimal control systems, let us consider first the linear closed-loop system

$$\dot{\mathbf{x}} = \mathbf{A}(t)\mathbf{x}(t) + \mathbf{B}(t)\mathbf{u}(t), \qquad \mathbf{x}(t_o) = \mathbf{x}_o, \qquad \mathbf{u}(t) = \mathbf{G}(t)\mathbf{x}(t), \tag{7.3-84}$$

which has been designed so that $\mathbf{x}(t_f) = \mathbf{0}$. To accomplish this, we may write the closed-loop system as

$$\dot{\mathbf{x}} = [\mathbf{A}(t) + \mathbf{B}(t)\mathbf{G}(t)]\mathbf{x}(t), \qquad \mathbf{x}(t_o) = \mathbf{x}_o \tag{7.3-85}$$

and note that, to satisfy the required terminal condition,

$$\mathbf{x}(t_f) = \mathbf{0} = \boldsymbol{\varphi}(t_f, t_o)\mathbf{x}(t_o). \tag{7.3-86}$$

Thus, we see that for nontrivial initial conditions the transition matrix must be equal to the null matrix for any t, $\boldsymbol{\varphi}(t_f, t) = \mathbf{0}$. This can be used to show that elements of the feedback matrix $\mathbf{G}(t)$ must become infinite as t approaches the final time t_f. We have previously noted this effect in our consideration of the linear regulator problem. It can be shown, and it is certainly quite reasonable, that sufficiently small changes in the system coefficient matrix $\mathbf{A}(t)$ or the control distribution matrix $\mathbf{B}(t)$ will not alter the fact that the system states will reach the origin at the terminal time if the feedback matrix $\mathbf{G}(t)$ is invariant. These comments certainly do not apply if a precalculated open-loop control for nominal parameters is used on a system with perturbed parameters in that, normally, we would never expect to reach the origin at the given final time under these circumstances.

EXAMPLE SENSITIVITY CALCULATIONS FOR VARIOUS METHODS: We have discussed a number of different methods for using sensitivity in optimal systems control. Several of the techniques are illustrated in the following example.

Example 7.3-2. We consider a system with dynamics represented by the equation

$$\dot{x} = -ax(t) + u(t), \qquad x(t_o) = x_o \tag{1}$$

where $x(t)$ and $u(t)$ are scalars, and a is a scalar plant parameter. If the system is to be driven to the origin, and the performance index is

$$J = \tfrac{1}{2}\int_{t_o}^{t_f} [x^2(t) + u^2(t)]\,dt, \tag{2}$$

we may compute the open-loop control and trajectory from the maximum principle:

$$\hat{u}_o(t) = \left\{\frac{a \sinh[\sqrt{1+a^2}(t_f - t)] - \sqrt{1+a^2}\cosh[\sqrt{1+a^2}(t_f - t)]}{\sinh\sqrt{1+a^2}(t_f - t_o)}\right\} x(t_o) \tag{3}$$

$$\hat{x}_o(t) = \left\{\frac{\sinh[\sqrt{1+a^2}(t_f - t)]}{\sinh[\sqrt{1+a^2}(t_f - t_o)]}\right\} x(t_o). \tag{4}$$

Similarly, we may obtain the closed-loop control from

$$u(t) = g(t)x(t), \qquad \hat{g}(t) = [a - \sqrt{1+a^2}]\{\coth[\sqrt{1+a^2}(t_f - t)]\}. \tag{5}$$

To simplify the computation, let us assume that $t_o = 0$, and $t_f = \infty$. In this event, we have for the open-loop control and trajectory

$$\begin{aligned} \hat{u}_o(t) &= [a - \sqrt{1+a^2}][\exp(-\sqrt{1+a^2}t)]x(0), \\ \hat{x}_o(t) &= [\exp(-\sqrt{1+a^2}t)]x(0), \end{aligned} \tag{6}$$

and for the closed-loop control

$$u(t) = gx(t), \qquad \hat{g} = a - \sqrt{1+a^2}. \tag{7}$$

We shall now assume that the plant parameter has changed to a value $\tilde{a} = a + \Delta a$. In this case the open-loop control and trajectory for the system $\dot{x} = -\tilde{a}x + u$ are, from Eq. (6),

$$u_o(t) = [a - \sqrt{1 + a^2}] \exp [(-\sqrt{1 + a^2}t)]x(0) \tag{8}$$

$$\tilde{x}_o(t) =$$

$$[\exp (-\tilde{a}t)]x(0) - [\exp (-\tilde{a}t) - \exp (-\sqrt{1 + a^2}t)]\left[\frac{(a - \sqrt{1 + a^2})}{(\tilde{a} - \sqrt{1 + a^2})}\right]x(0). \tag{9}$$

The controller we use for the closed-loop control is $u(t) = g(t)x(t)$ from Eq. (7), which yields the trajectory

$$x_c(t) = \{\exp [(\tilde{a} - a + \sqrt{1 + a^2})t]\} x(0). \tag{10}$$

The performance index, Eq. (2), is evaluated for the closed-loop control as

$$J(a, \tilde{a}) = \left[\frac{1 + a^2 - a\sqrt{1 + a^2}}{2(\tilde{a} - a + \sqrt{1 + a^2})}\right]x^2(0). \tag{11}$$

The performance index sensitivity vector is then computed as

$$\frac{\partial J}{\partial p} = \frac{\partial J}{\partial a} = \frac{1}{2}(\tilde{a} - a)\left[\frac{2a\sqrt{1 + a^2} - 1 - 2a^2}{(\tilde{a} - a + \sqrt{1 + a^2})^2\sqrt{1 + a^2}}\right]x^2(0), \tag{12}$$

and, as we expect, it is zero if $\tilde{a} = a$.

The sensitivity matrix of Cruz and Perkins is obtained from Eq. (7.3-46), where we use the parameters for this example, $A = -a$, $\tilde{A} = -\tilde{a}$, $B = \tilde{B} = 1$, and $F = a - \sqrt{1 + a^2}$, to obtain $S(s) = (s + \tilde{a})/(s + \tilde{a} - a + \sqrt{1 + a^2})$. Since the magnitude of $S(j\omega)$, $|S(j\omega)|$, is always less than unity, for all ω, we conclude that the sensitivity of the closed-loop system to variations in a is always less than that for the open-loop system.

We will now use the relative sensitivity approach of Rohrer and Sobral for this problem. The relative sensitivity, as defined by Eq. (7.3-50), is here (we recall that $J(\tilde{a}, \tilde{a})$ corresponds to the optimum performance index $J[u(t), p]$):

$$S^R(a, \tilde{a}) = \left[\frac{1 + a^2 - a\sqrt{1 + a^2}}{\tilde{a} - a + \sqrt{1 + a^2}}\right]\left[\frac{1 + \tilde{a}^2}{1 + \tilde{a}^2 - \tilde{a}\sqrt{1 + a^2}}\right] - 1. \tag{13}$$

Figure 7.3-5(a) illustrates the relative sensitivity versus $\tilde{a}$ for various values of a. Figure 7.3-5(b) illustrates the corresponding cost function, Eq. (11), $x(0) = 1$, versus $\tilde{a}$ for various values of a. The worth of the relative sensitivity curves for selecting a value of a, such that the cost function does not differ appreciably from the minimum value of the cost function as $\tilde{a}$ is varied, is apparent from this figure. For higher order systems the relative sensitivity curve often exhibits a more pronounced minimum than in this example. In this example the cost function does not exhibit an absolute minimum as parameters are varied. Often, of course, it does exhibit such a minimum.

To determine the performance index sensitivity vector as proposed by Pagurek, we first solve Eq. (7.3-74) which is, for this example,

$$\dot{P} = 2aP(t) + P^2(t) - 1, \qquad P(t_f) = S = \infty. \tag{14}$$

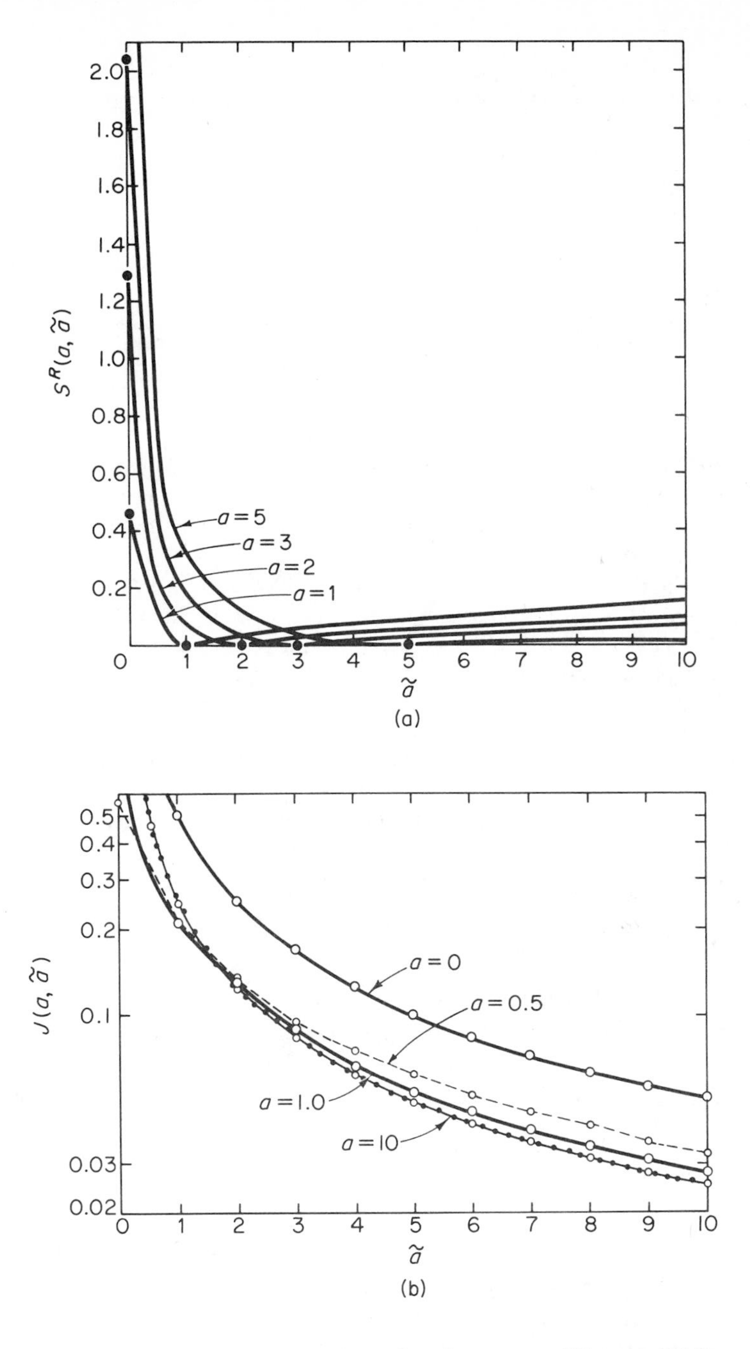

Fig. 7.3-5 Relative sensitivity and cost function curves of Example 7.3-2.

Since we have infinite terminal conditions (we are required to reach the origin at the terminal time), we solve instead the inverse Riccati equation

$$\dot{P}^{-1} = -2aP^{-1}(t) - 1 + P^{-2}(t), \qquad P^{-1}(t_f) = S^{-1} = 0. \tag{15}$$

Having determined $P(t)$, we solve Eq. (7.3-78) which is, for this example,

$$\dot{M} = 2[\tilde{a} + P(t)]M(t) - P^2(t) - 1, \qquad M(t_f) = S. \tag{16}$$

Again we must solve the inverse Riccati equation because of the infinite terminal condition. This is

$$\dot{M}^{-1} = 2[\tilde{a} + P(t)]M^{-1}(t) - P^2(t)M^{-2}(t) - M^{-2}(t), \qquad M^{-1}(t_f) = 0. \tag{17}$$

We finally solve Eq. (7.3-81) which is

$$\dot{\Gamma} = 2[a + P(t)]\Gamma(t) - 2P(t), \qquad \Gamma(t_f) = 0, \tag{18}$$

and then, at long last, we can determine the performance index sensitivity which is $\Gamma(t)x^2(t)/2$. All we have to do now is determine $x(t)$. It seems redundant for us to remark that this represents a lot of effort for even a first-order system.

For the infinite time interval case, we have $P = -a + \sqrt{1 + a^2}$, $M = (1 - a\sqrt{1 + a^2} + a^2)/(\tilde{a} - a + \sqrt{1 + a^2})$, and finally $\Gamma = (-a + \sqrt{1 + a^2})/\sqrt{1 + a^2}$. Thus we see that the performance index $\frac{1}{2}Mx^2(t_o)$ is precisely the same as that calculated before in Eq. (11). The Pagurek sensitivity index is not the same as that in Eq. (12) since in this case $\frac{1}{2}\Gamma x^2(t_o)$ is $-\partial J(a, \tilde{a})/\partial \tilde{a}$ evaluated at $\tilde{a} = a$, whereas in the previous case, it was $\partial J(a, \tilde{a})/\partial a$. Thus we are again faced with slightly different definitions of sensitivity.

To illustrate the effect of changing system parameters upon the terminal state, we will consider that the closed-loop control is determined from Eq. (5), whereas the actual system is perturbed to give

$$\dot{x} = -\tilde{a}x(t) + u(t), \qquad u(t) = [a - \sqrt{1 + a^2}]\{\coth[\sqrt{1 + a^2}(t_f - t)]\}x(t) \tag{19}$$

such that the complete closed-loop system is described by

$$\dot{x} = \{-\tilde{a} + [a - \sqrt{1 + a^2}]\coth[\sqrt{1 + a^2}(t_f - t)]\}x(t) = \alpha(t)x(t). \tag{20}$$

This equation has the solution $x(t) = \left\{\exp\left[\int_0^t \alpha(t)\,dt\right]\right\}x(0)$ and can be written as

$$x(t) = \left\{\frac{1 - \exp[(\tilde{a} - a + \sqrt{1 + a^2})(t_f - t)]}{1 - \exp[(\tilde{a} - a + \sqrt{1 + a^2})t_f]}\right\} \cdot \{\exp[(\tilde{a} - a + \sqrt{1 + a^2})t]\}x(0). \tag{21}$$

Thus we see that we do get the required final value, $x(t_f) = 0$, regardless of the value of the actual system parameter a. If we had used the open-loop control calculated for $A = -a$ and applied it to the system where $A = -\tilde{a}$, we would certainly not get $x(t_f) = 0$ unless $\tilde{a} = a$. Thus we have again demonstrated the value of the feedback solution.

SENSITIVITY TO CONTROL VARIATIONS: The sensitivity to variations in the plant parameters has been the object of much of the work on sensitivity in optimal control. However, Belanger [31] has investigated the effects of variations in the control. This is very useful in suboptimal control, for example, and also in studies of the sensitivity of the computation of the desired control.

Belanger considers both weak and strong variations in the control. For the weak or *continual* variation, the actual control differs from the desired control by an infinitesimal amount, $\boldsymbol{\delta}\mathbf{u}(t)$. Tolerances on the control can be set by limiting $\boldsymbol{\delta}\mathbf{u}(t)$. The strong or *intermittent* variation causes the actual control to differ from the desired control by large amounts, but only during infinitesimal intervals of time. The continual variation is applicable when the control is continuous, and the intermittent variation is useful for bang bang control. Here we will consider only the former case.

There are two effects of a variation in control. One result of a variation in the control would be the failure to hit a desired target or terminal state. The other effect would be variations in the value of the performance index. Both effects can be considered for the case in which the actual target has been replaced by an ideal target. This is useful since control tolerances necessary to hit the actual target can be determined and variation in cost calculated. For example, if the target were a small sphere, the control to hit a point at its center would be calculated, and then we would determine tolerances for this control such that we ensure that the sphere will always be reached. In all cases, we will find that the changes in cost function for the control variation assumed here are second-order, since the first-order effects are identically zero as a result of the necessary conditions.

The problem is as follows. We wish to transfer the system

$$\dot{\mathbf{x}} = \mathbf{f}[\mathbf{x}(t), \mathbf{u}(t), t], \qquad \mathbf{x}(t_o) = \mathbf{x}_o \tag{7.3-87}$$

to a terminal manifold or target set

$$\mathbf{N}[\mathbf{x}(t_f), t_f] = \mathbf{0} \tag{7.3-88}$$

where it is assumed that t_f is free. If t_f is specified, we may convert the problem to an unspecified terminal time problem by adding the equation $\dot{x}_{n+1} = 1$, $x_{n+1}(t_o) = t_o$, to the plant equations, and the relation $x_{n+1}(t_f) = t_f - t_o$ to the terminal manifold $\mathbf{N}$. A cost function is then postulated and the optimal control $\hat{\mathbf{u}}(t)$ determined. A tolerance is then placed on the size of the control deviation $\boldsymbol{\delta}\hat{\mathbf{u}}(t) = \hat{\mathbf{u}}(t) - \tilde{\mathbf{u}}(t)$. We desire to determine whether the terminal manifold can be reached using any control within the specified tolerance.

For a continuous control deviation from the optimal, we have for the state variable deviation $\boldsymbol{\delta}\hat{\mathbf{x}}(t) = \hat{\mathbf{x}}(t) - \tilde{\mathbf{x}}(t)$,

$$\boldsymbol{\delta}\dot{\hat{\mathbf{x}}} = -\frac{\partial \mathbf{f}[\hat{\mathbf{x}}(t), \hat{\mathbf{u}}(t), t]}{\partial \hat{\mathbf{x}}}\boldsymbol{\delta}\hat{\mathbf{x}}(t) - \frac{\partial \mathbf{f}[\hat{\mathbf{x}}(t), \hat{\mathbf{u}}(t)]}{\partial \hat{\mathbf{u}}}\boldsymbol{\delta}\hat{\mathbf{u}}(t), \qquad \boldsymbol{\delta}\hat{\mathbf{x}}(t_o) = 0. \tag{7.3-89}$$

If we perturb $\boldsymbol{\delta}\hat{\mathbf{u}}(t)$, we expect to change not only the trajectory, but also the time at which we arrive at the terminal manifold, since t_f is free. From Eq. (7.3-87) we have

$$\boldsymbol{\delta}\hat{\mathbf{x}}(\hat{t}_f + \delta t_f) = \boldsymbol{\delta}\hat{\mathbf{x}}(\hat{t}_f) + \mathbf{f}[\hat{\mathbf{x}}(\hat{t}_f), \hat{\mathbf{u}}(\hat{t}_f), \hat{t}_f]\,\delta t_f. \tag{7.3-90}$$

If $\tilde{\mathbf{x}}(\hat{t}_f + \delta t_f)$ is to be on the terminal manifold, we have for sufficiently small δt_f the approximation from Eq. (7.3-88) that

$$\left\{\frac{\partial \mathbf{N}[\hat{\mathbf{x}}(\hat{t}_f), \hat{t}_f]}{\partial \hat{\mathbf{x}}(\hat{t}_f)}\right\} \boldsymbol{\delta}\hat{\mathbf{x}}(\hat{t}_f + \delta t_f) = \mathbf{0}. \tag{7.3-91}$$

By combining Eqs. (7.3-90) and (7.3-91) we have

$$\left\{\frac{\partial \mathbf{N}[\hat{\mathbf{x}}(\hat{t}_f), \hat{t}_f]}{\partial \hat{\mathbf{x}}(\hat{t}_f)}\right\} \{\boldsymbol{\delta}\hat{\mathbf{x}}(\hat{t}_f) + \mathbf{f}[\hat{\mathbf{x}}(\hat{t}_f), \hat{\mathbf{u}}(\hat{t}_f), \hat{t}_f]\,\delta t_f\} = \mathbf{0}. \tag{7.3-92}$$

Generally, as we have seen in Chapter 3, the terminal manifold is such that we can obtain a relation between $\boldsymbol{\delta}\hat{\mathbf{x}}(\hat{t}_f)$ and δt_f, which allows us to determine the requirements for the terminal state to be on the terminal manifold.

We now inquire about the tolerance which must be imposed on the physical implementation of the control $\hat{\mathbf{u}}(t)$ such that all controls within tolerance take the system close to the terminal manifold, given that $\hat{\mathbf{u}}(t)$ takes the system there at time $\hat{t}_f$ starting at $\mathbf{x}(t_o)$. It is convenient to consider an "error" terminal manifold which is one-dimensional:

$$h[\boldsymbol{\delta}\hat{\mathbf{x}}(\hat{t}_f)] = 0. \tag{7.3-93}$$

This expression relates the ideal terminal manifold to an acceptable terminal manifold. Since the control variations $\boldsymbol{\delta}\hat{\mathbf{u}}(t)$ are small, this expression must be valid only in the vicinity of $\hat{\mathbf{x}}(\hat{t}_f)$. If $\tilde{\mathbf{x}}(\hat{t}_f + \delta t_f)$ is within this manifold, Eq. (7.3-93) is then approximately

$$h[\tilde{\mathbf{x}}(\hat{t}_f + \delta t_f) - \hat{\mathbf{x}}(\hat{t}_f)] = 0 = h\{\boldsymbol{\delta}\hat{\mathbf{x}}(\hat{t}_f) + \mathbf{f}[\hat{\mathbf{x}}(\hat{t}_f), \hat{\mathbf{u}}(\hat{t}_f), \hat{t}_f]\delta t_f\}. \tag{7.3-94}$$

It is this relation which yields $\boldsymbol{\delta}\hat{\mathbf{x}}(\hat{t}_f)$ in terms of δt_f.

Example 7.3-3. Let us consider the system $\dot{x}_1 = x_2(t)$, $\dot{x}_2 = u(t)$, $x_1(0) = 10$, $x_2(0) = 0$. The problem is to drive the system to the origin in one second. A control which accomplishes this is $u(t) = 120t - 60$. We will let h be the manifold $\delta\hat{x}_1^2(\hat{t}_f) + \delta\hat{x}_2^2(\hat{t}_f) = R^2$ and let $\tilde{t}_f = 1 + \delta t_f$ where δt_f is small. Thus we are saying that we will accept the system's performance if the norm of the terminal state is R units away from the origin. We now compute $\dot{\mathbf{x}}^T(\hat{t}_f) = [x_2(1), u(1)] = [0, 60]$ and see that $\boldsymbol{\delta}\hat{\mathbf{x}}(t_f)$ must lie in the set described by $|\delta\hat{x}_1(1)| \leq R$ since $\dot{\mathbf{x}}(t_f)$ is directed entirely in the x_2 direction, and since errors in the x_1 direction at time t_f cannot be corrected in time δt_f.

For the linearized system $\delta\dot{\hat{x}}_1 = \delta\hat{x}_2(t)$, $\delta\dot{\hat{x}}_2 = u(t)$, $\delta\hat{x}_1(0) = \delta\hat{x}_2(0) = 0$, we can use the maximum principle to compute the maximum $\delta\hat{x}_1(1)$ if the control is restricted such that $|\delta\hat{u}(t)| \leq K$. This turns out to be $|x_1(1)| \leq 0.5K$. We then use the fact that $|\delta\hat{x}_1(1)| \leq R$ to obtain $K \leq 2R$, and we thus conclude that the tolerance on the control must be $|\delta\hat{u}(t)| \leq 2R$ if we are to ensure the possibility of arrival on the terminal manifold $x_1^2 + x_2^2 = R$ at the terminal time. By a procedure similar to this, we can find the cost variation for the optimal trajectory. This we will leave as a problem for the interested reader who can easily show that the open-loop control given for this problem is the optimal open-loop control for the cost function $J = \frac{1}{2}\int_0^{t_f} u^2(t)\,dt$, where the terminal time is forced to be one second by the addition of the plant equation $\dot{x}_3 = 1$, $x_3(0) = 0$, and by the addition of the term $x_3(t_f) = 1$ to the terminal manifold which is just the origin of the x_1x_2 space.

7.4 Stability

We now conclude this chapter with a discussion of stability. The approach taken will be that of the second method of Lyapunov [38]. Lyapunov's first method is strictly concerned with local stability issues, where stable and unstable cases are separated on the basis of the stability or instability of linear approximations around equilibrium points. Further discussion of the first method can be found in [39, 40, 41]. The second method allows consideration of asymptotic stability in-the-large and global asymptotic stability as well as local stability. Other approaches to in-the-large and global asymptotic stability, such as the method of Popov and fixed-point theorems, are presented in [39, 40, 41]. An important advantage of Lyapunov's second method is that it allows the determination of the stability of a system without explicitly solving the system equations. We now begin our discussion of stability by stating several preliminary definitions.

Let

$$\dot{\mathbf{x}}(t) = \mathbf{f}[\mathbf{x}(t)] \tag{7.4-1}$$

describe an nth order autonomous system. The state $\mathbf{x}_e$ is said to be an *equilibrium point* if

$$\mathbf{0} = \mathbf{f}(\mathbf{x}_e);$$

that is, $\mathbf{x}_e$ is an equilibrium point for the system described by Eq. (7.4-1) if the system will not move away from $\mathbf{x}_e$ in the absence of a forcing function. The stability of a system is typically discussed relative to an equilibrium point. A system is said to be (Lyapunov) *stable* at the equilibrium point $\mathbf{x}_e$ if for any given $\epsilon > 0$ there exists a $\delta > 0$ such that $\|\mathbf{x}_o - \mathbf{x}_e\| < \delta$ implies that the solution of Eq. (7.4-1) with respect to initial condition $\mathbf{x}_o$, $\mathbf{x}(\cdot, \mathbf{x}_o)$, satisfies $\|\mathbf{x}(t, \mathbf{x}_o) - \mathbf{x}_e\| < \epsilon$, for all $t > t_o$, where $\mathbf{x}(t_o, \mathbf{x}_o) = \mathbf{x}_o$. That is, a system is stable with respect to the equilibrium point $\mathbf{x}_e$ if for any given ϵ-neighborhood of $\mathbf{x}_e$ there is a δ-neighborhood of $\mathbf{x}_e$ such that if the system is initially in the ϵ-neighborhood it will remain within the δ-neighborhood for all time. Similarly, $\mathbf{x}_e$ is *unstable* if there is an $\epsilon > 0$ such that there cannot be found a $\delta > 0$ which satisfies the conditions of the stability definition. An equilibrium point $\mathbf{x}_e$ is called *asymptotically stable* if $\mathbf{x}_e$ is stable and if there exists a δ such that if $\mathbf{x}_o$ is a member of a δ-neighborhood of $\mathbf{x}_e$, i.e., if $\mathbf{x}_o$ satisfies $||\mathbf{x}_o - \mathbf{x}_e|| < \delta$, then $\mathbf{x}(t, \mathbf{x}_o)$ converges to $\mathbf{x}_e$ as $t \longrightarrow \infty$. We say that an equilibrium point $\mathbf{x}_e$ of the system Eq. (7.4-1) is *asymptotically stable in-the-large* with respect to the bounded region $B \subset R^n$ if $\mathbf{x}_e$ is stable, and if every trajectory of the system having initial condition in B tends to $\mathbf{x}_e$ as $t \longrightarrow \infty$. When $B = R^n$, asymptotic stability in-the-large is called *global asymptotic stability*.

We now motivate our next definition with several remarks. The second method of Lyapunov is based on the simple intuitive fact that if the total energy of an autonomous physical system is monotonically decreasing, then the state of the system will tend toward an equilibrium state. This fact holds true because total energy, a nonnegative function of the state of the system, is minimum when the motion of the system ceases, which indicates that a stable equilibrium point has been achieved. In actuality, only the nonnegativity of the total energy function is of use in this argument; other nonnegative functions of the system state can also be useful in determining the stability or instability of a system. Our next definition characterizes such functions.

Let $V: R^n \longrightarrow R$ be continuous and have continuous first partial derivatives in a b-neighborhood $B, b > 0$, centered at the origin; i.e., $B = \{\mathbf{x} \in R^n : ||\mathbf{x}|| < b\}$. Assume V is positive definite in B; i.e., $V(0) = 0$ and $V(\mathbf{x}) > 0$ for all $\mathbf{x} \in B, \mathbf{x} \neq 0$. For simplicity, assume the coordinate system has been translated so that the origin is now the equilibrium point of Eq. (7.4-1) subject to investigation; that is, $\mathbf{f}(\mathbf{0}) = \mathbf{0}$. Assume that the derivative of V with respect to time, relative to Eq. (7.4-1) is negative semidefinite in B, i.e., $dV(\mathbf{x})/dt \leq \mathbf{0}$ for $\mathbf{x} \in B$, where

$$\frac{dV(\mathbf{x})}{dt} = \frac{\partial V}{\partial \mathbf{x}_1}\frac{d\mathbf{x}_1}{dt} + \cdots + \frac{\partial V}{\partial \mathbf{x}_n}\frac{d\mathbf{x}_n}{dt} = \left(\frac{\partial V}{\partial \mathbf{x}}\right)^T \mathbf{f}(\mathbf{x}).$$

A function V satisfying the above three properties is called a *Lyapunov function*. The results that follow in this section will be dependent on the existence or lack thereof of such a function.

7.4-1 Stability in-the-small

We now present four theorems due to Lyapunov which are sufficient conditions for local stability and instability for the system of Eq. (7.4-1). The proofs for these theorems are straightforward and are left as exercises.

Theorem 1 (First stability theorem of Lyapunov)
If a Lyapunov function can be found in some open neighborhood of the origin, then the origin is stable.

A stronger result is as follows:

Theorem 2 (Second stability theorem of Lyapunov)
Let V be a Lyapunov function in B, an open neighborhood of the origin. Furthermore, assume $dV/d\mathbf{x}$ is negative definite in B; i.e., $dV/d\mathbf{x} \leq \mathbf{0}$ for $\mathbf{x} \in B$ and $dV/d\mathbf{x} < \mathbf{0}$ for all $\mathbf{x} \in B, \mathbf{x} \neq \mathbf{0}$. Then, the origin is asymptotically stable.

Thus, the negative definite assumption on $dV/d\mathbf{x}$ in B implies that not only will the solution of Eq. (7.4-1) not venture too far away from the origin for reasonable initial conditions, but also will asymptotically approach the origin in the limit.

A significant drawback to the Lyapunov second method approach to system stability is that there is not a general method which easily allows us to determine whether or not a given function $V: R^n \longrightarrow R$ is Lyapunov. This observation, coupled with the fact that the above two theorems are only sufficient conditions, implies that if a desired Lyapunov function cannot be found, the system is not necessarily unstable; we, perhaps, have not been fortunate enough to select a satisfactory function. Fortunately, we can give sufficient conditions for local instability:

Theorem 3 (The first instability theorem of Lyapunov)
Let B be an open neighborhood of the origin. Assume there exists a continuous function $U: R^n \longrightarrow R$ with continuous first partial derivatives in B such that $U(\mathbf{0}) = \mathbf{0}$ and dU/dt is positive definite, but that U is neither negative definite nor negative semidefinite arbitrarily close to the origin. Then, the origin is unstable.

Another sufficient condition for instability is the following theorem:

Theorem 4 (The second instability theorem of Lyapunov)
Let B be an open neighborhood of the origin. Assume there exists a continuous function $U: R^n \longrightarrow R$ in B such that $U(\mathbf{0}) = 0$ and

$$\frac{dU(\mathbf{x})}{dt} = \alpha U(\mathbf{x}) + \omega(\mathbf{x}),$$

where $\alpha > 0$ and $\omega(\mathbf{x})$ is nonnegative definite on B. Assume further that U is neither negative definite nor negative semidefinite arbitrarily close to the origin. Then, the origin is unstable.

Example 7.4-1. The differential equation representing the equation of motion of a simple pendulum of mass M and length L is

$$\frac{d^2\theta}{dt^2} + \frac{Mg}{L} \sin \theta = 0$$

where g is the gravitational constant. We use the total energy of the system as the candidate V function by letting

$$V(\theta, \dot{\theta}) = \frac{1}{2}\dot{\theta}^2 + \frac{Mg}{L}(1 - \cos\theta).$$

Energy in a physical passive system must be positive definite, and $V(\theta, \dot{\theta})$ is positive definite (for $|\theta| < 2\pi$) as we expect. The time derivative is

$$\frac{dV(\theta, \dot{\theta})}{dt} = \dot{\theta}\ddot{\theta} + \frac{Mg}{L}\dot{\theta}\sin\theta = 0,$$

so $V(\mathbf{x})$ is a Lyapunov function and the origin is stable by Lyapunov's first stability theorem. We can determine that a limit cycle does exist for this simple pendulum. Addition of a damping arm to the differential equation for the pendulum will result in a negative semidefinite $dV(\mathbf{x})/dt$ for the same $V(\mathbf{x})$ as with no damping. The system then becomes asymptotically stable.

Example 7.4-2. Let us consider the first-order nonlinear system with a soft spring

$$\frac{dx}{dt} = -x + ax^3, a > 0$$

which has equilibrium points

$$x = 0, \pm a^{-1/2}.$$

We pick as a candidate $V(x)$ function

$$V(x) = \tfrac{1}{2}x^2,$$

so that

$$\frac{dV(x)}{dt} = x\dot{x} = -x^2 + ax^4.$$

The $V(x)$ function is always positive definite. The $\dot{V}$ function is negative definite for $ax^2 < 1$, and this is certainly true for small enough x. Thus the origin is asymptotically stable.

To examine behavior about the equilibrium point $x = a^{-1/2}$ we let

$$V(x) = \tfrac{1}{2}(x - a^{-1/2})^2$$

and get

$$\frac{dV(x)}{dt} = ax(x + a^{-1/2})(x - a^{-1/2})^2;$$

$\dot{V}(x)$ is positive in the vicinity of the equilibrium at $x = a^{-1/2}$. Thus we cannot conclude stability about $x = a^{-1/2}$. If we let

$$V(x) = -\frac{x^2}{2} + a\frac{x^4}{4} + \frac{1}{4a} = a\left[\frac{x^2}{2} - \frac{1}{2a}\right]^2$$

then

$$\frac{dV(x)}{dt} = (-x + ax^3)^2.$$

Now $\dot{V}(x)$ is certainly positive definite as is $V(x)$. By Lyapunov's first instability theorem the equilibrium point $x = a^{-1/2}$ must be unstable.

7.4-2
Stability in-the-large

Although the second method of Lyapunov is quite useful for determination of local stability, its primary advantage is in determining in-the-large and global asymptotic stability [42]. The following theorem gives sufficient conditions for a system to be asymptotically stable in-the-large.

Theorem 5
Let B be an open neighborhood of the origin that is bounded by k, i.e., $\mathbf{x} \in B$ implies $||\mathbf{x}|| \leq k$. Assume V is such that $V(\mathbf{x}) < k$ for all $\mathbf{x} \in R^n$. Further, assume that in B, V is a Lyapunov function with nonnegative definite time derivative dV/dt. Then, the origin is asymptotically stable in-the-large with respect to B.

When $B = R^n$, the system is globally asymptotically stable, and the following result can be given.

Theorem 6
Let V be a Lyapunov function such that dV/dt is negative definite on R^n. Further, assume $V(\mathbf{x}) \longrightarrow \infty$ as $||\mathbf{x}|| \longrightarrow \infty$. Then, the origin is globally asymptotically stable.

Example 7.4-3. Let us examine stability requirements for the system

$$\frac{dx}{dt} = -x - x^3$$

which has a single (real) equilibrium point at $x = 0$. If we let $V(x) = x^2$, then we easily determine that $\dot{V}(x) = -2x^2 - 2x^4$. Since $V(x)$ is positive definite, and $\dot{V}(x)$ is negative definite everywhere except at the equilibrium point, we conclude that the system is globally asymptotically stable.

Example 7.4-4. We now consider stability determination for the system

$$\frac{dx}{dt} = -x + x^3$$

which has equilibrium points at $x = 0, \pm 1$. We have seen from Example (7.4-2) that the origin is stable in-the-small whereas the equilibrium points at ± 1 are unstable. Now we wish to determine stability in-the-large properties of the system about the equilibrium point at the origin.

We first try the function $V(x) = x^2$ which leads to $\dot{V}(x) = -2x^2 + 2x^4$. Although $V(x)$ is certainly positive definite for all $x \neq 0$, $\dot{V}(x)$ is negative definite only for $|x| < 1$. There the sign of $\dot{V}(x)$ is negative such that $V(x)$ must decrease. The system is, therefore, stable for initial condition $x(t_o) < 1$.

We must generally be careful to determine stability regions according to the character of $\dot{V}(x)$ and $V(x)$. A simple example will illustrate the point. Suppose we use $V(x) = x^2 - 0.5x^4$ which results in $\dot{V}(x) = -2(-x + x^3)^2$.

Now $\dot{V}(x)$ is always negative so $V(x)$ must be monotone nonincreasing. Does $V(x)$, therefore, decrease to zero? In particular, does $V(x)$ decrease to zero for x such that $V(x) > 0$ or $|x| < \sqrt{2}$? The answer is no. For $|x| > 1$, the gradient of $V(x)$, $dV(x)/dx$, is $2x - 2x^3$, and for $|x| > 1$, this is negative. Now if $1 < |x| < \sqrt{2}$, the expression $\dot{V}(x)$ is negative, so $V(x)$ should decrease. Since $dV(x)/dx$ is negative for $|x| > 1$, the only way that $V(x)$ can decrease is for x to increase. This can and does happen with our particular $V(x)$ function.

This particular $V(x)$ function $V(x) = x^2 - 0.5x^4$ is everywhere less than a constant ϵ (here $\epsilon = 0.5$) only for $|x| < 1$. The true bounded region for this problem is thus $|x| < 1$.

7.4-3
Stability of linear systems

As is often true for other problems, a linearity assumption implies significantly stronger results for stability. In fact the linearity assumption will lead us to a necessary and sufficient condition for global asymptotic stability.

Consider the linear version of Eq. (7.4-1),

$$\dot{\mathbf{x}}(t) = \mathbf{A}\mathbf{x}(t). \tag{7.4-2}$$

Suppose we postulate a quadratic Lyapunov function

$$V(\mathbf{x}) = \mathbf{x}^T\mathbf{P}\mathbf{x}, \tag{7.4-3}$$

where $\mathbf{P}$ is symmetric and positive definite. Then, we easily obtain

$$\frac{dV(\mathbf{x})}{dt} = \left(\frac{\partial V(\mathbf{x})}{\partial \mathbf{x}}\right)^T \frac{d\mathbf{x}}{dt} = \mathbf{x}^T(\mathbf{A}^T\mathbf{P} + \mathbf{P}\mathbf{A})\mathbf{x}. \tag{7.4-4}$$

Let

$$\mathbf{Q} = -\mathbf{A}^T\mathbf{P} - \mathbf{P}\mathbf{A}, \tag{7.4-5}$$

which must also be a symmetric matrix. If $\mathbf{Q}$ is positive definite, it then follows from Theorem 6 that the system is clearly globally asymptotically stable.

For given matrices $\mathbf{A}$ and $\mathbf{P}$, $\mathbf{Q}$ is easily determined from Eq. (7.4-5). The reverse approach of assuming $\mathbf{Q}$ is positive definite and then inferring the properties of $\mathbf{P}$ which satisfy Eq. (7.4-5) is often quite useful. For example, if, for given positive definite $\mathbf{Q}$ ($\mathbf{Q} = \mathbf{I}$, for instance), the solution of Eq. (7.4-5) is also positive definite, then the linearity of the system and Theorem 6 imply that the system is globally asymptotically stable. On the other hand, if $\mathbf{Q}$ is assumed positive definite, but the solution of Eq. (7.4-5) is neither negative definite nor negative semidefinite, then the origin is unstable. Clearly, the characteristics of the solution of Eq. (7.4-5) for given $\mathbf{Q}$ are dependent on the properties of the $\mathbf{A}$ matrix. It can be shown that $\mathbf{P}$ and $\mathbf{Q}$ are positive definite if and only if $\mathbf{A}$ is stable; i.e., $\mathbf{A}$ has eigenvalues with no positive real parts. The following theorems result:

Theorem 7
The origin of the system $\dot{\mathbf{x}}(t) = \mathbf{A}\mathbf{x}(t)$ is globally asymptotically stable if and only if for any given positive definite, symmetric matrix $\mathbf{Q}$, there exists a positive definite, symmetric matrix $\mathbf{P}$ which satisfies Eq. (7.4-5).

Theorem 8
Assume that the origin is stable for $\dot{\mathbf{x}}(t) = \mathbf{A}\mathbf{x}(t)$. Then, there exists a unique Lyapunov function $V(\mathbf{x}) = \mathbf{x}^T\mathbf{P}\mathbf{x}$, where $\mathbf{P}$ is the solution of Eq. (7.4-5) for any positive definite, symmetric matrix $\mathbf{Q}$.

The stability of Eq. (7.4-2) can also be determined by the Routh-Hurwitz criterion. In fact we can derive the Routh-Hurwitz stability test in a relatively simple manner using the two previous theorems. In terms of computational effort, solution of Eq. (7.4-5) is roughly equivalent to determination of the eigenvalues of $\mathbf{A}$ and the associated eigenvectors. It might seem that this would be a preferred approach since we would then have the transient response available as well as knowledge of system stability. Solution of Eq. (7.4-5) gives us considerable insight [39, 40] into system properties so we should avoid this conclusion.

Let us examine some of these properties. For the linear system

$$\dot{\mathbf{x}}(t) = \mathbf{A}(t)\mathbf{x}(t), \qquad \mathbf{x}(0) = \mathbf{x}_o, \tag{7.4-6}$$

it turns out that solution of Eq. (7.4-5) also evaluates the integral

$$\int_0^\infty \mathbf{x}^T(t)\mathbf{Q}\mathbf{x}(t)\,dt = \mathbf{x}_0^T\mathbf{P}\mathbf{x}_o.$$

If $\mathbf{x}(t) = \mathbf{0}$ is a desired trajectory, then the integral serves to indicate the departure of $\mathbf{x}(t)$ from this desired trajectory. Thus $\mathbf{P}$ and $\mathbf{Q}$ serve as a performance measure. If an actual closed-loop system is described by

$$\dot{\mathbf{x}}(t) = \mathbf{F}\mathbf{x}(t) + \mathbf{B}\mathbf{u}(t)$$
$$\mathbf{u}(t) = \mathbf{K}\mathbf{x}(t),$$

then the solution of Eq. (7.4-5) with

$$\mathbf{A} = \mathbf{F} + \mathbf{B}\mathbf{K}$$

can be used to evaluate the performance of the feedback control law.

In this section we have indicated some system properties associated with stability. Our discussion is necessarily incomplete but hopefully adequate in supporting our development of optimum system concepts.

References

1. KALMAN, R.E., and BERTRAM, J.E., "A Unified Approach to the Theory of Sampling Systems." *J. Franklin Institute*, (1959), 405–25.

2. KALMAN, R.E., HO, Y.C., and NARENDRA, K.S., "Controllability of Linear Dynamical Systems." *Contributions to Differential Equations*, **1**, no. *2*, (1962), 189–213.

3. KALMAN, R.E., "Canonical Structure of Linear Dynamical Systems." *Proceedings Natl. Acad. Science*, **48**, no. *4*, (April 1962), 596–600.

4. KALMAN, R.E., "Mathematical Description of Linear Dynamical Systems." *J. Soc. Ind. Appl. Math.—Control Ser.*, Ser. A, **1**, no. *2*, (1963), 152–92.

5. KREINDLER, E., and SARACHIK, P.E., "On the Concepts of Controllability and Observability of Linear Systems." *IEEE Trans. Autom. Control*, **AC-9**, no. *2*, (1964), 129–36.

6. LEE, E.B., "On the Domain of Controllability for Linear Systems." *IRE Trans. Autom. Control*, **AC-8**, no. *2*, (April 1963), 172–73.

7. GILBERT, E.G., "Controllability and Observability in Multivariable Control Systems." *J. Soc. Ind. Appl. Math—Control Ser.*, Ser. A, **1**, no. *2*, (1963), 128–51.

8. POLAK, E., and WONG, E., *Notes for a First Course on Linear Systems.* Van Nostrand, New York, 1970.

9. DESOER, C.A., *Notes for a Second Course on Linear Systems.* Van Nostrand, New York, 1970.

10. BROCKETT, R.W., *Finite Dimensional Linear Systems.* Wiley, New York, 1970.

11. LEE, E.B., and MARKUS, L., *Foundations of Optimal Control Theory.* Wiley, New York, 1967.

12. NAYLOR, A.W., and SELL, G.R., *Linear Operator Theory in Engineering and Science.* Holt, New York, 1971.

13. CHOATE, W.C., and SAGE, A.P., "Operator Algebra for Differential Systems." *IEEE Trans. on System Science and Cybernetics*, **SSC-3**, no. *2*, (November 1967), 137–47.

14. BODE, W.H., *Network Analysis and Feedback Amplifier Design.* Van Nostrand, New York, 1945.

15. LINDORFF, D.P., "Sensitivity in Sampled Data Systems." *IRE Trans. Autom. Control*, **AC-8**, (April 1963), 120–24.

16. TOMOVIC, R., *Sensitivity Analysis of Dynamic Systems.* McGraw-Hill Book Co., New York, 1963.

17. TOMOVIC, R., "Modern Sensitivity Analysis." *IEEE Convention Record*, Part 6, (March 1965), 81–86.

18. HOROWITZ, I.M., *Synthesis of Feedback Systems.* Academic Press, New York, 1963.

19. VANNESS, J.E., BOYLE, J.M., and IMAD, F.P., "Sensitivities of Large, Multiple-loop Control Systems." *IEEE Trans. Autom. Control*, **AC-10**, (July 1965), 308–14.

20. SILJAK, D.D., and STOJIC, M.R., "Sensitivity Analysis of Self-Excited Nonlinear Oscillations." *IEEE Trans. Autom. Control*, **AC-10**, (October 1965), 413–20.

21. DORATO, P., "On Sensitivity in Optimal Control Systems." *IEEE Trans. Autom. Control*, **AC-8**, (July 1963), 256–57.

22. CRUZ, J.B., JR., and PERKINS, W.R., "A New Approach to the Sensitivity Problem in Multivariable Feedback Systems Design." *IEEE Trans. Autom. Control*, **AC-9**, (July 1964), 216–23.

23. ROHRER, R.A., and SOBRAL, JR., M., "Sensitivity Considerations in Optimal System Design." *IEEE Trans. Autom. Control*, **AC-10**, (January 1965), 45–48.

24. MACKO, D.S., and MESAROVIC, M., "Uncertainties and Optimal Control Approach to Feedback Control Problems." *Information and Control*, **8**, (1965), 468–92.

25. PAGUREK, P., "Sensitivity of the Performance of Optimal Control Systems to Plant Parameter Variations." *IEEE Trans. Autom. Control*, **AC-10**, (April 1965), 178–80.

26. DORATO, P., and KESTENBAUM, A., "Application of Game Theory to the Sensitivity Design of Optimal Systems." *IEEE Trans. Autom. Control*, **AC-12**, (February 1967), 85–87.

27. HOLTZMAN, J.M., and HORING, S., "The Sensitivity of Terminal Conditions of Optimal Control Systems to Parameter Variations." *IEEE Trans. Autom. Control*, **AC-10**, (October 1965), 420–26.

28. WITSENHAUSEN, H.S., "On the Sensitivity of Optimal Control Systems." *IEEE Trans. Autom. Control*, **AC-10**, (October 1965), 495–96.

29. SILJAK, D.D., and DORF, R.C., "On the Minimization of Sensitivity in Optimal Control Systems." *Proceedings 3rd Allerton Conf. on Circuit and System Theory*, University of Illinois, (October 1965), 225–29.

30. DORF, R.C., "System Sensitivity in the Time Domain." *Proceedings 3rd Allerton Conf. on Circuit and System Theory*, University of Illinois, (October 1965), 46–62.

31. BELANGER, P.R., "Some Aspects of Control Tolerances and First-Order Sensitivity in Optimal Control Systems." *IEEE Trans. Autom. Control*, **AC-11**, (January 1966), 77–83.

32. DORF, R.C., FARREN, M.C., and PHILLIPS, C.A., "Adaptive Sampling for Sampled-Data Control Systems." *IRE Trans. Autom. Control*, **AC-7**, (January 1962), 38–47.

33. TOMOVIC, R., and BEKEY, G.A., "Adaptive Sampling Based on Amplitude Sensitivity." *IEEE Trans. Autom. Control*, **AC-11**, (April 1966), 282–84.

34. BEKEY, G.A., and TOMOVIC, R., "Sensitivity of Discrete Systems to Variation of Sampling Interval." *IEEE Trans. Autom. Control*, **AC-11**, (April 1966), 284–87.

35. BENNETT, A.W., and SAGE, A.P., "Discrete System Parameter and Sampling

Interval Sensitivity." *Proceedings First Princeton Conf. on Infor. Sciences and Systems*, Princeton, New Jersey, (March 1967), 6–11.

36. BENNETT, A.W., and SAGE, A.P., "Discrete System Sensitivity and Variable Increment Sampling." *Proceedings Joint Autom. Control Conf.* (June 1967), 603–12.

37. CRUZ, J.B., JR., ed., *System Sensitivity Analysis.* Dowden, Stroudsburg, Pennsylvania, 1973.

38. LYAPUNOV, A.M., *Stability of Motion*, (English translation). Academic Press, New York, 1957.

39. ANDERSON, B.D.O., and MOORE, J.B., *Linear Optimal Control.* Prentice-Hall, Inc., Englewood Cliffs, New Jersey, 1971.

40. SCHULTZ, D.G., and MELSA, J.L., *State Functions and Linear Control Systems.* McGraw-Hill Book, Co., New York, 1967.

41. HSU, J.C., and MEYER, A.U., *Modern Control Principles and Applications.* McGraw-Hill Book Co., New York, 1968.

42. LASALLE, J.P., and LEFSCHETZ, S., *Stability by Lyapunov's Direct Method with Applications.* Academic Press, New York, 1961.

Problems

1. Determine whether or not the system described by

$$\dot{\mathbf{x}} = \mathbf{A}\mathbf{x}(t) + \mathbf{B}\mathbf{u}(t)$$

$$\mathbf{z}(t) = \mathbf{C}(t)\mathbf{x}(t)$$

is controllable and/or observable for the particular case in which

$$\mathbf{A} = \begin{bmatrix} a_{11} & a_{12} \\ a_{21} & a_{22} \end{bmatrix}, \qquad \mathbf{B} = \begin{bmatrix} b_1 \\ b_2 \end{bmatrix}, \qquad \mathbf{C} = [c_1, c_2].$$

By choosing specific numerical values, generate systems which are:

(a) state and output controllable, unobservable.
(b) state controllable and observable, not output controllable.
(c) observable and output controllable, not state controllable.
(d) state controllable, unobservable and not output controllable.
(e) output controllable, observable, not state controllable.

2. Show that a differential equation for the $\mathbf{W}(t_o, t_f)$ of Eq. (7.2-5) is

$$\frac{\partial \mathbf{W}(t_o, t_f)}{\partial t_o} = -\mathbf{B}(t_o)\mathbf{B}^T(t_o) + \mathbf{A}(t_o)\mathbf{W}(t_o, t_f) + \mathbf{W}(t_o, t_f)\mathbf{A}^T(t_o)$$

with initial condition $\mathbf{W}(t_f, t_f) = \mathbf{0}$. If you had to minimize $J = 0.5 \int_{t_o}^{t_f} \mathbf{u}^T(t)\mathbf{u}(t)\, dt$ for the system of Eq. (7.2-1) with $\mathbf{x}(t_o) = x_o$, $\mathbf{x}(t_f) = \mathbf{0}$, show that the minimum cost is given by the relation $J = 0.5\mathbf{x}^T(t_o)\mathbf{x}(t_o)$.

3. What is the continuous time equivalent to Problem 2 for observability? What is the discrete time equivalent?

4. Find the reachable zone as a function of t_f for the system

$$\dot{x}_1 = x_2(t), \qquad x_1(0) = 0$$
$$\dot{x}_2 = u(t), \qquad x_2(0) = 0$$

with

$$J = \frac{1}{2}\int_0^{t_f} u^2(t)\,dt = 1.$$

5. Derive the state and output controllability requirements as well as the observability requirement for the system

$$\dot{\mathbf{x}} = \mathbf{A}(t)\mathbf{x}(t) + \mathbf{B}(t)\mathbf{u}(t)$$
$$\mathbf{z}(t) = \mathbf{C}(t)\mathbf{x}(t) + \mathbf{D}(t)\mathbf{u}(t).$$

What simplifications can be made if **A**, **B**, **C**, and **D** are not functions of time?

6. Derive the continuous time observability requirements by finding conditions for existence of an optimum estimator to minimize

$$J = \frac{1}{2}\int_{t_o}^{t_f} ||\mathbf{z}(t) - \mathbf{c}(t)\mathbf{x}(t)||^2\,dt$$

subject to the constraint

$$\frac{d\mathbf{x}(t)}{dt} = \mathbf{A}(t)\mathbf{x}(t).$$

7. Show that a continuous time linear system

$$\frac{d\mathbf{x}(t)}{dt} = \mathbf{A}(t)\mathbf{x}(t), \qquad \mathbf{z}(t) = \mathbf{C}(t)\mathbf{x}(t)$$

is observable if the matrix $\mathbf{\Gamma}$ is of rank η where

$$\mathbf{\Gamma} = [\mathbf{\Gamma}_1, \mathbf{\Gamma}_2, \ldots, \mathbf{\Gamma}_\eta]$$
$$\mathbf{\Gamma}_1 = C^T(t)$$
$$\mathbf{\Gamma}_k = \frac{\partial \mathbf{\Gamma}_{k-1}}{\partial t} + \mathbf{A}^T(t)\mathbf{\Gamma}_{k-1}.$$

What do these results become for a constant coefficient linear system?

8. Let $\mathbf{M} = \mathbf{D}^T\mathbf{D}$. Show that:
   (a) $\mathbf{M} = \mathbf{M}^T$
   (b) $\mathbf{M}$ is positive semidefinite
   (c) the following conditions are equivalent:
      (i) $\mathbf{M}$ is positive definite
      (ii) $\mathbf{M}$ is nonsingular
      (iii) $\mathbf{D}$ has full rank.

9. Show that for the discrete time case, the observability Gramian $\mathbf{M}(k_o, k_f)$ has the following properties:
   (a) $\mathbf{M}(k_o, k_f)$ is symmetric
   (b) $\mathbf{M}(k_o, k_f)$ is nonnegative definite

(c)
$$\mathbf{M}(k, k_f) = \mathbf{C}^T(k)\mathbf{C}(k) + \mathbf{A}^T(k)\mathbf{M}(k+1, k_f)\mathbf{A}(k),$$
$$\mathbf{M}(k_f, k_f) = \mathbf{C}^T(k_f)\mathbf{C}(k_f)$$

(d)
$$\mathbf{M}(k_o, k_f) = \mathbf{M}(k_o, k) + \boldsymbol{\varphi}^T(k, k_o)\mathbf{M}(k, k_f)\boldsymbol{\varphi}(k, k_o).$$

10. For the continuous case, show that the observability Gramian $M(t_o, t_f)$:
(a) is symmetric
(b) is nonnegative definite
(c) satisfies

$$\frac{\partial \mathbf{M}(t, t_f)}{\partial t} = -\mathbf{A}^T(t)\mathbf{M}(t, t_f) - \mathbf{M}(t, t_f)\mathbf{A}(t) - \mathbf{C}^T(t)\mathbf{C}(t),$$
$$\mathbf{M}(t_f, t_f) = \mathbf{0}$$

(d) satisfies

$$\mathbf{M}(t_o, t_f) = \mathbf{M}(t_o, t) + \boldsymbol{\varphi}^T(t, t_o)\mathbf{M}(t, t_f)\boldsymbol{\varphi}(t, t_o).$$

11. Show that the continuous time controllability Gramian $\mathbf{W}(t_o, t_f)$:
(a) is symmetric
(b) is nonnegative definite
(c) satisfies

$$\frac{\partial \mathbf{W}(t, t_f)}{\partial t} = \mathbf{A}(t)\mathbf{W}(t, t_f) + \mathbf{W}(t, t_f)\mathbf{A}^T(t) - \mathbf{B}(t)\mathbf{B}^T(t),$$
$$\mathbf{W}(t_f, t_f) = \mathbf{0}$$

(d) satisfies

$$\mathbf{W}(t_o, t_f) = \mathbf{W}(t_o, t) + \boldsymbol{\varphi}(t_o, t)\mathbf{W}(t, t_f)\boldsymbol{\varphi}^T(t_o, t).$$

12. The matrix pair $(\mathbf{A}(\cdot), \mathbf{B}(\cdot))$ is said to be state controllable on $[t_o, t_f]$ if and only if $\mathbf{W}(t_o, t_f)$ is nonsingular. Similarly, the matrix pair $(\mathbf{A}(\cdot), \mathbf{C}(\cdot))$ is said to be observable on $[t_o, t_f]$ if and only if $\mathbf{M}(t_o, t_f)$ is nonsingular. Prove that $(\mathbf{A}(\cdot), \mathbf{B}(\cdot))$ is controllable if and only if $(\mathbf{A}^T(\cdot), \mathbf{C}^T(\cdot))$ is observable. This result is often referred to as the *duality theorem for controllability and observability.*

13. Determine the controllability requirement for a continuous time linear system by finding existence requirements for the optimum control $\mathbf{u}(t)$ which takes the linear system

$$\dot{\mathbf{x}}(t) = \mathbf{A}(t)\mathbf{x}(t) + \mathbf{B}(t)\mathbf{u}(t)$$

from $\mathbf{x}(t_o) = \mathbf{x}_o$ to $\mathbf{x}(t_f) = \mathbf{0}$ such as to minimize

$$J = \frac{1}{2}\int_{t_o}^{t_f} ||\mathbf{u}(t)||^2\, dt.$$

14. Show that an equivalent continuous time controllability requirement is that the matrix

$$\boldsymbol{\Gamma} = [\boldsymbol{\Gamma}_1, \boldsymbol{\Gamma}_2, \ldots, \boldsymbol{\Gamma}_n]$$

is of rank n where

$$\boldsymbol{\Gamma}_1(t) = \mathbf{B}(t)$$
$$\boldsymbol{\Gamma}_k(t) = \frac{\partial \boldsymbol{\Gamma}_{k-1}(t)}{dt} - \mathbf{A}(t)\boldsymbol{\Gamma}_{k-1}(t)$$

15. Determine the essential results of Sec. 7.1-2 for discrete time linear systems.

16. Determine the continuous parameter sensitivity vector for the second order, general, time-invariant system $\dot{\mathbf{x}} = \mathbf{A}\mathbf{x}(t) + \mathbf{b}u(t)$, where u is a unit step applied at time t_o.

17. Determine the discrete parameter sensitivity vector for a first difference approximation to Problem 16.

18. Determine the discrete system requirements for the sensitivity procedure of Dorato.

19. Determine the discrete system requirements for the sensitivity procedure of Rohrer and Sobral.

20. Obtain the discrete version of Eqs. (7.3-46) and (7.3-49).

21. Repeat the sensitivity calculations of Example 7.3-1 for the system and cost functions

(a) $\dot{x} = -ax + u, \qquad x(0) = x_o, \qquad J = \frac{1}{2}Sx^2(t_f) + \frac{1}{2}\int_{t_o}^{t_f} u^2(t)\,dt$

(b) $\dot{x}_1 = x_2, \qquad \dot{x}_2 = bu, \qquad x_1(0) = x_{10}, \qquad x_2(0) = x_{20},$

$$x_1(t_f) = x_2(t_f) = 0, \qquad J = \frac{1}{2}\int_0^{t_f} u^2(t)\,dt.$$

22. Determine t_f from Eq. (7.3-94) for Example 7.3-2.

23. In considering the cost variation for the control perturbation along the optimal trajectory in the sensitivity to control variations discussion, it is convenient to use

$$J = \int_0^{t_f} \phi[\mathbf{x}(t), \mathbf{u}(t), t]\,dt$$

as the cost function. If J^* is the minimum cost to transfer from $\mathbf{x}(t_o)$ to the ideal terminal manifold, and if J_T is the cost to transfer to the approximate manifold $h[\delta x(t_f)] = 0$, show that

$$J_T - J^* = \phi[\mathbf{x}^*(t_f), \mathbf{u}(t_f), t_f]\delta t_f - \boldsymbol{\lambda}^T(t_f)\boldsymbol{\delta}\mathbf{x}(t_f)$$

where $\boldsymbol{\lambda}(t)$ is the adjoint associated with the original control problem. Therefore, show that, since δt_f can be found in terms of $\boldsymbol{\delta}\mathbf{x}(t_f)$, $J_T - J^*$ can also be obtained.

24. Obtain Pagurek's sensitivity equation, Eq. (7.3-81), for the open-loop optimum linear regulator. Apply this result to Example (7.3-2). Consult [25] for details of the solution and assumed conditions, if necessary.

25. Show that, for sufficiently small changes in the system coefficient matrix $\mathbf{A}(t)$, or the control distribution matrix $\mathbf{B}(t)$, the system states will reach the origin at the terminal time if the feedback matrix $\mathbf{G}(t)$ of the closed-loop system of Eq. (7.3-84) is held invariant.

26. Determine bounds on max $(J_T - J^*)$ and min $(J_T - J^*)$ for Example 7.3-3.

27. Determine the local and global sampling interval sensitivity for the system of Example 7.3-1.

28. Prove the theorems stated in Sec. 7.4 concerning system stability and instability.

29. If $\dot{V}(\mathbf{x})$ is negative semidefinite, and you can show that no system solution can forever remain at those points at which $\dot{V}(\mathbf{x}) = 0$, show that the origin is at least asymptotically stable with respect to R_ϵ if $V(\mathbf{x})$ is a Lyapunov function.

30. For the equation $\mathbf{A}^T\mathbf{P} + \mathbf{PA} = -\mathbf{C}$, show that if $\mathbf{A}$ is stable and $\mathbf{C}$ is positive definite then $\mathbf{P}$ is also positive definite.

Optimum state estimation

8

We often wish to estimate the state of a system on the basis of statistically related data which may not provide enough information to determine the state of the system exactly. For example, we may desire to determine the position and velocity of an airborne object from radar tracking data, to distinguish the signal sent from the signal received in a communications link, or to evaluate a patient's health from his/her history of signs, symptoms, and laboratory test results. This chapter considers such state estimation problems. (The observation received by the observer is, for simplicity, assumed to be the sum of a linear transformation of the actual signal sent and the realization of a random variable modeling the noise in the system.) More specifically, our interest is to determine an estimate that is a linear function of all past observations which minimizes the expectation of the square of the error between the actual signal sent and the estimate of that signal. This problem has generally come to be called the problem of *linear least-squares estimation* or *linear filtering*.

Although there is a substantial history to this problem (for example, see the survey paper [1, Sec. II]), the first explicit solution for the linear filtering problem was given by Wiener in 1942 [2]. Unfortunately, the Wiener filter problem formulation is not well-suited to nonstationary recursive estimation problems where the signal to be estimated is the output of a dynamic time-varying linear system or a linearized nonlinear system driven by white noise. Accordingly, Kalman [3, 4, 5] modified Wiener's problem definition to a form particularly well-matched to state-space problem formulations generated by many modern physical problem needs. The result was the *Kalman filter*, which we consider, in Secs. 8.1, 8.2, and 8.3, in its continuous and discrete forms.

Although linear least-squares estimation is often viewed strictly as a particularly elegant application of the orthogonal projection theorem (see, for example, the development in [6]), several other proof techniques will be employed in the development of the results presented in this chapter. The reader interested in more advanced topics such as the extended Kalman filter and the error analysis of optimal and suboptimal filters is referred to [7]. A review and extensions of recent work on non-Riccati-based studies of estimation problems can be found in [8]. Nonlinear estimation is discussed in [7] and references contained therein. A discussion describing the relatively new martingale approach to estimation theory, which we will not pursue, can be found in [9] and [10].

For the case where the measurements are noise-free, a considerably less complex optimal observer, called the *Luenberger observer* can be obtained. This is the subject of Sec. 8.4.

8.1 State-space formulation for systems with random inputs and minimum error variance linear filtering

We develop the state-space approach to linear filtering in this chapter. A few mathematical preliminaries are necessary before we begin our study. In this section, we present a few results concerning the state-space formulation for linear nonstationary systems with nonstationary random inputs. We conclude our discussion with a simplified derivation of the optimum, in the sense of minimum error variance, unbiased linear filter with nonstationary random inputs. Appendix B summarizes pertinent facts concerning probability and stochastic processes used here. Reference [11] provides considerably greater coverage of this material.

Suppose we have a system described by

$$\dot{\mathbf{x}}(t) = \mathbf{A}(t)\mathbf{x}(t) + \mathbf{B}(t)\mathbf{w}(t), \tag{8.1-1}$$

where

$$\mathbf{x}^T(t) = [x_1, x_2, \ldots, x_n], \qquad \mathbf{w}^T(t) = [w_1, w_2, \ldots, w_m]$$

$$\mathbf{A}(t) = \begin{bmatrix} a_{11}(t) & a_{12}(t) & \ldots & a_{1n}(t) \\ a_{21}(t) & & & \\ \cdot & & & \\ \cdot & & & \\ \cdot & & & \\ a_{n1}(t) & \ldots & & a_{nn}(t) \end{bmatrix},$$

$$\mathbf{B}(t) = \begin{bmatrix} b_{11}(t) & b_{12}(t) & \dots & b_{1m}(t) \\ b_{21}(t) & & & \\ \cdot & & & \\ \cdot & & & \\ \cdot & & & \\ b_{n1}(t) & \dots & & b_{nm}(t) \end{bmatrix}.$$

We shall assume that both the characteristic matrix $\mathbf{A}(t)$ as well as the control distribution matrix $\mathbf{B}(t)$ are known. As demonstrated in Appendix A and Chapter 5, the solution to Eq. (8.1-1) is

$$\mathbf{x}(t) = \boldsymbol{\varphi}(t, t_o)\mathbf{x}(t_o) + \int_{t_o}^{t} \boldsymbol{\varphi}(t, \lambda)\mathbf{B}(\lambda)\mathbf{w}(\lambda)\, d\lambda, \qquad (8.1\text{-}2)$$

where $\boldsymbol{\varphi}(t, t_o)$ is the state transition matrix which must satisfy the relation

$$\frac{\partial \boldsymbol{\varphi}(t, t_o)}{\partial t} = \mathbf{A}(t)\boldsymbol{\varphi}(t, t_o) \qquad (8.1\text{-}3)$$

$$\boldsymbol{\varphi}(t_o, t_o) = \mathbf{I}. \qquad (8.1\text{-}4)$$

Now suppose we assume $\mathbf{w}(t)$ is a zero-mean white noise vector. This causes no loss in generality since we can use filtering to obtain any nonwhite noise as illustrated in Fig. 8.1-1.

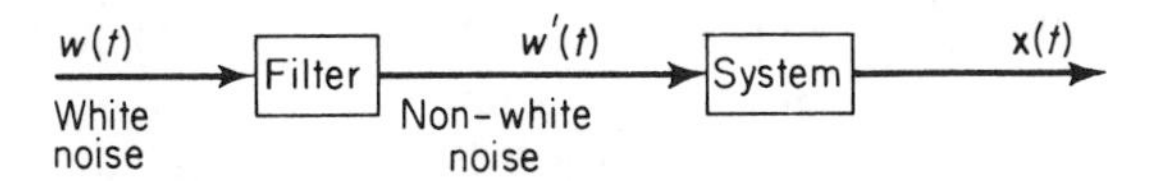

Fig. 8.1-1 Converting from white to nonwhite noise.

We now define the covariance matrix of a zero-mean random vector as

$$\mathbf{P}(t_1, t_2) = \mathcal{E}\{\mathbf{x}(t_1)\mathbf{x}^T(t_2)\}. \qquad (8.1\text{-}5)$$

If we form the inner product $\mathbf{x}^T\mathbf{x}$, we obtain a scalar. The form $\mathbf{x}\mathbf{x}^T$ is called the outer product of $\mathbf{x}$, and it is an $n \times n$ matrix, which is sometimes written $\mathbf{x} > < \mathbf{x}$. Substituting $\mathbf{x}(t)$ from Eq. (8.1-2) into the definition of Eq. (8.1-5), we obtain for the covariance matrix:

$$\mathbf{P}(t_1, t_2) = \mathcal{E}\{\boldsymbol{\varphi}(t_1, t_o)\mathbf{x}(t_o)\mathbf{x}^T(t_o)\boldsymbol{\varphi}^T(t_2, t_o)\} +$$
$$\mathcal{E}\left\{\int_{t_o}^{t_1}\int_{t_o}^{t_2} \boldsymbol{\varphi}(t_1, \lambda_1)\mathbf{B}(\lambda_1)\mathbf{w}(\lambda_1)\mathbf{w}^T(\lambda_2)\mathbf{B}^T(\lambda_2)\boldsymbol{\varphi}^T(t_2, \lambda_2)\, d\lambda_1\, d\lambda_2\right\}, \qquad (8.1\text{-}6)$$

and two other terms whose expected values are clearly zero. If we place the expected-value signs inside the integral, Eq. (8.1-6) becomes

$$\mathbf{P}(t_1, t_2) = \boldsymbol{\varphi}(t_1, t_o)\mathcal{E}\{\mathbf{x}(t_o)\mathbf{x}^T(t_o)\}\boldsymbol{\varphi}^T(t_2, t_o) +$$
$$\int_{t_o}^{t_1}\int_{t_o}^{t_2} \boldsymbol{\varphi}(t_1, \lambda_1)\mathbf{B}(\lambda_1)\mathcal{E}\{\mathbf{w}(\lambda_1)\mathbf{w}^T(\lambda_2)\}\mathbf{B}^T(\lambda_2)\boldsymbol{\varphi}^T(t_2, \lambda_2)\, d\lambda_1\, d\lambda_2. \qquad (8.1\text{-}7)$$

But we have defined

$$\mathcal{E}\{\mathbf{x}(t_o)\mathbf{x}^T(t_o)\} = \mathbf{P}(t_o, t_o)$$

which is normally written, because of the double t_o notation, as

$$\mathbf{P}(t_o, t_o) = \mathbf{P}(t_o). \tag{8.1-8}$$

In a similar fashion, we have

$$\mathcal{E}\{\mathbf{w}(\lambda_1)\mathbf{w}^T(\lambda_2)\} = \mathbf{Q}(\lambda_1)\delta(\lambda_1 - \lambda_2) \tag{8.1-9}$$

since we assumed nonstationary white noise inputs of zero-mean value. Here δ represents the impulse or Dirac delta function. Thus Eq. (8.1-7) becomes

$$\mathbf{P}(t_1, t_2) = \boldsymbol{\varphi}(t_1, t_o)\mathbf{P}(t_o)\boldsymbol{\varphi}^T(t_2, t_o) + \int_{t_o}^{t_1}\int_{t_o}^{t_2} \boldsymbol{\varphi}(t_1, \lambda_1)\mathbf{B}(\lambda_1)\mathbf{Q}(\lambda_1)\delta(\lambda_1 - \lambda_2)\mathbf{B}^T(\lambda_2)\boldsymbol{\varphi}^T(t_2, \lambda_2)\, d\lambda_1\, d\lambda_2. \tag{8.1-10}$$

This simplifies further to

$$\mathbf{P}(t_1, t_2) = \boldsymbol{\varphi}(t_1, t_o)\mathbf{P}(t_o)\boldsymbol{\varphi}^T(t_2, t_o) + \int_{t_o}^{\min(t_1, t_2)} \boldsymbol{\varphi}(t_1, \lambda_2)\mathbf{B}(\lambda_2)\mathbf{Q}(\lambda_2)\mathbf{B}^T(\lambda_2)\boldsymbol{\varphi}^T(t_2, \lambda_2)\, d\lambda_2. \tag{8.1-11}$$

Equation (8.1-11) gives the correlation of the outputs of a system with white noise inputs. We note that, to obtain the correlation matrix $\mathbf{P}(t_1, t_2)$, we must solve the matrix differential equation (8.1-3) for $\boldsymbol{\varphi}(t_1, t_o)$, which means solving n^2 differential equations. Then we must evaluate $n(n + 1)/2$ integrals for each $t_1 t_2$ combination in Eq. (8.1-11). This is a tedious job, to say the least. We are therefore usually satisfied with the mean-square value or variance matrix $\mathbf{P}(t, t)$, written $\mathbf{P}(t)$, which we obtain by setting $t_1 = t_2 = t$ in Eq. (8.1-11)

$$\mathbf{P}(t) = \boldsymbol{\varphi}(t, t_o)\mathbf{P}(t_o)\boldsymbol{\varphi}^T(t, t_o) + \int_{t_o}^{t} \boldsymbol{\varphi}(t, \lambda)\mathbf{B}(\lambda)\mathbf{Q}(\lambda)\mathbf{B}^T(\lambda)\boldsymbol{\varphi}^T(t, \lambda)\, d\lambda. \tag{8.1-12}$$

We have assumed that nonstationary white noise is present for each component of the input vector. This means that the correlation matrix for the noise used to derive Eq. (8.1-12) is

$$\mathbf{Q}(\lambda_1, \lambda_2) = \mathcal{E}\{\mathbf{w}(\lambda_1)\mathbf{w}^T(\lambda_2)\} = \mathbf{Q}(\lambda_1)\delta(\lambda_1 - \lambda_2) \tag{8.1-13}$$

where $\delta(\lambda)$ is the impulse function.

Equation (8.1-12), though a little simpler than Eq. (8.1-11), is still tedious to evaluate. Often, however, it simplifies somewhat. Suppose, for example, that we are only interested in one component of the state vector; that is, we desire only

$$z(t) = \mathbf{c}^T(t)\mathbf{x}(t), \tag{8.1-14}$$

where $\mathbf{c}(t)$ is a vector of the same dimension as $\mathbf{x}(t)$ such that $z(t)$ is a scalar. By definition, since $\mathcal{E}\{z(t)\} = 0$, the variance of $z(t)$ is

$$\begin{aligned}\operatorname{cov}\{z(t)\} &= \mathcal{E}\{z^2(t)\} = \mathcal{E}\{z(t)z^T(t)\} = \mathcal{E}\{\mathbf{c}^T(t)\mathbf{x}(t)\mathbf{x}^T(t)\mathbf{c}(t)\} \\ &= \mathbf{c}^T(t)\mathcal{E}\{\mathbf{x}(t)\mathbf{x}^T(t)\}\mathbf{c}(t) = \mathbf{c}^T(t)\mathbf{P}(t)\mathbf{c}(t).\end{aligned} \tag{8.1-15}$$

Therefore, from Eq. (8.1-12),

$$\text{cov}\{z(t)\} = \mathbf{c}^T(t)\boldsymbol{\varphi}(t, t_o)\mathbf{P}(t_o)\boldsymbol{\varphi}^T(t, t_o)\mathbf{c}(t) + \int_{t_o}^{t} \mathbf{c}^T(t)\boldsymbol{\varphi}(t, \lambda)\mathbf{B}(\lambda)\mathbf{Q}(\lambda)\mathbf{B}^T(\lambda)\boldsymbol{\varphi}^T(t, \lambda)\mathbf{c}(t)\, d\lambda. \tag{8.1-16}$$

We now define a row vector $\boldsymbol{\Gamma}^T(t, \lambda)$ such that

$$\boldsymbol{\Gamma}^T(t, \lambda) = \mathbf{c}^T(t)\boldsymbol{\varphi}(t, \lambda), \tag{8.1-17}$$

and we have

$$\text{cov}\{z(t)\} = \boldsymbol{\Gamma}^T(t, \lambda)\mathbf{P}(t_o)\boldsymbol{\Gamma}(t, \lambda) + \int_{t_o}^{t} \boldsymbol{\Gamma}^T(t, \lambda)\mathbf{B}(\lambda)\mathbf{Q}(\lambda)\mathbf{B}^T(\lambda)\boldsymbol{\Gamma}(t, \lambda)\, d\lambda \tag{8.1-18}$$

which is clearly a scalar. Even so, there are numerous practical difficulties. It is not often that we are able to determine the state transition matrix, and even when we can, the integral in Eq. (8.1-18) is still quite difficult to evaluate. Therefore it is desirable to consider an ancillary approach, that of converting the integral equation (8.1-12) to a differential equation. That this feat can be accomplished is one of the compelling advantages of the state-space approach. From Eq. (8.1-12) we have

$$\frac{d\mathbf{P}(t)}{dt} = \frac{\partial\boldsymbol{\varphi}(t, t_o)}{\partial t}\mathbf{P}(t_o)\boldsymbol{\varphi}^T(t, t_o) + \boldsymbol{\varphi}(t, t_o)\mathbf{P}(t_o)\frac{\partial\boldsymbol{\varphi}^T(t, t_o)}{\partial t} + \boldsymbol{\varphi}(t, t)\mathbf{B}(t)\mathbf{Q}(t)\mathbf{B}^T(t)\boldsymbol{\varphi}^T(t, t) + \int_{t_o}^{t} \frac{\partial\boldsymbol{\varphi}(t, \lambda)}{\partial t}\mathbf{B}(\lambda)\mathbf{Q}(\lambda)\mathbf{B}^T(\lambda)\boldsymbol{\varphi}^T(t, \lambda)\, d\lambda + \int_{t_o}^{t} \boldsymbol{\varphi}(t, \lambda)\mathbf{B}(\lambda)\mathbf{Q}(\lambda)\mathbf{B}^T(\lambda)\frac{\partial\boldsymbol{\varphi}^T(t, \lambda)}{\partial t}\, d\lambda. \tag{8.1-19}$$

If we use Eqs. (8.1-3), (8.1-4), and the transpose of Eq. (8.1-3)

$$\frac{\partial\boldsymbol{\varphi}^T(t, t_o)}{\partial t} = \boldsymbol{\varphi}^T(t, t_o)\mathbf{A}^T(t), \tag{8.1-20}$$

Eq. (8.1-19) becomes

$$\frac{d\mathbf{P}(t)}{dt} = \mathbf{A}(t)\boldsymbol{\varphi}(t, t_o)\mathbf{P}(t_o)\boldsymbol{\varphi}^T(t, t_o) + \boldsymbol{\varphi}(t, t_o)\mathbf{P}(t_o)\boldsymbol{\varphi}^T(t, t_o)\mathbf{A}^T(t) + \mathbf{B}(t)\mathbf{Q}(t)\mathbf{B}(t) + \mathbf{A}(t)\int_{t_o}^{t} \boldsymbol{\varphi}(t, \lambda)\mathbf{B}(\lambda)\mathbf{Q}(\lambda)\mathbf{B}^T(\lambda)\boldsymbol{\varphi}^T(t, \lambda)\, d\lambda + \int_{t_o}^{t} \boldsymbol{\varphi}(t, \lambda)\mathbf{B}(\lambda)\mathbf{Q}(\lambda)\mathbf{B}^T(\lambda)\boldsymbol{\varphi}^T(t, \lambda)\mathbf{A}^T(t)\, d\lambda. \tag{8.1-21}$$

When we use Eq. (8.1-12), the foregoing equation reduces simply to

$$\frac{d\mathbf{P}(t)}{dt} = \mathbf{A}(t)\mathbf{P}(t) + \mathbf{P}(t)\mathbf{A}^T(t) + \mathbf{B}(t)\mathbf{Q}(t)\mathbf{B}^T(t) \tag{8.1-22}$$

which we must solve subject to the initial conditions

$$\mathbf{P}(t_o) = \mathcal{E}\{\mathbf{x}(t_o)\mathbf{x}^T(t_o)\} = \mathbf{P}_o \tag{8.1-23}$$

where $\mathbf{P}_o$ is, of course, given as a part of the problem statement. If $\mathbf{x}(t_o)$ or $\mathbf{w}(t)$ have nonzero mean values, these mean values can be considered as deterministic and the response to them calculated in the normal way. Then we may use the superposition theorem to determine the response due to the deterministic portion plus the random zero-mean $\mathbf{x}(t_o)$ and $\mathbf{w}(t)$.

The significant fact here is that we have taken the convolution integral, which was almost impossible to evaluate, and converted it to a differential equation which can be solved on a digital computer. To solve Eq. (8.1-22), we must know $\mathbf{A}(t)$, $\mathbf{B}(t)$, and $\mathbf{P}_o$. However, we do not need to compute $\boldsymbol{\varphi}$, the state transition matrix. We note that the matrix differential equation (8.1-22) is linear and symmetric, in that $\mathbf{P}(t) = \mathbf{P}^T(t)$. Also $\mathbf{P}(t_2, t_1) = \boldsymbol{\varphi}(t_2, t_1)\mathbf{P}(t_1)$.

Example 8.1-1. Suppose we are given the constant coefficient system

$$\dot{x} = -\alpha x + \beta w$$

where x, w, α, and β are scalars, and α and β are constant with time. We wish to determine $p(t)$ given the initial covariance, $p_o = \mathcal{E}\{x^2(0)\} = p_o$, and given an input stationary white noise with correlation function $\mathcal{E}\{w(t)w(\tau)\} = q\delta(t - \tau)$. We note that the output random process will be nonstationary because the process is started at $t = 0$, not at $t = -\infty$, which would produce a stationary output random process. We must solve Eq. (8.1-22) with $A = -\alpha$, $B = \beta$, and $Q = q$, which yields for Eq. (8.1-22)

$$\dot{p} = -2\alpha p + \beta^2 q.$$

The solution to this is easily found to be

$$p(t) = p(0)e^{-2\alpha t} + \beta^2 q \frac{1 - e^{-2\alpha t}}{2\alpha}.$$

We see that the variance of the output is composed of two parts: one due to the initial variance, which decreases to zero exponentially, and a second due to the presence of noise, which grows to a constant value. The stationary value is given by $p(t \longrightarrow \infty) = \beta^2 q/2\alpha$.

Example 8.1-2. Suppose that we are given the system of Fig. 8.1-2 where the switch is closed at $t = 0$. The problem is basically the same as in the previous example with $\alpha = 0$, $\beta = 0$, and $p_o = 0$. Thus we have $\dot{p} = 1$ or $p(t) = t$, and we see that the variance of the output increases linearly with

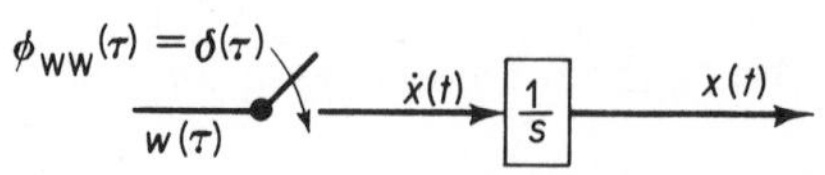

Fig. 8.1-2 Switched integrator for Example 8.1-2.

time. The output process of this example is often called a Wiener process if the input process is Gaussian.

It is possible to obtain a more direct derivation of the variance equation as follows.

For a zero-mean process $\mathcal{E}\{\mathbf{x}(t)\} = \mathbf{0}$, we have by definition,

$$\mathbf{P}(t) = \text{cov}\,\{\mathbf{x}(t), \mathbf{x}(t)\} = \mathcal{E}\{\mathbf{x}(t)\mathbf{x}^T(t)\}. \tag{8.1-24}$$

Upon differentiating this expression, we obtain

$$\dot{\mathbf{P}} = \mathcal{E}\{\dot{\mathbf{x}}(t)\mathbf{x}^T + \mathbf{x}(t)\dot{\mathbf{x}}^T(t)\}. \tag{8.1-25}$$

Substituting from the system equation (8.1-1) and its transpose

$$\dot{\mathbf{x}}^T(t) = \mathbf{x}^T(t)\mathbf{A}^T(t) + \mathbf{w}^T(t)\mathbf{B}^T(t), \tag{8.1-26}$$

we see that

$$\dot{\mathbf{P}} = \mathcal{E}\{\mathbf{A}(t)\mathbf{x}(t)\mathbf{x}^T(t) + \mathbf{B}(t)\mathbf{w}(t)\mathbf{x}^T(t) + \mathbf{x}(t)\mathbf{x}^T(t)\mathbf{A}^T(t) + \mathbf{x}(t)\mathbf{w}^T(t)\mathbf{B}^T(t)\}. \tag{8.1-27}$$

But $\mathcal{E}\{\mathbf{x}(t)\mathbf{x}^T(t)\} = \mathbf{P}(t)$; thus Eq. (8.1-27) becomes

$$\dot{\mathbf{P}} = \mathbf{A}(t)\mathbf{P}(t) + \mathbf{P}(t)\mathbf{A}^T(t) + \mathcal{E}\{\mathbf{B}(t)\mathbf{w}(t)\mathbf{x}^T(t) + \mathbf{x}(t)\mathbf{w}^T(t)\mathbf{B}^T(t)\}. \tag{8.1-28}$$

From our previous work, we have

$$\mathbf{x}(t) = \boldsymbol{\varphi}(t, t_o)\mathbf{x}(t_o) + \int_{t_o}^{t} \boldsymbol{\varphi}(t, \lambda)\mathbf{B}(\lambda)\mathbf{w}(\lambda)\,d\lambda.$$

Substituting this expansion into Eq (8.1-25) results in

$$\begin{aligned}\dot{\mathbf{P}} = {} & \mathbf{A}(t)\mathbf{P}(t) + \mathbf{P}(t)\mathbf{A}^T(t) + \\ & \mathcal{E}\left\{\mathbf{B}(t)\mathbf{w}(t)\left[\mathbf{x}^T(t_o)\boldsymbol{\varphi}^T(t, t_o) + \int_{t_o}^{t} \mathbf{w}^T(\lambda)\mathbf{B}^T(\lambda)\boldsymbol{\varphi}^T(t, \lambda)\,d\lambda\right]\right\} + \\ & \mathcal{E}\left\{\left[\boldsymbol{\varphi}(t, t_o)\mathbf{x}(t_o) + \int_{t_o}^{t} \boldsymbol{\varphi}(t, \lambda)\mathbf{B}(\lambda)\mathbf{w}(\lambda)\,d\lambda\right]\mathbf{w}^T(t)\mathbf{B}^T(t)\right\}\end{aligned} \tag{8.1-29}$$

or

$$\begin{aligned}\dot{\mathbf{P}} = {} & \mathbf{A}(t)\mathbf{P}(t) + \mathbf{P}(t)\mathbf{A}^T(t) + \mathbf{B}(t)\mathcal{E}\{\mathbf{w}(t)\mathbf{x}^T(t_o)\}\boldsymbol{\varphi}^T(t, t_o) + \\ & \int_{t_o}^{t} \mathbf{B}(t)\mathcal{E}\{\mathbf{w}(t)\mathbf{w}^T(\lambda)\}\mathbf{B}^T(\lambda)\boldsymbol{\varphi}^T(t, \lambda)\,d\lambda + \boldsymbol{\varphi}(t, t_o)\mathcal{E}\{\mathbf{x}(t_o)\mathbf{w}^T(t)\}\mathbf{B}^T(t) + \\ & \int_{t_o}^{t} \boldsymbol{\varphi}(t, \lambda)\mathbf{B}(\lambda)\mathcal{E}\{\mathbf{w}(\lambda)\mathbf{w}^T(t)\}\mathbf{B}^T(t)\,d\lambda.\end{aligned} \tag{8.1-30}$$

But $\mathcal{E}\{\mathbf{w}(t)\mathbf{x}^T(t_o)\} = 0$ and also $\mathcal{E}\{\mathbf{x}(t_o)\mathbf{w}^T(t)\} = 0$ since we assume no correlation between initial conditions and noise. Also, since we have assumed white noise inputs, we know further that $\mathcal{E}\{\mathbf{w}(t)\mathbf{w}^T(\lambda)\} = \mathbf{Q}(t)\delta(t - \lambda)$ and $\mathcal{E}\{\mathbf{w}(\lambda)\mathbf{w}^T(t)\} = \mathbf{Q}(\lambda)\delta(\lambda - t)$. Inserting these last four expressions into Eq. (8.1-30) gives

$$\begin{aligned}\dot{\mathbf{P}} = {} & \mathbf{A}(t)\mathbf{P}(t) + \mathbf{P}(t)\mathbf{A}^T(t) + \int_{t_o}^{t} \mathbf{B}(t)\mathbf{Q}(t)\delta(t - \lambda)\mathbf{B}^T(\lambda)\boldsymbol{\varphi}^T(t, \lambda)\,d\lambda + \\ & \int_{t_o}^{t} \boldsymbol{\varphi}(t, \lambda)\mathbf{B}(\lambda)\mathbf{Q}(\lambda)\delta(\lambda - t)\mathbf{B}^T(t)\,d\lambda.\end{aligned} \tag{8.1-31}$$

Now we can use the sifting property of the delta function to carry out the integrations. We know the delta function is an even function, $\delta(t - \lambda) = \delta(\lambda - t)$. Reasoning as before, we obtain, from Eq. (8.1-31), the variance equation

$$\dot{\mathbf{P}} = \mathbf{A}(t)\mathbf{P}(t) + \mathbf{P}(t)\mathbf{A}^T(t) + \mathbf{B}(t)\mathbf{Q}(t)\mathbf{B}^T(t) \tag{8.1-32}$$

which is solved with the initial condition $\mathbf{P}(t_o) = \mathbf{P}_o$.

We will continue this section with a development of the estimation algorithms for a linear minimum error variance unbiased filter. We will *not* make any assumptions in this section concerning the probability densities of the message and observation noise and the initial system state. Other derivations will be given in later sections which are valid for different error criteria where it is necessary to state explicitly these probability densities.

We assume that a state vector message model is generated by the linear differential equation

$$\dot{\mathbf{x}}(t) = \mathbf{F}(t)\mathbf{x}(t) + \mathbf{G}(t)\mathbf{w}(t) \tag{8.1-33}$$

where $\mathbf{w}(t)$ is zero-mean white noise. The first- and second-order moments of the white plant noise $\mathbf{w}(t)$ and the initial state vector are

$$\mathcal{E}\{\mathbf{w}(t)\} = \mathbf{0}, \qquad \operatorname{cov}\{\mathbf{w}(t), \mathbf{w}(\tau)\} = \mathbf{Q}(t)\,\delta(t - \tau)$$

$$\mathcal{E}\{\mathbf{x}(t_o)\} = \bar{\mathbf{x}}_o, \qquad \operatorname{var}\{\mathbf{x}(t_o)\} = \mathbf{P}_o$$

and

$$\operatorname{cov}\{\mathbf{x}(t), \mathbf{w}(\tau)\} = \mathbf{0}, \qquad \tau > t;$$

a noise-corrupted linearly modulated observation of the state vector

$$\mathbf{z}(t) = \mathbf{H}(t)\mathbf{x}(t) + \mathbf{v}(t) \tag{8.1-34}$$

is available for processing. $\mathbf{v}(t)$ is zero-mean white noise with

$$\mathcal{E}\{\mathbf{v}(t)\} = \mathbf{0}, \qquad \operatorname{cov}\{\mathbf{v}(t), \mathbf{v}(\tau)\} = \mathbf{R}(t)\delta(t - \tau)$$

and

$$\operatorname{cov}\{\mathbf{v}(t), \mathbf{w}(\tau)\} = \mathbf{0}, \qquad \operatorname{cov}\{\mathbf{v}(t), \mathbf{x}(\tau)\} = \mathbf{0} \qquad \text{for all } t, \tau.$$

As indicated we desire to construct a linear unbiased minimum error variance sequential estimator of the state vector $\mathbf{x}(t)$. We postulate a linear sequential filter of the form

$$\dot{\hat{\mathbf{x}}}(t) = \mathbf{A}(t)\hat{\mathbf{x}}(t) + \mathbf{K}(t)\mathbf{z}(t). \tag{8.1-35}$$

To make the filter unconditionally unbiased, we must have

$$\mathcal{E}\{\hat{\mathbf{x}}(t)\} = \mathcal{E}\{\mathbf{x}(t)\} = \bar{\mathbf{x}}(t). \tag{8.1-36}$$

We take the unconditional expectation of the postulated filter algorithm, Eq. (8.1-35), and obtain

$$\mathcal{E}\{\dot{\hat{\mathbf{x}}}(t)\} = \mathbf{A}(t)\mathcal{E}\{\hat{\mathbf{x}}(t)\} + \mathbf{K}(t)\mathcal{E}\{\mathbf{z}(t)\}. \tag{8.1-37}$$

We obtain, from Eq. (8.1-34),

$$\mathcal{E}\{\mathbf{z}(t)\} = \mathcal{E}\{\mathbf{H}(t)\mathbf{x}(t) + \mathbf{v}(t)\} = \mathbf{H}(t)\mathcal{E}\{\mathbf{x}(t_o)\}. \tag{8.1-38}$$

Thus we obtain, as a differential equation for the propagation of the mean of the system state,

$$\dot{\bar{\mathbf{x}}}(t) = [\mathbf{A}(t) + \mathbf{K}(t)\mathbf{H}(t)]\bar{\mathbf{x}}(t) \tag{8.1-39}$$

which follows directly from combining Eqs. (8.1-36) through (8.1-38). By taking the unconditional expectation of Eq. (8.1-33), we see that the mean of the system state vector must also propagate according to

$$\dot{\bar{\mathbf{x}}}(t) = \mathbf{F}(t)\bar{\mathbf{x}}(t) \tag{8.1-40}$$

since $\mathbf{w}(t)$ has zero-mean. By direct comparison of the foregoing two equations we see that one condition for an unbiased filter is that

$$\mathbf{A}(t) = \mathbf{F}(t) - \mathbf{K}(t)\mathbf{H}(t). \tag{8.1-41}$$

A second condition for the solution of Eqs. (8.1-37) or (8.1-39) and (8.1-40) to be the same is that

$$\hat{\mathbf{x}}(t_o) = \mathcal{E}\{\hat{\mathbf{x}}(t_o)\} = \mathcal{E}\{\mathbf{x}(t_o)\} = \bar{\mathbf{x}}_o. \tag{8.1-42}$$

We may also show that these two requirements are necessary but not sufficient for the filter to be unbiased so that

$$\mathcal{E}\{\mathbf{x}(t) \mid Z(t)\} = \mathcal{E}\{\hat{\mathbf{x}}(t) \mid Z(t)\} \tag{8.1-43}$$

where the conditioning variable $Z(t)$ represents all the data $\mathbf{z}(\tau)$ for $t_o \leq \tau \leq t$. When we satisfy these unconditionally unbiased requirements by imposing Eqs. (8.1-41) and (8.1-42) on the solution filter algorithm defined by Eq. (8.1-35), we have

$$\dot{\hat{\mathbf{x}}}(t) = \mathbf{F}(t)\hat{\mathbf{x}}(t) + \mathbf{K}(t)[\mathbf{z}(t) - \mathbf{H}(t)\hat{\mathbf{x}}(t)] \tag{8.1-44}$$

$$\hat{\mathbf{x}}(t_o) = \bar{\mathbf{x}}_o.$$

Our filter algorithm is arranged in this form for convenience. When $\mathbf{K}(t)$ is optimally selected to minimize the error variance, then the quantity $\mathbf{z}(t) - \mathbf{H}(t)\hat{\mathbf{x}}(t)$ turns out to be a white noise stochastic process called the *innovation process*, because it contains all of the new information in the observation $\mathbf{z}(t)$. This innovation has many important properties, which are discussed in advanced works on estimation theory [7]. We shall mention some of these in our later developments.

We now seek to complete the derivation by determining $\mathbf{K}(t)$ such that the filter has minimum error variance. We shall find $\mathbf{K}(t)$ such as to minimize

$$J(t) = \operatorname{tr}[\operatorname{var}\{\tilde{\mathbf{x}}(t)\}] = \operatorname{tr}[\mathbf{P}(t)] \tag{8.1-45}$$

where $\operatorname{tr}[\cdot]$ denotes the trace operation, and $\tilde{\mathbf{x}}(t)$ is the error in filtering which we will define as

$$\tilde{\mathbf{x}}(t) = \mathbf{x}(t) - \hat{\mathbf{x}}(t). \tag{8.1-46}$$

We will now obtain an expression for the error variance which we have defined by the symbol $\mathbf{P}(t)$. From the expression for the message model, Eq. (8.1-34), and the unconditionally unbiased postulated filter, Eq. (8.1-44), we have

$$\dot{\tilde{\mathbf{x}}}(t) = \dot{\mathbf{x}}(t) - \dot{\hat{\mathbf{x}}}(t) = \mathbf{F}(t)\mathbf{x}(t) - \mathbf{G}(t)\mathbf{w}(t) - \mathbf{F}(t)\hat{\mathbf{x}}(t) - \mathbf{K}(t)[\mathbf{z}(t) - \mathbf{H}(t)\hat{\mathbf{x}}(t)]$$

which becomes, using Eqs. (8.1-34) and (8.1-46),

$$\dot{\tilde{\mathbf{x}}}(t) = [\mathbf{F}(t) - \mathbf{K}(t)\mathbf{H}(t)]\tilde{\mathbf{x}}(t) + \mathbf{u}(t), \tag{8.1-47}$$

where

$$\mathbf{u}(t) = \mathbf{G}(t)\mathbf{w}(t) - \mathbf{K}(t)\mathbf{v}(t) \tag{8.1-48}$$

is a zero-mean white noise term with moments

$$\text{cov}\{\mathbf{u}(t), \mathbf{u}(\tau)\} = [\mathbf{G}(t)\mathbf{Q}(t)\mathbf{G}^T(t) + \mathbf{K}(t)\mathbf{R}(t)\mathbf{K}^T(t)]\,\delta(t - \tau),$$

and

$$\text{cov}\{\tilde{\mathbf{x}}(t), \mathbf{u}(\tau)\} = \mathbf{0} \qquad \text{for} \quad t_1 \geq t_2.$$

We may now write down by inspection the matrix differential equation for the propagation of the error variance. Analogous to the relationship between Eq. (8.1-1) and the associated variance relationship, Eq. (8.1-22), we have for the variance of Eq. (8.1-47),

$$\dot{\mathbf{P}}(t) = [\mathbf{F}(t) - \mathbf{K}(t)\mathbf{H}(t)]\mathbf{P}(t) - \mathbf{P}(t)[\mathbf{F}(t) - \mathbf{K}(t)\mathbf{H}(t)]^T + \mathbf{G}(t)\mathbf{Q}(t)\mathbf{G}^T(t) + \mathbf{K}(t)\mathbf{R}(t)\mathbf{K}^T(t). \tag{8.1-49}$$

This relation is important in its own right since it can be used to conduct studies of the effect of suboptimal gains $\mathbf{K}(t)$ upon the filter error variance. To solve this equation we need an initial condition. Since $\hat{\mathbf{x}}(t_o)$ is a deterministic quantity $\bar{\mathbf{x}}_o$, we have

$$\mathbf{P}(t_o) = \text{var}\{\mathbf{x}(t_o) - \hat{\mathbf{x}}(t_o)\} = \text{var}\{\mathbf{x}(t_o)\} = \mathbf{P}_o \tag{8.1-50}$$

as the needed initial condition.

Our basic objective now is to minimize the cost function J of Eq. (8.1-45) at each instant of time by proper choice of $\mathbf{K}(t)$. It turns out, for this specific problem, that this will be accomplished if we minimize

$$\frac{dJ(t)}{dt} = \text{tr}\,[\dot{\mathbf{P}}(t)] \tag{8.1-51}$$

where $\dot{\mathbf{P}}(t)$ is given by Eq. (8.1-49). In order to minimize this scalar error quantity $dJ(t)/dt$ by choice of the matrix $\mathbf{K}(t)$, it is necessary to use concepts from matrix calculus [12]. The reader can easily visualize the needed steps by analogy to the case where $\mathbf{K}(t)$ and $\mathbf{P}(t)$ are scalars. We have

$$\frac{d\text{tr}\,\{\mathbf{F}(t) - \mathbf{K}(t)\mathbf{H}(t)]\mathbf{P}(t)\}}{d\mathbf{K}(t)} = -\mathbf{P}(t)\mathbf{H}^T(t)$$

$$\frac{d\text{tr}\,\{\mathbf{P}(t)[\mathbf{F}(t) - \mathbf{K}(t)\mathbf{H}(t)]^T\}}{d\mathbf{K}(t)} = -\mathbf{P}(t)\mathbf{H}^T(t)$$

$$\frac{d\text{tr}\,\{\mathbf{K}(t)\mathbf{R}(t)\mathbf{K}^T(t)\}}{d\mathbf{K}(t)} = 2\mathbf{K}(t)\mathbf{R}(t)$$

such that Eq. (8.1-51) leads directly to the requirement that

$$\hat{\mathbf{K}}(t) = \mathbf{P}(t)\mathbf{H}^T(t)\mathbf{R}^{-1}(t). \tag{8.1-52}$$

Therefore we have

$$\dot{\mathbf{P}}(t) = \mathbf{F}(t)\mathbf{P}(t) + \mathbf{P}(t)\mathbf{F}^T(t) + \mathbf{G}(t)\mathbf{R}(t)\mathbf{G}^T(t) - \mathbf{P}(t)\mathbf{H}^T(t)\mathbf{R}^{-1}(t)\mathbf{H}(t)\mathbf{P}(t) \tag{8.1-53}$$

as the minimum error variance. This completes this simplified derivation of the optimum unbiased minimum error variance linear filter. Figure 8.1-3 should serve as a convenient summary of the pertinent results of this derivation of the optimum linear filter equations.

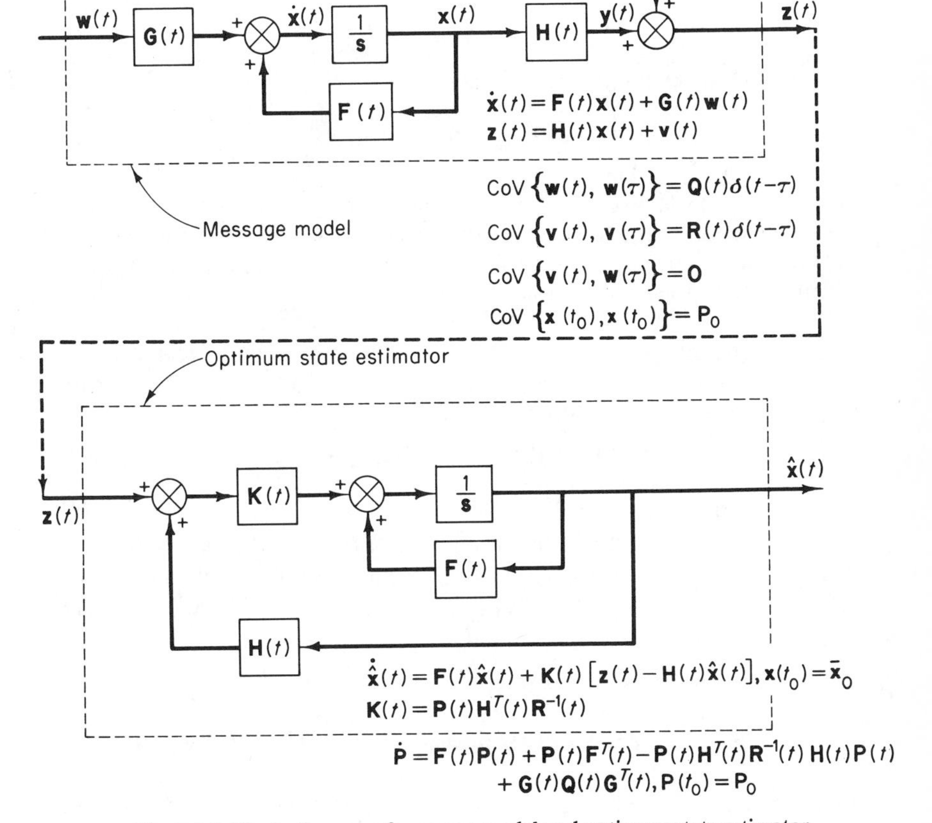

Fig. 8.1-3 Block diagram of message model and optimum state estimator.

Example 8.1-3. As a simple example, we consider the system given by the message model

$$\dot{x}(t) = 0, \qquad \mathcal{E}[x(0)] = 1, \qquad \text{var}\,[x(0)] = 10,$$

and the observation model

$$z(t) = 2x(t) + v(t), \qquad \mathcal{E}[v(t)] = 0, \qquad \text{cov}\,[v(t), v(\tau)] = 1\delta(t - \tau).$$

The problem is to determine the best minimum error variance filter so as to determine $x(t)$ from the measurements $z(t)$. The solution is given by our Kalman filter algorithms which can be determined for this specific example as follows:

1. First we identify $F = G = 0$, $H = 2$, $R = 1$, and $P(0) = 10$.
2. We solve the error variance equation for this problem which becomes

$$\dot{P} = -4P^2(t), \qquad P(0) = 10$$

and has the solution

$$P(t) = \frac{10}{1 + 40t}.$$

3. Thus, we see that

$$K(t) = \frac{20}{1 + 40t},$$

and the filter equation becomes

$$\dot{\hat{x}} = \frac{20}{1 + 40t}[z(t) - 2\hat{x}(t)], \qquad \hat{x}(0) = 1.$$

Before concluding this part of our discussion of the optimal estimator, we should note that, for the formulation given, the system model was assumed to be known exactly, the noise statistics are known perfectly, and the noise contaminates the measurement of every state.

These results may be used to establish several important facts. Among these are two versions of the orthogonal projection lemma, which indicate that for the optimum linear filter,

$$\operatorname{cov}\{\mathbf{z}(\lambda), \tilde{\mathbf{x}}(t)\} = \mathbf{0} \qquad \text{for all } t_o \leq \lambda \leq t \tag{8.1-54}$$

and

$$\operatorname{cov}\{\hat{\mathbf{x}}(t), \tilde{\mathbf{x}}(t)\} = \mathbf{0}; \tag{8.1-55}$$

these results follow directly from the characteristics of the optimum filter we have just obtained. This orthogonal projection lemma is quite an important one and can be proven to be valid under more general conditions than we have used here [7].

8.1-1 *Further properties of the minimum error variance linear filter*

In many cases, it will not be possible or desirable to use the correct Kalman gain $\mathbf{K}(t)$ in Eq. (8.1-44). In this case we may attempt to use an approximate suboptimal gain $\mathbf{K}_{so}(t)$ in Eq. (8.1-44) such that

$$\dot{\hat{\mathbf{x}}}_{so}(t) = \mathbf{F}(t)\hat{\mathbf{x}}_{so}(t) + \mathbf{K}_{so}(t)[\mathbf{z}(t) - \mathbf{H}(t)\hat{\mathbf{x}}_{so}(t)].$$

We may then define an error

$$\tilde{\mathbf{x}}_{so}(t) = \mathbf{x}(t) - \hat{\mathbf{x}}_{so}(t)$$

and show that the covariance matrix of this error, $\mathbf{E}(t) = \text{cov}\,[\tilde{\mathbf{x}}_{so}(t), \tilde{\mathbf{x}}_{so}(t)]$ where $\mathbf{D}(t) = \mathbf{F}(t) - \mathbf{K}_{so}(t)\mathbf{H}(t)$, is

$$\dot{\mathbf{E}} = \mathbf{D}(t)\mathbf{E}(t) + \mathbf{E}(t)\mathbf{D}^T(t) + \mathbf{K}_{so}(t)\mathbf{R}(t)\mathbf{K}_{so}^T(t) + \mathbf{G}(t)\mathbf{Q}(t)\mathbf{G}^T(t). \quad (8.1\text{-}56)$$

By use of Eq. (8.1-56), error analysis of optimal and suboptimal filters can be conducted. Equation (8.1-56) is, of course, entirely equivalent to Eq. (8.1-49).

Errors in Kalman filtering can arise from any of several sources other than use of a suboptimal Kalman gain. Among these errors are such things as:

1. An incorrect model for the system dynamics $\mathbf{F}(t)$, or the control distribution matrix $\mathbf{G}(t)$, or the measurement matrix $\mathbf{H}(t)$;
2. An incorrect estimate of the initial error variance; instead of the correct $\mathbf{P}(0)$, an incorrect $\mathbf{P}_c(0)$ is used;
3. An incorrect estimate of the time history of the covariance of the plant noise; an incorrect $\mathbf{Q}_c(t)$ is used rather than the correct $\mathbf{Q}(t)$;
4. An incorrect covariance matrix for the measurement noise; an incorrect $\mathbf{R}_c(t)$ is used rather than the correct $\mathbf{R}(t)$;
5. An intentionally incorrect Kalman gain $\mathbf{K}(t)$, which may be used in order to lessen the computational burden;
6. A combination of the above.

When any of these errors are present, the resulting Kalman filter is no longer an optimal one, but becomes suboptimal. We have just indicated the error determination for a suboptimal Kalman gain. Errors in $\mathbf{F}(t)$, $\mathbf{G}(t)$, and $\mathbf{H}(t)$ can be analyzed using the sensitivity methods of Chapter 7. Here we will consider errors 2, 3, and 4 above. If we designate the suboptimal state estimate by $\hat{\mathbf{x}}_a(t)$ and assume that the correct system characteristic matrix and control distribution matrix, as well as the measurement matrix, are used, then we may write the differential equation for the suboptimal estimator in the case of filtering as:

$$\dot{\hat{\mathbf{x}}}_a = \mathbf{F}(t)\hat{\mathbf{x}}_a(t) + \mathbf{K}_c(t)[\mathbf{z}(t) - \mathbf{H}(t)\hat{\mathbf{x}}_a(t)] \quad (8.1\text{-}57)$$

$$\mathbf{K}_c(t) = \mathbf{P}_c(t)\mathbf{H}^T(t)\mathbf{R}_c^{-1}(t) \quad (8.1\text{-}58)$$

$$\dot{\mathbf{P}}_c = \mathbf{F}(t)\mathbf{P}_c(t) + \mathbf{P}_c(t)\mathbf{F}^T(t) - \mathbf{P}_c(t)\mathbf{H}^T(t)\mathbf{R}_c^{-1}(t)\mathbf{H}(t)\mathbf{P}_c(t) + \mathbf{G}(t)\mathbf{Q}_c(t)\mathbf{G}^T(t). \quad (8.1\text{-}59)$$

In the equations above, $\mathbf{K}_c$ is the suboptimal Kalman gain, and $\mathbf{P}_c$ is the Riccati equation solved with the incorrect values of the covariance for the plant noise $\mathbf{Q}_c(t)$, the incorrect covariance of the observation noise $\mathbf{R}_c(t)$, and incorrect initial error variance $\mathbf{P}_c(0)$.

The actual covariance of the estimation error is not now obtained from the Riccati solution for $\mathbf{P}_c(t)$, but instead may be obtained from another

Riccati equation [7],

$$\dot{\mathbf{P}}_a = [\mathbf{F}(t) - \mathbf{K}_c(t)\mathbf{H}(t)]\mathbf{P}_a(t) + \mathbf{P}_a(t)[\mathbf{F}(t) - \mathbf{K}_c(t)\mathbf{H}(t)]^T + \mathbf{K}_c(t)\mathbf{R}(t)\mathbf{K}_c^T(t) + \mathbf{G}(t)\mathbf{Q}(t)\mathbf{G}^T(t), \qquad \mathbf{P}_a(0) = \mathbf{P}(0), \tag{8.1-60}$$

as we may easily show. Clearly, this equation represents actual errors in filtering and is a special form of Eq. (8.1-56). Here we define

$$\mathbf{P}_a(t) = \mathcal{E}\{[\hat{\mathbf{x}}_a(t) - \mathbf{x}(t)][\hat{\mathbf{x}}_a(t) - \mathbf{x}(t)]^T\},$$

and therefore we know the initial condition for Eq. (8.1-60). If the covariance of the errors for the suboptimal estimate is close to that obtained by the optimal estimate, then we conclude that the suboptimal estimator is a good one, despite the ignorance in knowledge of the initial error variance and the initial covariance matrices for the plant noise and the measurement noise. In order to discern the merits of the suboptimal estimator, it is convenient to talk of three related error measures. We will let $\mathbf{E}_{ca}(t)$ be the difference between the computed Riccati solution $\mathbf{P}_c(t)$ and the actual error covariance $\mathbf{P}_a(t)$. Also of interest is the difference between the error covariance matrix for the suboptimal estimate and the error covariance for the optimal Kalman filter; in other words, the difference between $\mathbf{P}_a(t)$ and $\mathbf{P}(t)$. Finally, we desire to compute the difference between the Riccati equation for the suboptimal estimate, which is not the variance of the errors, and the variance matrix for the actual optimum filter. Thus, we are interested in the three error quantities:

$$\mathbf{E}_{ca}(t) = \mathbf{P}_c(t) - \mathbf{P}_a(t) \tag{8.1-61}$$

$$\mathbf{E}_{ao}(t) = \mathbf{P}_a(t) - \mathbf{P}(t) \tag{8.1-62}$$

$$\mathbf{E}_{co}(t) = \mathbf{P}_c(t) - \mathbf{P}(t). \tag{8.1-63}$$

It is possible for us to show that the differential equations whose solutions represent these various error quantities are

$$\dot{\mathbf{E}}_{ca} = [\mathbf{F}(t) - \mathbf{K}_c(t)\mathbf{H}(t)]\mathbf{E}_{ca}(t) + \mathbf{E}_{ca}(t)[\mathbf{F}(t) - \mathbf{K}_c(t)\mathbf{H}(t)]^T + \mathbf{K}_c(t)\mathbf{R}_\epsilon(t)\mathbf{K}_c^T(t) + \mathbf{G}(t)\mathbf{Q}_\epsilon(t)\mathbf{G}^T(t) \tag{8.1-64}$$

$$\dot{\mathbf{E}}_{ao} = [\mathbf{F}(t) - \mathbf{K}_c(t)\mathbf{H}(t)]\mathbf{E}_{ao}(t) + \mathbf{E}_{ao}(t)[\mathbf{F}(t) - \mathbf{K}_c(t)\mathbf{H}(t)]^T + [\mathbf{K}_c(t)\mathbf{R}(t) - \mathbf{P}(t)\mathbf{H}^T(t)]\mathbf{R}(t)^{-1}[\mathbf{K}_c(t)\mathbf{R}(t) - \mathbf{P}(t)\mathbf{H}^T(t)]^T \tag{8.1-65}$$

$$\dot{\mathbf{E}}_{co} = [\mathbf{F}(t) - \mathbf{K}_c(t)\mathbf{H}(t)]\mathbf{E}_{co}(t) + \mathbf{E}_{co}(t)[\mathbf{F}(t) - \mathbf{K}_c(t)\mathbf{H}(t)]^T + \mathbf{E}_{co}(t)\mathbf{H}^T(t)\mathbf{R}_c^{-1}(t)\mathbf{H}(t)\mathbf{E}_{co}(t) + \mathbf{G}(t)\mathbf{Q}_\epsilon(t)\mathbf{G}^T(t) + \mathbf{P}(t)\mathbf{H}^T(t)[\mathbf{R}(t)\mathbf{R}_\epsilon^{-1}(t)\mathbf{R}(t) + \mathbf{R}(t)]^{-1}\mathbf{H}(t)\mathbf{P}(t), \tag{8.1-66}$$

where initial conditions follow directly from Eqs. (8.1-61), (8.1-62), (8.1-63), and

$$\mathbf{R}_\epsilon(t) = \mathbf{R}_c(t) - \mathbf{R}(t), \mathbf{Q}_\epsilon(t) = \mathbf{Q}_c(t) - \mathbf{Q}(t). \tag{8.1-67}$$

Equations (8.1-60) through (8.1-67) allow us to discern errors due to ignorance of the various covariance matrices representing plant and measurement noise and initial error variance.

Example 8.1-4. Let us consider a simple example wherein we desire to estimate an unknown constant. We assume the model

$$\dot{x} = 0, \qquad z = x + v, \qquad R = \sigma_v^2.$$

From Eqs. (8.1-29), (8.1-34), and (8.1-35) we easily obtain

$$\dot{\hat{x}} = K(t)[z - \hat{x}], \qquad \hat{x}(0) = 0, \qquad K(t) = \frac{p(t)}{\sigma_v^2}$$

$$\dot{p} = -\frac{p^2}{\sigma_v^2}; \qquad p(0) = p_0.$$

The variance equation has the solution

$$p(t) = \frac{p_0}{1 + p_0 t/\sigma_v^2}.$$

We note that in this example, $p(t) \longrightarrow 0$ as $t \longrightarrow \infty$; that is, x is estimated with zero error after sufficient observation time. Also, the gain $K(t)$ becomes zero as $t \longrightarrow \infty$. The implication of this is that less and less use is made of new measurements as time proceeds.

Now let us suppose that incorrect values of σ_v^2 and $p(0)$, σ_{vc}^2 and p_{oc}, are used in computation of the Kalman gain. Equation (8.1-44) for the error variance of error with the improper variance becomes

$$\dot{p}_a = -\frac{2p_{oc}}{\sigma_{vc}^2 + p_{oc}t} p_a(t) + \left[\frac{p_{oc}}{\sigma_{vc}^2 + p_{oc}t}\right]^2 \sigma_{vc}^2, \qquad p_a(0) = p_0.$$

Solution of this differential equation will allow us to determine the actual error variance which naturally will always be no less than the minimum error variance. Since we now know $p_a(t)$, $p(t)$, and $p_c(t)$, we can determine the errors of Eqs. (8.1-61), (8.1-62), and (8.1-63) by subtraction. Alternatively the given differential equations for these error quantities may be solved.

In a large number of practical applications, it is necessary to estimate states of systems which are defined by nonlinear differential equations. Thus, the first-order system of linear differential equations, which we have used previously, is not immediately applicable to many engineering problems of significance. In order to use the theory of the Kalman filter, the nonlinearity of the dynamic equations describing the system and also the output equations must be removed. In order to accomplish this, we may make use of a nominal solution of the nonlinear differential equations. This nominal solution must provide a good approximation to the actual system behavior; "good approximation" means that the difference between the nominal and the actual solutions can be described by a system of linear differential equations, which are commonly known as "linear perturbation equations."

There are several ways of effectively accomplishing this linearization. The method of linearization about a nominal trajectory has been developed in

Example 5.1-2. Here we discuss the classic method of linearization for application of the Kalman filter. This method can be conveniently summarized as follows:

1. We compute a nominal trajectory $\mathbf{x}^n(t)$ for a control $\mathbf{u}^n(t)$ using the equations of motion $\dot{\mathbf{x}} = \mathbf{f}(\mathbf{x}, \mathbf{u}, \mathbf{w}, t)$ and $\mathbf{z}(t) = \mathbf{h}(\mathbf{x}) + \mathbf{v}(t)$ to obtain $\dot{\mathbf{x}}^n = \mathbf{f}(\mathbf{x}^n, \mathbf{u}^n, \mathbf{0}, \mathbf{t})$, $\mathbf{x}^n(\mathbf{0}) = \hat{\mathbf{x}}(\mathbf{0})$.
2. We determine that the true trajectory $\mathbf{x}(t)$ equals $\mathbf{x}^n(t) + \Delta\mathbf{x}(t)$ and that the best estimate of the true trajectory is $\hat{\mathbf{x}}(t) = \mathbf{x}^n(t) + \Delta\hat{\mathbf{x}}(t)$, where $\Delta\hat{\mathbf{x}}(t)$ is the best estimate of the deviation.
3. We compute the nominal observation $\mathbf{z}^n(t)$ from $\mathbf{z}^n(t) = \mathbf{h}(\mathbf{x}^n)$ and then define the deviation in observation $\Delta\mathbf{z}$ by $\mathbf{z}(t) = \mathbf{z}^n(t) + \Delta\mathbf{z}(t)$.
4. Then we form $\Delta\dot{\mathbf{x}}$ and $\Delta\mathbf{z}(t)$ from (2) and (3) to obtain $\Delta\dot{\mathbf{x}} = \mathbf{f}(\mathbf{x}, \mathbf{u}^n, \mathbf{w}, t) - \mathbf{f}(\mathbf{x}^n, \mathbf{u}^n, \mathbf{0}, t)$ and $\Delta\mathbf{z}(t) = \mathbf{h}(\mathbf{x}) - \mathbf{h}(\mathbf{x}^n) + \mathbf{v}(t)$.
5. Finally, we linearize the nonlinear system equations about $\mathbf{x}^n(t)$ to obtain, after dropping all but the linear terms in a Taylor series expansion,

$$\Delta\dot{\mathbf{x}} = \mathbf{F}(t)\Delta\mathbf{x}(t) + \mathbf{G}(t)\mathbf{w}(t), \qquad \mathbf{F}(t) = \frac{\partial\mathbf{f}(\mathbf{x}^n, \mathbf{u}, \mathbf{0}, t)}{\partial\mathbf{x}^n},$$
$$\mathbf{G}(t) = \left.\frac{\partial\mathbf{f}(\mathbf{x}^n, \mathbf{u}^n, \mathbf{w}, t)}{\partial\mathbf{w}}\right|_{\mathbf{w}=0} \tag{8.1-68}$$

and linearize the nonlinear output equations about $\mathbf{x}^n(t)$ to obtain

$$\Delta\mathbf{z}(t) = \mathbf{H}(t)\Delta\mathbf{x}(t) + \mathbf{v}(t), \qquad \mathbf{H}(t) = \frac{\partial\mathbf{h}(\mathbf{x}^n)}{\partial\mathbf{x}^n}. \tag{8.1-69}$$

Then the linear theory is applied to compute $\Delta\hat{\mathbf{x}}(t)$ which leads to $\hat{\mathbf{x}}(t)$ from $\hat{\mathbf{x}}(t) = \mathbf{x}^n(t) + \Delta\hat{\mathbf{x}}(t)$. The $\mathbf{F}(t)$, $\mathbf{G}(t)$, and $\mathbf{H}(t)$ matrices of Eqs. (8.1-68) and (8.1-69) are used for this calculation.

Often the linearized filter solution that we have just obtained is quite acceptable. If however the actual $\mathbf{x}(t)$ deviates considerably from the nominal $\mathbf{x}^n(t)$, then a phenomenon known as *divergence* can occur. The extended Kalman filter can, at the expense of increased complexity, often prevent this divergence. If we expand the nonlinear differential equations in $\mathbf{f}$ and $\mathbf{h}$ about an assumed known optimum estimate $\mathbf{x}(t) = \hat{\mathbf{x}}(t)$ and the zero-mean noise $\mathbf{w}(t)$, we obtain for the message and observation model:

$$\dot{\mathbf{x}}(t) = \mathbf{f}(\mathbf{x}, \mathbf{u}, \mathbf{w}, t) \tag{8.1-70}$$

and

$$\mathbf{z}(t) = \mathbf{h}(\mathbf{x}, t) + \mathbf{v}(t), \tag{8.1-71}$$

the relations

$$\dot{\mathbf{x}}(t) = \mathbf{f}(\hat{\mathbf{x}}, \mathbf{u}, \mathbf{0}, t) + \frac{\partial\mathbf{f}(\hat{\mathbf{x}}, \mathbf{u}, \mathbf{0}, t)}{\partial\hat{\mathbf{x}}(t)}[\mathbf{x}(t) - \hat{\mathbf{x}}(t)] + \mathbf{G}(\hat{\mathbf{x}}, t)\mathbf{w}(t) \tag{8.1-72}$$

and

$$\mathbf{z}(t) = \mathbf{h}(\hat{\mathbf{x}}, t) + \frac{\partial\mathbf{h}(\hat{\mathbf{x}}, t)}{\partial\hat{\mathbf{x}}(t)}[\mathbf{x}(t) - \hat{\mathbf{x}}(t)] + \mathbf{v}(t), \tag{8.1-73}$$

where

$$\mathbf{G}(\hat{\mathbf{x}}, t) = \left.\frac{\partial \mathbf{f}(\hat{\mathbf{x}}, \mathbf{u}, \mathbf{w}, t)}{\partial \mathbf{w}(t)}\right|_{\mathbf{w}(t)=0}.$$

We may now repeat the derivation leading to the optimum linear filter. In this derivation it would be assumed that the estimate $\hat{\mathbf{x}}(t)$ in Eqs. (8.1-72) and (8.1-73) is a known quantity. It is a straightforward task to show that the optimum filter algorithms are now

$$\dot{\hat{\mathbf{x}}}(t) = \mathbf{f}(\hat{\mathbf{x}}, \mathbf{u}, \mathbf{0}, t) + \mathbf{K}(t)[\mathbf{z}(t) - \mathbf{h}(\hat{\mathbf{x}}, t)] \tag{8.1-74}$$

$$\mathbf{K}(t) = \mathbf{P}(t)\frac{\partial \mathbf{h}^T(\hat{\mathbf{x}}, t)}{\partial \hat{\mathbf{x}}(t)}\mathbf{R}^{-1}(t) \tag{8.1-75}$$

$$\dot{\mathbf{P}}(t) = \frac{\partial \mathbf{f}(\hat{\mathbf{x}}, \mathbf{u}, \mathbf{0}, t)}{\partial \hat{\mathbf{x}}(t)}\mathbf{P}(t) + \mathbf{P}(t)\frac{\partial \mathbf{f}^T(\hat{\mathbf{x}}, \mathbf{u}, \mathbf{0}, t)}{\partial \hat{\mathbf{x}}(t)} + \mathbf{G}(\hat{\mathbf{x}}, t)\mathbf{Q}(t)\mathbf{G}^T(\hat{\mathbf{x}}, t) - \mathbf{P}(t)\frac{\partial \mathbf{h}^T(\hat{\mathbf{x}}, t)}{\partial \hat{\mathbf{x}}(t)}\mathbf{R}^{-1}(t)\frac{\partial \mathbf{h}(\hat{\mathbf{x}}, t)}{\partial \hat{\mathbf{x}}(t)}\mathbf{P}(t) \tag{8.1-76}$$

with initial conditions

$$\hat{\mathbf{x}}(t_o) = \bar{\mathbf{x}}_o, \qquad \mathbf{P}(t_o) = \mathbf{P}_o. \tag{8.1-77}$$

Example 8.1-5. As a simple example of nonlinear estimation let us consider the system identification [13] problem in which we desire to estimate a system state $x(t)$ and a constant parameter $a(t)$ which evolves according to

$$\dot{x} = ax(t) + w(t), \qquad \dot{a} = 0,$$

where $w(t)$ is a zero-mean white plant noise term. We observe the state of the system after it has been corrupted by a zero-mean white noise term,

$$z(t) = x(t) + v(t).$$

The extended filter algorithms follow directly from Eqs. (8.1-76) through (8.1-77) in which we regard the state of the system as the augmented vector $\mathbf{x}^T = [x, a]$. We easily obtain

$$\dot{\hat{x}} = \hat{a}(t)\hat{x}(t) + r^{-1}(t)p_{11}(t)[z(t) - \hat{x}(t)]$$

$$\dot{\hat{a}} = r^{-1}(t)p_{12}(t)[z(t) - \hat{x}(t)]$$

$$\dot{p}_{11} = 2\hat{a}(t)p_{11}(t) + 2\hat{x}(t)p_{12}(t) - r^{-1}(t)p_{11}^2(t) + q(t)$$

$$\dot{p}_{12} = \hat{a}(t)p_{12}(t) + \hat{x}(t)p_{12}(t) - r^{-1}(t)p_{11}(t)p_{12}(t)$$

$$\dot{p}_{22} = -r^{-1}(t)p_{12}^2(t)$$

with initial conditions

$$\hat{x}(t_o) = \bar{x}_o$$

$$\hat{a}(t_o) = \bar{a}_o$$

$$p_{11}(t_o) = p_{110} = \text{var}\,\{x(t_o)\} = \text{var}\,\{\tilde{x}(t_o)\}$$

$$p_{12}(t_o) = p_{120} = \text{cov}\,\{x(t_o), a(t_o)\}$$

$$p_{22}(t_o) = p_{220} = \text{var}\,\{a(t_o)\} = \text{var}\,\{\tilde{a}(t_o)\}.$$

We note that the Riccati type equations for this problem involve the estimates themselves, and therefore the filter algorithms are more complicated than for an equivalent order linear problem. Also, to prevent a divergence problem, it turns out that we should model the constant parameter a as a Wiener process $\dot{a} = w_2(t)$ where $w_2(t)$ is white noise.

A full discussion of nonlinear filtering would take us far afield from our primary goal of a broad presentation of techniques for optimum systems control. Our limited discussion of nonlinear filtering has indicated that relatively simple extensions of the linear results will often suffice. References [7] and [13] and the references incorporated therein present much detail concerning nonlinear filtering. We will now turn our attention back to the linear case and derive some further important results that give us much insight into the behavior of the vitally important algorithms for linear filtering.

8.2 Further examination of the Kalman filter—continuous time case

Although our derivation of the Kalman filter equations in the previous section is correct and adequate for filter implementation, it is convenient here to rederive the filter equations utilizing a more fundamental approach. We then proceed to a discussion of duality and a further examination of practically important error analysis algorithms. We describe our estimation problem as follows.

We assume that the vector process $\mathbf{s}(t)$, representing the signal, can be described by

$$\mathbf{s}(t) = \mathbf{H}(t)\mathbf{x}(t), \tag{8.2-1}$$

where the state process $\{\mathbf{x}(t), t \in T\}$ satisfies

$$\dot{\mathbf{x}}(t) = \mathbf{F}(t)\mathbf{x}(t) + \mathbf{G}(t)\mathbf{w}(t), \tag{8.2-2}$$

and where the zero-mean plant noise $\mathbf{w}(t)$ is such that

$$\mathcal{E}\{\mathbf{w}(t)\mathbf{w}^T(\tau)\} = \mathbf{Q}(t)\,\delta(t-\tau). \tag{8.2-3}$$

We also specify that the zero-mean measurement noise $\mathbf{v}(t)$ must satisfy

$$\mathcal{E}\{\mathbf{v}(t)\mathbf{v}^T(\tau)\} = \mathbf{R}(t)\,\delta(t-\tau), \tag{8.2-4}$$

and that, for simplicity, it is uncorrelated with the plant noise such that

$$\mathcal{E}\{\mathbf{w}(t)\mathbf{v}^T(\tau)\} = \mathbf{0}. \tag{8.2-5}$$

Let the vector output again be described as

$$\mathbf{z}(t) = \mathbf{s}(t) + \mathbf{v}(t). \tag{8.2-6}$$

We assume that the zero-mean initial state $\mathbf{x}(t_o) = \mathbf{x}_o$ is such that

$$\mathcal{E}\{\mathbf{x}_o\mathbf{x}_o^T\} = \mathbf{P}(t_o) \tag{8.2-7}$$

and

$$\mathcal{E}\{\mathbf{v}(t)\mathbf{x}_o^T\} = \mathcal{E}\{\mathbf{w}(t)\mathbf{x}_o^T\} = 0, \qquad t \geq t_o. \tag{8.2-8}$$

The problem solved by the Kalman filter is the determination of a linear function of $\mathbf{Z}(t)$, $\hat{\mathbf{x}}(t)$, such that the criterion

$$J = \mathcal{E}\{[\mathbf{x}(t) - \hat{\mathbf{x}}(t)]^T[\mathbf{x}(t) - \hat{\mathbf{x}}(t)]\} \tag{8.2-9}$$

is minimized and $\hat{\mathbf{x}}(t)$ is an unbiased estimate of $\mathbf{x}(t)$, in that $\mathcal{E}\{\hat{\mathbf{x}}(t)\} = \mathcal{E}\{\mathbf{x}(t)\}$.

We define $\tilde{\mathbf{x}}(t) = \mathbf{x}(t) - \hat{\mathbf{x}}(t)$ which represents the estimation error at time t. We define the error variance matrix as $\mathbf{P}(t) \doteq \mathcal{E}\{\tilde{\mathbf{x}}(t)\tilde{\mathbf{x}}^T(t)\}$. We wish to determine functions $\mathbf{A}(t, \tau)$ and $\mathbf{B}(t, t_o)$ such that the linear estimate

$$\hat{\mathbf{x}}(t) = \mathbf{B}(t, t_o)\hat{\mathbf{x}}(t_o) + \int_{t_o}^{t} \mathbf{A}(t, \tau)\mathbf{z}(\tau)\, d\tau \tag{8.2-10}$$

will minimize $\mathbf{P}(t)$ with $\mathcal{E}\{\tilde{\mathbf{x}}(t)\} = \mathbf{0}$. Since

$$\tilde{\mathbf{x}}(t) = \mathbf{x}(t) - \hat{\mathbf{x}}(t) = \mathbf{x}(t) - \boldsymbol{B}(t, t_o)\hat{\mathbf{x}}(t_o) - \int_{t_o}^{t} \boldsymbol{A}(t, \tau)\mathbf{z}(\tau)\, d\tau, \tag{8.2-11}$$

it follows that:

$$\begin{aligned}\mathbf{P}(t) = \mathcal{E}\{\tilde{\mathbf{x}}(t)\tilde{\mathbf{x}}^T(t)\} = \mathcal{E}\{\mathbf{x}(t)\mathbf{x}^T(t)\} - \mathcal{E}\left\{\int_{t_o}^{t} \mathbf{x}(t)\mathbf{z}^T(\tau)\boldsymbol{A}^T(t, \tau)\, d\tau\right\} - \\ \mathcal{E}\left\{\int_{t_o}^{t} \boldsymbol{A}(t, \tau)\mathbf{z}(\tau)\mathbf{x}^T(t)\, d\tau\right\} + \mathcal{E}\left\{\int_{t_o}^{t}\int_{t_o}^{t} \boldsymbol{A}(t, \tau)\mathbf{z}(\tau)\mathbf{z}^T(\lambda)\boldsymbol{A}^T(t, \lambda)\, d\tau\, d\lambda\right\} + \\ \mathcal{E}\{\boldsymbol{B}(t, t_o)\hat{\mathbf{x}}(t_o)\hat{\mathbf{x}}^T(t_o)\boldsymbol{B}^T(t, t_o)\} - \\ \mathcal{E}\left\{\boldsymbol{B}(t, t_o)\hat{\mathbf{x}}(t_o)\left[\mathbf{x}^T(t) - \int_{t_o}^{t} \mathbf{z}^T(\tau)\boldsymbol{A}^T(t, \tau)\, d\tau\right]\right\} - \\ \mathcal{E}\left\{\left[\mathbf{x}(t) - \int_{t_o}^{t} \boldsymbol{A}(t, \tau)\mathbf{z}(\tau)\, d\tau\right]\hat{\mathbf{x}}^T(t_o)\boldsymbol{B}^T(t, t_o)\right\}.\end{aligned} \tag{8.2-12}$$

We take the indicated expected values, and note that, since we are going to obtain an optimum $\hat{\mathbf{x}}(t_o)$, $\hat{\mathbf{x}}(t_o)$ is not random. Use of the expectation of Eq. (8.2-11) in the last two terms of Eq. (8.2-12) yields, for zero-mean random processes,

$$\begin{aligned}\mathbf{P}(t) = \operatorname{cov}[\tilde{\mathbf{x}}(t), \tilde{\mathbf{x}}(t)] = \operatorname{cov}[\mathbf{x}(t), \mathbf{x}(t)] - \int_{t_o}^{t} \operatorname{cov}[\mathbf{x}(t), \mathbf{z}(\tau)]\boldsymbol{A}^T(t, \tau)\, d\tau - \\ \int_{t_o}^{t} \boldsymbol{A}(t, \tau)\operatorname{cov}[\mathbf{z}(\tau), \mathbf{x}(t)]\, d\tau + \int_{t_o}^{t}\int_{t_o}^{t} \boldsymbol{A}(t, \tau)\operatorname{cov}[\mathbf{z}(\tau), \mathbf{z}(\lambda)]\boldsymbol{A}^T(t, \lambda)\, d\tau\, d\lambda - \\ \boldsymbol{B}(t, t_o)\hat{\mathbf{x}}(t_o)\hat{\mathbf{x}}^T(t_o)\boldsymbol{B}^T(t, t_o).\end{aligned} \tag{8.2-13}$$

Now we apply the basic variational calculus: In Eq. (8.2-13), let

$$\begin{aligned}\boldsymbol{A}(t, \tau) &= \hat{\boldsymbol{A}}(t, \tau) + \epsilon\boldsymbol{\eta}(t, \tau) \\ \boldsymbol{B}(t, \tau) &= \hat{\boldsymbol{B}}(t, \tau) + \epsilon\mathbf{v}(t, \tau) \\ \hat{\mathbf{x}}(t_o) &= \hat{\mathbf{x}}_o(t_o) + \epsilon\boldsymbol{\Gamma}(t_o).\end{aligned} \tag{8.2-14}$$

We must then solve

$$\left.\frac{\partial \operatorname{cov}[\tilde{\mathbf{x}}(t, \tilde{\mathbf{x}}(t)]}{\partial \epsilon}\right|_{\epsilon=0} = \mathbf{0} \tag{8.2-15}$$

for $\hat{\boldsymbol{A}}(t, \tau)$ and $\hat{\mathbf{x}}_o(t_o)$. Equation (8.2-15) must be true independent of $\mathbf{v}(t, \tau)$,

$\boldsymbol{\eta}(t, \tau)$, and $\boldsymbol{\Gamma}(t_o)$. Solving Eq. (8.2-15), we obtain:

$$-\int_{t_o}^{t} \boldsymbol{\eta}(t, \tau) \operatorname{cov}[\mathbf{z}(\tau), \mathbf{x}(t)]\, d\tau + \int_{t_o}^{t}\int_{t_o}^{t} \boldsymbol{\eta}(t, \tau) \operatorname{cov}[\mathbf{z}(\tau), \mathbf{z}(\lambda)]\hat{\boldsymbol{A}}^T(t, \lambda)\, d\tau\, d\lambda +$$
$$\mathbf{v}(t, t_o)\hat{\mathbf{x}}_o(t_o)\hat{\mathbf{x}}_o^T(t_o)\hat{\boldsymbol{B}}^T(t, t_o) + \hat{\boldsymbol{B}}(t, t_o)[\hat{\mathbf{x}}_o(t_o)\boldsymbol{\Gamma}^T(t_o) + \boldsymbol{\Gamma}(t_o)\hat{\mathbf{x}}_o^T(t_o)]\hat{\boldsymbol{B}}^T(t, t_o) -$$
$$\int_{t_o}^{t} \operatorname{cov}[\mathbf{x}(t), \mathbf{z}(\tau)]\boldsymbol{\eta}^T(t, \tau)\, d\tau + \int_{t_o}^{t}\int_{t_o}^{t} \hat{\boldsymbol{A}}(t, \tau) \operatorname{cov}[\mathbf{z}(\tau), \mathbf{z}(\lambda)]\boldsymbol{\eta}^T(t, \lambda)\, d\tau\, d\lambda -$$
$$\hat{\boldsymbol{B}}(t, t_o)\hat{\mathbf{x}}_o(t_o)\hat{\mathbf{x}}_o^T(t_o)\mathbf{v}^T(t, t_o) = \mathbf{0}. \tag{8.2-16}$$

We note that the last three terms in Eq. (8.2-16) are simply the transpose of the first three. Thus we can set either pair equal to zero. Thus, we have

$$\mathbf{0} = \int_{t_o}^{t} \left\{\operatorname{cov}[\mathbf{x}(t), \mathbf{z}(\lambda)] - \int_{t_o}^{t} \hat{\boldsymbol{A}}(t, \tau) \operatorname{cov}[\mathbf{z}(\tau), \mathbf{z}(\lambda)]\, d\tau\right\} \boldsymbol{\eta}^T(t, \lambda)\, d\lambda +$$
$$\hat{\boldsymbol{B}}(t, t_o)\hat{\mathbf{x}}_o(t_o)\hat{\mathbf{x}}_o^T(t_o)\mathbf{v}^T(t, t_o) \tag{8.2-17}$$

independent of $\boldsymbol{\eta}(t, \lambda)$ and $\mathbf{v}(t, t_o)$. But in order to make the middle term of Eq. (8.2-16) zero for arbitrary $\boldsymbol{\Gamma}(t)$, we must have $\hat{\mathbf{x}}_o(t_o) = \mathbf{0}$. Thus the optimum filter is characterized by either of the equivalent requirements

$$\operatorname{cov}[\mathbf{x}(t), \mathbf{z}(\lambda)] = \int_{t_o}^{t} \hat{\boldsymbol{A}}(t, \tau) \operatorname{cov}[\mathbf{z}(\tau), \mathbf{z}(\lambda)]\, d\tau, \qquad t_o < \lambda < t \tag{8.2-18}$$

$$\operatorname{cov}\{\tilde{\mathbf{x}}(t), z(\lambda)\} = \mathbf{0}, \qquad \operatorname{cov}\{\tilde{\mathbf{x}}(t), \hat{\mathbf{x}}(t)\} = \mathbf{0} \tag{8.2-19}$$

each of which is a statement of the orthogonal projection lemma.

As stated, we wish to reduce this matrix Wiener-Hopf integral equation of Eq. (8.2-18) to a differential equation. We accomplish this by taking the time derivative of both sides of Eq. (8.2-18). Operating first on the left side,

$$\begin{aligned}\frac{\partial}{\partial t}\{\operatorname{cov}[\mathbf{x}(t), \mathbf{z}(\lambda)]\} &= \operatorname{cov}[\dot{\mathbf{x}}(t), \mathbf{z}(\lambda)] \\ &= \operatorname{cov}[\mathbf{F}(t)\mathbf{x}(t) + \mathbf{G}(t)\mathbf{w}(t), \mathbf{z}(\lambda)] \\ &= \mathbf{F}(t) \operatorname{cov}[\mathbf{x}(t), \mathbf{z}(\lambda)] + \mathbf{G}(t) \operatorname{cov}[\mathbf{w}(t), \mathbf{z}(\lambda)].\end{aligned} \tag{8.2-20}$$

From Eq. (8.2-18), we see that λ is restricted such that $t_o < \lambda < t$. Thus since $t > \lambda$ always, $\operatorname{cov}[\mathbf{w}(t), \mathbf{z}(\lambda)] = \mathbf{0}$. This simply states that there is no correlation between the input at time t and the output at some time $\lambda < t$. With this relation, Eq. (8.2-20) becomes

$$\frac{\partial}{\partial t}\{\operatorname{cov}[\mathbf{x}(t), \mathbf{z}(\lambda)]\} = \mathbf{F}(t) \operatorname{cov}[\mathbf{x}(t), \mathbf{z}(\lambda)]. \tag{8.2-21}$$

Now consider the time derivative of the right-hand side of Eq. (8.2-18):

$$\frac{\partial}{\partial t}\left\{\int_{t_o}^{t} \hat{\boldsymbol{A}}(t, \tau) \operatorname{cov}[\mathbf{z}(\tau), \mathbf{z}(\lambda)]\, d\tau\right\} = \hat{\boldsymbol{A}}(t, t) \operatorname{cov}[\mathbf{z}(t), \mathbf{z}(\lambda)] +$$
$$\int_{t_o}^{t} \frac{\partial}{\partial t}\{\hat{\boldsymbol{A}}(t, \tau) \operatorname{cov}[\mathbf{z}(\tau), \mathbf{z}(\lambda)]\}\, d\tau. \tag{8.2-22}$$

Equating the result of the differentiations, we now have

$$\mathbf{F}(t)\operatorname{cov}[\mathbf{x}(t), \mathbf{z}(\lambda)] = \hat{\mathcal{A}}(t, t)\operatorname{cov}[\mathbf{z}(t), \mathbf{z}(\lambda)] + \int_{t_0}^{t} \frac{\partial}{\partial t}\{\hat{\mathcal{A}}(t, \tau)\operatorname{cov}[\mathbf{z}(\tau), \mathbf{z}(\lambda)]\}\, d\tau. \tag{8.2-23}$$

We then note the following equalities:

$$\operatorname{cov}[\mathbf{x}(t), \mathbf{z}(\lambda)] = \mathcal{E}\{\mathbf{x}(t)[\mathbf{H}(\lambda)\mathbf{x}(\lambda) + \mathbf{v}(\lambda)]^T\} = \operatorname{cov}\{\mathbf{x}(t), \mathbf{x}(\lambda)\}\mathbf{H}^T(\lambda) \tag{8.2-24}$$

$$\begin{aligned}\operatorname{cov}[\mathbf{z}(\tau), \mathbf{z}(\lambda)] &= \mathcal{E}\{[\mathbf{H}(\tau)\mathbf{x}(\tau) + \mathbf{v}(\tau)][\mathbf{H}(\lambda)\mathbf{x}(\lambda) + \mathbf{v}(\lambda)]^T\} \\ &= \mathbf{H}(\tau)\operatorname{cov}\{\mathbf{x}(\tau), \mathbf{x}(\lambda)\}\mathbf{H}^T(\lambda) + \mathbf{R}(\tau)\,\delta(\lambda - \tau)\end{aligned} \tag{8.2-25}$$

$$\operatorname{cov}[\mathbf{x}(t), \mathbf{x}(\lambda)] = \boldsymbol{\varphi}(t, \gamma)\operatorname{cov}\{\mathbf{x}(\gamma), \mathbf{x}(\gamma)\}\boldsymbol{\varphi}^T(\lambda, \gamma) + \int_{\lambda}^{t} \boldsymbol{\varphi}(t, \gamma)\mathbf{G}(\gamma)\mathbf{Q}(\gamma)\mathbf{G}^T(\gamma)\boldsymbol{\varphi}^T(\lambda, \gamma)\, d\gamma \tag{8.2-26}$$

where Eqs. (8.2-24) and (8.2-25) follow directly from the definition of $\mathbf{x}$ and $\mathbf{z}$. Equation (8.2-26) is a restatement of Eq. (8.1-11). It is convenient to define some of the terms in Eq. (8.2-10) in slightly different form. Thus we let

$$\begin{aligned}\hat{\mathbf{x}}(t) &= \int_{t_0}^{t} \hat{\mathcal{A}}(t, \tau)\mathbf{z}(\tau)\, d\tau \\ \hat{\mathcal{A}}(t, t) &= \mathbf{K}(t) \\ \mathbf{K}(t) &= \boldsymbol{\Xi}(t)\mathbf{H}^T(t)\mathbf{R}^{-1}(t).\end{aligned} \tag{8.2-27}$$

We notice that our derivation of the filter equations is following a path quite similar to the development of the previous section.

In order to determine the differential equations for the optimum filter, we examine Eq. (8.2-23) in detail. If we use Eq. (8.2-18) we see that the term to the left of the equals sign in Eq. (8.2-23) may be written as

$$\mathbf{F}(t)\operatorname{cov}[\mathbf{x}(t), \mathbf{z}(\lambda)] = \mathbf{F}(t)\int_{t_0}^{t} \hat{\mathcal{A}}(t, \tau)\operatorname{cov}[\mathbf{z}(\tau), \mathbf{z}(\lambda)]\, d\tau. \tag{8.2-28}$$

Since $\mathbf{z}(t) = \mathbf{H}(t)\mathbf{x}(t) + \mathbf{v}(t)$, we may substitute for the first term to the right of the equals sign in Eq. (8.2-23) the expression

$$\hat{\mathcal{A}}(t, t)\operatorname{cov}[\mathbf{z}(t), \mathbf{z}(\lambda)] = \hat{\mathcal{A}}(t, t)\mathbf{H}(t)\operatorname{cov}[\mathbf{x}(t), \mathbf{z}(\lambda)] + \hat{\mathcal{A}}(t, t)\mathbf{R}(t)\,\delta(t - \lambda). \tag{8.2-29}$$

But, for $\lambda < t$, we know that $\delta(t - \lambda) = 0$. Thus when we substitute the foregoing two relations in Eq. (8.2-23), we obtain, using Eq. (8.2-18),

$$\int_{t_0}^{t} \left\{\mathbf{F}(t)\hat{\mathcal{A}}(t, \tau) - \frac{\partial}{\partial t}[\hat{\mathcal{A}}(t, \tau)] - \hat{\mathcal{A}}(t, t)\mathbf{H}(t)\hat{\mathcal{A}}(t, \tau)\right\}\operatorname{cov}[\mathbf{z}(\tau), \mathbf{z}(\lambda)]\, d\tau = \mathbf{0} \tag{8.2-30}$$

for all $t_0 \leq \lambda < t$. This will be satisfied only if the term premultiplying the cov [$\mathbf{z}(\tau)$, $\mathbf{z}(\lambda)$] term is zero. Therefore we must have as the differential equation for the impulse response of the optimum filter

$$\mathbf{F}(t)\hat{\mathbf{A}}(t,\tau) - \frac{\partial}{\partial t}\hat{\mathbf{A}}(t,\tau) - \hat{\mathbf{A}}(t,t)\mathbf{H}(t)\hat{\mathbf{A}}(t,\tau) = \mathbf{0} \qquad (8.2\text{-}31)$$

for all $t_o \leq \tau < t$. By combining Eq. (8.2-31) together with Eq. (8.2-27) and its derivative, we immediately obtain the differential equation for the optimal state estimator

$$\dot{\hat{\mathbf{x}}} = \mathbf{F}(t)\hat{\mathbf{x}}(t) + \mathbf{K}(t)[\mathbf{z}(t) - \mathbf{H}(t)\hat{\mathbf{x}}(t)]. \qquad (8.2\text{-}32)$$

From Eqs. (8.2-24) through (8.2-26), we obtain, with the definitions of Eqs. (8.2-13) and (8.2-27), $\mathbf{\Xi} = \mathbf{P}$, and the variance equation

$$\dot{\mathbf{P}} = \mathbf{F}(t)\mathbf{P}(t) + \mathbf{P}(t)\mathbf{F}^T(t) - \mathbf{P}(t)\mathbf{H}^T(t)\mathbf{R}^{-1}(t)\mathbf{H}(t)\mathbf{P}(t) + \mathbf{G}(t)\mathbf{Q}(t)\mathbf{G}^T(t) \qquad (8.2\text{-}33)$$

which describes the error variance in estimation. Also, the initial conditions are†

$$\hat{\mathbf{x}}(t_o) = \mathbf{0} \qquad (8.2\text{-}34)$$

$$\mathbf{P}(t_o) = \mathcal{E}[\tilde{\mathbf{x}}(t_o)\tilde{\mathbf{x}}^T(t_o)] = \text{cov}\,[\tilde{\mathbf{x}}(t_o), \tilde{\mathbf{x}}(t_o)]. \qquad (8.2\text{-}35)$$

We can construct a block diagram of the optimum filter, as shown in Fig. 8.1-3, which serves as a summary of the most pertinent results of this section concerning the derivation of the Kalman filter.

We now examine the process $\tilde{\mathbf{z}}(t)$, called the *innovations process*, where we define

$$\tilde{\mathbf{z}}(t) = \mathbf{z}(t) - \mathbf{H}(t)\hat{\mathbf{x}}(t) = \mathbf{H}(t)\tilde{\mathbf{x}}(t) + \mathbf{v}(t). \qquad (8.2\text{-}36)$$

The realization of the innovations process at time t can be thought of as the datum containing new information about the state; hence, the sample paths of the innovations process and the output process contain essentially the same information.

An interesting and useful property of the innovations process, which always has zero-mean, is that

$$\text{cov}\,[\tilde{\mathbf{z}}(t), \tilde{\mathbf{z}}(\tau)] = \text{cov}\,[\mathbf{v}(t), \mathbf{v}(\tau)] = \mathbf{R}(t)\,\delta(t-\tau). \qquad (8.2\text{-}37)$$

(Additionally, it can be shown [7, 14] that if $\mathbf{v}(t)$ is Gaussian, then so is $\tilde{\mathbf{z}}(t)$, implying that $\mathbf{v}(t)$ and $\tilde{\mathbf{z}}(t)$ can be thought of as essentially the same stochastic process.) This property and the fact which follows from Eqs. (8.2-4) and

---

†We are assuming zero mean random processes and so obtain as a consequence of this assumption $\hat{\mathbf{x}}(t_o) = \mathcal{E}\{\mathbf{x}(t_o)\} = \mathbf{0}$. If the expected value of $\mathbf{x}(t_o)$ is non zero, we can show that we should use $\hat{\mathbf{x}}(t_o) = \mathcal{E}\{\mathbf{x}(t_o)\} = \bar{\mathbf{x}}_o$. Finally, if $\mathbf{x}(t_o)$ is known, say $\mathbf{x}_o^k$, we use $\hat{\mathbf{x}}(t_o) = \mathbf{x}_o^k$ and $\mathbf{P}(t_o) = \mathbf{0}$ as the initial conditions for the optimum filter and the variance equations.

(8.2-6) that the filter process can be represented as

$$\dot{\hat{\mathbf{x}}}(t) = \mathbf{F}(t)\hat{\mathbf{x}}(t) + \mathbf{K}(t)\tilde{\mathbf{z}}(t) \tag{8.2-38}$$

$$\mathbf{z}(t) = \mathbf{H}(t)\hat{\mathbf{x}}(t) + \tilde{\mathbf{z}}(t), \tag{8.2-39}$$

imply trivially that Eq. (8.2-32) is the desired estimation formula. Several other associated results are discussed in [1]. The potential of an innovations approach to estimation is most evident in the derivation of various smoothing algorithms, which is discussed in [7].

We note that the best filter determined above has been constrained to be linear in the data, as was true in Sec. 8.1, and is, therefore, not necessarily the best filter in the class of all nonlinear functions. For a rather general class of criteria containing the minimum error variance criterion as a special case, the expected mean of the state, conditioned on the data, is the best estimate [7, 14]. For the case where the stochastic processes $\mathbf{w}(t)$ and $\mathbf{v}(t)$ and the random variable $\mathbf{x}_o$ are all Gaussian, then the expected mean of the state conditioned on the data is linear in the data. As a result, the best linear filter is also the best filter out of the class of all nonlinear functions for the Gaussian case.

We now show that the time-invariant version of the estimation problem defined earlier in this section is equivalent to, or is the dual of, a deterministic control problem in the sense that both have the same optimal cost and control. Specifically, we show that the dual of the time-invariant optimal estimation problem is the time-invariant optimal regulator problem previously discussed in Chapters 4 and 5. The time-varying case is discussed in [16]. Our presentation follows that of Astrom [17].

We are given the system or message model

$$\dot{\mathbf{x}}(t) = \mathbf{F}\mathbf{x}(t) + \mathbf{G}\mathbf{w}(t) \tag{8.2-40}$$

and the observation model

$$\mathbf{z}(t) = \mathbf{H}\mathbf{x}(t) + \mathbf{v}(t), \tag{8.2-41}$$

where both noise inputs are zero-mean and satisfy

$$\operatorname{cov}[\mathbf{w}(t), \mathbf{w}(\tau)] = \mathbf{Q}\,\delta(t - \tau)$$

$$\operatorname{cov}[\mathbf{v}(t), \mathbf{v}(\tau)] = \mathbf{R}\,\delta(t - \tau)$$

$$\operatorname{cov}[\mathbf{v}(t), \mathbf{w}(\tau)] = \mathbf{0}.$$

We wish to find an unbiased estimate of $\mathbf{x}(t)$, $\hat{\mathbf{x}}(t)$, which is linear in $Z(t)$ and which minimizes $\mathcal{E}\{\|\tilde{\mathbf{x}}(t)\|_{\mathbf{S}}^2\}$, where $\tilde{\mathbf{x}}(t) - \mathbf{x}(t) - \hat{\mathbf{x}}(t)$, the estimation error. For simplicity we assume there exists a column vector $\boldsymbol{l}$ such that $\mathbf{S} = \boldsymbol{l}\boldsymbol{l}^T$. Hence, we wish to minimize

$$J = \mathcal{E}\{[\boldsymbol{l}^T\tilde{\mathbf{x}}(t)]^2\} = \mathcal{E}\{\|\tilde{\mathbf{x}}(t)\|_{\mathbf{S}}^2\} \tag{8.2-42}$$

for all t.

Since the estimate is linear, $\mathbf{l}^T\hat{\mathbf{x}}(t)$ is of the form

$$\mathbf{l}^T\hat{\mathbf{x}}(t) = \mathbf{b}^T\hat{\mathbf{x}}(t_o) - \int_{t_o}^{t} \mathbf{a}^T(\tau)\mathbf{z}(\tau)\,d\tau, \tag{8.2-43}$$

where we wish to find $\mathbf{a}(t)$ and $\mathbf{b}$ which minimize Eq. (8.2-42). To do this we let $\mathbf{x}^*(t)$ be the function satisfying

$$\dot{\mathbf{x}}^*(t) = -\mathbf{F}^T\mathbf{x}^*(t) - \mathbf{H}^T\mathbf{a}(t) \tag{8.2-44}$$

with terminal condition $\mathbf{x}^*(t_f) = \mathbf{l}$, where $[t_o, t_f]$ is the fixed time interval of interest. We note that

$$\begin{aligned}
\mathbf{l}^T\mathbf{x}(t_f) &= \mathbf{x}^{*T}(t_f)\mathbf{x}(t_f) \\
&= \mathbf{x}^{*T}(t_o)\mathbf{x}(t_o) + \int_{t_o}^{t_f} d[\mathbf{x}^{*T}(\tau)\mathbf{x}(\tau)] \\
&= \mathbf{x}^{*T}(t_o)\mathbf{x}(t_o) + \int_{t_o}^{t_f} [-\mathbf{a}^T(\tau)\mathbf{H}\mathbf{x}(\tau)\,d\tau + \mathbf{x}^{*T}(\tau)\mathbf{G}\mathbf{w}(\tau)\,d\tau].
\end{aligned} \tag{8.2-45}$$

This result follows immediately from the relation

$$\begin{aligned}
d[\mathbf{x}^{*T}(\tau)\mathbf{x}(\tau)] &= d\mathbf{x}^{*T}(\tau)\,\mathbf{x}(\tau) + \mathbf{x}^{*T}(\tau)\,d\mathbf{x}(\tau) \\
&= [-\mathbf{F}^T\mathbf{x}^*(\tau)\,d\tau - \mathbf{H}^T\mathbf{a}(\tau)\,d\tau]^T\mathbf{x}(\tau) \\
&\quad + \mathbf{x}^{*T}(\tau)[\mathbf{F}\mathbf{x}(\tau)\,d\tau + \mathbf{G}\mathbf{w}(\tau)\,d\tau] \\
&= -\mathbf{a}^T(\tau)\mathbf{H}\mathbf{x}(\tau)\,d\tau + \mathbf{x}^{*T}(\tau)\mathbf{G}\mathbf{w}(\tau)\,d\tau.
\end{aligned}$$

It then follows that

$$\begin{aligned}
\mathbf{l}^T\tilde{\mathbf{x}}(t_f) = \mathbf{l}^T\mathbf{x}(t_f) - \mathbf{l}^T\hat{\mathbf{x}}(t_f) = \mathbf{x}^{*T}(t_o)\mathbf{x}(t_o) - \mathbf{b}^T\hat{\mathbf{x}}(t_o) + \\
\int_{t_o}^{t_f} [\mathbf{x}^{*T}(\tau)\mathbf{G}\mathbf{w}(\tau) + \mathbf{a}^T(\tau)\mathbf{v}(\tau)]\,d\tau.
\end{aligned} \tag{8.2-46}$$

We wish that our estimate be unbiased for all $t \in [t_o, t_f]$. Thus we wish to insure that the unconditional expectation of $\mathbf{l}^T\tilde{\mathbf{x}}(t_f)$ is zero for arbitrary but fixed t_f; $\mathbf{w}(t)$ and $\mathbf{v}(t)$ are now zero-mean white noise terms. Taking the expectation of both sides of Eq. (8.2-46), we assume, therefore, that $\mathbf{x}^*(t_o) = \mathbf{b}$, and $\hat{\mathbf{x}}(t_o) = \mathcal{E}\{\mathbf{x}(t_o)\}$. From Eq. (8.2-46), it follows that the criterion of Eq. (8.2-42) becomes, in the context of the variables $\mathbf{x}^*$ and $\mathbf{a}$,

$$J = \mathcal{E}\{\|\tilde{\mathbf{x}}(t_f)\|_{\mathbf{S}}^2 = \|\mathbf{x}^*(t_o)\|_{\mathbf{P}(t_o)}^2 + \int_{t_o}^{t_f} [\|\mathbf{x}^*(\tau)\|_{\mathbf{GQG}_{\tau}^T}^2 + \|a(\tau)\|_{\mathbf{R}}^2]\,d\tau. \tag{8.2-47}$$

Thus, determining the minimum of Eq. (8.2-47) subject to the constraint Eq. (8.2-44) is equivalent to the state estimation problem. We have now established the duality relationship between estimation and control.

Example 8.2-1. Let us consider the problem of optimum state estimation for the message and noise model illustrated in Fig. 8.2-1 and described

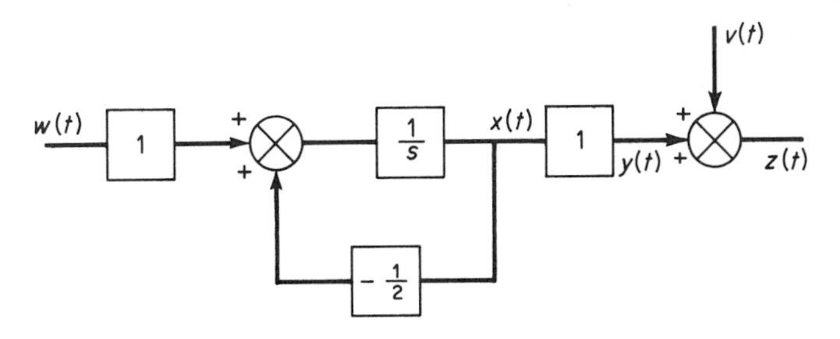

Fig. 8.2-1 Message and noise model for Example 8.2-1.

by the equations:

$$\dot{x} = -\tfrac{1}{2}x + w(t)$$
$$z = x(t) + v(t)$$
$$\operatorname{cov}[w(t), w(\tau)] = Q(t)\,\delta(t-\tau) = 2\delta(t-\tau)$$
$$\operatorname{cov}[v(t), v(\tau)] = R(t)\,\delta(t-\tau) = \tfrac{1}{4}\delta(t-\tau)$$
$$\operatorname{cov}[\tilde{x}(t_o), \tilde{x}(t_o)] = 0$$
$$t_o = 0.$$

Thus we see that the estimation equations are

$$\dot{\hat{x}} = -\tfrac{1}{2}\hat{x} + 4p(t)[z(t) - \tilde{x}(t)], \qquad \hat{x}(0) = 0$$
$$\frac{dp(t)}{dt} = -p(t) + 4p^2(t) + 2, \qquad p(0) = 0.$$

The Kalman gain for the estimator is given by

$$K(t) = p(t)H^TR^{-1} = 4p(t).$$

The determination of $P(t)$ (which we shall write as $p(t)$ since it is a scalar in this, our specific problem) involves solution of the Riccati equation, which is also the error-variance equation

$$\dot{p} = -p - 4p^2 + 2. \tag{1}$$

The solution of this equation will be obtained by two methods. The first will be to guess the form of the solution and choose the unspecified parameters to make it satisfy the differential equation. The second method will be to convert the nonlinear differential equation into two linear differential equations. By solving the latter, we will obtain $p(t)$. Of course, a computer could be used, but for this simple problem it is informative to solve the variance equation analytically.

For the first method, we try the "solution"

$$p(t) = \alpha + \beta \tanh(\gamma t + \phi) = \alpha + \beta \tanh\theta \tag{2}$$

$$\frac{dp}{dt} = \beta\dot{\theta}(1 + \tanh^2\theta) = \beta\gamma(1 - \tanh^2\theta). \tag{3}$$

Substituting Eqs. (2) and (3) into Eq. (1), we obtain

$$\beta\gamma(1 - \tanh^2\theta) = -\alpha - \beta\tanh\theta - 4(\alpha^2 + 2\alpha\beta\tanh\theta + \beta^2\tanh^2\theta) + 2. \tag{4}$$

Collecting terms, we have

$$\beta\gamma + \alpha + 4\alpha^2 - 2 + \beta(1 + 8\alpha)\tanh\theta + \beta(4\beta - \gamma)\tanh^2\theta = 0. \tag{5}$$

With $\beta = 0$ not allowed, Eq. (5) can hold for all θ if and only if $\alpha = -\frac{1}{8}$, $\beta = \gamma/4$. Substituting these values into Eq. (5), we get $\gamma^2 = \frac{33}{4}$. Therefore, $\gamma = \pm\sqrt{33}/2$, $\beta = \pm\sqrt{33}/8$. Choosing the positive signs, we have finally

$$p(t) = -\frac{1}{8} + \frac{1}{8}\sqrt{33}\tanh\left(\frac{\sqrt{33}}{2}t + \phi\right). \tag{6}$$

Since $p(0) = 0$, we have

$$p(t) = -\frac{1}{8} + \frac{1}{8}\sqrt{33}\tanh\left(\frac{\sqrt{33}}{2}t + \tanh^{-1}\frac{1}{\sqrt{33}}\right). \tag{7}$$

Using Eq. (5) we then have

$$K(t) = -\frac{1}{2} + \frac{1}{2}\sqrt{33}\tanh\left(\frac{\sqrt{33}}{2}t + \tanh^{-1}\frac{1}{\sqrt{33}}\right).$$

For the second method, it is easy to show that, in general [5], if

$$\mathbf{\Theta}(t, t_o) = \begin{bmatrix} \mathbf{\theta}_{11}(t, t_o) & \mathbf{\theta}_{12}(t, t_o) \\ \mathbf{\theta}_{21}(t, t_o) & \mathbf{\theta}_{22}(t, t_o) \end{bmatrix} \tag{8}$$

is the transition matrix for the solution of the equation

$$\dot{\boldsymbol{\xi}} = \mathbf{A}(t)\boldsymbol{\xi}(t) \tag{9}$$

where

$$\boldsymbol{\xi}^T(t) = [\mathbf{x}^T(t), \boldsymbol{\lambda}^T(t)] \tag{10}$$

$$\mathbf{A}(t) = \begin{bmatrix} -\mathbf{F}^T(t) & \mathbf{H}^T(t)\mathbf{R}^{-1}(t)\mathbf{H}(t) \\ \mathbf{G}(t)\mathbf{Q}(t)\mathbf{G}^T(t) & \mathbf{F}(t) \end{bmatrix} \tag{11}$$

then,

$$\mathbf{P}(t) = [\mathbf{\theta}_{21}(t, t_o) + \mathbf{\theta}_{22}(t, t_o)\mathbf{P}(t_o)][\mathbf{\theta}_{11}(t, t_o) + \mathbf{\theta}_{12}(t, t_o)\mathbf{P}(t_o)]^{-1} \tag{12}$$

is the solution to the Riccati equation,

$$\dot{\mathbf{P}} = \mathbf{F}(t)\mathbf{P}(t) + \mathbf{P}(t)\mathbf{F}^T(t) - \mathbf{P}(t)\mathbf{H}^T(t)\mathbf{R}^{-1}(t)\mathbf{H}(t)\mathbf{P}(t) + \mathbf{G}(t)\mathbf{Q}(t)\mathbf{G}^T(t). \tag{13}$$

For this second method, we let $\boldsymbol{\xi}^T = [x(t), \lambda(t)]$. Thus the canonic differential equations for the optimal estimator become

$$\dot{\boldsymbol{\xi}} = \mathbf{A}(t)\boldsymbol{\xi}(t) = \begin{bmatrix} \frac{1}{2} & 4 \\ 2 & -\frac{1}{2} \end{bmatrix}\boldsymbol{\xi}(t). \tag{14}$$

Solving for the eigenvalues of A, we have

$$\lambda_1, \lambda_2 = -\frac{\sqrt{33}}{2}, \frac{+\sqrt{33}}{2} \tag{15}$$

and the corresponding eigenvectors

$$\mathbf{u}_1 = \begin{bmatrix} 1 \\ -\frac{1}{8}(\sqrt{33} + 1) \end{bmatrix}, \quad \mathbf{u}_2 = \begin{bmatrix} 1 \\ \frac{1}{8}(\sqrt{33} - 1) \end{bmatrix}. \tag{16}$$

Defining what is frequently called the "modal matrix" corresponding to $\mathbf{A}$ as

$$\mathbf{M} = \begin{bmatrix} 1 & 1 \\ -\frac{1}{8}(\sqrt{33}+1) & \frac{1}{8}(\sqrt{33}-1) \end{bmatrix} \tag{17}$$

and calculating its inverse

$$\mathbf{M}^{-1} = \begin{bmatrix} \dfrac{\sqrt{33}-1}{2\sqrt{33}} & \dfrac{-4}{\sqrt{33}} \\ \dfrac{\sqrt{33}+1}{2\sqrt{33}} & \dfrac{4}{\sqrt{33}} \end{bmatrix}, \tag{18}$$

we are now prepared to make the change of variable $\boldsymbol{\zeta} = \mathbf{M}^{-1}\boldsymbol{\xi}$ so that the following differential equation applies

$$\dot{\boldsymbol{\zeta}} = \text{diag}\left\{-\frac{\sqrt{33}}{2}, \frac{\sqrt{33}}{2}\right\}\boldsymbol{\zeta}. \tag{19}$$

The solution of Eq. (19) is obtained by inspection as

$$\boldsymbol{\zeta}(t) = \begin{bmatrix} e^{-\sqrt{33}t/2} & 0 \\ 0 & e^{+\sqrt{33}t/2} \end{bmatrix}\boldsymbol{\zeta}(0) = \text{diag}\,\{e^{-\sqrt{33}t/2}, e^{+\sqrt{33}t/2}\}\boldsymbol{\zeta}(0). \tag{20}$$

Using the abbreviations $\epsilon_1 = \epsilon_2(t) = \exp[-\sqrt{33}t/2]$ and $\epsilon_2 = \epsilon_2(t) = \exp[\sqrt{33}t/2]$, we can write the solution to the original differential equation as

$$\boldsymbol{\xi} = \mathbf{M}\begin{bmatrix} \epsilon_1 & 0 \\ 0 & \epsilon_2 \end{bmatrix}\mathbf{M}^{-1}\boldsymbol{\xi}(0) = \begin{bmatrix} \theta_{11} & \theta_{12} \\ \theta_{21} & \theta_{22} \end{bmatrix}\boldsymbol{\xi}(0). \tag{21}$$

Carrying out the indicated matrix multiplications, we obtain

$$\boldsymbol{\xi} = \begin{bmatrix} \sqrt{33}\cosh\dfrac{\sqrt{33}}{2}t + \sinh\dfrac{\sqrt{33}}{2}t & 8\sinh\dfrac{\sqrt{33}}{2}t \\ 4\sinh\dfrac{\sqrt{33}}{2}t & \sqrt{33}\cosh\dfrac{\sqrt{33}}{2}t - \sinh\dfrac{\sqrt{33}}{2}t \end{bmatrix}\boldsymbol{\xi}(0). \tag{22}$$

Because $\mathbf{P}(0) = p(0) = 0$ for this specific problem, Eq. (12) reduces to $p(t) = \theta_{21}/\theta_{11}$. Substituting from Eq. (22) for θ_{21} and θ_{11}, we obtain

$$p(t) = \frac{4\sinh(\sqrt{33}/2)t}{\sqrt{33}\cosh(\sqrt{33}/2)t + \sinh(\sqrt{33}/2)t} = \frac{(4/\sqrt{33})\tanh(\sqrt{33}/2)t}{1 + (1/\sqrt{33})\tanh(\sqrt{33}/2)t}.$$

Factoring out $\sqrt{33}/8$, then adding and subtracting $1/\sqrt{33}$ to and from the remaining terms, we have, after use of a trigonometric equality,

$$p(t) = \frac{\sqrt{33}}{8}\left[\frac{-1}{\sqrt{33}} + \tanh\left(\frac{\sqrt{33}}{2}t + \tanh^{-1}\frac{1}{\sqrt{33}}\right)\right] \tag{23}$$

which is exactly the same as Eq. (7). As is readily apparent, the effort involved in solving this simple Riccati equation is such as to suggest that the analytical solution of a Riccati equation of higher order than the first would be prohibitive. The final estimator equations are therefore

$$\dot{\hat{x}} = -\tfrac{1}{2}\hat{x} + K(t)[z(t) - \hat{x}(t)], \qquad \hat{x}(0) = 0$$

$$K(t) = -\frac{1}{2} + \frac{1}{2}\sqrt{33}\tanh\left(\frac{\sqrt{33}}{2}t + \tanh^{-1}\frac{1}{\sqrt{33}}\right).$$

Figure 8.2-2 illustrates the message and noise model as well as the optimum filter. Figure 8.2-3 illustrates the time-varying gain plot. The mean-square error is one-fourth the value of $K(t)$ at any given t since $K(t) = 4p(t)$.

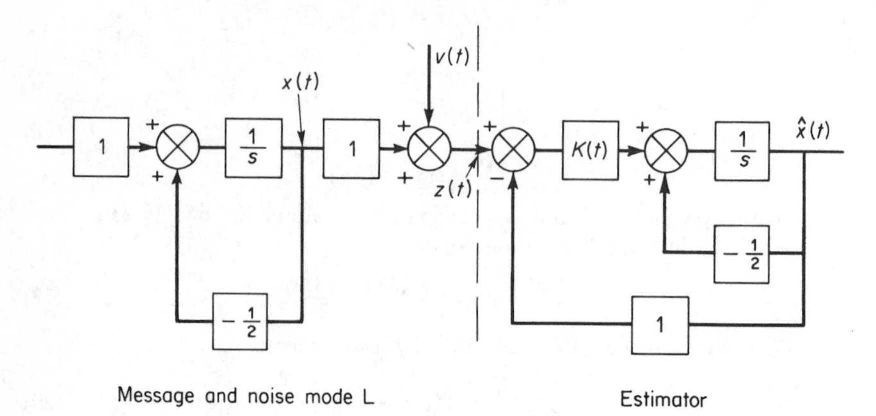

Fig. 8.2-2 Optimal estimator of Example 8.2-1.

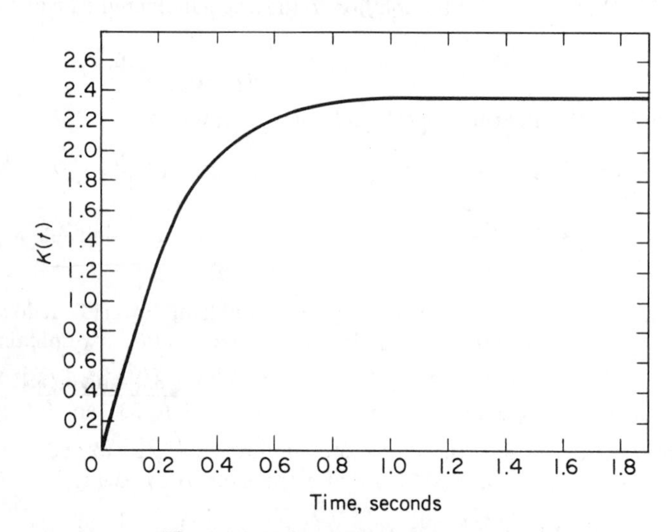

Fig. 8.2-3 Kalman gain for Example 8.2-1.

Example 8.2-2. Consider a dynamic system in which the acceleration is white noise, a problem typical of aerospace guidance. Position is observed after it is contaminated by white noise. The process equation is therefore

$$\dot{\mathbf{x}} = \mathbf{F}(t)\mathbf{x} + \mathbf{G}(t)w(t) = \begin{bmatrix} 0 & 1 \\ 0 & 0 \end{bmatrix}\mathbf{x}(t) + \begin{bmatrix} 0 \\ 1 \end{bmatrix}w(t),$$

where $\mathbf{x}^T(t) = [x_1(t), x_2(t)]$ and $w(t)$ is a zero-mean scalar, Gaussian white noise, with a constant variance which shall be taken equal to 1 for this example. The measurement is given by

$$z(t) = [1 \quad 0]\,\mathbf{x} + v(t) = \mathbf{H}(t)\mathbf{x} + v(t),$$

where z is a scalar and $v(t)$ is a zero-mean scalar, Gaussian white noise, with a constant variance of 16. We will further assume that

$$\mathbf{P}(t_o) = \mathbf{P}(0) = \begin{bmatrix} 1 & 0 \\ 0 & 0 \end{bmatrix}.$$

Application of Eqs. (8.2-41) and (8.2-42) to this example leads immediately to the final estimation equations:

$$\dot{\hat{x}}_1 = \hat{x}_2 + \tfrac{1}{16}p_{11}(t)[z(t) - \hat{x}_1(t)], \qquad \hat{x}_1(0) = 0$$

$$\dot{\hat{x}}_2 = \tfrac{1}{16}p_{12}(t)[z(t) - \hat{x}_1(t)], \qquad \hat{x}_2(0) = 0$$

$$\dot{p}_{11} = 2p_{12} - \frac{p_{11}^2}{16}, \qquad p_{11}(0) = 1$$

$$\dot{p}_{12} = p_{22} - \frac{p_{11}p_{12}}{16}, \qquad p_{12}(0) = 0$$

$$\dot{p}_{22} = \frac{-p_{12}^2}{16} + 1, \qquad p_{22}(0) = 0$$

$$\mathbf{K}(t) = \begin{bmatrix} K_1(t) \\ K_2(t) \end{bmatrix} = \tfrac{1}{16}\begin{bmatrix} p_{11}(t) \\ p_{12}(t) \end{bmatrix}.$$

Figure 8.2-4 illustrates the block diagram of the optimum filter and the Kalman gains resulting from a computer solution of the Riccati or variance equation.

Example 8.2-3. Suppose we have the message and noise model

$$\frac{dx}{dt} = -(1 + \cos t)x + w(t)$$

$$z(t) = x(t) + v(t)$$

$$\operatorname{cov}[w(t), w(\tau)] = \delta(t - \tau) = \operatorname{cov}[v(t), v(\tau)]$$

$$\operatorname{cov}[w(t), v(\tau)] = 0 = \operatorname{cov}[\tilde{x}(t_o), \tilde{x}(t_o)].$$

From the equations for the optimal estimate, we have for this specific problem that

$$\dot{\hat{x}} = -(1 + \cos t)\hat{x} + p(t)[z - \hat{x}(t)], \qquad \hat{x}(0) = 0$$

$$\dot{p} = -2(1 + \cos t)p(t) - p^2(t) + 1, \qquad p(0) = 0$$

are the desired equations for the optimal estimate. The solution of the error-variance equation is given in Fig. 8.2-5. For this particular case, we see that the initial error variance is zero and does not become constant for large time. This is due to the nonstationary nature of the message model. We can modify our problem slightly by letting $R(t) = e^{-0.1t}$ in which case the estimation equations become

$$\dot{\hat{x}} = -(1 + \cos t)\hat{x}(t) + e^{0.1t}p(t)[z(t) - \hat{x}(t)], \qquad \hat{x}(0) = 0$$

$$\dot{p} = -2(1 + \cos t)p(t) - e^{0.1t}p^2(t) + 1, \qquad p(0) = 0.$$

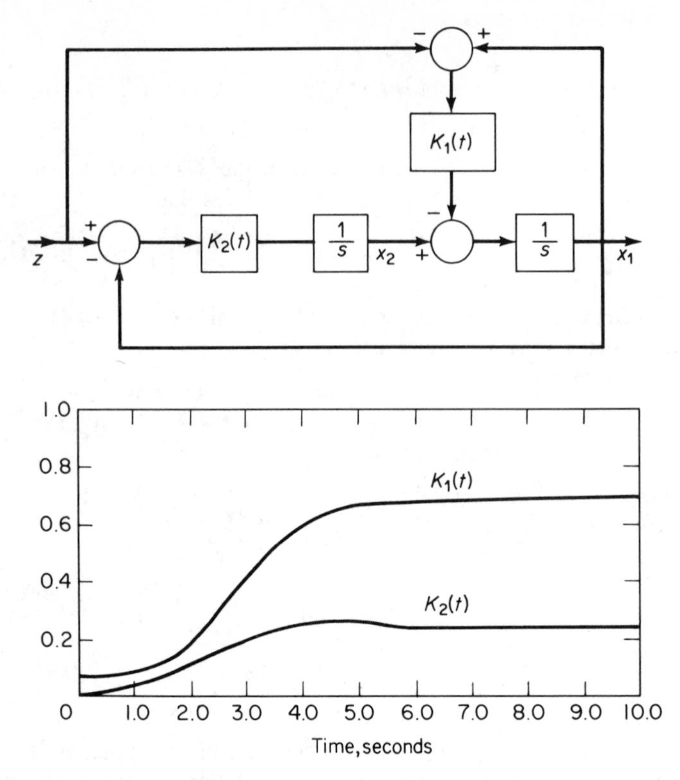

Fig. 8.2-4 Optimum state estimation filter and Kalman gains for Example 8.2-2.

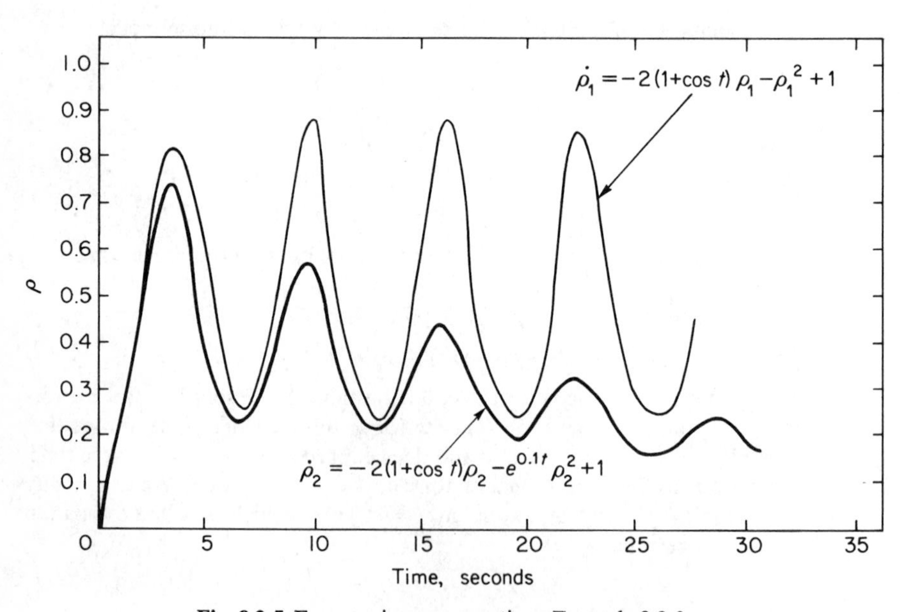

Fig. 8.2-5 Error variance versus time, Example 8.2-3.

In this case, the error variance becomes zero for large enough time, since the mean-square value of the measurement noise decreases with time. Figure 8.2-5 also illustrates the solution of the variance equation for this case.

8.2-1 *Error analysis algorithms*

The optimality of the Kalman filter is clearly dependent on how correctly and precisely the system, subject to observation and estimation, has been modeled. Correct and precise modeling, however, is generally an impossible task. Hence, for most applications, we must be concerned with the effects of parameter errors and variations on the accuracy of estimation. Such concerns aid in the modeling process by determining the accuracy that the model and its parameters must satisfy in order to meet specified estimation accuracy requirements. In this subsection, we determine an algorithm for the difference between the modeled error variance and the actual error variance following the development in [18]. This expands upon our treatment in the previous section.

We present an approach to error analysis in which the system estimation error is separated into optimum and nonoptimum error components. These two types of errors are shown to be orthogonal. The result of this separation is the derivation of a set of "delta" algorithms which directly compute the difference between the true system mean-squared error, the true error variance, and the optimum system mean-squared error, or optimum error variance. These delta equations simplify the computational problem when the optimum and true system covariances are large compared with their difference.

The difference is obtained directly and not by subtracting one large covariance from the other, since the latter method makes large numerical errors a likely occurrence.

We assume the equations

$$\dot{\bar{\mathbf{x}}} = \bar{\mathbf{F}}\bar{\mathbf{x}} + \bar{\mathbf{G}}\bar{\mathbf{w}} \tag{8.2-48}$$

$$\mathbf{z} = \bar{\mathbf{H}}\bar{\mathbf{x}} + \bar{\mathbf{v}} \tag{8.2-49}$$

represent our imprecise model of the physical system under consideration which is precisely modeled by the equations

$$\dot{\mathbf{x}} = \mathbf{F}\mathbf{x} + \mathbf{G}\mathbf{w} \tag{8.2-50}$$

$$\mathbf{z} = \mathbf{H}\mathbf{x} + \mathbf{v}. \tag{8.2-51}$$

Throughout the remainder of this subsection, we will eliminate all explicit dependence on time for notational brevity. The Kalman filter based on the model of Eqs. (8.2-48) and (8.2-49) is given by

$$\dot{\hat{\bar{\mathbf{x}}}} = \bar{\mathbf{F}}\hat{\bar{\mathbf{x}}} + \bar{\mathbf{K}}(\mathbf{z} - \bar{\mathbf{H}}\hat{\bar{\mathbf{x}}})$$

where

$$\bar{\mathbf{K}} = \bar{\mathbf{P}}\bar{\mathbf{H}}^T\bar{\mathbf{R}}^{-1} \tag{8.2-52}$$

and

$$\dot{\bar{\mathbf{P}}} = \bar{\mathbf{F}}\bar{\mathbf{P}} + \bar{\mathbf{P}}\bar{\mathbf{F}}^T - \bar{\mathbf{P}}\bar{\mathbf{H}}^T\bar{\mathbf{R}}^{-1}\bar{\mathbf{H}}\bar{\mathbf{P}} + \bar{\mathbf{G}}\bar{\mathbf{Q}}\bar{\mathbf{G}}^T. \tag{8.2-53}$$

If we define $\tilde{\hat{\mathbf{x}}} = \mathbf{x} - \hat{\mathbf{x}}$, $\Delta\mathbf{F} = \mathbf{F} - \bar{\mathbf{F}}$, and $\Delta\mathbf{H} = \mathbf{H} - \bar{\mathbf{H}}$, then algebraic manipulation implies that $\tilde{\hat{\mathbf{x}}}$ satisfies

$$\dot{\tilde{\hat{\mathbf{x}}}} = (\bar{\mathbf{F}} - \bar{\mathbf{K}}\bar{\mathbf{H}})\tilde{\hat{\mathbf{x}}} + (\Delta\mathbf{F} - \bar{\mathbf{K}}\,\Delta\mathbf{H})\mathbf{x} + \mathbf{G}\mathbf{w} - \bar{\mathbf{K}}\mathbf{v}.$$

Let $\hat{\mathbf{x}}^o$ be the Kalman filter associated with the precise model. It then can be shown with further algebraic manipulation that $\tilde{\hat{\mathbf{x}}} = \tilde{\mathbf{x}}^o + \tilde{\mathbf{x}}^a + \tilde{\mathbf{x}}^b$, where

$$\dot{\tilde{\mathbf{x}}}^o = (\mathbf{F} - \mathbf{K}\mathbf{H})\tilde{\mathbf{x}}^o + \mathbf{G}\mathbf{w} - \mathbf{K}\mathbf{v} \tag{8.2-54}$$

$$\dot{\tilde{\mathbf{x}}}^a = (\mathbf{F} - \mathbf{K}\mathbf{H})\tilde{\mathbf{x}}^a + \Delta\mathbf{K}[\mathbf{H}\tilde{\mathbf{x}}^o + \mathbf{v}] \tag{8.2-55}$$

$$\dot{\tilde{\mathbf{x}}}^b = (\mathbf{F} - \mathbf{K}\mathbf{H})\tilde{\mathbf{x}}^b + (\Delta\mathbf{F} - \mathbf{K}\,\Delta\mathbf{H})\hat{\mathbf{x}}^o, \tag{8.2-56}$$

and where it is easily shown that $\tilde{\mathbf{x}}^o = \mathbf{x} - \hat{\mathbf{x}}^o$. It is seen that $\tilde{\mathbf{x}}^o$ in Eq. (8.2-54) is uncorrelated with $\tilde{\mathbf{x}}^b$ in Eq. (8.2-56) since the optimum error $\tilde{\mathbf{x}}^o$ is uncorrelated with $\hat{\mathbf{x}}^o$. Thus,

$$\mathcal{E}\{\tilde{\mathbf{x}}^o\tilde{\mathbf{x}}^{bT}\} = \operatorname{cov}\{\tilde{\mathbf{x}}^o, \tilde{\mathbf{x}}^b\} = \mathbf{0}. \tag{8.2-57}$$

Note that the driving function for $\tilde{\mathbf{x}}^a$ in Eq. (8.2-55) is the innovation $\tilde{\mathbf{z}} = \mathbf{H}\tilde{\mathbf{x}}^o + \mathbf{v} = \mathbf{z} - \mathbf{H}\hat{\mathbf{x}}^o$. We recall that two properties of the zero-mean innovation process are

$$\mathcal{E}\{\tilde{\mathbf{x}}^o\tilde{\mathbf{z}}^T\} = \mathbf{0}$$

and

$$\mathcal{E}\{\tilde{\mathbf{z}}\tilde{\mathbf{z}}^T\} = \mathcal{E}\{\mathbf{v}, \mathbf{v}^T\} = \mathbf{R}\,\delta(t - \tau). \tag{8.2-58}$$

From Eqs. (8.2-55) and (8.2-58), it follows that

$$\mathcal{E}\{\tilde{\mathbf{x}}^o\tilde{\mathbf{x}}^{aT}\} = \mathbf{0}. \tag{8.2-59}$$

Thus, the estimation error associated with the precise systems model is orthogonal to the remaining portions of the estimation error associated with the imprecise systems model, which implies the result

$$\mathcal{E}\{\tilde{\mathbf{x}}^o\hat{\hat{\mathbf{x}}}^T\} = \operatorname{cov}\{\tilde{\mathbf{x}}^o, \hat{\hat{\mathbf{x}}}\} = \mathbf{0}. \tag{8.2-60}$$

We note that since $\hat{\hat{\mathbf{x}}}$ is necessarily a linear transformation of the data, and $\tilde{\mathbf{x}}^o$ represents the error between the actual state, subject to estimation, and the best estimate in the linear vector space generated by the data, Eq. (8.2-59) also results from a direct application of the orthogonal projection theorem.

Examination of Eqs. (8.2-54) and (8.2-55) implies that $\tilde{\mathbf{x}}^o$ and $\tilde{\mathbf{x}}^a$ are zero-mean processes; however, this is not true of $\tilde{\mathbf{x}}^b$ due to the fact that $\tilde{\mathbf{x}}^b$ in Eq. (8.2-56) is subject to a nonzero-mean forcing function $\hat{\mathbf{x}}^o$. The process $\tilde{\mathbf{x}}^b$ actually represents that portion of the error due to the state amplitude feeding through the dynamic modeling errors of the system. Thus,

$$\tilde{\mathbf{x}}^b = \tilde{\mathbf{x}}^e + \mathbf{u}_{\tilde{x}} \tag{8.2-61}$$

such that

$$\dot{\tilde{\mathbf{x}}}^e = (\bar{\mathbf{F}} - \bar{\mathbf{K}}\bar{\mathbf{H}})\tilde{\mathbf{x}}^e + (\Delta\bar{\mathbf{F}} - \bar{\mathbf{K}}\,\Delta\mathbf{H})\hat{\mathbf{x}}^e \tag{8.2-62}$$

and

$$\dot{\boldsymbol{\mu}}_{\tilde{x}} = (\bar{\mathbf{F}} - \bar{\mathbf{K}}\bar{\mathbf{H}})\boldsymbol{\mu}_{\tilde{x}} + (\Delta\mathbf{F} - \bar{\mathbf{K}}\,\Delta\mathbf{H})\boldsymbol{\mu}_{x} \tag{8.2-63}$$

where $\boldsymbol{\mu}_x$ is the system mean, $\boldsymbol{\mu}_{\tilde{x}}$ is the mean value of the estimation error, $\hat{\mathbf{x}}^e$ is the part of $\hat{\mathbf{x}}^o$ with *a priori* zero-mean, and $\tilde{\mathbf{x}}^e$ represents the zero-mean part of the state dependent error. We note that Eq. (8.2-63) is a deterministic equation. Thus, $\hat{\mathbf{x}}^e$ and $\boldsymbol{\mu}_x$ satisfy the equations

$$\dot{\hat{\mathbf{x}}}^e = \mathbf{F}\hat{\mathbf{x}}^e + \mathbf{K}\tilde{\mathbf{z}} \tag{8.2-64}$$

$$\dot{\boldsymbol{\mu}}_x = \mathbf{F}\boldsymbol{\mu}_x \tag{8.2-65}$$

where we note that

$$\hat{\mathbf{x}}^o = \hat{\mathbf{x}}^e + \boldsymbol{\mu}_x. \tag{8.2-66}$$

Algebraic manipulation shows that

$$\tilde{\mathbf{x}}^d = \tilde{\mathbf{x}}^a + \tilde{\mathbf{x}}^e \tag{8.2-67}$$

satisfies

$$\dot{\tilde{\mathbf{x}}}^d = (\bar{\mathbf{F}} - \bar{\mathbf{K}}\bar{\mathbf{H}})\tilde{\mathbf{x}}^d + (\Delta\mathbf{F} - \bar{\mathbf{K}}\,\Delta\mathbf{H})\hat{\mathbf{x}}^e + \Delta\mathbf{K}\tilde{\mathbf{z}}. \tag{8.2-68}$$

Since $\tilde{\mathbf{x}}^o$ and $\tilde{\mathbf{x}}^a$ are uncorrelated $\tilde{\mathbf{x}}^o$ and $\tilde{\mathbf{x}}^b$ are also uncorrelated. This implies that the total system error variance $\mathbf{V}_{\tilde{x}}$ satisfies

$$\begin{aligned} \mathbf{V}_{\tilde{x}} &= \mathcal{E}\{(\tilde{\mathbf{x}} - \boldsymbol{\mu}_{\tilde{x}})(\tilde{\mathbf{x}} - \boldsymbol{\mu}_{\tilde{x}})^T\} \\ &= \mathcal{E}\{(\tilde{\mathbf{x}}^o + \tilde{\mathbf{x}}^d)(\tilde{\mathbf{x}}^o + \tilde{\mathbf{x}}^d)^T\} \\ &= \mathcal{E}\{\tilde{\mathbf{x}}^o\tilde{\mathbf{x}}^{oT}\} + \mathcal{E}\{\tilde{\mathbf{x}}^d\tilde{\mathbf{x}}^{dT}\}. \end{aligned} \tag{8.2-69}$$

Thus, if we define $\Delta\mathbf{V}_{\tilde{x}} = \mathbf{V}_{\tilde{x}} - \mathbf{P}$, it follows that

$$\Delta\mathbf{V}_{\tilde{x}} = \mathcal{E}\{\tilde{\mathbf{x}}^d\tilde{\mathbf{x}}^{dT}\}. \tag{8.2-70}$$

Use of Eq. (8.2-68) then implies that $\Delta\mathbf{V}_{\tilde{x}}$ satisfies

$$\begin{aligned} \Delta\dot{\mathbf{V}}_{\tilde{x}} = (\bar{\mathbf{F}} - \bar{\mathbf{K}}\bar{\mathbf{H}})\,\Delta\mathbf{V}_{\tilde{x}} + \mathbf{V}_{\tilde{x}}(\bar{\mathbf{F}} - \bar{\mathbf{K}}\bar{\mathbf{H}})^T + (\Delta\mathbf{F} - \bar{\mathbf{K}}\,\Delta\mathbf{H})\,\Delta\mathbf{V}_c + \\ \Delta\mathbf{V}_c^T(\Delta\mathbf{F} - \bar{\mathbf{F}}\,\Delta\mathbf{H})^T + \Delta\mathbf{K}\mathbf{R}\,\Delta\mathbf{K} \end{aligned} \tag{8.2-71}$$

where we define $\Delta\mathbf{V}_c = \mathcal{E}\{\hat{\mathbf{x}}^e\tilde{\mathbf{x}}^{dT}\}$. Further algebraic manipulation shows that $\Delta\mathbf{V}_c$ satisfies

$$\Delta\dot{\mathbf{V}}_c = \mathbf{F}\,\Delta\mathbf{V}_c + \Delta\mathbf{V}_c(\bar{\mathbf{F}} - \bar{\mathbf{K}}\bar{\mathbf{H}})^T + \mathbf{K}\mathbf{R}\,\Delta\mathbf{K}^T + \Delta\mathbf{V}_x(\Delta\mathbf{F} - \bar{\mathbf{K}}\,\Delta\mathbf{H}) \tag{8.2-72}$$

where

$$\Delta\dot{\mathbf{V}}_x = \mathbf{F}\,\Delta\mathbf{V}_x + \Delta\mathbf{V}_x\,\mathbf{F}^T + \mathbf{K}\mathbf{R}\mathbf{K}. \tag{8.2-73}$$

Define $\mathbf{V}_{\tilde{x}} = \mathbf{P} + \Delta\mathbf{V}_{\tilde{x}}$, $\mathbf{V}_c = \mathbf{P} + \Delta\mathbf{V}_c$, and $\mathbf{V}_x = \mathbf{P} + \Delta\mathbf{V}_x$. Then $\mathbf{V}_{\tilde{x}}$, $\mathbf{V}_c$, and $\mathbf{V}_{\tilde{x}}$ satisfy

$$\begin{aligned} \dot{\mathbf{V}}_{\tilde{x}} = (\bar{\mathbf{F}} - \bar{\mathbf{K}}\bar{\mathbf{H}})\mathbf{V}_{\tilde{x}} + \mathbf{V}_{\tilde{x}}(\bar{\mathbf{F}} - \bar{\mathbf{K}}\bar{\mathbf{H}})^T + \mathbf{V}_c^T(\Delta\mathbf{F} - \bar{\mathbf{K}}\,\Delta\mathbf{H})^T + \\ (\Delta\mathbf{F} - \bar{\mathbf{K}}\,\Delta\mathbf{H})\mathbf{V}_c + \mathbf{G}\mathbf{Q}\mathbf{G}^T + \bar{\mathbf{K}}\mathbf{R}\bar{\mathbf{K}}, \end{aligned} \tag{8.2-74}$$

$$\dot{\mathbf{V}}_c = \mathbf{F}\mathbf{V}_c + \mathbf{V}_c(\bar{\mathbf{F}} - \bar{\mathbf{K}}\bar{\mathbf{H}})^T + \mathbf{V}_x(\Delta\mathbf{F} - \bar{\mathbf{K}}\,\Delta\mathbf{H})^T + \mathbf{G}\mathbf{Q}\mathbf{G}^T, \tag{8.2-75}$$

and

$$\dot{\mathbf{V}}_x = \mathbf{F}\mathbf{V}_x + \mathbf{V}_x\mathbf{F}^T + \mathbf{G}\mathbf{Q}\mathbf{G}^T. \tag{8.2-76}$$

We often wish to describe the difference in the modeled error variance and the actual system error variance. This is

$$\Delta\bar{\mathbf{V}}_{\tilde{x}} = \mathbf{V}_{\tilde{x}} - \bar{\mathbf{V}}_{\tilde{x}}. \tag{8.2-77}$$

It follows from Eq. (8.2-74) that

$$\begin{aligned}\dot{\bar{\mathbf{V}}}_{\tilde{x}} + \Delta\dot{\bar{\mathbf{V}}}_{\tilde{x}} = (\bar{\mathbf{F}} - \bar{\mathbf{K}}\bar{\mathbf{H}})\bar{\mathbf{V}}_{\tilde{x}} + \bar{\mathbf{V}}_{\tilde{x}}(\bar{\mathbf{F}} - \bar{\mathbf{K}}\bar{\mathbf{H}})^T + \bar{\mathbf{G}}\bar{\mathbf{Q}}\bar{\mathbf{G}}^T + \bar{\mathbf{K}}\bar{\mathbf{R}}\bar{\mathbf{K}}^T +\\ (\bar{\mathbf{F}} - \bar{\mathbf{K}}\bar{\mathbf{H}})\,\Delta\bar{\mathbf{V}}_{\tilde{x}} + \Delta\bar{\mathbf{V}}_{\tilde{x}}(\bar{\mathbf{F}} - \bar{\mathbf{K}}\bar{\mathbf{H}})^T + \mathbf{V}_c^T(\Delta\mathbf{F} - \bar{\mathbf{K}}\,\Delta\mathbf{H})^T +\\ (\Delta\mathbf{F} - \bar{\mathbf{K}}\,\Delta\mathbf{H})\mathbf{V}_c + (\mathbf{G}\mathbf{Q}\mathbf{G}^T - \bar{\mathbf{G}}\bar{\mathbf{Q}}\bar{\mathbf{G}}^T) + \bar{\mathbf{K}}(\mathbf{R} - \bar{\mathbf{R}})\bar{\mathbf{K}}^T.\end{aligned} \tag{8.2-78}$$

But $\bar{\mathbf{V}}_{\tilde{x}} = \bar{\mathbf{P}}$, given in Eq. (8.2-53), which represents the first six terms on the right-hand side of Eq. (8.2-78). Thus, we have

$$\begin{aligned}\Delta\dot{\bar{\mathbf{V}}}_{\tilde{x}} = (\bar{\mathbf{F}} - \bar{\mathbf{K}}\bar{\mathbf{H}})\,\Delta\bar{\mathbf{V}}_{\tilde{x}} + \bar{\mathbf{V}}_{\tilde{x}}(\bar{\mathbf{F}} - \bar{\mathbf{K}}\bar{\mathbf{H}})^T + (\Delta\mathbf{F} - \bar{\mathbf{K}}\,\Delta\mathbf{H})\mathbf{V}_c +\\ \mathbf{V}_c^T(\Delta\mathbf{F} - \bar{\mathbf{K}}\,\Delta\mathbf{H})^T + (\mathbf{G}\mathbf{Q}\mathbf{G}^T - \bar{\mathbf{G}}\bar{\mathbf{Q}}\bar{\mathbf{G}}^T) + \bar{\mathbf{K}}(\mathbf{R} - \bar{\mathbf{R}})\bar{\mathbf{K}}^T\end{aligned} \tag{8.2-79}$$

which is our desired algorithm for the difference between the modeled error variance and the actual error variance. We note that if $\Delta\mathbf{F} - \bar{\mathbf{K}}\,\Delta\mathbf{H} = \mathbf{0}$, then Eq. (8.2-79) is all that is required to determine $\Delta\bar{\mathbf{V}}_{\tilde{x}}$. If $\Delta\mathbf{F} - \bar{\mathbf{K}}\,\Delta\mathbf{H} \neq \mathbf{0}$, it is necessary to compute $\mathbf{V}_c$ from Eq. (8.2-74) which in turn requires computation of $\mathbf{V}_x$ from Eq. (8.2-76).

We wish to emphasize that Eqs. (8.2-71), (8.2-74), and (8.2-79) are variance or change in variance equations only and, in general, do not represent all the system error. They represent the total error only for zero initial value for the mean of x and in the absence of bias inputs. Unlike optimum estimation, the estimation process with modeling errors is, in general, no longer unbiased, and the bias error propagation is obtained from Eq. (8.2-63) which in turn depends on Eq. (8.2-65). This bias error must be added in to obtain $\mathbf{P}_{\tilde{x}}$, the total mean-squared error of the system.

$$\mathbf{P}_{\tilde{x}} = \mathbf{V}_{\tilde{x}} + \boldsymbol{\mu}_{\tilde{x}}\boldsymbol{\mu}_{\tilde{x}}^T = \mathbf{V}_{\tilde{x}}^o + \Delta\mathbf{V}_{\tilde{x}} + \boldsymbol{\mu}_{\tilde{x}}\boldsymbol{\mu}_{\tilde{x}}' = \bar{\mathbf{V}}_{\tilde{x}} + \Delta\bar{\mathbf{V}}_{\tilde{x}} + \boldsymbol{\mu}_{\tilde{x}}\boldsymbol{\mu}_{\tilde{x}}^T. \tag{8.2-80}$$

The mean-squared error may also be obtained directly by use of Eqs. (8.2-71) (8.2-72), (8.2-73), and (8.2-77) provided proper initial conditions are applied. However, inclusion of nonzero-mean initial conditions indicates that $\mathbf{V}_{\tilde{x}}$ will represent the mean-squared error, not the error variance.

The initial conditions for the variance equations are:

$$\mathbf{V}_{\tilde{x}}(t_o) = \mathbf{V}_c(t_o) = \mathbf{V}_{\mathbf{x}}(t_o) = \mathbf{V}_{\tilde{x}}^o(t_o) = \operatorname{var}[\mathbf{x}(t_o) - \boldsymbol{\mu}_{\mathbf{x}}(t_o)]$$

$$\boldsymbol{\mu}_{\mathbf{x}}(t_o) = \mathcal{E}[\mathbf{x}(t_o)]$$

$$\Delta\mathbf{V}_{\tilde{x}}(t_o) = \Delta\mathbf{V}_c(t_o) = \Delta\mathbf{V}_{\mathbf{x}}(t_o) = \mathbf{0}.$$

The initial values of $\boldsymbol{\mu}_{\tilde{x}}(t_o)$, $\bar{\boldsymbol{\mu}}_{\mathbf{x}}(t_o)$, $\bar{\mathbf{V}}_{\tilde{x}}(t_o)$ and $\Delta\bar{\mathbf{V}}_{\tilde{x}}(t_o)$ are, of course, left to the discretion of the investigator, except that

$$\bar{\mathbf{V}}_{\tilde{x}}(t_o) + \Delta\bar{\mathbf{V}}_{\tilde{x}}(t_o) = \mathbf{V}_{\tilde{x}}(t_o)$$

$$\boldsymbol{\mu}_{\mathbf{x}}(t_o) = \bar{\boldsymbol{\mu}}_{\mathbf{x}}(t_o) + \boldsymbol{\mu}_{\tilde{x}}(t_o). \tag{8.2-81}$$

For direct calculation of the system mean-squared error, the initial conditions are the same except that

$$\mathbf{V}_{\mathbf{x}}(t_o) = \mathbf{V}^o_{\tilde{\mathbf{x}}}(t_o) + \boldsymbol{\mu}_{\mathbf{x}}(t_o)\boldsymbol{\mu}^T_{\mathbf{x}}(t_o)$$
$$\Delta\mathbf{V}_{\mathbf{x}}(t_o) = \boldsymbol{\mu}_{\mathbf{x}}(t_o)\boldsymbol{\mu}^T_{\mathbf{x}}(t_o).$$

The initial conditions for $\boldsymbol{\mu}_{\tilde{\mathbf{x}}}(t_o)$, $\bar{\boldsymbol{\mu}}_{\mathbf{x}}(t_o)$, $\bar{\mathbf{V}}_{\tilde{\mathbf{x}}}(t_o)$ and $\Delta\bar{\mathbf{V}}_{\tilde{\mathbf{x}}}(t_o)$ are again at the discretion of the user, with the restrictions of Eq. (8.2-81).

Further results concerning this development are presented in [18]. In particular a much more general case is considered in which nonzero-mean and correlated noises are considered. Also, the system equation may contain a known deterministic input. Several approximate error analysis algorithms are also determined, and a sensitivity analysis is conducted.

Since the Kalman filter itself requires the solution of a fairly complex matrix Riccati equation, it is not surprising that estimation error analysis algorithms are also quite complex. To help in accessing the magnitude of the task of error analysis using different approaches, Table 8.2-1 is presented. Each column lists the parameter computation required for a particular approach in the order in which the algorithms might be processed. The number to the right of the parameter to be computed indicates the equation number which applies to that computation.

The $\Delta\mathbf{V}_{\tilde{\mathbf{x}}}$ approach, presented in the first column, requires the most parameters to be computed, ten, but also provides exact computation of the most often required parameter, the difference between the optimum error covariance and the true system error. The $\Delta\bar{\mathbf{V}}_{\tilde{\mathbf{x}}}$ approach, given by column 2, requires a calculation of eight parameters, omitting the calculation of the optimum covariance and the optimum Kalman gain. It provides direct computation of the difference between the modeled error variance and the actual

Table 8.2-1
EQUATIONS REQUIRED FOR ERROR ANALYSIS FOR VARIOUS APPROACHES

$\Delta V_{\tilde{\mathbf{x}}}$ approach	$\Delta\bar{\mathbf{V}}_{\tilde{\mathbf{x}}}$ approach	$\mathbf{V}_{\tilde{\mathbf{x}}}$ approach	$\tilde{\mathbf{x}}^b$ approach
$\bar{\mathbf{P}}$ (8.2-53)	$\bar{\mathbf{P}}$ (8.2-53)	$\bar{\mathbf{P}}$ (8.2-53)	$\bar{\mathbf{P}}$ (8.2-53)
$\bar{\mathbf{K}}$ (8.2-52)	$\bar{\mathbf{K}}$ (8.2-52)	$\bar{\mathbf{K}}$ (8.2-52)	$\bar{\mathbf{K}}$ (8.2-52)
$\mathbf{V}^o_{\tilde{\mathbf{x}}}$ (8.1-53)	*$\mathbf{V}_{\mathbf{x}}$ (8.2-76)	*$\mathbf{V}_{\mathbf{x}}$ (8.2-76)	*$\Delta\mathbf{V}_{\mathbf{x}}$ (8.2-73)
$\mathbf{K}$ (8.1-52)	*$\mathbf{V}_{\mathbf{c}}$ (8.2-75)	*$\mathbf{V}_{\mathbf{c}}$ (8.2-75)	*$\tilde{\mathbf{x}}^b$ (8.2-56)
*$\Delta\mathbf{V}_{\mathbf{x}}$ (8.2-73)	$\Delta\bar{\mathbf{V}}_{\tilde{\mathbf{x}}}$ (8.2-79)	$\mathbf{V}_{\tilde{\mathbf{x}}}$ (8.2-73)	$\mathbf{P}_{\tilde{\mathbf{x}}}$ (8.2-80)
*$\Delta\mathbf{V}_{\mathbf{c}}$ (8.2-72)	$\mathbf{Y}_{\mathbf{x}}$ (8.2-65)	$\mathbf{Y}_{\mathbf{x}}$ (8.2-65)	
$\Delta\mathbf{V}_{\tilde{\mathbf{x}}}$ (8.2-71)	$\mathbf{Y}_{\tilde{\mathbf{x}}}$ (8.2-63)	$\mathbf{Y}_{\tilde{\mathbf{x}}}$ (8.2-63)	
$\mathbf{Y}_{\mathbf{x}}$ (8.2-65)	$\mathbf{P}_{\tilde{\mathbf{x}}}$ (8.2-80)	$\mathbf{P}_{\tilde{\mathbf{x}}}$ (8.2-80)	
$\mathbf{Y}_{\tilde{\mathbf{x}}}$ (8.2-63)			
$\mathbf{P}_{\tilde{\mathbf{x}}}$ (8.2-80)			

error variance. This does not disclose how far the system is from optimum. However, if the parameter change is small, then the modeled error variance will be quite close to the optimum error variance even though the actual error variance may diverge considerably. Thus, in many practical situations the $\Delta\bar{\mathbf{V}}_{\tilde{x}}$ approach may be the most attractive. The $\mathbf{V}_{\tilde{x}}$ approach shown in column 3 provides direct computation of the actual variance, $\mathbf{V}_{\tilde{x}}$. Usually one of the first two approaches is preferred. The last column provides the "conditional bias" approach. This approach is considerably simpler computationally than other approaches but provides approximate answers applicable only to cases in which only the system models $\bar{\mathbf{F}}$ and $\bar{\mathbf{H}}$ are in error. In this approach only two matrix Riccati equations must be solved, whereas the exact approaches require solution of four or more equations.

The parameters indicated in this table are those required when errors in all parameters are to be considered. When there are no errors in $\bar{\mathbf{F}}$ and $\bar{\mathbf{H}}$ parameters, equations indicated by * are not required. We will now consider three examples. More detailed results from these examples are presented in [18].

Example 8.2-4. We now consider a system, illustrated in Fig. 8.2-6, with scalar state vector x described by

$$\dot{x}(t) = -ax(t) + w(t)$$
$$z(t) = x(t) + v(t),$$

where an error in the value of a only will be assumed, the modeled value being $\bar{a}$. Only the steady state solution of this example will be discussed. The modeled estimation equations are:

$$\dot{\bar{\hat{x}}} = -\bar{a}\bar{\hat{x}} + \bar{k}(\bar{z} - \bar{\hat{x}})$$
$$\bar{k} = \bar{p}/r$$
$$\dot{\bar{p}} = -2\bar{a}\bar{p} + q - \bar{p}^2/r.$$

For convenience, define

$$\bar{y} = \sqrt{1 + q/(r\bar{a}^2)}, \qquad y = \sqrt{1 + q/(ra^2)}$$

such that under steady state conditions

$$\bar{k} = (\bar{y} - 1)\bar{a} \qquad k = (y - 1)a$$
$$\bar{P} = (\bar{y} - 1)\bar{a}r \qquad P = (y - 1)ar.$$

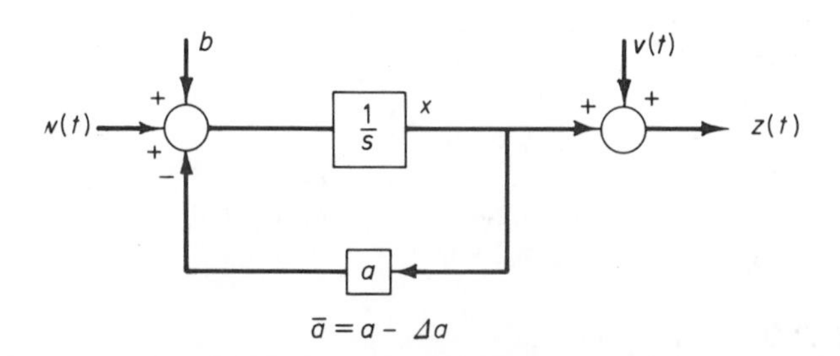

Fig. 8.2-6 Model for Example 8.2-4.

We note further that

$$\Delta V_x = k^2 r/(2a)$$

$$\Delta V_c = \frac{kr}{(a + \bar{a}\bar{y})}\left(\Delta k - \frac{\Delta ak}{2a}\right)$$

$$\Delta V_{\tilde{x}} = \frac{r}{\bar{a}\bar{y}}\frac{(\Delta k)^2}{2} - \frac{k\,\Delta a}{(a + \bar{a}y)}\,\Delta k - \frac{\Delta ak}{2a}.$$

For small $\Delta a = (a - \bar{a})$, a good approximation for $\Delta k = (k - \bar{k})$ is

$$\Delta k = k - \bar{k} = (y - 1)a - (\bar{y} - 1)\bar{a} \cong \Delta a(y - 1).$$

Substituting this approximation into the expression for $\Delta V_{\tilde{x}}$ implies that all terms of $\Delta V_{\tilde{x}}$ are second- or higher-order terms of Δa.

Example 8.2-5. This example demonstrates divergence due to modeling errors. The "conditional bias" approximation is shown to give good results. It is further demonstrated that large differences between modeling and optimum variances do not imply large differences between actual and optimum variances. Conversely, small differences between modeling and optimum variances do not imply small differences between actual and optimum variances.

For this example we consider Fig. 8.2-7. The actual parameters $\mathbf{w}(t)$ and $\mathbf{v}(t)$ are as described in the previous example. The variable parameters are $\bar{\mathbf{Q}}$, $\bar{\mathbf{R}}$, and $\bar{a}$. The system equations are

$$\dot{\mathbf{x}} = \mathbf{F}\mathbf{x} + \mathbf{w}$$

$$y = \mathbf{H}\mathbf{x} + v$$

with the actual system described by:

$$\mathbf{x} = \begin{bmatrix} x_1 \\ x_2 \end{bmatrix}; \quad \mathbf{F} = \begin{bmatrix} -0.5 & 1 \\ 0 & 0 \end{bmatrix}; \quad \mathbf{H} = [1 \quad 0]$$

$$\mathbf{Q} = \begin{bmatrix} 0 & 0 \\ 0 & 10 \end{bmatrix}; \quad R = 1.0.$$

The parameter deviations are:

$$\Delta\mathbf{F} = \begin{bmatrix} \Delta a & 0 \\ 0 & 0 \end{bmatrix}; \quad \Delta\mathbf{Q} = \begin{bmatrix} 0 & 0 \\ 0 & \Delta Q \end{bmatrix}; \quad \Delta\mathbf{R} = \Delta R.$$

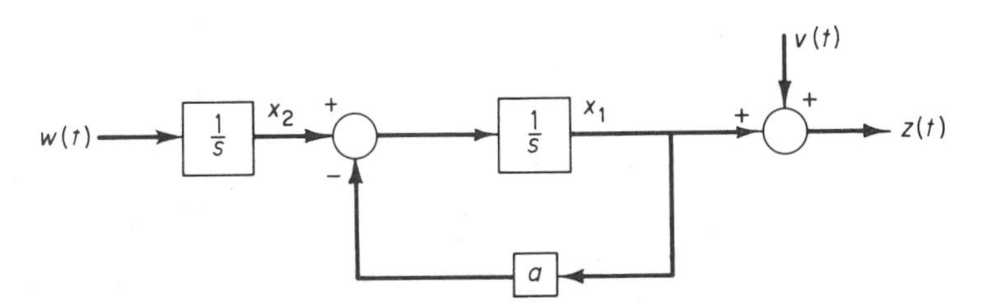

Fig. 8.2-7 Model for Example 8.2-5.

The following numerical values are assumed,

$$\Delta Q = 5, -5, -10$$
$$\Delta R = 0.5, -0.5, -1.0$$
$$\Delta a = -0.005, 0.005, 0.01.$$

Only one deviation was made per experiment. The first error source to be considered is an error in the system time constant a. Because of the integrator, the state variances $\mathbf{V_x}$ for both states 1 and 2, not the error variances $\mathbf{V_{\tilde{x}}}$ continue to increase without bound as t increases. The large variance of state 1 then affects the system error variance through any variation Δa in the system time constant. Since the state variance of x_1 can get very large, it is to be expected that the error variances of the system due to Δa variation might also get quite large. However, since the state x_1 appears directly at the output, its error variance remains small; state x_2 error variance grows without bound as time increases. The deviation of the error variance for state 1 from the optimum $\Delta V_{\tilde{x}11}$ reaches a steady state value of about 2.4×10^{-5} in about five sec. On the other hand $\Delta V_{\tilde{x}22}$ keeps increasing without bound as time increases. This fact is demonstrated in Fig. 8.2-8 which shows the effects of

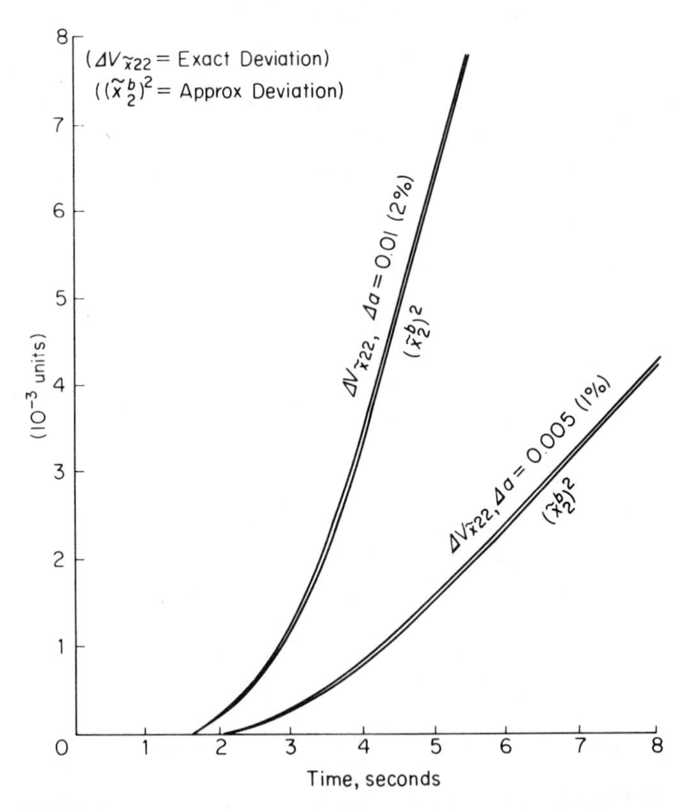

Fig. 8.2-8 Deviation from optimum state 2 error variance due to variation in system time constant.

1 and 2 percent variation in a. Also, apparent from Fig. 8.2-8 is that $(\tilde{x}_2^b)^2$ provides an excellent approximation to $\Delta V_{\tilde{x}22}$. Although $(\tilde{x}_1^b)^2$ does not do a good job of approximating $\Delta V_{\tilde{x}11}$, the error in computing $V_{\tilde{x}11}$ is very small because the deviation is small.

Figure 8.2-9 shows the deviation of the actual error variance from the modeled state 2 error variance due to variation in the system time constant. We note that this figure shows curves of the deviation of the actual error variance from the modeled or computed variances whereas Fig. 8.2-8 shows the deviation of the actual error variance from the optimum.

For this example, the difference between the modeled error variance and the optimum error variance is very small, as would be expected since the parameter variation is very small. However, a small difference between optimum and modeled error variance does not indicate the effect that the parameter variation has on the actual error variance, and it is the actual error variance which is of critical importance in an actual design situation.

The effect on modeled and actual error variance of the states for variation in the plant noise variance is next considered. The modeled noise variance is 5, 15, and 20; or 1/2, 3/2, and 2 times the actual plant noise variance. Variations of this magnitude cause the modeled error variance to deviate considerably from optimum while the deviation of the actual error variance is much less. This is in contrast to the results due to plant dynamics variations. The

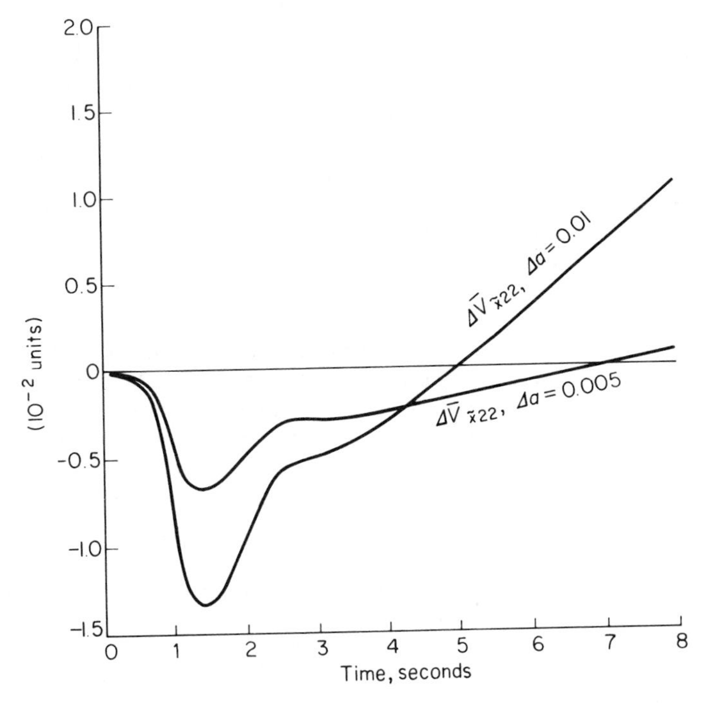

Fig. 8.2-9 Deviation from modeled state 2 error variance due to variation in system time constant.

reason is that the variance of the input plant noise is used in the estimation procedure only for the computation of the Kalman gain, and fortunately, the error variances are relatively insensitive to changes in the Kalman gain. The effect on the modeled and actual error variance for variations in plant measurement noise is that the variation between the optimum and the modeled error variances is much greater than that between the optimum and the actual error variances. This is fortunate in that it is the actual error variance which is of concern, and thus the conclusion is reached that, for this example, the computed or modeled error variance is a poor indicator of the actual error variance.

Example 8.2-6. The last example concerns the system whose block diagram is given by Fig. 8.2-10. The system equations and the numerical

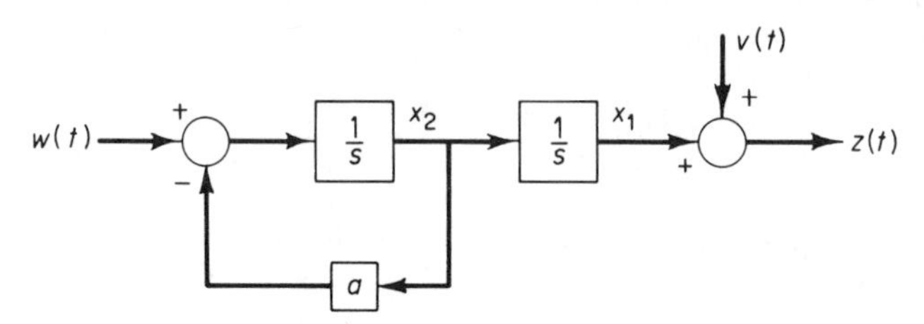

Fig. 8.2-10 Model for Example 8.2-6.

value of the constants a, Q, and R are the same as in the previous example. The overall system transfer function is also the same. However, we note that here the integration occurs after the time lag, so that the state 2 variance remains bounded. It is this state variance that feeds back through the parameter a, so that the system error variance does not diverge for system time constant variation as it did in the previous example. This is because, even though the variance of state 1 gets very large, it is uncoupled from the variations of system time constant a. That is, only ΔV_{x22} affects $\Delta \mathbf{V}_{\tilde{\mathbf{x}}}$, and it is bounded. For this system the $(\tilde{x}^b)^2$ approximation of $\Delta \mathbf{V}_{\tilde{\mathbf{x}}}$ is not good for either state. Here $(\tilde{x}^b)^2$ is 0.11×10^{-2} compared to the actual deviation of 0.24×10^{-2} for state 1. For state 2 the $(\tilde{x}^b)^2$ approximation was 0.47×10^{-2} for the actual deviation. However, the approximation of $\mathbf{V}_{\tilde{\mathbf{x}}}$ using $(\tilde{x}^b)^2$ is quite good, simply because $\Delta \mathbf{V}_{\tilde{\mathbf{x}}}$ remains quite small. We thus reach the important conclusion that all state variable models of a given transfer function are not equivalent when modeling errors are present.

8.3 Kalman filter—discrete-time case

In this section, we derive the discrete version of the Kalman filter determined for the continuous case in Sec. 8.2. We note that, historically, discrete-time Kalman filter results were actually obtained before the continuous results. Motivation for the discrete-time version of the Kalman filter was in part due to the system theory activity in the field of sampled-data systems which developed during the 50s as a result of the use of digital computers in control and communications links. Fundamental work for this problem is due to Kalman [3], Ho and Lee [19], Lee [20], and others.

We wish to consider the development of a linear unbiased minimum error variance algorithm for discrete-time estimation. Consider the vector difference equation

$$\mathbf{x}(k+1) = \boldsymbol{\varphi}(k+1, k)\mathbf{x}(k) + \boldsymbol{\Gamma}(k)\mathbf{w}(k) \tag{8.3-1}$$

and the observation equation

$$\mathbf{z}(k) = \mathbf{H}(k)\mathbf{x}(k) + \mathbf{v}(k), \tag{8.3-2}$$

where

$$\mathcal{E}\{\mathbf{w}(k)\} = \mathcal{E}\{\mathbf{v}(k)\} = \mathbf{0}$$
$$\mathcal{E}\{\mathbf{x}(k_o)\} = \bar{\mathbf{x}}(k_o)$$
$$\text{var}\,\{\mathbf{x}(k_o)\} = \mathbf{P}(k_o)$$
$$\mathcal{E}\{\mathbf{w}(k)\mathbf{w}^T(j)\} = \mathbf{Q}(k)\,\delta(k-j)$$
$$\mathcal{E}\{\mathbf{v}(k)\mathbf{v}^T(j)\} = \mathbf{R}(k)\,\delta(k-j)$$

and

$$\text{cov}\,\{\mathbf{v}(k), \mathbf{x}(k_o)\} = \text{cov}\,\{\mathbf{w}(k), \mathbf{x}(k_o)\} = \text{cov}\,\{\mathbf{v}(k), \mathbf{w}(j)\} = \mathbf{0}.$$

Again, the assumption that $\text{cov}\,\{\mathbf{v}(k), \mathbf{w}(j)\} = \mathbf{0}$ is made for simplicity. Here the δ symbol represents the Kronecker delta function.

The message and observation model for our discrete time problem is illustrated in Fig. 8.3-1. It is a straightforward problem for us to show that algorithms which describe the propagation of the mean and variance of $\mathbf{x}(k)$

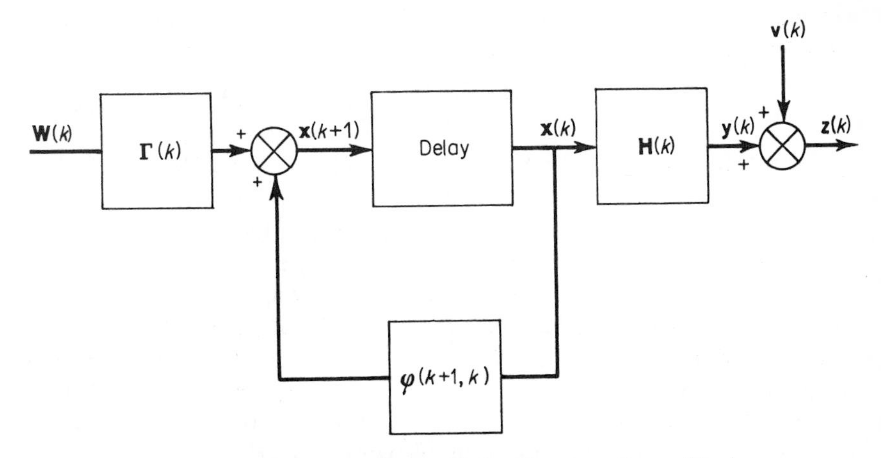

Fig. 8.3-1 Message and noise model for discrete optimum filtering.

are:

$$\mathcal{E}\{\mathbf{x}(k+1)\} = \bar{\mathbf{x}}(k+1) = \boldsymbol{\varphi}(k+1, k)\bar{\mathbf{x}}(k) \tag{8.3-3}$$

and also

$$\begin{aligned} \text{var}\,\{\mathbf{x}(k+1)\} = \mathbf{P}(k+1) &= \mathcal{E}\{[\mathbf{x}(k+1) - \bar{\mathbf{x}}(k+1)][\mathbf{x}(k+1) - \bar{\mathbf{x}}(k+1)]^T\} \\ &= \text{var}\,\{\boldsymbol{\varphi}(k+1, k)[\mathbf{x}(k - \bar{\mathbf{x}}(k)] + \boldsymbol{\Gamma}(k)\mathbf{w}(k)\} \\ &= \boldsymbol{\varphi}(k+1, k)\mathbf{P}(k)\boldsymbol{\varphi}^T(k+1, k) + \boldsymbol{\Gamma}(k)\mathbf{Q}(k)\boldsymbol{\Gamma}^T(k) \end{aligned} \tag{8.3-4}$$

where $\bar{\mathbf{x}}(k_o)$ and $\mathbf{P}(k_o)$ are given as the initial condition vectors and matrices. It is also easy to show that

$$\mathcal{E}\{\mathbf{x}(k+1) \,|\, \mathbf{x}(k)\} = \boldsymbol{\varphi}(k+1, k)\mathbf{x}(k) \tag{8.3-5}$$

$$\text{var}\,\{\mathbf{x}(k+1) \,|\, \mathbf{x}(k)\} = \boldsymbol{\Gamma}(k)\mathbf{Q}(k)\boldsymbol{\Gamma}^T(k). \tag{8.3-6}$$

A random process may be generalized from the case of t discrete to the continuous-time case in the usual way by

$$\frac{d\mathbf{x}(t)}{dt} = \lim_{T \to 0} \frac{\mathbf{x}(t = kT) - \mathbf{x}(t + T = \overline{k+1}T)}{T}.$$

It is easy for us to show that this changes the variance equation from

$$\mathbf{P}(k+1) = \boldsymbol{\varphi}(k+1, k)\mathbf{P}(k)\boldsymbol{\varphi}^T(k+1, k) + \boldsymbol{\Gamma}(k)\mathbf{Q}(k)\boldsymbol{\Gamma}^T(k)$$

to the continuous variance equation

$$\dot{\mathbf{P}} = \mathbf{A}(t)\mathbf{P}(t) + \mathbf{P}(t)\mathbf{A}^T(t) + \mathbf{B}(t)\mathcal{Q}(t)\mathbf{B}^T(t),$$

and the system equation from

$$\mathbf{x}(k+1) = \boldsymbol{\varphi}(k+1, k)\mathbf{x}(k) + \boldsymbol{\Gamma}(k)\mathbf{w}(k)$$

to the continuous system equation

$$\dot{\mathbf{x}} = \mathbf{A}(t)\mathbf{x}(t) + \mathbf{B}(t)\mathbf{w}(t),$$

where cov $\{\mathbf{w}(t)\} = \mathbf{Q}(t)/T$ such that, as T becomes zero, we have a Dirac δ function in the covariance of $\mathbf{w}(t)$ and

$$\mathbf{A}(t) = \lim_{T \to 0} \frac{\mathbf{I} - \boldsymbol{\varphi}(t+T, t)}{T}, \qquad \mathbf{B}(t) = \lim_{T \to 0} \frac{\boldsymbol{\Gamma}(t)}{T}, \tag{8.3-7}$$

$$\text{cov}\,[\mathbf{w}(t_1), \mathbf{w}(t_2)] = \mathcal{Q}(t_1)\,\delta(t_1 - t_2), \qquad \mathcal{Q}(t) = \lim_{\substack{T \to 0 \\ kT \to t}} T\mathbf{Q}(kT). \tag{8.3-8}$$

The other pertinent continuous expressions for means and variances may also be obtained in a direct fashion from the discrete case.

We wish to determine a function $\hat{\mathbf{x}}(k+1 \,|\, k)$ having the following characteristics: it is linearly dependent on the data sequence $\boldsymbol{Z}(k) = \{\mathbf{z}(k_o + 1), \ldots, \mathbf{z}(k)\}$ and the initial mean $\bar{\mathbf{x}}(t_o)$; it is conditionally and unconditionally unbiased in that $\mathcal{E}\{\hat{\mathbf{x}}(k+1 \,|\, k) \,|\, \boldsymbol{Z}(k)\} = \mathcal{E}\{\mathbf{x}(k+1) \,|\, \boldsymbol{Z}(k)\}$ and $\mathcal{E}\{\hat{\mathbf{x}}(k+1 \,|\, k)\} = \mathcal{E}\{\mathbf{x}(k+1)\}$, and it minimizes var $\{\tilde{\mathbf{x}}(k+1 \,|\, k) \,|\, \boldsymbol{Z}(k)\}$ with respect to the set of all such linear functions dependent on the pair $(\boldsymbol{Z}(k), \bar{\mathbf{x}}(k_o))$, where

$$\tilde{\mathbf{x}}(k+1 \,|\, k) = \mathbf{x}(k+1) - \hat{\mathbf{x}}(k+1 \,|\, k) \tag{8.3-9}$$

is called the one-stage prediction estimation error. The function $\hat{\mathbf{x}}(k+1 \mid k)$ is called the optimum linear one-stage predictor and will be used later to derive the optimum linear filter.

We again appeal to the orthogonal projection theorem and related results presented in Appendix B to determine a recursive equation for $\hat{\mathbf{x}}(k+1 \mid k)$. The linear minimum error variance of $\mathbf{x}$, given a linear observation space Z, is given by the orthogonal projection of $\mathbf{x}$ onto Z, that is $\hat{\mathbf{x}} = \hat{\mathcal{E}}\{\mathbf{x} \mid \mathbf{z}\}$. We use the symbol $\hat{\mathcal{E}}$ rather than $\mathcal{E}$ because the linear minimum error variance estimator is generally not the true conditional mean. If we assume all random variables are Gaussian, then $\hat{\mathcal{E}}\{\mathbf{x} \mid \mathbf{z}\} = \mathcal{E}\{\mathbf{x} \mid \mathbf{z}\}$. In the non-Gaussian case the estimator which we will obtain in this development is only the best linear estimator and not necessarily the best estimator.

Let the orthogonal sequence $\{\boldsymbol{\alpha}_i\}_{i=1}^m$ form a basis for Z. Then by Lemma B2 of Appendix B,

$$\hat{\mathcal{E}}\{\mathbf{x} \mid Z\} = \sum_{i=1}^{m} \mathcal{E}\{\mathbf{x}\boldsymbol{\alpha}_i^T\}\mathcal{E}\{\boldsymbol{\alpha}_i\boldsymbol{\alpha}_i^T\}^{-1}\boldsymbol{\alpha}_i. \tag{8.3-10}$$

From this same result, it also follows that if $\boldsymbol{\alpha}^*$ is orthogonal to Z, that is $\mathcal{E}\{Z^T\alpha^*\} = 0$ where $\alpha^* = \{\alpha_1, \alpha_2, \ldots, \alpha_m\}$, then

$$\hat{\mathcal{E}}\{\mathbf{x} \mid Z, \boldsymbol{\alpha}^*\} = \hat{\mathcal{E}}\{\mathbf{x} \mid Z\} + \hat{\mathcal{E}}\{\mathbf{x} \mid \boldsymbol{\alpha}^*\}. \tag{8.3-11}$$

We proceed by induction in the development of a recursive equation for $\hat{\mathbf{x}}(k+1 \mid k)$ by assuming $\hat{\mathbf{x}}(k \mid k-1)$ is known and computing $\mathbf{x}(k+1 \mid k)$ in terms of $\mathbf{x}(k \mid k-1)$ and $\mathbf{z}(k)$. In order to use Eq. (8.3-11), we must determine an $\boldsymbol{\alpha}^*$ which is orthogonal to $Z(k-1)$ since in general $\mathbf{z}(k)$ is not orthogonal to $Z(k-1)$. It is easily shown that the innovation

$$\tilde{\mathbf{z}}(k) = \mathbf{z}(k) - \mathbf{H}(k)\hat{\mathbf{x}}(k \mid k-1) \tag{8.3-12}$$

is orthogonal to $Z(k-1)$. It can also be easily shown that the linear vector spaces generated by $\{\mathbf{z}(k), Z(k-1)\}$ and $\{\tilde{\mathbf{z}}(k), Z(k-1)\}$ are identical.

Therefore, from Eq. (8.3-11), we see that

$$\begin{aligned}\hat{\mathbf{x}}(k+1 \mid k) &= \hat{\mathcal{E}}\{\mathbf{x}(k+1) \mid Z(k)\} \\ &= \hat{\mathcal{E}}\{\mathbf{x}(k+1) \mid \{\tilde{\mathbf{z}}(k), Z(k-1)\}\} \\ &= \hat{\mathcal{E}}\{\mathbf{x}(k+1) \mid Z(k-1)\} + \hat{\mathcal{E}}\{\mathbf{x}(k+1) \mid \tilde{\mathbf{z}}(k)\}. \end{aligned} \tag{8.3-13}$$

It follows from Eq. (8.3-1) and the assumptions on the noise processes $\mathbf{w}(k)$ and $\mathbf{v}(k)$ that

$$\hat{\mathcal{E}}\{\mathbf{x}(k+1) \mid Z(k-1)\} = \boldsymbol{\varphi}(k+1 \mid k)\hat{\mathbf{x}}(k \mid k-1). \tag{8.3-14}$$

We may also show that $\hat{\mathbf{x}}(k+1 \mid k) = \boldsymbol{\varphi}(k+1 \mid k)\hat{\mathcal{E}}\{\mathbf{x}(k) \mid \mathbf{z}(k)\}$, a result which will be of use later. Thus, the first term on the right-hand side of Eq. (8.3-13) may be represented in terms of $\hat{\mathbf{x}}(k \mid k-1)$.

We now wish to examine the second term on the right-hand side of Eq. (8.3-13). It follows from Eq. (8.3-10) that

$$\hat{\mathcal{E}}\{\mathbf{x}(k+1) \mid \tilde{\mathbf{z}}(k)\} = \mathcal{E}\{\mathbf{x}(k+1)\tilde{\mathbf{z}}^T(k)\}[\mathcal{E}\{\tilde{\mathbf{z}}(k)\tilde{\mathbf{z}}^T(k)\}]^{-1}\tilde{\mathbf{z}}(k). \tag{8.3-15}$$

Noting that $\mathbf{z}(k) = \mathbf{H}(k)\mathbf{x}(k|k-1) + \mathbf{v}(k)$, we obtain

$$\begin{aligned}
&\mathcal{E}\{\mathbf{x}(k+1)\tilde{\mathbf{z}}^T(k)\} \\
&= \mathcal{E}\{[\boldsymbol{\varphi}(k+1|k)\mathbf{x}(k) + \boldsymbol{\Gamma}(k)\mathbf{w}(k)] \times [\mathbf{H}(k)\tilde{\mathbf{x}}(k|k-1) + \mathbf{v}(k)]^T\} \\
&= \boldsymbol{\varphi}(k+1|k)\,\mathcal{E}\{\mathbf{x}(k)\tilde{\mathbf{x}}^T(k|k-1)\}\mathbf{H}^T(k) \\
&\quad + \boldsymbol{\varphi}(k+1|k)\,\mathcal{E}\{\mathbf{x}(k)\mathbf{v}^T(k)\} \\
&\quad + \boldsymbol{\Gamma}(k)\,\mathcal{E}\{\mathbf{w}(k)\tilde{\mathbf{x}}^T(k|k-1)\}\mathbf{H}^T(k) \\
&\quad + \boldsymbol{\Gamma}(k)\,\mathcal{E}\{\mathbf{w}(k)\mathbf{v}^T(k)\} \\
&= \boldsymbol{\varphi}(k+1|k)\,\mathcal{E}\{\mathbf{x}(k)\tilde{\mathbf{x}}^T(k|k-1)\}\mathbf{H}^T(k),
\end{aligned} \tag{8.3-16}$$

where all other terms are zero due to the assumptions on the noise processes. Eq. (8.3-9) implies that:

$$\begin{aligned}
&\mathcal{E}\{\mathbf{x}(k)\tilde{\mathbf{x}}^T(k|k-1)\} \\
&= \mathcal{E}\{\hat{\mathbf{x}}(k|k-1)\tilde{\mathbf{x}}^T(k|k-1)\} + \mathcal{E}\{\tilde{\mathbf{x}}(k|k-1)\tilde{\mathbf{x}}^T(k|k-1)\} \\
&= \mathcal{E}\{\tilde{\mathbf{x}}(k|k-1)\tilde{\mathbf{x}}^T(k|k-1)\}
\end{aligned} \tag{8.3-17}$$

where the term $\mathcal{E}\{\hat{\mathbf{x}}(k|k-1)\tilde{\mathbf{x}}^T(k|k-1)\}$ is zero by the orthogonal projection theorem. If

$$\boldsymbol{\Sigma}\,(k|k-1) = \mathcal{E}\{\tilde{\mathbf{x}}(k|k-1)\tilde{\mathbf{x}}^T(k|k-1),$$

then it follows from Eqs. (8.3-16) and (8.3-17) that

$$\mathcal{E}\{\mathbf{x}(k+1)\tilde{\mathbf{z}}^T(k)\} = \boldsymbol{\varphi}(k+1|k)\,\boldsymbol{\Sigma}\,(k|k-1)\mathbf{H}^T(k). \tag{8.3-18}$$

Similarly, it is easily shown that

$$\mathcal{E}\{\tilde{\mathbf{z}}(k)\tilde{\mathbf{z}}^T(k)\} = \mathbf{H}(k)\,\boldsymbol{\Sigma}\,(k|k-1)\mathbf{H}^T(k) + \mathbf{R}(k). \tag{8.3-19}$$

Substituting Eqs. (8.3-18) and (8.3-19) into Eq. (8.3-15) implies that

$$\begin{aligned}
\hat{\mathcal{E}}\{\mathbf{x}(k+1)|\tilde{\mathbf{z}}(k)\} &= \boldsymbol{\varphi}(k+1|k)\,\boldsymbol{\Sigma}\,(k|k-1)\mathbf{H}^T(k) \times \\
&\quad [\mathbf{H}(k)\,\boldsymbol{\Sigma}\,(k|k-1)\mathbf{H}^T(k) + \mathbf{R}(k)]^{-1}\tilde{\mathbf{z}}(k).
\end{aligned} \tag{8.3-20}$$

Eqs. (8.3-14) and (8.3-20), when submitted into Eq. (8.3-13), yield

$$\begin{aligned}
\hat{\mathbf{x}}(k+1|k) &= \boldsymbol{\varphi}(k+1|k)\hat{\mathbf{x}}(k|k-1) + \boldsymbol{\varphi}(k+1|k)\,\boldsymbol{\Sigma}\,(k|k-1)\mathbf{H}^T(k) \times \\
&\quad [\mathbf{H}(k)\,\boldsymbol{\Sigma}\,(k|k-1)\mathbf{H}^T(k) + \mathbf{R}(k)]^{-1} \times [\mathbf{z}(k) - \mathbf{H}(k)\hat{\mathbf{x}}(k|k-1)],
\end{aligned} \tag{8.3-21}$$

which is the desired recursive equation for the linear minimum variance estimator. Often Eq. (8.3-21) is written as

$$\begin{aligned}
\hat{\mathbf{x}}(k+1|k) &= \boldsymbol{\varphi}(k+1|k)\hat{\mathbf{x}}(k|k-1) + \\
&\quad \mathcal{K}(k)[\mathbf{z}(k) - \mathbf{H}(k)\hat{\mathbf{x}}(k|k-1)],
\end{aligned} \tag{8.3-22}$$

where the Kalman gain $\mathcal{K}(k)$ is given by

$$\begin{aligned}
\mathcal{K}(k) &= \boldsymbol{\varphi}(k+1|k)\,\boldsymbol{\Sigma}\,(k|k-1)\mathbf{H}^T(k) \times \\
&\quad [\mathbf{H}(k)\,\boldsymbol{\Sigma}\,(k|k-1)\mathbf{H}^T(k) + \mathbf{R}(k)]^{-1}.
\end{aligned} \tag{8.3-23}$$

We see from Eq. (8.3-23) that calculation of the Kalman gain requires determination of $\boldsymbol{\Sigma}(k \mid k-1)$ which we now show can also be described recursively. It is easily shown that $\tilde{\mathbf{x}}(k+1 \mid k)$ satisfies the recursive equation

$$\tilde{\mathbf{x}}(k+1 \mid k) = \boldsymbol{\phi}(k+1, k)\tilde{\mathbf{x}}(k \mid k-1) + \boldsymbol{\Gamma}(k)\mathbf{w}(k) - \mathcal{K}(k)\tilde{\mathbf{z}}(k). \quad (8.3\text{-}24)$$

Clearly, $\mathbf{w}(k)$ is independent, and hence orthogonal, to both $\tilde{\mathbf{x}}(k \mid k-1)$ and $\tilde{\mathbf{z}}(k)$. Recalling that

$$\tilde{\mathbf{z}}(k) = \mathbf{H}(k)\tilde{\mathbf{x}}(k \mid k-1) + \mathbf{v}(k),$$

we also note that

$$\mathcal{E}\{\tilde{\mathbf{z}}(k)\tilde{\mathbf{x}}^T(k \mid k-1)\} = \mathbf{H}(k)\boldsymbol{\Sigma}(k \mid k-1).$$

It then follows from the definition of $\boldsymbol{\Sigma}(k+1 \mid k)$ and Eqs. (8.3-19), (8.3-22), and the foregoing, that $\boldsymbol{\Sigma}(k+1 \mid k)$ satisfies the recursive equation

$$\begin{aligned}\boldsymbol{\Sigma}(k+1 \mid k) = {} & \boldsymbol{\varphi}(k+1 \mid k)\boldsymbol{\Sigma}(k \mid k-1)\boldsymbol{\varphi}^T(k+1 \mid k) + \boldsymbol{\Gamma}(k)\mathbf{Q}(k)\boldsymbol{\Gamma}^T(k) + \\ & \mathcal{K}(k)[\mathbf{H}(k)\boldsymbol{\Sigma}(k \mid k-1)\mathbf{H}^T(k) + \mathbf{R}(k)]\mathcal{K}^T(k) - \\ & \boldsymbol{\varphi}(k+1 \mid k)\boldsymbol{\Sigma}(k \mid k-1)\mathbf{H}^T(k)\mathcal{K}^T(k) - \\ & \mathcal{K}(k)\mathbf{H}(k)\boldsymbol{\Sigma}(k \mid k-1)\boldsymbol{\varphi}^T(k+1 \mid k). \qquad (8.3\text{-}25)\end{aligned}$$

There are several other ways of describing the above recursive relationship. For example, substitution of $\mathcal{K}(k)$, defined in Eq. (8.3-23), into Eq. (8.3-25) implies, after some algebraic manipulation, that

$$\begin{aligned}\boldsymbol{\Sigma}(k+1 \mid k) = {} & \boldsymbol{\varphi}(k+1 \mid k)\boldsymbol{\Sigma}(k \mid k-1)\boldsymbol{\varphi}^T(k+1 \mid k) - \\ & \boldsymbol{\varphi}(k+1 \mid k)\boldsymbol{\Sigma}(k \mid k-1)\mathbf{H}^T(k) \times [\mathbf{H}(k)\boldsymbol{\Sigma}(k \mid k-1)\mathbf{H}^T(k) + \mathbf{R}(k)]^{-1} \times \\ & \mathbf{H}(k)\boldsymbol{\Sigma}(k \mid k-1)\boldsymbol{\varphi}^T(k+1 \mid k) + \boldsymbol{\Gamma}(k)\mathbf{Q}(k)\boldsymbol{\Gamma}^T(k), \qquad (8.3\text{-}26)\end{aligned}$$

which is known as the *Riccati equation*. Other useful descriptions of this recursive relationship can be found in [7] and [15].

Eq. (8.3-22) represents a recursive description of the linear minimum error variance single-stage predictor which requires knowledge of the minimum error variance presented in the recursive Eqs. (8.3-25) and (8.3-26). We note that neither of the recursive equations for $\boldsymbol{\Sigma}(k+1 \mid k)$ requires knowledge of $Z(k)$, and hence $\boldsymbol{\Sigma}(k+1 \mid k)$ can be computed before the data appear. The optimal predictor, however, must be computed on-line due to its direct dependence on the data. We conclude this development of the recursive equations for $\hat{\mathbf{x}}(k+1 \mid k)$ and $\boldsymbol{\Sigma}(k+1 \mid k)$ by noting that the required initial conditions for the recursive equations are

$$\hat{\mathbf{x}}(k_o+1 \mid k_o) = \boldsymbol{\varphi}(k_o+1, k_o)\bar{\mathbf{x}}(k_o) \qquad (8.3\text{-}27)$$

and

$$\boldsymbol{\Sigma}(k_o+1 \mid k_o) = \boldsymbol{\varphi}(k_o+1, k_o)\mathbf{P}(k_o)\boldsymbol{\varphi}^T(k_o+1, k_o) + \boldsymbol{\Gamma}(k_o)\mathbf{Q}(k_o)\boldsymbol{\Gamma}^T(k_o). \qquad (8.3\text{-}28)$$

Discussions of duality and innovations for the discrete case are found in the problems at the end of this chapter.

We often, but not invariably, prefer determining $\hat{\mathbf{x}}(k \mid k) = \hat{\mathbf{x}}(k) = \hat{\mathcal{E}}\{\mathbf{x}(k) \mid \boldsymbol{Z}(k)\}$ to the one-stage predictor $\hat{\mathbf{x}}(k+1 \mid k)$. We have noted previously that these are related by

$$\hat{\mathbf{x}}(k+1 \mid k) = \boldsymbol{\varphi}(k+1, k)\hat{\mathbf{x}}(k). \tag{8.3-29}$$

Substituting the right-hand side of Eq. (8.3-29) into Eq. (8.3-21) and premultiplying both sides by $\boldsymbol{\varphi}^{-1}(k+1, k) = \boldsymbol{\varphi}(k, k+1)$, we have

$$\hat{\mathbf{x}}(k) = \boldsymbol{\varphi}(k, k-1)\hat{\mathbf{x}}(k-1) + \mathbf{K}(k)[\mathbf{z}(k) - \mathbf{H}(k)\boldsymbol{\varphi}(k, k-1)\hat{\mathbf{x}}(k-1)] \tag{8.3-30}$$

with initial condition $\hat{\mathbf{x}}(k_o) = \bar{\mathbf{x}}(k_o)$, where

$$\mathbf{K}(k) = \boldsymbol{\varphi}(k, k+1)\boldsymbol{\mathcal{K}}(k). \tag{8.3-31}$$

This optimal filter is displayed in Fig. 8.3-2. It is easily shown, using the basic approach illustrated above in determining $\hat{\mathbf{x}}(k+1 \mid k)$, that $\mathbf{x}(k)$ is the linear function of the pair $(\boldsymbol{Z}(k), \mathbf{x}_o)$ that minimizes $\mathcal{E}\{[\mathbf{x}(k) - \hat{\mathbf{x}}(k)][\mathbf{x}(k) - \hat{\mathbf{x}}(k)]^T\}$.

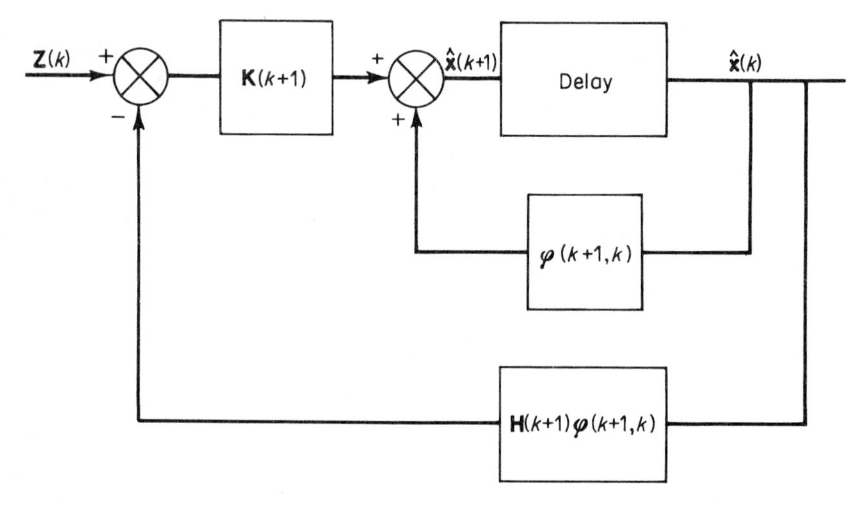

Fig. 8.3-2 Block diagram of optimum discrete filter.

Example 8.3-1. Let us consider discrete state estimation for the scalar process

$$x(k+1) = \varphi x(k) + w(k)$$
$$z(k) = x(k) + v(k),$$

where r and q are constant and $P(k_o) = P_o$. From Eqs. (8.3-21) and (8.3-22),

$$\hat{x}(k+1 \mid k) = \varphi \hat{x}(k \mid k-1) + K(k)[z(k) - \hat{x}(k \mid k-1)],$$

where

$$\mathcal{K}(k) = [\Sigma\,(k \mid k-1) + r]\varphi\,\Sigma\,(k \mid k-1)$$

and

$$\Sigma\,(k+1 \mid k) = \frac{\varphi^2\,\Sigma\,(k \mid k-1)r}{\Sigma\,(k \mid k-1) + r} + q.$$

It then follows from Eqs. (8.3-30) and (8.3-31) that

$$\hat{x}(k) = \varphi\hat{x}(k-1) + K(k)[z(k) - \varphi\hat{x}(k-1)]$$

where
$$K(k) = \varphi^2(\Sigma\,(k|k-1) + r)\,\Sigma\,(k|k-1).$$

As is again apparent, the two sets of estimation equations are nearly the same, being related by $\hat{x}(k+1|k) = \varphi\hat{x}(k)$ and $K(k) = \varphi\mathcal{K}(k)$.

Example 8.3-2. Assume that the system described by Eqs. (8.3-1) and (8.3-2) is such that $\mathbf{w}(k) = \mathbf{0}$ for all k. We wish to minimize the criterion

$$J = \sum_{k=k_o}^{k_f} ||\mathbf{z}(k) - \mathbf{H}(k)\hat{\mathbf{x}}(k)||^2_{\mathbf{R}(k)}$$

with respect to the sequence $\hat{\mathbf{x}}(k_o), \ldots, \hat{\mathbf{x}}(k_f)$ subject to the constraint

$$\hat{\mathbf{x}}(k+1) = \boldsymbol{\varphi}(k+1, k)\hat{\mathbf{x}}(k),$$

where $\mathbf{R}(k)$ is positive definite and where $\hat{\mathbf{x}}(k)$ has available $\boldsymbol{Z}(k_f)$ or $\mathbf{z}(k_o), \ldots, \mathbf{z}(k_f)$, $k = k_o, \ldots, k_f$. Equivalently, we wish to select $\hat{\mathbf{x}}(k_o)$ so that the criterion

$$J = \sum_{k=k_o}^{k_f} ||\mathbf{z}(k) - \mathbf{H}(k)\boldsymbol{\varphi}(k, k_o)\hat{\mathbf{x}}(k_o)||^2_{\mathbf{R}(k)} \tag{1}$$

is minimized, where $\boldsymbol{\varphi}(k, k_o)$ satisfies

$$\boldsymbol{\varphi}(k, k_o) = \boldsymbol{\varphi}(k, t)\boldsymbol{\varphi}(t, k_o),\ \boldsymbol{\varphi}(k, k) = \mathbf{I},$$

where $\mathbf{I}$ is the identity matrix, and $\boldsymbol{\varphi}(k+1, k)$ is given by Eq. (8.3-1). To perform this (unconstrained) minimization, it is necessary [and sufficient due to the convexity of Eq. (1)] that an optimizing $\hat{\mathbf{x}}(k_o)$ satisfy $\partial J/\partial\hat{\mathbf{x}}(k_o) = \mathbf{0}$. This implies from Eq. (1) that

$$\sum_{k=k_o}^{k_f} \boldsymbol{\varphi}^T(k, k_o)\mathbf{H}^T(k)\mathbf{R}(k)[\mathbf{z}(k) - \mathbf{H}(k)\boldsymbol{\varphi}(k, k_o)\hat{\mathbf{x}}(k_o)] = 0. \tag{2}$$

We then note that $\hat{\mathbf{x}}(k_o)$ can be removed from the summation and Eq. (2) can be rewritten as

$$\sum_{k=k_o}^{k_f} \boldsymbol{\varphi}^T(k, k_o)\mathbf{H}^T(k)\mathbf{R}(k)\mathbf{z}(k) = \sum_{k=k_o}^{k_f} \boldsymbol{\varphi}^T(k, k_o)\mathbf{H}^T(k)\mathbf{R}(k)\mathbf{H}(k)\boldsymbol{\varphi}(k, k_o)\hat{\mathbf{x}}(k_o).$$

We remark that the coefficient of $\hat{\mathbf{x}}(k_o)$ is the observability Gramian when $\mathbf{R}(k) = \mathbf{I}$ for all k. Thus, results from Sec. 7.1 state that this problem has a unique solution if and only if the system is observable on $[k_o, k_f]$ if and only if the coefficient of $\hat{\mathbf{x}}(k_o)$ is nonsingular.

8.3-1
Other approaches to the discrete-time optimum linear filter

We now consider another development of the Kalman filter algorithms in which we make explicit assumptions concerning probabilities of random terms in the message and observation models.

We have defined the problem of state estimation as that of determining the "best" estimate $\hat{\mathbf{x}}(t)$ of a system state vector $\mathbf{x}(t)$ from measurements $\boldsymbol{Z}(t_1)$ of the state vector. This definition provides the basis upon which we may distinguish three basic types of estimation. If t is less than t_1, the process is termed *interpolation* or *smoothing* [7]. If t is equal to t_1, the process is termed *filtering*. If t is greater than t_1, we say that we are accomplishing *extrapolation* or *prediction*. The work in this text will be concerned with the filtering problem.

Two other features of this definition should be noted. The first is the use of the term "best" estimate. Clearly, the "bestness" criterion depends upon the particular problem and must be specified for any given problem. The other factor to be considered is the effect of system inputs and disturbances on the estimation process. It is impossible to identify the state of a system without some knowledge of the system inputs and disturbances. Thus, the definition implies a knowledge of the system inputs and at least some statistical information concerning the input noise $\mathbf{w}(t)$ and the output measurement error $\mathbf{v}(t)$.

It is important to recognize the probabilistic nature of state estimation in the presence of random noise. In probabilitistic terms, all the information necessary to estimate the value of a random variable is contained in the probability density functions of that variable. That is, once the conditional probabilities of a random process are known, it is theoretically possible to answer any statistical question concerning that process. Specifically, we are able to compute the conditional means and the conditional variances at any given time. In view of the statistical nature of the estimation process, the definition of estimation can be restated as follows: Given the observed values of $\mathbf{z}(t)$ for $t \leq t_1$, find the conditional probability of $\mathbf{x}(t)$ given $\mathbf{z}(t)$, or in symbols, $p[(\mathbf{x}(t) \mid \mathbf{z}(t), t \leq t_1]$.

In terms of probability theory, we can distinguish various "degrees" or "orders" of estimation, depending upon the statistical knowledge required. Thus, we may define:

1. *Zero-Order Estimation:* This requires knowledge of only the first moments. It implies a nondynamical, nonstochastic model.
2. *First-Order Estimation:* This type of estimation requires knowledge of the first and second moments. It implies a dynamical model and yields the best linear estimate.
3. *Higher-Order Estimation:* This requires knowledge of higher-order statistics. We should remark that very little is known at present about higher-order estimation.
4. *Complete Estimation:* This requires knowledge of statistics of all orders. For a Gaussian process, complete estimation requires a knowledge of only the first and second moments. This, plus the fact that many processes are Gaussian, makes the problem we are treating in this section all important.

On the basis of these classifications, it can be seen that optimal estimation is difficult to attain except for the special case when either the random process is Gaussian or only a "linear" estimate is required. A linear estimate is one which is obtained as a linear combination of the observed values. The "bestness" criterion employed in the case of a linear estimator is that of minimization of mean-square error. That is, an optimal estimate of $\mathbf{x}(t)$ is determined so as to minimize the quantity $\mathcal{E}\{\|\mathbf{x}(t) - \hat{\mathbf{x}}(t)\|^2\}$. This criterion is also referred to as "least-squares." As previously indicated, for the Gaussian case, the best linear estimate is also the best minimum mean-square estimate. That is, for a Gaussian process, no nonlinear filter is superior to the best linear filter in the mean-square error sense.

The probabilistic viewpoint also provides us with a precise definition of the term "optimal estimate." This is especially important since the adjective, optimal, is frequently difficult to define with precision. In terms of probablity theory, an optimal estimate is one that utilizes all available statistical information concerning the process. Since very little statistical information is usually available in a practical application, the ability to estimate despite the lack of information is important. It is also important, however, to recognize the fact that, if more detailed statistical information is available, the least-squares estimation process will not necessarily provide the best possible estimate.

One of the most widely used methods for state estimation is the Kalman computational algorithm for Wiener filtering. This method does not necessarily assume Gaussian statistics. In our previous efforts, we assumed a linear filter and derived the minimum mean-square error estimator without the Gaussian assumption. The following development utilizes the Bayesian approach to deriving the Wiener-Kalman equations for a single stage of a discrete process using Gaussian statistics. We will then extend the result to the case of a multistage process.

The observations (measurements) $\mathbf{z}(k)$ consist of linear combinations of the signals $\mathbf{x}(k)$ corrupted by random noise $\mathbf{v}(k)$. The signal $\mathbf{x}(k)$ and the noise $\mathbf{v}(k)$ are uncorrelated Gaussian-Markov random vector variables. Since $\mathbf{x}(k)$ and $\mathbf{v}(k)$ are Gaussian-Markov processes, and since $\mathbf{z}(k)$ is a linear function of $\mathbf{x}(k)$ and $\mathbf{v}(k)$, then $\mathbf{z}(k)$ is also Gaussian-Markov. Also, it is clear that

$$\mathcal{E}\{\mathbf{z}(k)\} = \mathbf{H}(k)\mathcal{E}\{\mathbf{x}(k)\} = \mathbf{H}(k)\bar{\mathbf{x}}(k) \tag{8.3-32}$$

$$\begin{aligned}\operatorname{cov}\{\mathbf{z}(k)\} &= \operatorname{cov}\{(\mathbf{Hx} + \mathbf{v})\} \\ &= \mathbf{H}(k)\mathbf{P}(k)\mathbf{H}^T(k) + \mathbf{R}(k).\end{aligned} \tag{8.3-33}$$

Therefore, the probability density functions of $\mathbf{x}(k)$, $\mathbf{v}(k)$, and $\mathbf{z}(k)$ are, respectively (here we drop the stage variable k since we consider a single-stage process):

$$p_1(\mathbf{x}) = c_1 \exp\{-0.5(\mathbf{x} - \bar{\mathbf{x}})^T\mathbf{P}_0^{-1}(\mathbf{x} - \bar{\mathbf{x}})\} \tag{8.3-34}$$

$$p_2(\mathbf{v}) = c_2 \exp\{-0.5(\mathbf{z} - \mathbf{Hx})^T\mathbf{R}^{-1}(\mathbf{z} - \mathbf{Hx})\} \tag{8.3-35}$$

$$p_3(\mathbf{z}) = c_3 \exp\{-0.5(\mathbf{z} - \mathbf{H}\bar{\mathbf{x}})^T[\mathbf{HP}_0\mathbf{H}^T + \mathbf{R}]^{-1}(\mathbf{z} - \mathbf{H}\bar{\mathbf{x}})\} \tag{8.3-36}$$

where the c's can certainly be evaluated but do not contribute anything to our present development. In this formulation of the estimation problem, the optimal estimate of $\mathbf{x}(k)$ is taken as that value of $\mathbf{x}(k)$ which maximizes the conditional probability of $\mathbf{x}(k)$, given the measurements $\mathbf{z}(k)$. Since $\mathbf{x}(k)$ and $\mathbf{v}(k)$ are uncorrelated, their joint probability density function is

$$p_4(\mathbf{x}, \mathbf{v}) = p_1(\mathbf{x})p_2(\mathbf{v}). \tag{8.3-37}$$

If we equate incremental probabilities, the joint probabilities are related by the equation

$$p_5(\mathbf{x}, \mathbf{z})\, d\mathbf{x}\, d\mathbf{z} = p_4(\mathbf{x}, \mathbf{v})\, d\mathbf{x}\, d\mathbf{v} \tag{8.3-38}$$

where $d\mathbf{x} = dx_1 dx_2, \ldots dx_n$, an n-dimensional volume element. For fixed $\mathbf{x}$, $d\mathbf{v} = d\mathbf{z}$, such that

$$p_5(\mathbf{x}, \mathbf{z}) = p_4(\mathbf{x}, \mathbf{v}). \tag{8.3-39}$$

And, from Eqs. (8.3-37) and (8.3-38),

$$p_5(\mathbf{x}, \mathbf{z}) = p_1(\mathbf{x})p_2(\mathbf{v}) = p_1(\mathbf{x})p_2(\mathbf{z} - \mathbf{H}\mathbf{x}). \tag{8.3-40}$$

To proceed further, we find it convenient to use *Bayes' rule* which states,

$$p(\mathbf{x} \mid \mathbf{z}) = \frac{p(\mathbf{x}, \mathbf{z})}{p(\mathbf{z})} = \frac{p(\mathbf{z} \mid \mathbf{x})p(\mathbf{x})}{p(\mathbf{z})} \tag{8.3-41}$$

where $p(\mathbf{x})$ is the *a priori* density function of $\mathbf{x}$, and $p(\mathbf{z})$ is the density function of the measurements. Thus, $p(\mathbf{x} \mid \mathbf{z})$ can be obtained if the joint density $p(\mathbf{x}, \mathbf{z})$ is given or can be computed or if the conditional density function $p(\mathbf{z} \mid \mathbf{x})$ is given or can be computed. In this case, the joint density function is available, Eq. (8.3-40), and thus Bayes' rule yields for the conditional probability of $\mathbf{x}(t)$, given $\mathbf{z}(t)$,

$$p_6(\mathbf{x} \mid \mathbf{z}) = \frac{p_1(\mathbf{x})p_2(\mathbf{z} - \mathbf{H}\mathbf{x})}{p_3(\mathbf{z})}. \tag{8.3-42}$$

It is this quantity which we wish to maximize.† A maximum of $p_6(\mathbf{x} \mid \mathbf{z})$ requires:

$$\frac{\partial p_6}{\partial \mathbf{x}} = \mathbf{0}, \qquad \frac{\partial}{\partial \mathbf{x}}\left[\frac{\partial p_6}{\partial \mathbf{x}}\right] = \text{negative definite.} \tag{8.3-43}$$

So we calculate

$$\frac{\partial p_6}{\partial \mathbf{x}} = \frac{\left[p_1(\mathbf{x})\frac{\partial}{\partial \mathbf{x}}[p_2(\mathbf{z} - \mathbf{H}\mathbf{x})] + \left[\frac{\partial}{\partial \mathbf{x}}p_1(\mathbf{x})\right]p_2(\mathbf{z} - \mathbf{H}\mathbf{x})\right]}{p_3(\mathbf{z})} = 0. \tag{8.3-44}$$

Equation (8.3-44) requires that

†We will call the $\hat{\mathbf{x}}$ which maximizes Eq. (8.3-42) the maximum likelihood estimate. The classical maximum likelihood estimate results when $p[\mathbf{z} \mid \mathbf{x}]$ is maximized with respect to $\hat{\mathbf{x}}$. We are obtaining a Bayesian maximum likelihood estimator, a maximum *a posteriori* estimate.

$$p_1(\mathbf{x}) \frac{\partial}{\partial \mathbf{x}}[p_2(\mathbf{z} - \mathbf{H}\mathbf{x})] = -\left\{\frac{\partial}{\partial \mathbf{x}}[p_1(\mathbf{x})]\right\} p_2(\mathbf{z} - \mathbf{H}\mathbf{x}). \qquad (8.3\text{-}45)$$

But

$$\frac{\partial}{\partial \mathbf{x}}[p_1(\mathbf{x})] = -\mathbf{P}_0^{-1}(\hat{\mathbf{x}} - \bar{\mathbf{x}})p_1(\mathbf{x}) \qquad (8.3\text{-}46)$$

$$\frac{\partial}{\partial \mathbf{x}}[p_2(\mathbf{z} - \mathbf{H}\mathbf{x})] = \mathbf{H}^T\mathbf{R}^{-1}(\mathbf{z} - \mathbf{H}\mathbf{x})p_2(\mathbf{z} - \mathbf{H}\mathbf{x}). \qquad (8.3\text{-}47)$$

Therefore, by cancelling the scalar probability density function $p_2(\mathbf{z} - \mathbf{H}\mathbf{x})$ we obtain

$$\mathbf{H}^T\mathbf{R}^{-1}(\mathbf{z} - \mathbf{H}\hat{\mathbf{x}}) = \mathbf{P}_0^{-1}(\hat{\mathbf{x}} - \bar{\mathbf{x}}). \qquad (8.3\text{-}48)$$

Upon rearranging terms and making use of Eq. (8.3-45), we see that

$$\mathbf{H}^T\mathbf{R}^{-1}\mathbf{z} - \mathbf{H}^T\mathbf{R}^{-1}\mathbf{H}\hat{\mathbf{x}} = \mathbf{P}_0^{-1}\hat{\mathbf{x}} - \mathbf{P}_0^{-1}\bar{\mathbf{x}}$$

$$\mathbf{P}_0^{-1}\hat{\mathbf{x}} + \mathbf{H}^T\mathbf{R}^{-1}\mathbf{H}\hat{\mathbf{x}} = \mathbf{H}^T\mathbf{R}^{-1}\mathbf{z} + \mathbf{P}_0^{-1}\bar{\mathbf{x}}$$

$$(\mathbf{P}_0^{-1} + \mathbf{H}^T\mathbf{R}^{-1}\mathbf{H})\hat{\mathbf{x}} = \mathbf{H}^T\mathbf{R}^{-1}(\mathbf{z} - \mathbf{H}\bar{\mathbf{x}}) + (\mathbf{P}_0^{-1} + \mathbf{H}^T\mathbf{R}^{-1}\mathbf{H})\bar{\mathbf{x}}$$

$$\hat{\mathbf{x}} = \bar{\mathbf{x}} + (\mathbf{P}_0^{-1} + \mathbf{H}^T\mathbf{R}^{-1}\mathbf{H})^{-1}\mathbf{H}^T\mathbf{R}^{-1}(\mathbf{z} - \mathbf{H}\bar{\mathbf{x}}).$$

Or, by defining the matrix $\mathbf{P}$ we may write

$$\hat{\mathbf{x}} = \bar{\mathbf{x}} + \mathbf{P}\mathbf{H}^T\mathbf{R}^{-1}(\mathbf{z} - \mathbf{H}\bar{\mathbf{x}}) \qquad (8.3\text{-}49)$$

$$\mathbf{P} = (\mathbf{P}_0^{-1} + \mathbf{H}^T\mathbf{R}^{-1}\mathbf{H})^{-1}. \qquad (8.3\text{-}50)$$

Finally, by using the matrix inversion lemma,† we have

$$\hat{\mathbf{x}} = \bar{\mathbf{x}} + \mathbf{P}\mathbf{H}^T\mathbf{R}^{-1}(\mathbf{z} - \mathbf{H}\bar{\mathbf{x}}) \qquad (8.3\text{-}51)$$

$$\mathbf{P} = \mathbf{P}_0 - \mathbf{P}_0\mathbf{H}^T(\mathbf{H}\mathbf{P}_0\mathbf{H}^T + \mathbf{R})^{-1}\mathbf{H}\mathbf{P}_0. \qquad (8.3\text{-}52)$$

If $\mathbf{R}$ is of lower dimension than $\mathbf{P}$, Eq. (8.3-52) is much less demanding computationally than Eq. (8.3-50).

For the multistage case, we can show that $p[\mathbf{x}(k) \mid Z(k)]$ is Gaussian with

$$\mathcal{E}\{\mathbf{x}(k) \mid Z(k)\} = \hat{\mathbf{x}}(k) \qquad (8.3\text{-}53)$$

$$\text{cov}\{\mathbf{x}(k) \mid Z(k)\} = \mathbf{P}(k) \qquad (8.3\text{-}54)$$

where $p[\mathbf{x}(k) \mid Z(k)]$ indicates $p[\mathbf{x}(k) \mid \mathbf{z}(k_o), \mathbf{z}(k_1), \ldots, \mathbf{z}(k-1), \mathbf{z}(k)]$. The joint density of the white Gaussian-Markov noises is

$$p[\mathbf{w}(k), \mathbf{v}(k+1) \mid \mathbf{x}(k), \mathbf{z}(k)] = p[\mathbf{w}(k), \mathbf{v}(k+1)] = p[\mathbf{w}(k)]p[\mathbf{v}(k+1)] \qquad (8.3\text{-}55)$$

with the assumed means and covariances

$$\mathcal{E}\{\mathbf{w}(k)\} = \mathcal{E}\{\mathbf{v}(k+1)\} = \mathbf{0}$$

$$\text{cov}\{\mathbf{w}(k)\} = \mathbf{Q}(k)$$

$$\text{cov}\{\mathbf{v}(k)\} = \mathbf{R}(k),$$

---

†In Appendix D, we present a proof of this very useful matrix inversion lemma.

where the process is defined by the vector difference equation

$$\mathbf{x}(k+1) = \boldsymbol{\varphi}(k+1, k)\mathbf{x}(k) + \boldsymbol{\Gamma}(k)\mathbf{w}(k) \tag{8.3-56}$$

$$\mathbf{z}(k) = \mathbf{H}(k)\mathbf{x}(k) + \mathbf{v}(k). \tag{8.3-57}$$

We can demonstrate that [19]

1. $p[\mathbf{x}(k+1) \mid \boldsymbol{Z}(k)]$ is Gaussian with

$$\mathcal{E}\{\mathbf{x}(k+1) \mid \boldsymbol{Z}(k)\} = \boldsymbol{\varphi}(k+1, k)\hat{\mathbf{x}}(k)$$

$$\operatorname{cov}\{\mathbf{x}(k+1) \mid \boldsymbol{Z}(k)\} = \boldsymbol{\varphi}(k+1, k)\mathbf{P}(k)\boldsymbol{\varphi}^T(k+1, k) + \boldsymbol{\Gamma}(k)\mathbf{Q}(k)\boldsymbol{\Gamma}^T(k) = \mathbf{N}(k+1).$$

2. $p[\mathbf{z}(k+1) \mid \boldsymbol{Z}(k)]$ is Gaussian with

$$\mathcal{E}\{\mathbf{z}(k+1) \mid \boldsymbol{Z}(k)\} = \mathbf{H}(k+1)\boldsymbol{\varphi}(k+1, k)\hat{\mathbf{x}}(k)$$

$$\operatorname{cov}\{\mathbf{z}(k+1) \mid \boldsymbol{Z}(k)\} = \mathbf{H}(k+1)\mathbf{N}(k+1)\mathbf{H}^T(k+1) + \mathbf{R}(k+1).$$

3. $p[\mathbf{z}(k+1) \mid \mathbf{x}(k+1)]$ is Gaussian with

$$\mathcal{E}\{\mathbf{z}(k+1) \mid \mathbf{x}(k+1)\} = \mathbf{H}(k+1)\mathbf{x}(k+1)$$

$$\operatorname{cov}[\mathbf{z}(k+1) \mid \mathbf{x}(k+1)] = \mathbf{R}(k+1).$$

We note that we are in effect setting $\hat{\mathbf{x}}(k) = \mathcal{E}\{\mathbf{x}(k) \mid \boldsymbol{Z}(k)\}$ which is the conditional mean estimate when we let $\mathcal{E}\{\mathbf{x}(k+1) \mid \boldsymbol{Z}(k)\} = \boldsymbol{\varphi}(k+1, k)\hat{\mathbf{x}}(k)$ in 1. However we do not (yet) know that $\hat{\mathbf{x}}(k)$ is an optimal estimate. Thus far, it is just a parameter.

We are now directly in a position to determine the optimum filter by maximizing the expression $p[\mathbf{x}(k+1) \mid \boldsymbol{Z}(k+1)]$ and its Bayesian equivalent

$$p[\mathbf{x}(k+1) \mid \boldsymbol{Z}(k+1)] = \frac{p[\mathbf{z}(k+1) \mid \mathbf{x}(k+1)]p[\mathbf{x}(k+1) \mid \boldsymbol{Z}(k)]}{p[\mathbf{z}(k+1) \mid \boldsymbol{Z}(k)]}. \tag{8.3-58}$$

This expression is more convenient to use than Bayes' rule. It follows directly from the Markov assumption in that:

$$\begin{aligned} p[\mathbf{x}(k+1) \mid \boldsymbol{Z}(k+1)] &= \frac{p[\mathbf{x}(k+1), \boldsymbol{Z}(k+1)]}{p[\boldsymbol{Z}(k+1)]} \\ &= \frac{p[\mathbf{x}(k+1), \mathbf{z}(k+1) \mid \boldsymbol{Z}(k)]p[\boldsymbol{Z}(k)]}{p[\mathbf{z}(k+1) \mid \boldsymbol{Z}(k)]p[\boldsymbol{Z}(k)]} \\ &= \frac{p[\mathbf{z}(k+1) \mid \mathbf{x}(k+1)]p[\mathbf{x}(k+1) \mid \boldsymbol{Z}(k)]}{p[\mathbf{z}(k+1)\, \boldsymbol{Z}(k)]}. \end{aligned}$$

By a procedure essentially the same as for the single-stage case, we may show that

$$p[\mathbf{x}(k+1) \mid \boldsymbol{Z}(k+1)] = C_7 \exp\{-\tfrac{1}{2}[\mathbf{x}(k+1) - \hat{\mathbf{x}}(k+1)]^T\mathbf{P}^{-1}(k+1)[\mathbf{x}(k+1) - \hat{\mathbf{x}}(k+1)]\}, \tag{8.3-59}$$

where:

$$\hat{\mathbf{x}}(k+1) = \boldsymbol{\varphi}(k+1, k)\hat{\mathbf{x}}(k) + \mathbf{N}(k+1)\mathbf{H}^T(k+1) \times [\mathbf{H}(k+1)\mathbf{N}(k+1)\mathbf{H}^T(k+1) + \mathbf{R}(k+1)]^{-1} \times [\mathbf{z}(k+1) - \mathbf{H}(k+1)\boldsymbol{\varphi}(k+1, k)\hat{\mathbf{x}}(k)] \quad (8.3\text{-}60)$$

$$\mathbf{N}(k+1) = \boldsymbol{\varphi}(k+1, k)\mathbf{P}(k)\boldsymbol{\varphi}^T(k+1, k) + \boldsymbol{\Gamma}(k)\mathbf{Q}(k)\boldsymbol{\Gamma}^T(k) \quad (8.3\text{-}61)$$

$$\mathbf{P}^{-1}(k+1) = \mathbf{N}^{-1}(k+1) + \mathbf{H}^T(k+1)\mathbf{R}^{-1}(k+1)\mathbf{H}(k+1) \quad (8.3\text{-}62)$$

$$\mathbf{P}(k+1) = \mathbf{N}(k+1) - \mathbf{N}(k+1)\mathbf{H}^T(k+1) \times [\mathbf{H}(k+1)\mathbf{N}(k+1)\mathbf{H}^T(k+1) + \mathbf{R}(k+1)]^{-1} \times [\mathbf{H}(k+1)\mathbf{N}(k+1)]. \quad (8.3\text{-}63)$$

The first two equations are also the estimation equations since they maximize Eq. (8.3-58). These estimation equations, which are sequential in nature and thus capable of being processed in an on-line digital computer, yield the maximum likelihood estimate of $\mathbf{x}(k)$ for the special case of a discrete linear system with Gaussian statistics. Figure 8.3-1 illustrates a block diagram of the message model, and Fig. 8.3-2 shows a block diagram of the optimum discrete filter for this problem.

It should be noted that $\mathbf{N}(k)$ is not the covariance matrix of the error $[\mathbf{x}(k) - \hat{\mathbf{x}}(k)]$ given measurements up to and including $\mathbf{z}(k)$. From Eq. (8.3-54) we see that $\mathbf{P}(k) = \operatorname{cov}\{\mathbf{x}(k) \mid \mathbf{Z}(k)\}$. Then from Eq. (8.3-60) we see that the estimate $\hat{\mathbf{x}}$ may be rewritten as

$$\hat{\mathbf{x}}(k+1) = \boldsymbol{\varphi}(k+1, k)\hat{\mathbf{x}}(k) + \mathbf{K}(k+1)[\mathbf{z}(k+1) - \mathbf{H}(k+1)\boldsymbol{\varphi}(k+1, k)\hat{\mathbf{x}}(k)], \quad (8.3\text{-}63)$$

where

$$\mathbf{K}(k+1) = \mathbf{N}(k+1)\mathbf{H}^T(k+1)[\mathbf{H}(k+1)\mathbf{N}(k+1)\mathbf{H}^T(k+1) + \mathbf{R}(k+1)]^{-1}$$

$$\mathbf{K}(k+1) = \mathbf{P}(k+1)\mathbf{H}^T(k+1)\mathbf{R}^{-1}(k+1). \quad (8.3\text{-}64)$$

Certainly, $\hat{\mathbf{x}}(k+1)$ is Gaussian-Markov, and we can show that the mean and covariance of $\hat{\mathbf{x}}(k+1)$ in Eq. (8.3-63) are:

$$\mathcal{E}\{\hat{\mathbf{x}}(k+1)\} = \boldsymbol{\varphi}(k+1, k)\mathcal{E}\{\hat{\mathbf{x}}(k)\} = \boldsymbol{\varphi}(k+1, k)\hat{\mathbf{x}}(k) \quad (8.3\text{-}65)$$

$$\operatorname{cov}\{\hat{\mathbf{x}}(k+1)\} = \boldsymbol{\varphi}(k+1, k)\operatorname{cov}\{\hat{\mathbf{x}}(k)\}\boldsymbol{\varphi}^T(k+1, k) + \mathbf{K}(k+1) \times [\mathbf{H}(k+1)\mathbf{N}(k+1)\mathbf{H}^T(k+1) + \mathbf{R}(k+1)]\mathbf{K}^T(k+1). \quad (8.3\text{-}66)$$

These values follow directly when we recall from Eq. (8.3-53) that $\hat{\mathbf{x}}(k+1) = \mathcal{E}\{\mathbf{x}(k+1) \mid \mathbf{Z}(k+1)\}$. They also follow from Eqs. (8.3-63) and (8.3-64).

The difference equation for the estimation error can be obtained if we subtract Eq. (8.3-60) from Eq. (8.3-56):

$$\tilde{\mathbf{x}}(k+1) = \mathbf{x}(k+1) - \hat{\mathbf{x}}(k+1) = \boldsymbol{\varphi}(k+1, k)\tilde{\mathbf{x}}(k) + \boldsymbol{\Gamma}(k)\mathbf{w}(k) -$$
$$\mathbf{K}(k+1)[\mathbf{H}(k+1)\boldsymbol{\varphi}(k+1, k)\tilde{\mathbf{x}}(k) + \mathbf{v}(k+1) + \mathbf{H}(k+1)\boldsymbol{\Gamma}(k)\mathbf{w}(k)], \tag{8.3-67}$$

where var $\{\mathbf{z}(k) - \mathbf{H}(k)\boldsymbol{\varphi}(k, k-1)\hat{\mathbf{x}}(k-1)\} = \mathbf{H}(k)\mathbf{N}(k)\mathbf{H}^T(k) + \mathbf{R}(k)$.

The probability density of this error is Gaussian. The covariance matrix is

$$\text{cov}\,\{\tilde{\mathbf{x}}(k+1)\} = \mathbf{N}(k+1) - \mathbf{K}(k+1)\mathbf{H}(k+1)\mathbf{N}(k+1) = \mathbf{P}(k+1). \tag{8.3-68}$$

These solutions are the discrete version of the continuous Wiener-Kalman filter equations which we derived earlier in this chapter. Often $\mathbf{N}(k+1)$ is spoken of as the error covariance matrix. Actually it is an extrapolated error covariance at stage $k+1$ based upon the actual error covariance matrix $\mathbf{P}(k)$ at stage k.

Since $\mathbf{v}(k)$ is a zero-mean white random sequence, it is possible to show that the prediction solution is given by

$$\hat{\mathbf{x}}(k+r \mid k+1) = \boldsymbol{\varphi}(k+r, k+1)\hat{\mathbf{x}}(k+1). \tag{8.3-69}$$

For the case where the transition matrix is constant, this reduces to

$$\hat{\mathbf{x}}(k+r \mid k+1) = \boldsymbol{\varphi}^{r-1}\hat{\mathbf{x}}(k+1) \tag{8.3-70}$$

where the symbol $\hat{\mathbf{x}}(k+r \mid k+1)$ means $\hat{\mathbf{x}}$ at stage $k+r$, given information up to stage $k+1$. It is also possible to develop the analogous solution for smoothing [7].

It should be noted that the previous solutions are not those first obtained by Kalman [3]. In his original paper, Kalman assumed that we wished to estimate $\mathbf{x}(k+1)$, given information up to but not including the $\mathbf{z}(k+1)$st measurement. In that case, it is desired to maximize the expression $p[\mathbf{x}(k+1) \mid \boldsymbol{Z}(k)]$. We used this expression earlier in developing our Kalman filter algorithms. This probability was Gaussian, with

$$\mathcal{E}\{\mathbf{x}(k+1) \mid \boldsymbol{Z}(k)\} = \boldsymbol{\varphi}(k+1, k)\hat{\mathbf{x}}(k) \tag{8.3-71}$$

$$\text{cov}\,\{\mathbf{x}(k+1) \mid \boldsymbol{Z}(k)\} = \mathbf{N}(k+1) =$$
$$\boldsymbol{\varphi}(k+1, k)\mathbf{P}(k)\boldsymbol{\varphi}^T(k+1, k) + \boldsymbol{\Gamma}(k)\mathbf{Q}(k)\boldsymbol{\Gamma}^T(k). \tag{8.3-72}$$

Therefore we have

$$p[\mathbf{x}(k+1) \mid \boldsymbol{Z}(k)] = \frac{1}{(2\pi)^{n/2}|\mathbf{N}(k+1)|^{1/2}} \cdot$$
$$\exp\{-\tfrac{1}{2}[\mathbf{x}(k+1) - \boldsymbol{\varphi}(k+1, k)\hat{\mathbf{x}}(k)]^T\mathbf{N}^{-1}(k+1)[\mathbf{x}(k+1) -$$
$$\boldsymbol{\varphi}(k+1, k)\hat{\mathbf{x}}(k)]\}. \tag{8.3-73}$$

This expression is maximized by

$$\hat{\mathbf{x}}(k+1 \mid k) = \boldsymbol{\varphi}(k+1, k)\hat{\mathbf{x}}(k \mid k) \tag{8.3-74}$$

which is the best predicted estimate of $\hat{\mathbf{x}}(k+1)$. This could have been obtained from Eq. (8.3-69). Thus, we will use $\hat{\mathbf{x}}(k)$ to represent the estimate of $\mathbf{x}$ at stage k with observations including stage k and $\hat{\mathbf{x}}(k \mid k_1)$ to represent the (filtered) estimate at stage k with observations up to and including stage k_1. For consistency, we will also use $\mathbf{\Sigma}(k+1 \mid k) = \mathbf{N}(k+1)$.

If Eq. (8.3-74) is incorporated into our estimation equations, we obtain

$$\hat{\mathbf{x}}(k+1 \mid k) = \boldsymbol{\varphi}(k+1, k)\hat{\mathbf{x}}(k \mid k-1) + \mathbf{K}'(k)[\mathbf{z}(k) - \mathbf{H}(k)\hat{\mathbf{x}}(k \mid k-1)] \tag{8.3-75}$$

$$\mathbf{K}'(k) = \boldsymbol{\varphi}(k+1, k)\,\mathbf{\Sigma}(k \mid k-1)\mathbf{H}^T(k)[\mathbf{H}(k)\,\mathbf{\Sigma}(k \mid k-1)\mathbf{H}^T(k) + \mathbf{R}(k)]^{-1} \tag{8.3-76}$$

$$\begin{aligned}\mathbf{\Sigma}(k+1 \mid k) = \boldsymbol{\varphi}(k+1, k)\{\mathbf{\Sigma}(k \mid k-1) - \mathbf{\Sigma}(k \mid k-1)\mathbf{H}^T(k) \times \\ [\mathbf{H}(k)\,\mathbf{\Sigma}(k \mid k-1)\mathbf{H}^T(k) + \mathbf{R}(k)]^{-1}\mathbf{H}(k)\,\mathbf{\Sigma}(k \mid k-1)\}\boldsymbol{\varphi}^T(k+1, k) + \\ \mathbf{\Gamma}(k)\mathbf{Q}(k)\mathbf{\Gamma}^T(k).\end{aligned} \tag{8.3-77}$$

Close examination will reveal that these equations are very similar to but not quite the same as the previous Wiener-Kalman equations. The reason for the difference is basically the predictive term. $\mathbf{\Sigma}$ is the covariance matrix of the error in the estimate for this case. It is apparent that both types of filters are "optimum" with but slightly different optimization criteria.

Example 8.3-3. Let us consider discrete state estimation for the scalar process where,

$$x(k+1) = \varphi x(k) + w(k)$$

$$z(k) = x(k) + v(k),$$

where r and q are constant and $P(k_o) = P_o$. For the problem of maximizing $p[x(k+1) \mid Z(k+1)]$, we have from Eqs. (8.3-60) and (8.3-61)

$$\hat{x}(k+1) = \varphi\hat{x}(k) + n(k+1)[n(k+1) + r]^{-1}[z(k+1) - \varphi\hat{x}(k)]$$

$$n(k+1) = \varphi^2 p(k) + q$$

where the error variance is obtained from Eq. (8.3-63) as

$$\begin{aligned} p(k+1) &= n(k+1) - n(k+1)[n(k+1) + r]^{-1}n(k+1) \\ &= [n^{-1}(k+1) + r^{-1}]^{-1}. \end{aligned}$$

The foregoing two equations may be combined to yield

$$n(k+1) = \frac{\varphi^2 r n(k)}{r + n(k)} + q.$$

Thus, the filter equations may be written as

$$\hat{x}(k+1) = \varphi\hat{x}(k) + r^{-1}p(k+1)[z(k+1) - \varphi\hat{x}(k)]$$

$$p(k+1) = \frac{r[\varphi^2 p(k) + q]}{\varphi^2 p(k) + q + r}.$$

For the problem of maximizing $p[x(k+1) \mid Z(k)]$, we have from Eqs. (8.3-75) through (8.3-77)

$$\hat{x}(k+1\,|\,k) = \varphi\hat{x}(k\,|\,k-1) + K'(k)[z(k) - \hat{x}(k\,|\,k-1)],$$

where

$$K'(k) = \varphi\,\Sigma\,(k\,|\,k-1)[\Sigma\,(k\,|\,k-1) + r]^{-1}$$

$$\Sigma\,(k+1\,|\,k) = \varphi^2\{\Sigma\,(k\,|\,k-1) -$$

$$\Sigma\,(k\,|\,k-1)[\Sigma\,(k\,|\,k-1) + r]^{-1}\,\Sigma\,(k\,|\,k-1)\} + q.$$

As is apparent again, the two sets of estimation equations are nearly the same. They are related by $\hat{x}(k+1\,|\,k) = \varphi\hat{x}(k)$ and $N(k) = \Sigma\,(k\,|\,k-1)$.

Example 8.3-4. Earlier in the chapter we considered estimation for the process defined by the scalar differential system

$$\dot{x} = -\tfrac{1}{2}x + w(t), \qquad Q = 2, \qquad p(0) = 0$$

$$z = x(t) + v(t), \qquad R = \tfrac{1}{4},$$

and we obtained the estimation equations

$$\dot{\hat{x}} = -\tfrac{1}{2}\hat{x} + 4p(t)[z(t) - \hat{x}(t)], \qquad \hat{x}(0) = 0$$

$$\dot{p} = -p(t) - 4p^2(t) + 2, \qquad p(0) = 0.$$

The difference equations we obtain by taking a first difference approximation to these estimation equations are

$$\hat{x}(k+1) = \left(1 - \frac{T}{2}\right)\hat{x}(k) + 4TP(k)[z(k) - \hat{x}(k)], \qquad \hat{x}(0) = 0$$

$$p(k+1) = (1-T)p(k) - 4Tp^2(k) + 2T, \qquad p(0) = 0.$$

Other equally valid but different first difference or higher-order difference approximations may, of course, be obtained. The Wiener-Kalman discrete estimation equations which minimize $p[x(k+1)\,|\,\boldsymbol{Z}(k+1)]$ may be obtained from a first difference approximation to the original process equations. This yields

$$x(k+1) = \left(1 - \frac{T}{2}\right)x(k) + Tw(k), \qquad Q = \frac{2}{T}, \qquad R = \frac{1}{4T}$$

$$z(k) = x(k) + v(k), \qquad p(0) = 0.$$

The estimation equations are therefore

$$\hat{x}(k+1) = \left(1 - \frac{T}{2}\right)\hat{x}(k) + \frac{n(k+1)}{n(k+1) + 1/4T}\left[z(k+1) - \left(1 - \frac{T}{2}\right)\hat{x}(k)\right]$$

$$n(k+1) = \left(1 - \frac{T}{2}\right)^2 p(k) + 2T,$$

where from Eq. (8.3-62)

$$p(k) = \frac{n(k)}{1 + 4Tn(k)}.$$

These may be rewritten as

$$\hat{x}(k+1) = \left(1 - \frac{T}{2}\right)\hat{x}(k) + 4Tp(k+1)\left[z(k+1) - \left(1 - \frac{T}{2}\right)\hat{x}(k)\right],$$

where

$$p(k+1) = \frac{(1 - T/2)^2 p(k) + 2T}{(1 - T/2)^2 4Tp(k) + (1 + 8T^2)}.$$

The estimation equations which yield the one stage prediction solution are

$$\hat{x}(k+1 \mid k) = \left(1 - \frac{T}{2}\right)\hat{x}(k \mid k-1) + K'(k)[z(k) - x(k \mid k-1)]$$

$$K'(k) = \frac{(1 - T/2)\,\Sigma\,(k \mid k-1)}{\Sigma\,(k \mid k-1) + 1/4T}$$

$$\Sigma\,(k+1 \mid k) = \left(1 - \frac{T}{2}\right)^2 \left\{\Sigma\,(k \mid k-1) - \frac{\Sigma^2\,(k \mid k-1)}{\Sigma\,(k \mid k-1) + 1/4T}\right\} + 2T.$$

For T small, these last equations can be easily approximated by

$$\hat{x}(k+1 \mid k) = \left(1 - \frac{T}{2}\right)\hat{x}(k \mid k-1) + K'(k)[z(k) - \hat{x}(k \mid k-1)]$$

$$K'(k) = 4T\Sigma\,(k \mid k-1)$$

$$\Sigma\,(k+1 \mid k) \approx (1 - T)\,\Sigma\,(k \mid k-1) - 4T\Sigma^2\,(k \mid k-1) + 2T.$$

These are exactly the same as the discrete estimation equations which were obtained by a difference approximation to the continuous estimation equations. An effect similar to this was noted in Chapter 6 on the discrete maximum principle. The comments in that chapter on discrepancy between the two approaches applies to the problem at hand almost without modification.

The behavior of the solution of Eq. (8.3-26) for the time-invariant case as the time of observation grows long is of some theoretical importance. The asymptotic behavior and the existence and uniqueness of the solution of the continuous variance equation is discussed in [5, 7, 21]. We can show that under suitable hypotheses $\Sigma\,(k \mid k-1)$ is uniformly bounded and converges to a finite solution as k grows large. The sufficient condition common to both of these results is the observability of the system Eqs. (8.3-1) and (8.3-2). Such a result has considerable intuitive appeal since observability implies that the data received provides information about the present state of all elements of the state vector. Such information, in turn, causes the covariance of the state estimation error to be bounded.

To show that the observability of $(\boldsymbol{\varphi}, \mathbf{H})$ implies that $\Sigma\,(k \mid k-1)$ is uniformly bounded as $k \longrightarrow \infty$, we can determine a suboptimal unbiased admissible estimator. We postulate an unbiased estimator that is linear in the past observations, which is uniformly bounded in k. The error covariance of this suboptimal estimator will then necessarily bound $\Sigma\,(k \mid k-1)$ from above for all k, implying the result. The existence of a solution for $\Sigma\,(k \mid k-1)$ is guaranteed if $(\boldsymbol{\varphi}, \mathbf{H})$ is observable. Our intuition is correct in this case. Proofs and extensions of these results can be found in [7], [21], and [22].

8.4
Reconstruction of state variables from output variables—observers

In previous sections, we were concerned with the estimation of system state variables from noisy observation of output variables. We developed the theory of the Kalman filter in order to cope with problems of this type. In many cases, however, there is no measurement noise present. In this case a limiting form of the Kalman filter could, of course, be used. [$\mathbf{R}^{-1}$ does not exist.] However, as we shall see, it is possible to develop a state estimator for the measurement noise-free case with considerably less complexity than the Kalman filter.

There are two solutions to the problem posed. Certainly, if the system dynamics are known, a model of the system can be constructed which has all of its state variables directly measurable. Any input signal is applied to the model as well as to the system. Thus, even if the state vector of the original system cannot be measured, the model's state variables, which are equivalent to those of the system, are available. It is also necessary to set the correct initial conditions on the model. The principal objection to this solution is that it is not often possible to know the system dynamics exactly, and disturbance inputs which enter the system but not the model may cause disastrous errors. Another method consists of differentiating the output variables a number of times and combining these terms to form the unmeasured state variables. The poor noise characteristics and the difficulty of building good differentiators generally make this method impractical. We shall now indicate two alternative methods for state variable reconstruction, one due to Kalman [4] and the other due to Luenberger [23, 24].

8.4-1
Reconstruction of all system state vectors

Suppose that we have a linear observable differential system described by

$$\dot{\mathbf{x}} = \mathbf{A}(t)\mathbf{x}(t) + \mathbf{B}(t)\mathbf{u}(t), \qquad \mathbf{x}(t_o) = \mathbf{x}_o \tag{8.4-1}$$

$$\mathbf{z}(t) = \mathbf{C}(t)\mathbf{x}(t). \tag{8.4-2}$$

If it happens that $\mathbf{C}(t)$ is nonsingular, the state vector $\mathbf{x}(t)$ may easily be determined from $\mathbf{x}(t) = \mathbf{C}^{-1}(t)\mathbf{z}(t)$. If we are not this fortunate, we may attempt to let the output vector $\mathbf{z}(t)$ serve as the input to a linear observer described by

$$\dot{\boldsymbol{\xi}} = \mathbf{D}(t)\boldsymbol{\xi}(t) + \mathbf{E}(t)\mathbf{u}(t) + \mathbf{F}(t)\mathbf{x}(t), \qquad \boldsymbol{\xi}(t_o) = \boldsymbol{\xi}_o. \tag{8.4-3}$$

We shall attempt to adjust the observer such that

$$\boldsymbol{\xi}(t) = \mathbf{T}\mathbf{x}(t), \tag{8.4-4}$$

where $\mathbf{T}$ is a constant and nonsingular matrix. If we premultiply Eq. (8.4-1) by $\mathbf{T}$ and differentiate Eq. (8.4-4) with respect to time, we have from Eq. (8.4-3),

$$\mathbf{D}(t)\mathbf{T}\mathbf{x}(t) + \mathbf{E}(t)\mathbf{u}(t) + \mathbf{F}(t)\mathbf{x}(t) = \mathbf{T}\mathbf{A}(t)\mathbf{x}(t) + \mathbf{T}\mathbf{B}(t)\mathbf{u}(t). \tag{8.4-5}$$

Thus we see that, for nontrivial $\mathbf{x}(t)$ and $\mathbf{u}(t)$, we must set

$$\mathbf{D}(t)\mathbf{T} - \mathbf{T}\mathbf{A}(t) + \mathbf{F}(t) = \mathbf{0} \tag{8.4-6}$$

$$\mathbf{E}(t) = \mathbf{T}\mathbf{B}(t). \tag{8.4-7}$$

If $\mathbf{D}(t)$ and $\mathbf{A}(t)$ have no common eigenvalues, it can be shown that $\mathbf{T}$ in Eq. (8.4-6) will have a unique solution for specified $\mathbf{D}(t)$, $\mathbf{A}(t)$, and $\mathbf{C}(t)$. Therefore, to design the observer, we may specify $\mathbf{D}(t)$ or $\mathbf{T}$ which, in turn, will determine $\mathbf{E}(t)$. $\mathbf{A}(t)$, $\mathbf{B}(t)$, $\mathbf{C}(t)$, and $\mathbf{F}(t)$ are unalterable here. If we premultiply Eq. (8.4-1) by $\mathbf{T}$ and subtract the result from Eq. (8.4-3), we have

$$\dot{\boldsymbol{\xi}} - \mathbf{T}\dot{\mathbf{x}} = \mathbf{D}(t)\boldsymbol{\xi}(t) + [\mathbf{E}(t) - \mathbf{T}\mathbf{B}(t)]\mathbf{u}(t) + [\mathbf{F}(t) - \mathbf{T}\mathbf{A}(t)]\mathbf{x}(t). \tag{8.4-8}$$

When we use Eqs. (8.4-6) and (8.4-7), the foregoing equation reduces to

$$\dot{\boldsymbol{\xi}} - \mathbf{T}\dot{\mathbf{x}} = \mathbf{D}(t)[\boldsymbol{\xi}(t) - \mathbf{T}\mathbf{x}(t)] \tag{8.4-9}$$

which has the solution

$$\boldsymbol{\xi}(t) = \mathbf{T}\mathbf{x}(t) + \boldsymbol{\varphi}(t, t_o)[\boldsymbol{\xi}(t_o) - \mathbf{T}\mathbf{x}(t_o)] \tag{8.4-10}$$

where

$$\frac{\partial \boldsymbol{\varphi}(t, t_o)}{\partial t} = \mathbf{D}(t)\boldsymbol{\varphi}(t, t_o), \qquad \boldsymbol{\varphi}(t, t) = \mathbf{I}. \tag{8.4-11}$$

We see from Eq. (8.4-10) that we do get the required result $\boldsymbol{\xi}(t) = \mathbf{T}\mathbf{x}(t)$, but that we must set the initial condition vector of the observer so that the result $\boldsymbol{\xi}(t) = \mathbf{T}\mathbf{x}(t)$ is obtained. This is

$$\boldsymbol{\xi}(t_o) = \mathbf{T}\mathbf{x}(t_o). \tag{8.4-12}$$

It is apparent we must use caution here for, in an actual case, it will seldom be possible to set these initial conditions correctly. Thus, it is mandatory that $\boldsymbol{\varphi}(t, t_o)$ be the transition matrix of a stable system or that the eigenvalues of $\mathbf{D}(t)$ all have negative real parts. If the real parts of the eigenvalues of $\mathbf{D}(t)$ are quite negative, then any transient due to an error in the setting will normally die out quickly and lead to a good observer. We use the word "normally" since, for a time-varying system, negative real eigenvalues do not guarantee system stability.

From the form of Eq. (8.4-4), we know that $\mathbf{T}$ must be an $n \times n$ matrix [$\boldsymbol{\xi}(t)$ and $\mathbf{x}(t)$ must be n vectors] if we are to be able to invert $\mathbf{T}$ to obtain $\mathbf{x}(t)$. It is desirable to avoid this inversion, and thus we inquire into the possibilities of setting $\mathbf{T} = \mathbf{I}$, an $n \times n$ identity matrix. In this case, we have from Eqs. (8.4-6) and (8.4-7)

$$\mathbf{D}(t) = \mathbf{A}(t) - \mathbf{F}(t), \qquad \mathbf{E}(t) = \mathbf{B}(t), \qquad \mathbf{F}(t) = \mathbf{G}(t)\mathbf{C}(t), \tag{8.4-13}$$

such that the observer is described by

$$\dot{\boldsymbol{\xi}} = [\mathbf{A}(t) - \mathbf{F}(t)]\boldsymbol{\xi}(t) + \mathbf{B}(t)\mathbf{u}(t) + \mathbf{F}(t)\mathbf{x}(t) \tag{8.4-14}$$

and can be realized in block diagram form as shown in Fig. 8.4-1. This figure indicates one method of reconstruction of all state vectors from the measurement of as few as a single component of the state variables. The observation $\mathbf{z}(t)$ must, of course, be observable if we are to be able to reconstruct all state variables. We also note that there is additional freedom not mentioned thus far. The $\mathbf{C}(t)$ matrix is given in the problem specification, but the $\mathbf{F}(t)$ matrix is not, and we naturally ask how we should pick this matrix. [Actually, we pick a $\mathbf{G}(t)$ matrix such that $\mathbf{F}(t) = \mathbf{G}(t)\mathbf{C}(t)$.] The answer is that we pick the $\mathbf{F}(t)$ matrix so that the transition matrix of Eq. (8.4-10) decays to zero very rapidly. We may reduce this aspect of the problem to an optimal control problem and obtain its solution [4]. Clearly, the observer of Fig. 8.4-1 is just a Kalman filter with a known input $\mathbf{u}(t)$, where $\mathbf{G}(t)$ represents the Kalman gain which we are free to adjust since there is no measurement or plant noise.

We may apply exactly the same procedure to the discrete case. We have a discrete system

$$\mathbf{x}(k+1) = \mathbf{A}(k)\mathbf{x}(k) + \mathbf{B}(k)\mathbf{u}(k), \qquad \mathbf{x}(k_o) = \mathbf{x}_o \tag{8.4-15}$$

$$\mathbf{z}(k) = \mathbf{C}(k)\mathbf{x}(k) \tag{8.4-16}$$

and desire to determine an observer of the form

$$\boldsymbol{\xi}(k+1) = \mathbf{D}(k)\boldsymbol{\xi}(k) + \mathbf{E}(k)\mathbf{u}(k) + \mathbf{G}(k)\mathbf{z}(k), \qquad \boldsymbol{\xi}(k_o) = \boldsymbol{\xi}_o \tag{8.4-17}$$

such that $\mathbf{x}(k) = \boldsymbol{\xi}(k)$. This is possible if for $\mathbf{F}(k) = \mathbf{G}(k)\mathbf{C}(k)$

$$\mathbf{D}(k) = \mathbf{A}(k) - \mathbf{F}(k), \qquad \mathbf{E}(k) = \mathbf{B}(k), \qquad \mathbf{x}(k_o) = \boldsymbol{\xi}(k_o) \tag{8.4-18}$$

which are exactly the same requirements as for the continuous system. The block diagram realization of the discrete observer is precisely the same as that shown for the continuous system in Fig. 8.4-1 except that the integrators

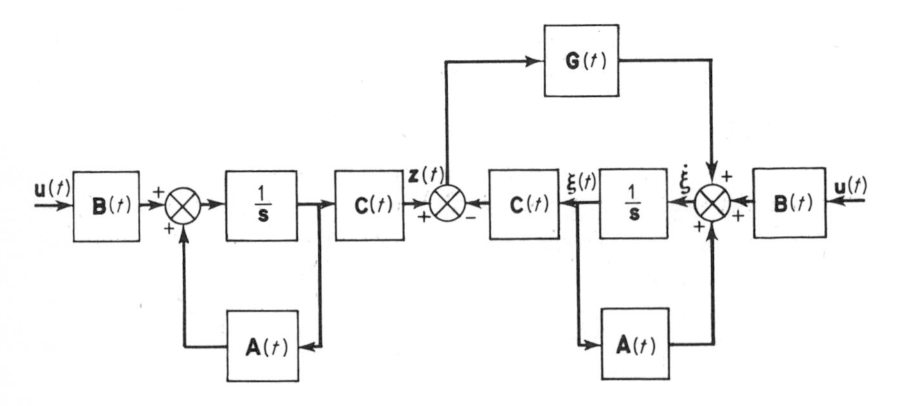

Fig. 8.4-1 Block diagram realization of an observer.

are replaced by one-period delays. A possible disadvantage to this method is that all state variables are produced by the observer, whereas some (at least one) are already available and some are not needed. We will discover a method for reproducing only desired state variables in the next subsection.

Example 8.4-1. Let us consider the problem of constructing an observer for the system

$$\dot{x} = x_2(t), \qquad \dot{x}_2 = u(t), \qquad z(t) = x_1(t)$$

which is written in state vector notation as

$$\dot{\mathbf{x}} = \mathbf{A}\mathbf{x}(t) + \mathbf{b}u = \begin{bmatrix} 0 & 1 \\ 0 & 0 \end{bmatrix}\mathbf{x}(t) + \begin{bmatrix} 0 \\ 1 \end{bmatrix}u(t)$$

$$z(t) = \mathbf{C}(t)\mathbf{x}(t) = [1 \quad 0]\mathbf{x}(t).$$

From Eq. (8.4-14) we see that the observer is characterized by

$$\begin{bmatrix} \dot{\xi}_1 \\ \dot{\xi}_2 \end{bmatrix} = \begin{bmatrix} -g_1(t) & 1 \\ -g_2(t) & 0 \end{bmatrix}\begin{bmatrix} \xi_1(t) \\ \xi_2(t) \end{bmatrix} + \begin{bmatrix} 0 \\ u(t) \end{bmatrix} + \begin{bmatrix} g_1(t) \\ g_2(t) \end{bmatrix} x_1(t),$$

where $g_1(t)$ and $g_2(t)$ must be selected such as to ensure good transient response for errors in the correct setting of the initial condition observer vector $\boldsymbol{\xi}(t_o) = \mathbf{x}(t_o)$.

Figure 8.4-2 illustrates a block diagram of the observer for this example. It is reasonable to let g_1 and g_2 be constant functions such that the eigenvalues of the observer are $2s = -g_1 \pm \sqrt{g_1^2 - 4g_2}$. For stability of the observer, we must make g_1 and g_2 greater than zero. A reasonable value for g_2 would be $g_1^2/4$ such that the observer has two real eigenvalues at $-g_1/2$. The output of the observer is then, from Eq. (8.4-10),

$$\boldsymbol{\xi}(t) = \mathbf{x}(t) + \boldsymbol{\varphi}(t, t_o)[\boldsymbol{\xi}(t_o) - \mathbf{x}(t_o)],$$

where

$$\boldsymbol{\varphi}(t, t_o) = \begin{bmatrix} 1 - \dfrac{g_1(t - t_o)}{2} & t - t_o \\ \dfrac{-g_1^2(t - t_o)}{4} & 1 + \dfrac{g_1(t - t_o)}{2} \end{bmatrix} \exp\left[\frac{-g_1(t - t_o)}{2}\right].$$

Thus we see that for this simple example, the larger we make g_1 (with $g_2 = g_1^2/4$), the faster is the error decay for incorrect initial condition setting of the observer.

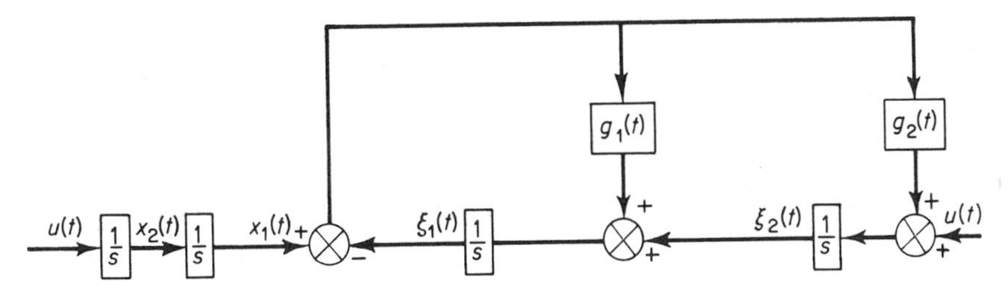

Fig. 8.4-2 Model and observer for Example 8.4-1.

8.4-2
State reconstruction with observers of low order

In the example just completed, we were given $z(t) = x_1(t)$ and we determined an observer to reconstruct the original state vector $\mathbf{x}(t)$. Some of the information contained in the observer is redundant. Specifically, we obtain $x_1(t)$ as an observer output, whereas $x_1(t)$ is already available as the system output. Let us see if we can reconstruct the $x_2(t)$ state variable by an observer represented by a first-order differential equation. We will then return to some more general remarks concerning low-order observers.

Example 8.4-2. We assume that we are given the system and output represented by

$$\dot{\mathbf{x}} = \mathbf{A}(t)\mathbf{x}(t) + \mathbf{B}(t)u(t) = \begin{bmatrix} 0 & 1 \\ 0 & 0 \end{bmatrix}\mathbf{x}(t) + \begin{bmatrix} 0 \\ 1 \end{bmatrix}u(t)$$

$$z(t) = \mathbf{C}(t)\mathbf{x}(t) = [1 \quad 0]\mathbf{x}(t)$$

and attempt to construct an observer which is a first-order system with dynamics given by

$$\dot{\xi} = D(t)\xi(t) + \mathbf{F}(t)\mathbf{x}(t) + E(t)u(t) = -3\xi(t) + [1, 0]\mathbf{x}(t) + Eu(t).$$

We will again try to obtain $\xi(t) = \mathbf{T}\mathbf{x}(t)$ where $\mathbf{T}$ will now be a 1×2 matrix which will, of course, not be invertible. Certainly Eq. (8.4-6) will still hold and $\mathbf{TA} - D\mathbf{T} - \mathbf{F} = \mathbf{0}$. However, it is now necessary that, in almost all cases, we restrict $\mathbf{A}$, $\mathbf{B}$, and $\mathbf{C}$ to be constant in time.

We have already specified $\mathbf{A}$, D, and $\mathbf{F}$ such that we may now solve for $\mathbf{T}$ from the foregoing, which becomes

$$\mathbf{T}\begin{bmatrix} 0 & 1 \\ 0 & 0 \end{bmatrix} + 3\mathbf{T} - [1 \quad 0] = \mathbf{0}$$

such that $\mathbf{T} = [\frac{1}{3}, -\frac{1}{9}]$. Then since we have $E(t) = \mathbf{TB}(t)$ from Eq. (8.4-7), we have for this problem $E = -\frac{1}{9}$. So now we have

$$\xi(t) = \mathbf{T}\mathbf{x}(t) = \frac{x_1(t)}{3} - \frac{x_2(t)}{9}.$$

We observe $x_1(t)$ so that it is a trivial matter to combine $\xi(t)$ and $x_1(t)$ to obtain $x_2(t)$ as illustrated in Fig. 8.4-3.

Notice that we specified the dynamics in this example and then solved for the resulting linear change of variable relating $\xi(t)$ and $\mathbf{x}(t)$, $\xi(t) = \mathbf{T}\mathbf{x}(t)$. We could specify other dynamics for the observer and obtain a different linear change of variable.

We can make the observer arbitrarily fast by increasing g_1 without limit in Example 8.4-1 or by increasing the scalar $-D(t)$ in this example. In practice, this is not possible because sensitivity of the observer output to slight changes in the system dynamics becomes a serious problem for very fast observers.

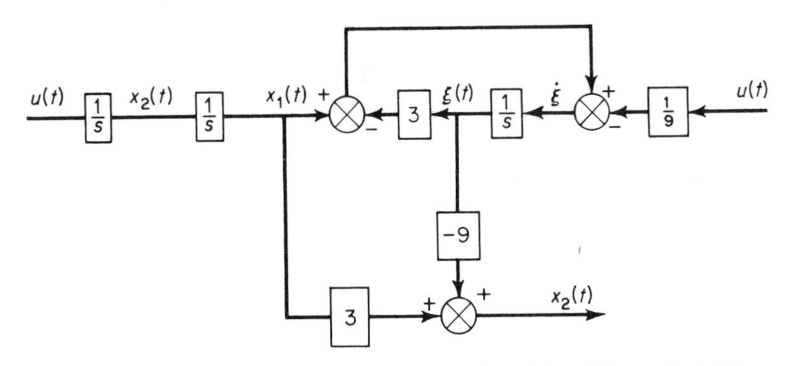

Fig. 8.4-3 A low order observer for Example 8.4-1 and Example 8.4-2.

Consider now the problem of obtaining an observer for an nth-order constant-coefficient system which has m outputs. We let the m outputs drive an $(n - m)$th-order system with $(n - m)$ outputs which will be part of the observer. If we can design the second system such that the $(n - m)$ outputs, together with the m outputs of the nth-order system, are each independent linear combinations of the original state variables, we may then find these original state variables by simple matrix operations only. The resulting system block diagram is as shown in Fig. 8.4-4.

In doing this, we realize that all of the mathematics of the preceding section applies except that, in the equation $\boldsymbol{\xi}(t) = \mathbf{T}\mathbf{x}(t)$, $\mathbf{T}$ is an $(n - m) \times (m)$ matrix and so is not invertible. Thus, we shall adjoin to the $\boldsymbol{\xi}(t)$ vector components, each of which is an independent linear combination of the components of $\mathbf{x}(t)$, such that we have $\boldsymbol{\xi}'(t) = \mathbf{T}'\mathbf{x}(t)$, where $\mathbf{T}'$ is now an invertible $n \times n$ matrix. In order for us to do this, it is required that:

(a) the eigenvalues of the observer must be distinct and different from those of the original system.

(b) the original system and the observer must be completely observable.

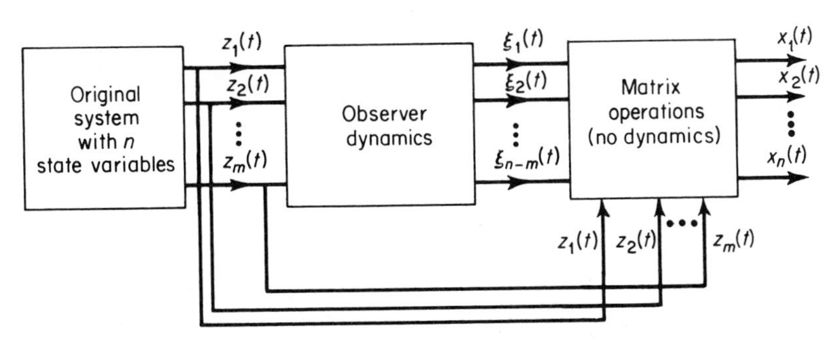

Fig. 8.4-4 Observer of low order.

(c) the observer must be of at least the $(n - m)$th order.
(d) the observer dynamics, as well as those of the system, must be time-invariant.

We note further that, if it is possible to separate a system into two or more connected subsystems, observers for the subsystems will be easier to design than observers for the system as a whole.

It is convenient to summarize what we have done thus far. We consider the plant model given by

$$\dot{\mathbf{x}}(t) = \mathbf{A}\mathbf{x}(t) + \mathbf{B}\mathbf{u}(t); \tag{8.4-19}$$

the measurement model given by

$$\mathbf{z}(t) = \mathbf{C}\mathbf{x}(t); \tag{8.4-20}$$

and the general linear state estimator given by

$$\dot{\hat{\mathbf{x}}} = \mathbf{K}\hat{\mathbf{x}}(t) + \mathbf{L}\mathbf{z}(t) + \mathbf{B}\mathbf{u}. \tag{8.4-21}$$

The error and error rate of the estimator are given by

$$\mathbf{e}(t) = \mathbf{x}(t) - \hat{\mathbf{x}}(t) \tag{8.4-22}$$

and

$$\begin{aligned}\dot{\mathbf{e}}(t) &= \dot{\mathbf{x}}(t) - \dot{\hat{\mathbf{x}}}(t)\\ &= \mathbf{A}\mathbf{x}(t) - \mathbf{K}\hat{\mathbf{x}}(t) - \mathbf{L}\mathbf{z}(t)\\ &= \mathbf{A}\mathbf{x}(t) - \mathbf{K}\hat{\mathbf{x}}(t) - \mathbf{L}\mathbf{C}\mathbf{x}(t) + \mathbf{K}\mathbf{x}(t) - \mathbf{K}\mathbf{x}(t)\\ &= [\mathbf{A}(t) - \mathbf{K}(t) - \mathbf{L}\mathbf{C}(t)]\mathbf{x}(t) + \mathbf{K}(t)[\mathbf{x}(t) - \hat{\mathbf{x}}(t)]\\ &= [\mathbf{A}(t) - \mathbf{K}(t) - \mathbf{L}\mathbf{C}(t)]\mathbf{x}(t) + \mathbf{K}\mathbf{e}(t).\end{aligned} \tag{8.4-23}$$

To ensure that $\mathbf{x}$ is unbiased for all time, we need $\mathcal{E}\{\mathbf{e}\} = \mathcal{E}[\dot{\mathbf{e}}] = \mathbf{0}$. Taking the expectation of Eq. (8.4-23) and equating it to zero, we determine that it is necessary that

$$\mathbf{K}(t) = \mathbf{A}(t) - \mathbf{L}\mathbf{C}(t). \tag{8.4-24}$$

Substituting Eq. (8.4-24) into (8.4-21) yields

$$\begin{aligned}\dot{\hat{\mathbf{x}}}(t) &= [\mathbf{A}(t) - \mathbf{L}\mathbf{C}(t)]\hat{\mathbf{x}}(t) + \mathbf{L}\mathbf{z}(t) + \mathbf{B}\mathbf{u}(t)\\ &= [\mathbf{A}(t) - \mathbf{L}\mathbf{C}(t)]\hat{\mathbf{x}}(t) + \mathbf{L}\mathbf{H}\mathbf{x}(t) + \mathbf{B}\mathbf{u}(t)\\ &= \mathbf{A}\hat{\mathbf{x}}(t) + \mathbf{L}\mathbf{C}[\mathbf{x}(t) - \hat{\mathbf{x}}(t)] + \mathbf{B}\mathbf{u}(t)\end{aligned} \tag{8.4-25}$$

which represents an unbiased observer with arbitrary dynamics. This observer structure possesses a degree of redundancy, however, since an estimate is constructed for the entire state vector although a part of the state is already available by direct measurement. To eliminate this redundacy, a lower ordered observer can be constructed. A theorem presented by Luenberger [24] states that a state observer of order n–m can be constructed having arbitrary eigenvalues for any nth-order completely controllable linear time-invariant system having m linearly independent outputs. Gopinath [25] has developed a particularly simple approach for constructing the reduced-order identity observer, which is now presented.

Consider the partitioned plant model where $\mathbf{x}_1$ is measured exactly and $\mathbf{x}_2$ is to be estimated:

$$\dot{\mathbf{x}}_1(t) = \mathbf{A}_{11}\mathbf{x}_1(t) + \mathbf{A}_{12}\mathbf{x}_2(t) + \mathbf{B}_1\mathbf{u}(t) \tag{8.4-26}$$

$$\dot{\mathbf{x}}_2(t) = \mathbf{A}_{21}\mathbf{x}_1(t) + \mathbf{A}_{22}\mathbf{x}_2(t) + \mathbf{B}_2\mathbf{u}(t). \tag{8.4-27}$$

If $\mathbf{x}_1$ is measured, we then have available the following information about $\mathbf{x}_2$, since $\dot{\mathbf{x}}_1$ and $\mathbf{u}$ are then also known:

$$\dot{\mathbf{x}}_2(t) = \mathbf{A}_{21}\mathbf{x}_1(t) + \mathbf{A}_{22}\mathbf{x}_2(t) + \mathbf{B}_2\mathbf{u}(t) \tag{8.4-28}$$

and

$$\mathbf{z}(t) = \mathbf{A}_{12}\mathbf{x}_2(t) = \dot{\mathbf{x}}_1(t) - \mathbf{A}_{11}\mathbf{x}_1(t) - \mathbf{B}_1\mathbf{u}(t). \tag{8.4-29}$$

Equations (8.4-28) and (8.4-29) are analogous to Eq. (8.4-19) and (8.4-20), respectively, with the following associations:

$$\mathbf{Ax}(t) \sim \mathbf{A}_{22}\mathbf{x}_2(t)$$

$$\mathbf{Bu}(t) \sim \mathbf{A}_{21}\mathbf{x}_1(t) + \mathbf{B}_2\mathbf{u}(t)$$

$$\mathbf{Hx}(t) \sim \mathbf{A}_{12}\mathbf{x}_2(t).$$

Thus, by inspection of Eq. (8.4-25) we can write the reduced-order observer structure for $\mathbf{x}_2$ as

$$\begin{aligned}\dot{\hat{\mathbf{x}}}_2 &= \mathbf{A}_{22}\hat{\mathbf{x}}_2 + \mathbf{LA}_{12}(\mathbf{x}_2 - \hat{\mathbf{x}}_2) + \mathbf{A}_{21}\mathbf{x}_1 + \mathbf{B}_2\mathbf{u} \\ &= (\mathbf{A}_{22} - \mathbf{LA}_{12})\hat{\mathbf{x}}_2 + \mathbf{LA}_{12}\mathbf{x}_2 + \mathbf{A}_{21}\mathbf{x}_1 + \mathbf{B}_2\mathbf{u}.\end{aligned} \tag{8.4-30}$$

Now $\mathbf{A}_{12}\mathbf{x}_2$ is given by Eq. (8.4-29) which upon substitution into the above gives

$$\begin{aligned}\dot{\hat{\mathbf{x}}}_2 &= (\mathbf{A}_{22} - \mathbf{LA}_{12})\hat{\mathbf{x}}_2 + \mathbf{L}(\dot{\mathbf{x}}_1 - \mathbf{A}_{11}\mathbf{x}_1 - \mathbf{B}_1\mathbf{u}) + \mathbf{A}_{21}\mathbf{x}_1 + \mathbf{B}_2\mathbf{u} \\ &= (\mathbf{A}_{22} - \mathbf{LA}_{12})\hat{\mathbf{x}}_2 + (\mathbf{A}_{21} - \mathbf{LA}_{11})\mathbf{x}_1 + \mathbf{L}\dot{\mathbf{x}}_1 - \mathbf{LB}_1\mathbf{u} + \mathbf{B}_2\mathbf{u}.\end{aligned} \tag{8.4-31}$$

In order to eliminate the differentiation of $\mathbf{x}_1$ we let

$$\hat{\mathbf{x}}_2 = \mathbf{y} + \mathbf{Lx}_1. \tag{8.4-32}$$

Then

$$\begin{aligned}\dot{\mathbf{y}} &= \dot{\hat{\mathbf{x}}}_2 - \mathbf{L}\dot{\mathbf{x}}_1 \\ &= (\mathbf{A}_{22} - \mathbf{LA}_{12})\hat{\mathbf{x}}_2 + (\mathbf{A}_{21} - \mathbf{LA}_{11})\mathbf{x}_1 + \mathbf{B}_2\mathbf{u} - \mathbf{LB}_1\mathbf{u} \\ &= (\mathbf{A}_{22} - \mathbf{LA}_{12})\mathbf{y} + [(\mathbf{A}_{21} - \mathbf{LA}_{11}) + (\mathbf{A}_{22} - \mathbf{LA}_{12})\mathbf{L}]\mathbf{x}_1 \\ &\quad + (\mathbf{B}_2 - \mathbf{LB}_1)u\end{aligned} \tag{8.4-33}$$

with

$$\mathbf{y}(0) = \mathbf{y}_o. \tag{8.4-34}$$

Example 8.4-3. As a simple example, we consider the system given by

$$\begin{bmatrix}\dot{x}_1 \\ \dot{x}_2\end{bmatrix} = \begin{bmatrix}-2 & 1 \\ 0 & -1\end{bmatrix}\begin{bmatrix}x_1 \\ x_2\end{bmatrix} + \begin{bmatrix}0 \\ 1\end{bmatrix}u$$

and

$$z = x_1.$$

The problem is to determine an observer structure so that x_2 can be estimated from measurements made of x_1. The solution is given by (8.4-32) and (8.4-33) which can be determined as follows:

1. First we identify

$$A_{11} = -2, \quad A_{12} = 1, \quad A_{21} = 0, \quad A_{22} = -1,$$
$$B_1 = 0, \quad B_2 = 1.$$

2. From Eqs. (8.4-32) and (8.4-33), $\hat{x}_2$ can be written as

$$\hat{x}_2 = y + Lx_1,$$
$$\dot{y} = -(1 + L)y + (L - L^2)x_1 + u.$$

3. If L is set arbitrarily to the value of 2, we obtain

$$\hat{x}_2 = y + 2x_1$$

and

$$\dot{y} = -3y - 2x_1 + u; \qquad y(0) = y_o.$$

4. We note that $e = x_2 - \hat{x}_2 = x_2 - y - 2x_1$, and:

$$\begin{aligned} \dot{e} &= \dot{x}_2 - \dot{y} - 2\dot{x}_1 \\ &= -3x_2 + 3y + 6x_1 \\ &= -3(x_2 - y - 2x_1) \\ &= -3e. \end{aligned}$$

Thus, we see that $e \longrightarrow 0$ as $t \longrightarrow \infty$.

Before concluding this development of the observer-estimator problem, it should be noted that for the formulation given, the system model was assumed to be known exactly and there was no input noise forcing function. These restrictions can be removed but we will not consider these extensions here.

References

1. KAILATH, T., "A View of Three Decades of Linear Filtering Theory." *IEEE Trans. Infor. Theory*, **IT-20**, (March 1974), 146–81.

2. WIENER, N., *Extrapolation, Interpolation and Smoothing of Stationary Time Series, with Engineering Applications*. Technology Press and Wiley, New York, 1949.

3. KALMAN, R.E., "A New Approach to Linear Filtering and Prediction Problems." *Trans. AME, J. Basic Eng.*, 82, (March 1960), 34–45.

4. KALMAN, R.E., "New Methods of Weiner Filtering Theory." J.L. Bogdanoff and F. Kozin, eds., *Proceedings First Symposium Eng. Appl. of Random Function Theory and Probability*. Wiley, New York, 1963.

5. KALMAN, R.E., and BUCY, R.S., "New Results in Linear Filtering and Prediction Theory." *Trans. ASME, Ser. D, J. Basic Eng.*, 83, (December 1961), 95–107.

6. LUENBERGER, D.G., *Optimization by Vector Space Methods.* Wiley, New York, 1969.

7. SAGE, A.P., and MELSA, J.L., *Estimation Theory with Applications to Communications and Control.* McGraw-Hill Book Co., New York, 1971.

8. MORF, M., and KAILATH, T., "Square-Root Algorithms for Least-Squares Estimation." *IEEE Trans. Autom. Control,* **AC-20**, (August 1975), 487–97.

9. SEGALL, A., "Stochastic Processes in Estimation Theory." ESL-P-588, Electronic Systems Laboratory, M.I.T., January 1975.

10. WONG, E., "Recent Progress in Stochastic Processes—A Survey." *IEEE Trans. Infor. Theory,* **IT-19**, (May 1973), 262–75.

11. MELSA, J.L., and SAGE, A.P., *An Introduction to Probability and Stochastic Processes.* Prentice-Hall, Inc., Englewood Cliffs, New Jersey, 1973.

12. ATHANS, M., "The Matrix Minimum Principle." *Information and Control,* **11**, (1968), 592–606.

13. SAGE, A.P., and MELSA, J.L., *System Identification.* Academic Press, New York, 1971.

14. SHERMAN, S., "Non Mean–Square Error Criteria." *Trans. IRE,* **IT-4**, (1958), 125–26.

15. KAILATH, T., and FROST, P., "An Innovations Approach to Least Squares Estimation—Part II: Linear Smoothing in Additive White Noise." *IEEE Trans. Autom. Control,* **AC-13**, (December 1968), 655–60.

16. RHODES, I.B., "A Tutorial Introduction to Estimation and Filtering." *IEEE Trans. Autom. Control,* **AC-16**, (1971), 688–706.

17. ASTROM, K.J., *Introduction to Stochastic Control Theory.* Academic Press, New York, 1970.

18. BROWN, R.J., and SAGE, A.P., "Error Analysis of Modeling and Bias Errors in Continuous Time State Estimation." *Automatica,* **7**, (1971), 577–90.

19. HO, Y.C., and LEE, R.C.K., "A Bayesian Approach to Problems in Stochastic Estimation and Control." *IEEE Trans. Autom. Control,* **AC-9**, (1964), 333–39.

20. LEE, R.C.K., *Optimal Estimation, Identification and Control.* M.I.T. Press, Cambridge, Massachusetts, 1964.

21. KALMAN, R.E., and BERTRAM, J.E., "Control System Analysis and Design via the Second Method of Liapunov, I, Continuous-Time Systems." *Trans. ASME, J. Basic Eng.*, **82D**, (1960), 371–93.

22. KUSHNER, H., *Introduction to Stochastic Control.* Holt, New York, 1971.

23. LUENBERGER, D.G., "Observers for Multivariate Systems." *IEEE Trans. Autom. Control,* **AC-11**, (1966), 190–97.

24. LUENBERGER, D.G., "An Introduction to Observers." *IEEE Trans. Autom. Control,* **AC-16,** (1971), 596–602.

25. GOPINATH, B., "On the Control of Linear Multiple Input-Output Systems." *The Bell System Tech. J.* **50,** (March 1971), 1063–80.

Problems

1. For the block diagram shown,

$$\mathcal{E}\{w_1(t)w_1(\tau)\} = 0$$

$$\mathcal{E}\{\mathbf{v}(t)\mathbf{v}^T(\tau)\} = \begin{bmatrix} 1 & 0 \\ 0 & 1 \end{bmatrix} \delta(t - \tau)$$

$$\mathbf{P}(0) = \begin{bmatrix} 4 & 0 \\ 0 & 4 \end{bmatrix}$$

$$\mathcal{E}\{w_1(t)\} = \mathcal{E}\{v_1(t)\} = \mathcal{E}\{v_2(t)\} = 0.$$

Find the optimal estimator equations and determine the error variance matrix as a function of time.

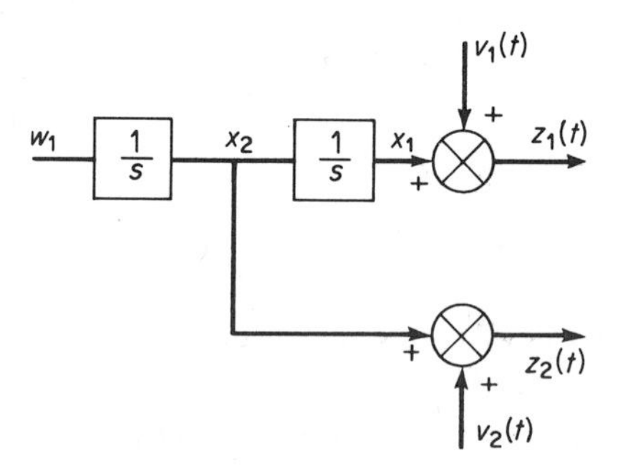

2. For the block diagram shown,

$$\mathcal{E}\{w_1(t)w_1(\tau)\} = \delta(t - \tau)$$

$$\mathcal{E}\{v_1(t)v_1(\tau)\} = \tfrac{1}{4}\delta(t - \tau)$$

$$\mathbf{P}(0) = \begin{bmatrix} 1 & 0 \\ 0 & 1 \end{bmatrix}.$$

Find the optimal estimator equations and determine the error variance matrix as a function of time. If $\mathcal{E}\{w_1(t)w_1(\tau)\} = \exp[-|t - \tau|]$, what do the estimator and variance equations become?

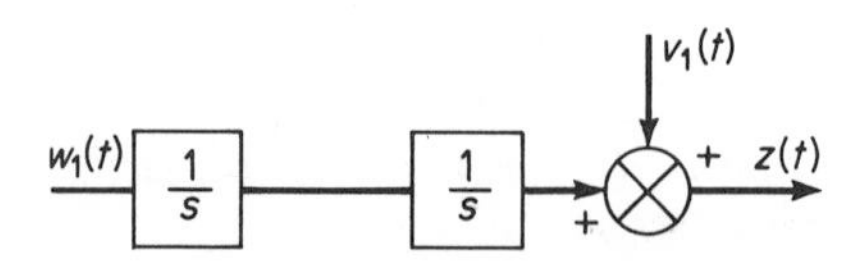

3. Using the method of solution presented in Example 8.2-1, find the solution for the Kalman gain of the general first-order linear constant message model; i.e., one where

$$F = f_{11} \qquad H = 1 \qquad R(t) = r_{11}$$

$$G = 1 \qquad Q(t) = q_{11} \qquad p(0) = p_o.$$

4. Determine the configuration of the dual problems for Problems 1 and 2.

5. Find the "stationary" solutions to Problem 2 by setting $dp/dt = 0$ in the Riccati variance equation.

6. Derive the variance equation (8.2-42), by the method outlined in Sec. 8.2.

7. Show that the solution of the variance equation, Eq. (8.2-18), for constant system

$$\dot{\mathbf{P}} = -\mathbf{A}^T\mathbf{P}(t) - \mathbf{P}(t)\mathbf{A} + Q, \qquad \mathbf{P}(t_o) = \mathbf{P}_o$$

is

$$\mathbf{P}(t) = \mathbf{P}_o e^{\mathbf{A}(t-t_o)} + \int_{t_o}^{t} e^{\mathbf{A}^T(\tau-t)}\mathbf{Q}e^{\mathbf{A}(\tau-t)}\,d\tau.$$

What is the similar result for nonstationary systems?

8. Derive Eq. (12) in Example 8.2-1. Obtain the matrix Riccati equation from Eq. (12). Does this same result hold for any $\mathbf{A}(t)$ matrix of the form

$$\mathbf{A}(t) = \begin{bmatrix} \mathbf{F}(t) & \mathbf{\Gamma}(t) \\ \mathbf{\Xi}(t) & -\mathbf{F}^T(t) \end{bmatrix}?$$

What is the matrix Riccati equation for this $\mathbf{A}(t)$?

9. Determine the Kalman gain $K(t)$ and the differential equation describing $\hat{x}(t)$ in terms of $z(t)$ for the case where:

$$\dot{x}(t) = w(t)$$

$$z(t) = x(t) + v(t)$$

$$q = r = 1, \qquad \mathcal{E}\{w(0)\} = \mathcal{E}\{v(t)\} = 0, \qquad P(0) = 1.$$

10. Determine the Kalman filter $\hat{\mathbf{x}}(t)$ and Kalman gain $\mathbf{K}(t)$ for the cases:
(a) $\mathbf{H}(t) = \mathbf{I}$, the identity matrix, and $\mathbf{R}(t) = \mathbf{0}$.
(b) $\mathbf{H}(t) = \mathbf{0}$.
How are these cases special?

11. Show that cov $[\tilde{\mathbf{x}}(t), \hat{\mathbf{x}}(t)] = \mathbf{0}$.

12. Find $\mathbf{P}(j, k)$ for a linear system

$$\mathbf{P}(j, k) = \mathcal{E}\{[\mathbf{x}(j) - \bar{\mathbf{x}}(j)][\mathbf{x}(k) - \bar{\mathbf{x}}(k)]^T\}$$

$$\mathbf{x}(k+1) = \boldsymbol{\varphi}(k+1, k)\mathbf{x}(k).$$

13. For the block diagram of Fig. 8.3-1, find $\mathcal{E}\{\mathbf{y}(k)\}$, cov $\{\mathbf{y}(k)\}$ and the mean and variance of $p[\mathbf{y}(k+1), \mathbf{y}(k)]$.

14. The discrete system,

$$\mathbf{x}(k+1) = \boldsymbol{\varphi}(k+1, k)\mathbf{x}(k) + \mathbf{\Gamma}(k)\mathbf{w}(k),$$

may be written as

$$\frac{\mathbf{x}(k+1)-\mathbf{x}(k)}{T}=\frac{\boldsymbol{\varphi}(k+1,k)-\mathbf{I}}{T}\mathbf{x}(k)+\frac{\boldsymbol{\Gamma}(k)\mathbf{w}(k)}{T}.$$

As the sample period becomes zero, this may be written as

$$\frac{d\mathbf{x}(t)}{dt}=\mathbf{A}(t)\mathbf{x}(t)+\mathbf{B}(t)\mathbf{w}'(t).$$

Show that this transformation, applied to the discrete variance equation,

$$\operatorname{var}\{\mathbf{x}(k+1)\}=\mathbf{P}(k+1)=\boldsymbol{\varphi}(k+1,k)\mathbf{P}(k)\boldsymbol{\varphi}^T(k+1,k)+\boldsymbol{\Gamma}(k)\mathbf{Q}(k)\boldsymbol{\Gamma}^T(k),$$

gives us the continuous variance equation

$$\dot{\mathbf{P}}(t)=\mathbf{A}(t)\mathbf{P}(t)+\mathbf{P}(t)\mathbf{A}^T(t)+\mathbf{B}(t)\mathbf{Q}(t)\mathbf{B}^T(t).$$

Define all parameters.

15. Rederive Eq. (8.2-79) for the case where $\operatorname{cov}\{\mathbf{w}(t),\mathbf{v}(\tau)\}=\mathbf{S}(t)\,\delta(t-\tau)$.

16. Find the variance of the output state vector for the system described by

$$\mathbf{x}(k+1)=\boldsymbol{\varphi}\mathbf{x}(k)+\boldsymbol{\Gamma}(k)\mathbf{w}(k),$$

where

$$\mathbf{x}^T=[x_1,x_2],\qquad \boldsymbol{\varphi}=\begin{bmatrix}-0.5 & -0.5\\ 1 & 0\end{bmatrix},\qquad \boldsymbol{\Gamma}=\begin{bmatrix}1\\0\end{bmatrix}$$

$$\mathcal{E}\{w(k)\}=0,\qquad \mathcal{E}\{\mathbf{x}(0)\}=\begin{bmatrix}0\\0\end{bmatrix}$$

$$\operatorname{cov}\{w(k)\}=1,\qquad \operatorname{cov}\{\mathbf{x}(0)\}=\mathbf{0}.$$

17. Consider the problem of discrete state estimation for the message model,

$$\mathbf{x}(k+1)=\boldsymbol{\varphi}\mathbf{x}(k)+\boldsymbol{\Gamma}\mathbf{w}(k)$$
$$\mathbf{z}(k)=\mathbf{H}\mathbf{x}(k)+\mathbf{v}(k),$$

where

$$\boldsymbol{\varphi}=\begin{bmatrix}1 & 1\\ 0 & 1\end{bmatrix},\qquad \boldsymbol{\Gamma}=\begin{bmatrix}1 & 0\\ 0 & 1\end{bmatrix},\qquad \mathbf{H}=[1\quad 0]$$

$$\mathbf{Q}=\begin{bmatrix}q_{11} & 0\\ 0 & q_{22}\end{bmatrix},\qquad R=r_{11},\qquad \mathbf{P}(0)=\begin{bmatrix}0 & 0\\ 0 & 0\end{bmatrix}.$$

Find the two sets of equations for the optimal state estimation filters. Compare the resulting two sets of computational algorithms.

18. Show the $\operatorname{cov}[\hat{\mathbf{x}}(k),\tilde{\mathbf{x}}(k)]=\mathbf{0}$.

19. Assume that $\operatorname{cov}\{\mathbf{v}(k),\mathbf{w}(j)\}=\mathbf{S}(k)\,\delta(k-j)$. Determine the effect of this generalizing assumption on $\hat{\mathbf{x}}(k)$, $\hat{\mathbf{x}}(k+1\,|\,k)$, and their associated Kalman gains.

20. Show that the innovation $\tilde{\mathbf{z}}(k)$ is orthogonal to $\boldsymbol{Z}(k-1)$.

21. Justify the statement below Eq. (8.3-15).

22. Derive Eq. (8.3-18).

23. Develop the discrete version of the duality result presented in Sec. 8.2.

24. Determine cov $[\tilde{\mathbf{z}}(k), \tilde{\mathbf{z}}(j)]$, the covariance of the innovations for the discrete case, where $\tilde{\mathbf{z}}(k) = \mathbf{z}(k) - \hat{\mathbf{z}}(k \mid k-1)$.

25. Show that it is possible to build an $(n-1)$th-order observer by the method of Sec. 8.4-1. Consider a system

$$\dot{\mathbf{x}} = \mathbf{A}(t)\mathbf{x}(t) + \mathbf{B}(t)\mathbf{u}(t)$$

$$z(t) = x_1(t)$$

and find the equations for an observer such that

$$\boldsymbol{\xi}^T(t) = [x_2, x_3, x_4, \ldots, x_n].$$

Note, however, that the dynamics of the observer are fixed and may well be totally unsatisfactory.

26. Design a second-order observer for the system

$$\begin{bmatrix} \dot{x}_1 \\ \dot{x}_2 \end{bmatrix} = \begin{bmatrix} -2 & 1 \\ 0 & -1 \end{bmatrix} \begin{bmatrix} x_1(t) \\ x_2(t) \end{bmatrix} + \begin{bmatrix} 0 \\ 1 \end{bmatrix} u(t), \qquad z(t) = x_1(t).$$

Illustrate the effect on the observer of changing the parameter -2 to $-2 + \epsilon$, where ϵ is a small number.

27. Design a first-order observer for the system of Problem 26. Again illustrate the effect on the observer of the parameter change. Which observer is most satisfactory in this regard?

28. Design two first-order observers for the system shown, where the outputs x_1 and x_3 only are observable.

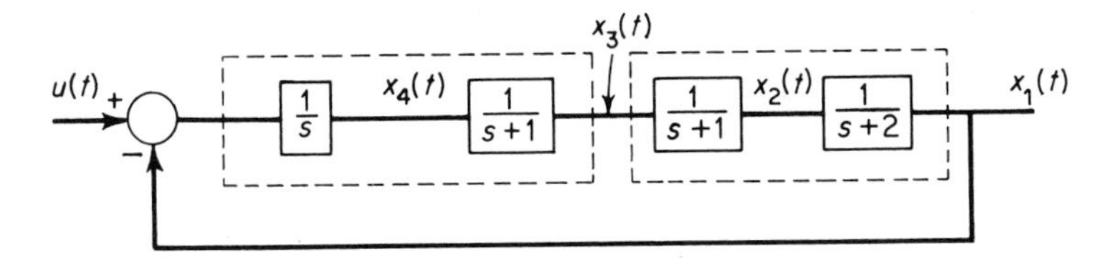

9 Combined estimation and control—the linear quadratic Gaussian problem

Chapters 2 through 7 have considered the optimal control and properties of deterministic models of physical systems. Chapter 8 was concerned with the problem of optimal state estimation for stochastic descriptions of physical systems. Invariably, however, physical systems subject to control are at best difficult to model deterministically and often only noisy observations of some of the state variables of systems are available. Thus, a more physically realistic control model would often be a stochastic process model of the physical system, where we would make decisions or determine controls on the basis of imperfect measurements of the system. Both state estimation and control would then be involved in the decision-making process. We examine these notions more explicitly in the following section by considering a general discrete control problem. Section 9.2 considers a specific example of the problem discussed in Sec. 9.1, the *Linear Quadratic Gaussian* (LQG) *problem*. The continuous time LQG problem is then examined in Sec. 9.3. These problems have been treated by Kalman [1], Joseph and Tou [2], Gunckel and Franklin [3], and others. More recent treatments of the LQG problem can be found in [4, 5, 6]. The final section considers extensions of concepts presented earlier in the chapter.

9.1 Problem formulation—a general discussion

We let the equation

$$\mathbf{x}(k+1) = \mathbf{f}[\mathbf{x}(k), \mathbf{u}(k), \mathbf{w}(k), k], \qquad k = k_0, \ldots, k_f - 1 \tag{9.1-1}$$

describe the evolution of the state of a system, where $\mathbf{u}(k)$ is the control applied at time k, $\mathbf{w}(k)$ is the random variable modeling system uncertainties, and $\mathbf{x}(k_0)$ is a random variable modeling the uncertainty of the initial state vector. We assume that the measurement at time k is obtained from

$$\mathbf{z}(k) = \mathbf{g}[\mathbf{x}(k), \mathbf{u}(k), \mathbf{v}(k), k], \tag{9.1-2}$$

where $\mathbf{v}(k)$ is the random variable modeling the uncertainties in state measurement. We note that the control $\mathbf{u}(k)$ is allowed to directly influence both the state trajectory and the observation quality. The sequences $\mathbf{x}(k_o), \ldots, \mathbf{x}(k_f)$ and $\mathbf{u}(k_o), \ldots, \mathbf{u}(k_f - 1)$ are assumed to accrue cost $\mathcal{L}[\mathbf{x}(k_o), \ldots, \mathbf{x}(k_f), \mathbf{u}(k_o), \ldots, \mathbf{u}(k_f - 1)]$, which is (under reasonable assumptions) a random variable. The performance index which we wish to minimize is

$$J = \mathcal{E}\{\mathcal{L}[\mathbf{x}(k_o), \ldots, \mathbf{x}(k_f), \mathbf{u}(k_o), \ldots, \mathbf{u}(k_f - 1)]\}. \tag{9.1-3}$$

We remark that the control sequence $\mathbf{u}(k_o), \ldots, \mathbf{u}(k_f - 1)$ represents the control values chosen by the controller. The controller chooses such a sequence on the basis of knowledge of the system structure and available data. Intuitively, the minimum of the performance index is dependent on what the controller knows when decisions are made. We will assume throughout this chapter that a control law or decision-making rule is allowed to depend on the following data:

1. the joint probability density function p of the random variables $\mathbf{x}(k_o), \mathbf{w}(k_o), \ldots, \mathbf{w}(k_f - 1), \mathbf{v}(k_o), \ldots, \mathbf{v}(k_f - 1)$,
2. the functional relationship of the system dynamics $\mathbf{f}$,
3. the functional relationship of the system measurement $\mathbf{g}$,
4. the cost function $\mathcal{L}$,
5. all past and present measurements, $\boldsymbol{Z}(k) = [\mathbf{z}(k_o), \ldots, \mathbf{z}(k)]$, at each time of control setting determination, and
6. all past control values, $\boldsymbol{U}(k) = [\mathbf{u}(k_o), \ldots, \mathbf{u}(k - 1)]$, at each time of control setting determination.

We let $\boldsymbol{\Xi} = (p, \mathbf{f}, \mathbf{g}, \mathcal{L})$, the structure of the system. We then say that a control law $\mathbf{u}$ is admissible if and only if

$$\mathbf{u}(k) = \boldsymbol{\alpha}[\boldsymbol{\Xi}, \boldsymbol{Z}(k), \boldsymbol{U}(k), k]. \tag{9.1-4}$$

Thus, we allow the controller to base the kth decision, $k - k_o, \ldots, k_f - 1$, on knowledge of the system structure over the entire length of the control horizon $\{k_o, \ldots, k_f\}$, all past and present measurements, and all past control settings. The latter two assumptions require perfect memory of the controller. We note, however, that the controller is not clairvoyant in that no future measurements are assumed known and no present and future control settings are known. The assumption that $\boldsymbol{\Xi}$ is known over the entire control horizon implies that the controller is always aware of the informational value and availability of future measurements and has knowledge of the effect of future control settings on the cost and on the future states and measurements. Such a definition of admissibility is called a *closed-loop* [7] or *truly feedback* [8] *policy*. Other policies are described in [7]. The objective in solving the control problem is to determine an admissible control law $\boldsymbol{\alpha}^*$ such that $J(\boldsymbol{\alpha}^*) \leq J(\boldsymbol{\alpha})$, for all other admissible control laws $\boldsymbol{\alpha}$.

We may show that assumptions 1 through 5 include assumption 6, i.e.,

$$\mathbf{u}(t) = \boldsymbol{\alpha}[\boldsymbol{Z}(t), t], \tag{9.1-5}$$

by substituting Eq. (9.1-5) for $t = k_o, \ldots, k-1$ into Eq. (9.1-4) and using a standard induction argument to obtain:

$$\begin{aligned}
\mathbf{u}(k) &= \boldsymbol{\alpha}[\boldsymbol{Z}(k), \mathbf{u}(k_o), \ldots, \mathbf{u}(k-1), k] \\
&= \boldsymbol{\alpha}\{\boldsymbol{Z}(k), \boldsymbol{\alpha}(k_o), \ldots, \boldsymbol{\alpha}[\boldsymbol{Z}(k-1), k-1], k\} \\
&= \boldsymbol{\alpha}[\boldsymbol{Z}(k), k].
\end{aligned}$$

We remark that the present control setting may affect both the future state trajectory, through Eq. (9.1-1), and the future state uncertainty, due to Eq. (9.1-2). Even if $\mathbf{g}$ is independent of the present control setting, the dependence on future measurements is affected by the future state trajectory. Hence, the controller performs the dual role of controlling the quality of future state measurements, through Eq. (9.1-2), as well as the future state trajectory. This intercoupling of estimation and control, two roles that are possibly conflicting, is often called the *dual effect* [9].

An obvious difficulty in solving the previous problem is that as k grows, so does the pair $[\boldsymbol{Z}(k), \boldsymbol{U}(k)]$, which can quickly become unmanageably large. For the case where Eq. (9.1-3) is additive, i.e., when

$$\mathcal{L} = \theta[\mathbf{x}(k_f)] + \sum_{k=k_o}^{k_f-1} \phi[\mathbf{x}(k), \mathbf{u}(k), k], \tag{9.1-6}$$

it is sufficient to know $[\boldsymbol{Z}(k), \boldsymbol{U}(k)]$ through the conditional density of $\mathbf{x}(k)$, conditioned on $[\boldsymbol{Z}(k), \boldsymbol{U}(k)]$, which is $p[\mathbf{x}(k) \mid \boldsymbol{Z}(k), \boldsymbol{U}(k)]$, [10]. This probability density can be determined recursively using Bayes' rule. Unfortunately, $p[\mathbf{x}(k) \mid \boldsymbol{Z}(k), \boldsymbol{U}(k)]$ is usually infinite dimensional and hence of little practical value. (An important exception is the case where $\mathbf{x}$ and $\mathbf{z}$ have finite state-spaces [11].) If certain assumptions on $\boldsymbol{\Xi}$ are satisfied, then our stochastic control problem is the linear quadratic Gaussian (LQG) problem, which will be formulated and analyzed in the sections to follow. Although it is often the case that the aforementioned assumptions on $\boldsymbol{\Xi}$ are not satisfied, the LQG problem may still be of use. For example, if the desired deterministic trajectory and associated control law are known, then (linear) perturbation equations, representing the dynamics of the errors between the actual and ideal trajectories and controls, are easily determined from Eq. (9.1-1), and control results associated with the LQG problem can be applied to minimize second order effects [12]. If, however, the system is very nonlinear and the noise level is high due to inherent uncertainties which imply that a deterministic trajectory is unrealistic, then other near-optimal control strategies must be considered [13]. We now proceed to the discrete time version of the LQG problem.

9.2 The LQG problem—discrete case

We consider the system

$$\mathbf{x}(k+1) = \mathbf{A}(k)\mathbf{x}(k) + \mathbf{B}(k)\mathbf{u}(k) + \mathbf{w}(k) \tag{9.2-1}$$

$$\mathbf{z}(k) = \mathbf{C}(k)\mathbf{x}(k) + \mathbf{v}(k), \tag{9.2-2}$$

where $\mathbf{x}(k)$ is the state n-vector, $\mathbf{u}(k)$ is the control m-vector, and $\mathbf{z}(k)$ is the observation r-vector. The processes $\{\mathbf{w}(k), k = k_o, k_o + 1, \ldots\}$ and $\{\mathbf{v}(k), k = k_o, k_o + 1, \ldots\}$ are sequences of zero-mean, independent Gaussian random variables with covariances,

$$\operatorname{cov}\{\mathbf{w}(k), \mathbf{w}(j)\} = \mathbf{Q}_2(k)\,\delta(k - j), \tag{9.2-3}$$

$$\operatorname{cov}\{\mathbf{v}(k), \mathbf{v}(j)\} = \mathbf{R}_2(k)\,\delta(k, j), \tag{9.2-4}$$

$$\operatorname{cov}\{\mathbf{w}(k), \mathbf{v}(j)\} = \mathbf{0}. \tag{9.2-5}$$

These processes, corresponding respectively to models of plant noise and measurement noise, are assumed to be independent of the normal random n-vector $\mathbf{x}(k_o)$, which has mean $\bar{\mathbf{x}}(k_o)$ and variance $\mathbf{P}_2(k_o)$. We also assume that $\mathbf{P}_2(k_o)$ and $\mathbf{Q}_2(k)$ are nonnegative definite, and $\mathbf{R}_2(k)$ is positive definite for all $k = k_o, k_o + 1, \ldots$. We wish to find an admissible control law (as defined in Sec. 9.1) that minimizes the criterion

$$J = \mathcal{E}\left\{ \|\mathbf{x}(k_f)\|_{\mathbf{S}}^2 + \sum_{k=k_o}^{k_f-1} \left(\|\mathbf{x}(k)\|_{\mathbf{Q}_1(k)}^2 + \|\mathbf{u}(k)\|_{\mathbf{R}_1(k)}^2 \right) \right\} \tag{9.2-6}$$

with respect to the set of all admissible control laws, where $\mathbf{S}$ and $\mathbf{Q}_1(k)$ are assumed nonnegative definite and (without loss of generality) symmetric, and $\mathbf{R}_1(k)$ is positive definite and symmetric, for all $k = k_o, k_o + 1, \ldots, k_f$.

We will solve this problem using dynamic programming. We define

$$\tilde{\Gamma}_k[Z(k-1)] = \min_{\mathbf{u}(k)} \ldots$$

$$\min_{\mathbf{u}(k_f-1)} \mathcal{E}\left\{ \|\mathbf{x}(k_f)\|_{\mathbf{S}}^2 + \sum_{t=k}^{k_f-1} \left(\|\mathbf{x}(t)\|_{\mathbf{Q}_1(t)}^2 + \|\mathbf{u}(t)\|_{\mathbf{R}_1(t)}^2 \right) \,\middle|\, Z(k-1) \right\} \tag{9.2-7}$$

which is the minimum expected cost to be accrued between time k to a fixed final time k_f, given measurement sequence $Z(k-1)$. Noting that $\mathbf{x}(k)$ and $\mathbf{u}(k)$ are not (causally) affected by $\mathbf{u}(k+1), \ldots, \mathbf{u}(k_f-1)$, we see that Eq. (9.2-7) can be written as

$$\tilde{\Gamma}_k[Z(k-1)] = \min_{\mathbf{u}(k)} \mathcal{E}\Big\{[\|\mathbf{x}(k)\|^2_{\mathbf{Q}_1(k)} + \|\mathbf{u}(k)\|^2_{\mathbf{R}_1(k)}] +$$
$$\min_{\mathbf{u}(k+1)} \cdots \min_{\mathbf{u}(k_f-1)} \Big[\|\mathbf{x}(k_f)\|^2_{\mathbf{S}} + \sum_{t=k+1}^{k_f-1} (\|\mathbf{x}(t)\|^2_{\mathbf{Q}_1(t)} + \|\mathbf{u}(t)\|^2_{\mathbf{R}_1(t)})\Big] \Big| Z(k-1)\Big\} =$$
$$\min_{\mathbf{u}(k)} \Big[\mathcal{E}\Big\{\|\mathbf{x}(k)\|^2_{\mathbf{Q}_1(k)} + \|\mathbf{u}(k)^2_{\mathbf{R}_1(k)}\|Z(k-1)\Big\} +$$
$$\mathcal{E}\Big\{\min_{\mathbf{u}(k+1)} \cdots \min_{\mathbf{u}(k_f-1)} \mathcal{E}\Big[\|\mathbf{x}(k_f)\|^2_{\mathbf{S}} + \sum_{t=k+1}^{k_f-1} (\|\mathbf{x}(t)\|^2_{\mathbf{Q}_1(t)} +$$
$$\|\mathbf{u}(t)\|^2_{\mathbf{R}_1(t)}) \,|\, Z(k)\Big] \Big| Z(k-1)\Big\}\Big]$$

such that finally we have

$$\tilde{\Gamma}_k[Z(k-1)] = \min_{\mathbf{u}(k)} [\mathcal{E}\{\|\mathbf{x}(k)\|^2_{\mathbf{Q}_1(k)} + \|\mathbf{u}(k)\|^2_{\mathbf{R}_1(k)} \,|\, Z(k-1)\} +$$
$$\mathcal{E}\{\mathcal{E}\{\tilde{\Gamma}_{k+1}[Z(k)] \,|\, Z(k)\} \,|\, Z(k-1)\}]. \tag{9.2-8}$$

From Eq. (9.2-7) we immediately see that the terminal condition for Eq. (9.2-8) is

$$\tilde{\Gamma}_{k_f}[Z(k_f-1)] = \mathcal{E}\{\|x(k_f)\|^2_{\mathbf{S}} \,|\, Z(k_f-1)\}. \tag{9.2-9}$$

We will eventually wish to show that both the cost-to-go function $\tilde{\Gamma}$ and the optimal control are dependent on measurement data only through the one step predictor of the state. In proving this result, it will be particularly useful for us to note that $\mathcal{E}\{\mathbf{x}^T\mathbf{M}\mathbf{x}\}$ is of the form

$$\mathcal{E}\{\mathbf{x}^T\mathbf{M}\mathbf{x}\} = \bar{\mathbf{x}}^T\mathbf{M}\bar{\mathbf{x}} + \operatorname{tr}\{\mathbf{M} \operatorname{cov}[\mathbf{x}]\}, \tag{9.2-10}$$

where $\mathbf{x}$ is a random variable having mean $\bar{\mathbf{x}}$. To prove Eq. (9.2-10), we note that

$$\mathcal{E}\{\mathbf{x}^T\mathbf{M}\mathbf{x}\} = \mathcal{E}\{(\mathbf{x}-\bar{\mathbf{x}})^T\mathbf{M}(\mathbf{x}-\bar{\mathbf{x}})\} + \bar{\mathbf{x}}^T\mathbf{M}\bar{\mathbf{x}},$$
$$(\mathbf{x}-\bar{\mathbf{x}})^T\mathbf{M}(\mathbf{x}-\bar{\mathbf{x}}) = \operatorname{tr}\{(\mathbf{x}-\bar{\mathbf{x}})^T\mathbf{M}(\mathbf{x}-\bar{\mathbf{x}})\}$$
$$= \operatorname{tr}\{\mathbf{M}(\mathbf{x}-\bar{\mathbf{x}})^T(\mathbf{x}-\bar{\mathbf{x}})\}.$$

Thus, we see that

$$\mathcal{E}\{(\mathbf{x}-\bar{\mathbf{x}})^T\mathbf{M}(\mathbf{x}-\bar{\mathbf{x}})\} = \operatorname{tr}\{\mathbf{M}\mathcal{E}\{(\mathbf{x}-\bar{\mathbf{x}})(\mathbf{x}-\bar{\mathbf{x}})^T\}\},$$

and Eq. (9.2-10) follows directly.

We now wish to show that for each $k = k_f, k_f - 1, \ldots, k_o$, there exists a function $\mathbf{\Gamma}_k$, dependent on $\hat{\mathbf{x}}(k|k-1)$, a matrix $\mathbf{M}(k)$, and a scalar $m(k)$, such that $\tilde{\Gamma}_k[Z(k-1)] = \mathbf{\Gamma}_k[\hat{\mathbf{x}}(k|k-1)]$. To do this we assume a relationship of the form

$$\mathbf{\Gamma}_k(\mathbf{x}) = \mathbf{x}^T\mathbf{M}(\mathbf{k})\mathbf{x} + m(k). \tag{9.2-11}$$

The procedure will also show that the optimal control is dependent on $Z(k-1)$ at time k only through $\hat{\mathbf{x}}(k|k-1)$. Hence the process $\{\hat{\mathbf{x}}(k|k-1), k = k_o, k_o + 1, \ldots\}$ is what is called a *sufficient statistic* [10] for the discrete LQG problem. We note that Eqs. (9.2-9) and (9.2-10) imply that

$$\tilde{\Gamma}_{k_f}[Z(k_f - 1)] = \hat{\mathbf{x}}^T(k_f|k_f - 1)\mathbf{S}\hat{\mathbf{x}}(k_f|k_f - 1) + \operatorname{tr}\{\mathbf{S}\,\mathbf{\Sigma}\,(k_f|k_f - 1)\}, \tag{9.2-12}$$

where the error variance expression $\mathbf{\Sigma}\,(k_f|k_f - 1)$ is defined by

$$\mathbf{\Sigma}\,(k_f|k_f - 1) = \operatorname{cov}\{\mathbf{x}(t_f)|Z(k_f - 1)\} = \operatorname{cov}\{\tilde{\mathbf{x}}(k_f|k_f - 1)\}, \tag{9.2-13}$$

and where the error term $\tilde{\mathbf{x}}(k_f|k_f - 1)$ is given in terms of the conditional mean estimate as

$$\tilde{\mathbf{x}}(k_f|k_f - 1) = \mathbf{x}(k_f) - \mathcal{E}\{\mathbf{x}(k_f)|Z(k_f - 1)\} = \mathbf{x}(k_f) - \hat{\mathbf{x}}(k_f|k_f - 1). \tag{9.2-14}$$

To interrelate Eq. (9.2-11) to Eq. (9.2-12) we see that $\mathbf{M}(k_f) = \mathbf{S}$, and $m(k_f) = \operatorname{tr}\{\mathbf{S}\,\mathbf{\Sigma}\,(k_f|k_f - 1)\}$. Thus Eq. (9.2-12) may be written in terms of Eq. (9.2-11), which now becomes

$$\Gamma_k[\hat{\mathbf{x}}(k|k - 1)] = \hat{\mathbf{x}}(k|k - 1)\mathbf{M}(k)\hat{\mathbf{x}}(k|k - 1) + \mathbf{m}(k). \tag{9.2-15}$$

We also assume that Eq. (9.2-11) holds true for $k = k + 1$ so that we have

$$\begin{aligned}\tilde{\Gamma}_{k+1}[Z(k)] &= \Gamma_{k+1}[\hat{\mathbf{x}}(k + 1|k)] \\ &= \hat{\mathbf{x}}^T(k + 1|k)\mathbf{M}(k + 1)\hat{\mathbf{x}}(k + 1|k) + m(k + 1).\end{aligned} \tag{9.2-16}$$

Our derivation of the Bayes maximum likelihood estimator in Sec. 8.3 indicated that the mean of the state conditioned on the observation was the Bayes maximum likelihood estimator. If we extend this derivation of the optimum discrete time estimator to include a known control input into the message model equation, it is trivial to show that the filter algorithm becomes

$$\hat{\mathbf{x}}(k + 1|k) = \mathbf{A}(k)\hat{\mathbf{x}}(k|k - 1) + \mathbf{B}(k)\mathbf{u}(k) + \mathbf{K}(k)\tilde{\mathbf{z}}(k|k - 1), \tag{9.2-17}$$

where $\mathbf{K}(k)$ is the Kalman gain, and the innovation $\tilde{\mathbf{z}}(k|k - 1)$ is defined as

$$\tilde{\mathbf{z}}(k|k - 1) = \mathbf{z}(k) - \mathbf{H}(k)\,\mathcal{E}\{\mathbf{x}(k)|\mathbf{Z}(k - 1)\} = \mathbf{z}(k) - \mathbf{H}(k)\hat{\mathbf{x}}(k|k - 1) \tag{9.2-18}$$

and has zero-mean and covariance

$$\operatorname{var}\{\tilde{\mathbf{z}}(k)\} = \mathbf{C}(k)\,\mathbf{\Sigma}\,(k|k - 1)\mathbf{C}^T(k) + \mathbf{R}_2(k). \tag{9.2-19}$$

Thus,

$$\mathcal{E}\{\hat{\mathbf{x}}(k + 1|k)|Z(k - 1)\} = \mathbf{A}(k)\hat{\mathbf{x}}(k|k - 1) + \mathbf{B}(k)\mathbf{u}(k) \tag{9.2-20}$$

and

$$\operatorname{cov}[\hat{\mathbf{x}}(k + 1|k)|Z(k - 1)] = \mathbf{K}(k)[\mathbf{C}(k)\,\mathbf{\Sigma}\,(k|k - 1)\mathbf{C}^T(k) + \mathbf{R}_2(k)]\mathbf{K}^T(k). \tag{9.2-21}$$

Substitution of Eqs. (9.2-10) and (9.2-16) into Eq. (9.2-8) yields, after some elementary manipulation:

$$\tilde{\Gamma}_k[Z(k-1)] = \min_{\mathbf{u}(k)} \{\hat{\mathbf{x}}^T(k|k-1)\mathbf{Q}_1(k)\hat{\mathbf{x}}(k|k-1) +$$
$$\text{tr}[\mathbf{Q}_1(k)\boldsymbol{\Sigma}(k|k-1)] + \mathbf{u}^T(k)\mathbf{R}_1(k)\mathbf{u}(k) + [\mathbf{A}(k)\hat{\mathbf{x}}(k|k-1) +$$
$$\mathbf{B}(k)\mathbf{u}(k)]^T\mathbf{M}(k+1)[\mathbf{A}(k)\hat{\mathbf{x}}(k|k-1) + \mathbf{B}(k)\mathbf{u}(k)] +$$
$$\text{tr}[\mathbf{M}(k+1)\mathbf{K}(k)[\mathbf{C}(k)\boldsymbol{\Sigma}(k|k-1)\mathbf{C}^T(k) + \mathbf{R}_2(k)]\mathbf{K}^T(k)] + m(k+1)\} =$$
$$\min_{\mathbf{u}(k)} \{\hat{\mathbf{x}}^T(k|k-1)[\mathbf{A}^T(k)\mathbf{M}(k+1)\mathbf{A}(k) + \mathbf{Q}_1(k) - \mathbf{L}^T(k)[\mathbf{R}_1(k) +$$
$$\mathbf{B}^T(k)\mathbf{M}(k+1)\mathbf{B}(k)]\mathbf{L}(k)]\hat{\mathbf{x}}(k|k-1) + [\mathbf{u}(k) + \mathbf{L}(k)\mathbf{x}(k|k-1)]^T[\mathbf{R}_1(k) +$$
$$\mathbf{B}^T(k)\mathbf{M}(k+1)\mathbf{B}(k)][\mathbf{u}(k) + \mathbf{L}(k)\hat{\mathbf{x}}(k|k-1)] + m(k)\} =$$
$$\hat{\mathbf{x}}^T(k|k-1)\mathbf{M}(k)\hat{\mathbf{x}}(k|k-1) + \mathbf{m}(k), \tag{9.2-22}$$

where

$$\mathbf{L}(k) = [\mathbf{R}_1(k) + \mathbf{B}^T(k)\mathbf{M}(k+1)\mathbf{B}(k)]^{-1}\mathbf{B}^T(k)\mathbf{M}(k+1)\mathbf{A}(k), \tag{9.2-23}$$

$$m(k) = m(k+1) + \text{tr}[\mathbf{Q}_1(k)\boldsymbol{\Sigma}(k|k-1)] +$$
$$\text{tr}[\mathbf{M}(k+1)\mathbf{K}(k)[\mathbf{C}(k)\boldsymbol{\Sigma}(k|k-1)\mathbf{C}^T(k) + \mathbf{R}_2(k)]\mathbf{K}^T(k)] \tag{9.2-24}$$

and

$$\mathbf{M}(k) = \mathbf{A}^T(k)\mathbf{M}(k+1)\mathbf{A}(k) + \mathbf{Q}_1(k) -$$
$$\mathbf{L}^T(k)[\mathbf{R}_1(k) + \mathbf{B}^T(k)\mathbf{M}(k+1)\mathbf{B}(k)]\mathbf{L}(k), \tag{9.2-25}$$

and where the optimal feedback control is obtained as a linear function of the conditional mean estimate

$$\mathbf{u}(k) = -\mathbf{L}(k)\hat{\mathbf{x}}(k|k-1). \tag{9.2-26}$$

These results are correct if the matrix $[\mathbf{R}_1(k) + \mathbf{B}^T(k)\mathbf{M}(k+1)\mathbf{B}(k)]$ is positive definite. To prove that $[\mathbf{R}_1(k) + \mathbf{B}^T(k)\mathbf{M}(k+1)\mathbf{B}(k)]$ is positive definite, it is sufficient to show that the nonnegative definiteness of $\mathbf{M}(k+1)$ implies the nonnegative definiteness of $\mathbf{M}(k)$, since by assumption $\mathbf{M}(k_f) = \mathbf{S}$ is nonnegative definite. Substitution of Eq. (9.2-23) into Eq. (9.2-25) implies that

$$\mathbf{M}(k) = \mathbf{A}^T(k)\mathbf{M}(k+1)[\mathbf{A}(k) - \mathbf{B}(k)\mathbf{L}(k)] + \mathbf{Q}_1(k). \tag{9.2-27}$$

It follows directly from Eq. (9.2-23) that

$$\mathbf{R}_1(k)\mathbf{L}(k) = \mathbf{B}^T(k)\mathbf{M}(k+1)[\mathbf{A}(k) - \mathbf{B}(k)\mathbf{L}(k)]. \tag{9.2-28}$$

Combining Eqs. (9.2-27) and (9.2-28) shows that

$$\mathbf{M}(k) = [\mathbf{A}(k) - \mathbf{B}(k)\mathbf{L}(k)]^T\mathbf{M}(k+1)[\mathbf{A}(k) - \mathbf{B}(k)\mathbf{L}(k)] +$$
$$\mathbf{L}^T(k)\mathbf{R}_1(k)\mathbf{L}(k) + \mathbf{Q}_1(k), \tag{9.2-29}$$

and our result follows directly. We note that the assumed symmetry of $\mathbf{S}$ and Eq. (9.2-22) imply that $\mathbf{M}(k)$ is also symmetric for all k.

The form of the optimal control for the LQG problem given in Eq. (9.2-26) is of particular interest. We note that the optimal control is linear in the unbiased, minimum error variance estimator of the state. Since the optimal control depends on the data $\mathbf{Z}(k-1)$ only through $\hat{\mathbf{x}}(k|k-1)$, it is said to have the *separation property*. Further examination of $\mathbf{L}(k)$ and Example 6.2-1 implies that the linear operator acting on the current state which produces the current optimal control for the discrete linear regulator is identical to $\mathbf{L}(k)$. Thus, it is sufficient to compute the matrix $\mathbf{L}(k)$, as in Example 6.2-1, for the completely deterministic linear regulator and then use the single-stage predictor resulting from the Kalman filter to determine the optimal control for the LQG problem. In general, a control is said to have the *certainty equivalence property* if the optimal control has the same form as the deterministic version of the problem where the actual state of the system subject to control is replaced by the single-stage predictor. Hence, the optimal control for the LQG problem has both the separation property and the certainty equivalence property. A stochastic optimal control problem having the separation property but not the more general certainty equivalence property is presented in [14]. Further discussion of this can also be found in [7]. We also note that computation of $\mathbf{L}(k)$ given in Eq. (9.2-16) does not depend on the covariance matrices of the noise processes $\mathbf{Q}_2(k)$ and $\mathbf{R}_2(k)$; thus, the optimal control for the LQG problem does not require the available knowledge regarding the quality of future observations.

Our results can be used to show that there exists a completely observed LQG problem which is control-wise equivalent to the LQG problem formulated above. Consider the system described by Eq. (9.2-14), which we assume is completely observed, and the cost function

$$J' = \mathcal{E}\Big\{\|\hat{\mathbf{x}}(k_f|k_f-1)\|_{\mathbf{S}}^2 + \sum_{k=k_o}^{k_f-1}(\|\hat{\mathbf{x}}(k|k-1)\|_{\mathbf{Q}_1(k)}^2 + \|\mathbf{u}(k)\|_{\mathbf{R}_1(k)}^2)\,|\,\hat{\mathbf{x}}(k_o)\Big\}. \tag{9.2-30}$$

A procedure identical to the development of the LQG problem shows the equivalence (in the sense that both problems have identical optimal controls) of this problem and the more general LQG problem.

Intuitively, we feel that the more uncertainty involved with a stochastic optimum systems control problem, the less likely a controller will be able to satisfactorily control the system and, hence, the greater the expected cost. The minimum cost, which from Eqs. (9.2-7) and (9.2-12), is

$$J_{\min} = \hat{\mathbf{x}}^T(k_o|k_o-1)\mathbf{M}(k_o)\hat{\mathbf{x}}(k_o|k_o-1) + m(k_o), \tag{9.2-31}$$

can be expressed in a form so that our intuition is easily validated. The optimal cost is [6, p. 282]

$$J = \bar{\mathbf{x}}^T(k_o)\mathbf{M}(k_o)\bar{\mathbf{x}}(k_o) + \operatorname{tr}[\mathbf{M}(k_o)\mathbf{P}_2(k_o)] + \sum_{k=k_o}^{k_f-1}\operatorname{tr}[\mathbf{M}(k+1)\mathbf{Q}_2(k)] + \sum_{k=k_o}^{k_f-1}\operatorname{tr}[\boldsymbol{\Sigma}(k|k-1)\mathbf{L}^T(k)[\mathbf{B}^T(k)\mathbf{M}(k+1)\mathbf{B}(k) + \mathbf{R}_1(k)]\mathbf{L}(k)], \tag{9.2-32}$$

where the last three terms in Eq. (9.2-32) reflect additional costs associated with various uncertainties. The second term is the cost due to the uncertainty of the initial state. The third term represents the cost related to the uncertainty of future values of the state. The fourth term reflects cost resulting from effects of future uncertainty in the state estimation. Further interpretation of these costs can be found in [6, 12].

Our development has been necessarily long, so it is convenient to summarize our results. The system subject to estimation and control is given by

$$\mathbf{x}(k+1) = \mathbf{A}(k)\mathbf{x}(k) + \mathbf{B}(k)\mathbf{u}(k) + \mathbf{w}(k)$$
$$\mathbf{z}(k) = \mathbf{C}(k)\mathbf{x}(k) + \mathbf{v}(k),$$

where $\mathbf{w}(k)$ and $\mathbf{v}(k)$ are zero-mean uncorrelated white Gaussian random processes with variance $\mathbf{Q}_2(k)$ and $\mathbf{R}_2(k)$. The random variable $\mathbf{x}(k_o)$ is Gaussian with mean $\bar{\mathbf{x}}(k_o)$, covariance $\mathbf{P}_2(k_o)$, and is uncorrelated with $\mathbf{w}(k)$ and $\mathbf{v}(k)$. The problem is to minimize

$$J = \mathcal{E}\left\{\|\mathbf{x}(k_f)\|^2_{\mathbf{S}} + \sum_{k=k_o}^{k_f-1} [\|\mathbf{x}(k)\|^2_{\mathbf{Q}_1(k)} + \|\mathbf{u}(k)\|^2_{\mathbf{R}_1(k)}]\right\},$$

subject to previously described conditions on the system equations.

The optimal control is given by

$$\mathbf{u}(k) = -\mathbf{L}(k)\hat{\mathbf{x}}(k|k-1)$$

where the gain $\mathbf{L}(k)$ is determined from

$$\mathbf{L}(k) = [\mathbf{R}_1(k) + \mathbf{B}^T(k)\mathbf{M}(k+1)\mathbf{B}(k)]^{-1}\mathbf{B}^T(k)\mathbf{M}(k+1)\mathbf{A}(k);$$

$\mathbf{M}(k)$ is obtained by solving the discrete matrix Riccati equation

$$\mathbf{M}(k) = \mathbf{A}^T(k)\mathbf{M}(k+1)\mathbf{A}(k) + \mathbf{Q}_1(k) - \mathbf{L}^T(k)[\mathbf{R}_1(k) + \mathbf{B}^T(k)\mathbf{M}(k+1)\mathbf{B}(k)]\mathbf{L}(k),$$

with terminal condition

$$\mathbf{M}(k_f) = \mathbf{S}.$$

The optimum estimate is obtained from

$$\hat{\mathbf{x}}(k+1|k) = \mathbf{A}(k)\hat{\mathbf{x}}(k|k-1) + \mathbf{B}(k)\mathbf{u}(k) + \mathbf{K}(k)\tilde{\mathbf{z}}(k),$$

with initial condition

$$\hat{\mathbf{x}}(k_o+1|k_o) = \mathbf{A}(k_o)\bar{\mathbf{x}}(k_o).$$

The Kalman gain $\mathbf{K}(k)$ is obtained from

$$\mathbf{K}(k) = \mathbf{A}(k)\,\boldsymbol{\Sigma}(k|k-1)\mathbf{C}^T(k)[\mathbf{C}(k)\,\boldsymbol{\Sigma}(k|k-1)\mathbf{C}^T(k) + \mathbf{R}_2(k)]^{-1},$$

where the error variance equation propagates according to

$$\boldsymbol{\Sigma}(k+1|k)\mathbf{A} = (k)\,\boldsymbol{\Sigma}(k|k-1)\mathbf{A}^T(k) + \mathbf{A}(k)\,\boldsymbol{\Sigma}(k|k-1)\mathbf{C}^T(k)[\mathbf{C}(k)\,\boldsymbol{\Sigma}(k|k-1)\mathbf{C}^T(k) + \mathbf{R}_2(k)]^{-1} \times \mathbf{C}(k)\,\boldsymbol{\Sigma}(k|k-1)\mathbf{A}^T(k) + \mathbf{Q}_2(k),$$

with initial condition

$$\boldsymbol{\Sigma}(k_o+1|k_o) = \mathbf{A}(k_o)\mathbf{P}_2(k_o)\mathbf{A}^T(k_o) + \mathbf{Q}_2(k_o).$$

9.3 The LQG problem—continuous time case

We now examine the continuous time version of the LQG problem. Consider a linear system described by

$$\dot{\mathbf{x}}(t) = \mathbf{A}(t)\mathbf{x}(t) + \mathbf{B}(t)\mathbf{u}(t) + \mathbf{w}(t) \tag{9.3-1}$$

$$\mathbf{z}(t) = \mathbf{C}(t)\mathbf{x}(t) + \mathbf{v}(t), \tag{9.3-2}$$

where the state vector $\mathbf{x}(t)$ is an n-vector, the control vector $\mathbf{u}(t)$ is an m-vector, and $\mathbf{z}(t)$ is the observation r-vector. Assume that the initial condition $\mathbf{x}(t_o)$ is Gaussian with mean $\bar{\mathbf{x}}(t_o)$ and symmetric nonnegative definite covariance matrix $\mathbf{P}_2(t_o) = \mathbf{P}_{20}$, and assume that the noise processes $\{\mathbf{w}(t), t \in [t_o, t_f]\}$ and $\{\mathbf{v}(t), t \in [t_o, t_f]\}$ are zero-mean white Gaussian such that

$$\operatorname{cov}[\mathbf{w}(t), \mathbf{w}(\tau)] = \mathbf{Q}_2(t)\,\delta(t-\tau)$$

and

$$\operatorname{cov}[\mathbf{v}(t), \mathbf{v}(\tau)] = \mathbf{R}_2(t)\,\delta(t-\tau).$$

Further, we assume that $\mathbf{x}(t_o)$, $\{\mathbf{w}(t)\}$, and $\{\mathbf{v}(t)\}$ are independent.

Analogous to the discrete case, we let the criterion for optimality be

$$J = \mathcal{E}\left\{\|\mathbf{x}(t_f)\|^2_{\mathbf{S}} + \int_{t_o}^{t_f} (\|\mathbf{x}(\tau)\|^2_{\mathbf{Q}_1(\tau)} + \|\mathbf{u}(\tau)\|^2_{\mathbf{R}_1(\tau)})\,d\tau\right\} \tag{9.3-3}$$

which we wish to minimize. The matrices $\mathbf{R}_1(t)$ and $\mathbf{R}_2(t)$ are assumed to be positive definite symmetric; $\mathbf{S}$, $\mathbf{Q}_1(t)$, and $\mathbf{Q}_2(t)$ are assumed to be nonnegative definite and symmetric.

We now modify our definition of admissibility for controls, which we have presented in Sec. 9.1, to the continuous time case. Let $Z(t) = \{\mathbf{z}(s), t_o \le s \le t\}$. A control function is admissible if

$$\mathbf{u}(t) = \boldsymbol{\alpha}[t, Z(t)], \qquad t \in [t_o, t_f] \tag{9.3-4}$$

where $\boldsymbol{\alpha}$ must (for technical reasons) be reasonably smooth.† We wish to determine an admissible control function that minimizes the criterion of Eq. (9.3-3) with respect to the set of all admissible control functions.

We now use the well-known relation (for instance, see [16, p.162])

$$\mathcal{E}\{\mathcal{E}\{\mathbf{X}\,|\,\mathbf{Y}\}\} = \mathcal{E}\{\mathbf{X}\} \tag{9.3-5}$$

to describe Eq. (9.3-3) in a form more suitable for the analysis to follow. Equations (9.3-3) and (9.3-5) imply that

$$J = \mathcal{E}\{\mathcal{E}[\|\mathbf{x}(t_f)\|^2_{\mathbf{S}}\,|\,Z(t_f)]\} + \mathcal{E}\left\{\int_{t_o}^{t_f} \mathcal{E}\{\|\mathbf{x}(\tau)\|^2_{\mathbf{Q}_1(\tau)}\,|\,Z(\tau)\}\,d\tau\right\} + \mathcal{E}\left\{\int_{t_o}^{t_f} \|\mathbf{u}(\tau)\|^2_{\mathbf{R}_1(\tau)}\,d\tau\right\}. \tag{9.3-6}$$

†More precisely, $\boldsymbol{\alpha}(t, f_1)$ satisfies the Lipschitz condition $\|\boldsymbol{\alpha}(t, \mathbf{f}_1) - \boldsymbol{\alpha}(t, \mathbf{f}_2)\| \le \beta\|\mathbf{f}_1 - \mathbf{f}_2\|$ for continuous r-vector functions $\mathbf{f}_i\colon [t_o, t_f] \longrightarrow R$, $i = 1, 2$, where β is a constant [15, 5].

Eqs. (9.2-10) and (9.3-6) then imply that

$$J = \mathcal{E}\left\{\|\hat{\mathbf{x}}(t_f)\|_{\mathbf{S}}^2 + \int_{t_o}^{t_f} [\|\hat{\mathbf{x}}(\tau)\|_{\mathbf{Q}_1(\tau)}^2 + \|\mathbf{u}(\tau)\|_{\mathbf{R}_1(\tau)}^2]\, d\tau\right\} +$$

$$\mathcal{E}\left\{\mathrm{tr}\left[\mathbf{S}\mathbf{P}_2(t_f) + \int_{t_o}^{t_f} \mathbf{Q}_1(\tau)\mathbf{P}_2(\tau)\, d\tau\right]\right\}, \tag{9.3-7}$$

where

$$\hat{\mathbf{x}}(t) = \mathcal{E}\{\mathbf{x}(t) \mid Z(t)\}$$

and

$$\mathbf{P}_2(t) = \mathcal{E}\{[\mathbf{x}(t) - \hat{\mathbf{x}}(t)][\mathbf{x}(t) - \hat{\mathbf{x}}(t)]^T \mid Z(t)\}.$$

It is now appropriate to determine recursive equations for $\hat{\mathbf{x}}(t)$ and $\mathbf{P}_2(t)$. We recall that for the $\mathbf{u}(t) = \mathbf{0}$ case, such equations were derived in Sec. 8.2 for the case of the best linear estimator. It can be shown [17] that, just as in the discrete time case in Sec. 8.3, the best linear filter is the conditional mean filter for the case where the statistics are Gaussian. Extension of the results in Sec. 8.2 to include the control input demonstrates that $\hat{\mathbf{x}}(t)$ and $\mathbf{P}_2(t)$ satisfy

$$\dot{\hat{\mathbf{x}}}(t) = \mathbf{A}(t)\hat{\mathbf{x}}(t) + \mathbf{B}(t)\mathbf{u}(t) + \mathbf{K}(t)[\mathbf{z}(t) - \mathbf{C}(t)\hat{\mathbf{x}}(t)] \tag{9.3-8}$$

$$\hat{\mathbf{x}}(t_o) = \bar{\mathbf{x}}(t_o),$$

where

$$\mathbf{K}_2(t) = \mathbf{P}_2(t)\mathbf{C}^T(t)\mathbf{R}_2^{-1}(t), \tag{9.3-9}$$

and

$$\dot{\mathbf{P}}_2(t) = \mathbf{A}(t)\mathbf{P}_2(t) + \mathbf{P}_2(t)\mathbf{A}^T(t) - \mathbf{P}_2(t)\mathbf{C}^T(t)\mathbf{R}_2^{-1}(t)\mathbf{C}(t)\mathbf{P}_2(t) + \mathbf{Q}_2(t), \tag{9.3-10}$$

where $\mathbf{P}_2(t_o) = \mathbf{P}_{20}$. Interestingly, we note, by analogy to the discrete case, that the estimation error covariance $\mathbf{P}_2(t)$ is independent of the measurements and the control settings and is deterministic. This implies that the last term on the right-hand side of Eq. (9.3-7) is not affected by the expectation operator and will not affect choice of the control. Hence, minimization of J is equivalent to minimizing

$$J' = \mathcal{E}\left\{\|\hat{\mathbf{x}}(t_f)\|_{\mathbf{S}}^2 + \int_{t_o}^{t_f} (\|\hat{\mathbf{x}}(\tau)\|_{\mathbf{Q}_1(\tau)}^2 + \|\mathbf{u}(\tau)\|_{\mathbf{R}_1(\tau)}^2)\, d\tau\right\}. \tag{9.3-11}$$

We recall from Sec. 8.2 that the innovations process $\{\tilde{\mathbf{z}}(t)\}$, $\tilde{\mathbf{z}}(t) = \mathbf{z}(t) - \mathbf{C}(t)\hat{\mathbf{x}}(t)$ is zero-mean and Gaussian, having covariance $\mathcal{E}\{\tilde{\mathbf{z}}(t)\tilde{\mathbf{z}}^T(\tau)\} = \mathbf{R}_2(t)\delta(t - \tau)$ (see also [5, 17]).

Now we note that Eqs. (9.3-8) through (9.3-11) constitute a linear regulator problem in optimum control. We therefore have an equivalent control problem: minimize Eq. (9.3-11), subject to Eq. (9.3-8), with respect to the set of all admissible control laws. We note that the equivalent problem assumes perfect observations of the state process, where the state is now $\hat{\mathbf{x}}(t)$. Following the dynamic programming approach used in Sec. 9.2, (see also [5, 15, 16]), we can show that the optimal admissible control which minimizes Eq. (9.3-11) subject to Eq. (9.3-8), is

$$\mathbf{u}^*(t) = -\mathbf{R}_1^{-1}(t)\mathbf{B}^T(t)\mathbf{P}_1(t)\hat{\mathbf{x}}(t) = \mathbf{K}_1(t)\hat{\mathbf{x}}(t) \tag{9.3-12}$$

$$\dot{\mathbf{P}}_1(t) = -\mathbf{P}_1(t)\mathbf{A}(t) - \mathbf{A}^T(t)\mathbf{P}_1(t) + \mathbf{P}_1(t)\mathbf{B}(t)\mathbf{R}_1^{-1}(t)\mathbf{B}^T(t)\mathbf{P}_1(t) - \mathbf{Q}_1(t), \tag{9.3-13}$$

where $\mathbf{P}_1(t_f) = \mathbf{S}$. The expected optimal cost to be accrued between time t and final time t_f is then

$$J'(\hat{\mathbf{x}}(t), t) = \|\hat{\mathbf{x}}(t)\|^2_{\mathbf{P}_1(t)} + \operatorname{tr}\left\{\int_t^{t_f} \mathbf{K}_2(\tau)\mathbf{R}_2(\tau)\mathbf{K}_2^T(\tau)\mathbf{P}_1(\tau)\,d\tau\right\}.$$

Thus, the optimal expected cost to be accrued over the problem horizon is

$$J = \|\bar{\mathbf{x}}(t_o)\|^2_{\mathbf{P}_1(t_o)} + \operatorname{tr}\left[\int_{t_o}^{t_f} \mathbf{K}_2(\tau)\mathbf{R}_2(\tau)\mathbf{K}_2^T(\tau)\mathbf{P}_1(\tau)\,d\tau\right] + \operatorname{tr}\left[\mathbf{S}\mathbf{P}_2(t_f) + \int_{t_o}^{t_f} \mathbf{Q}_1(\tau)\mathbf{P}_2(\tau)\,d\tau\right], \tag{9.3-14}$$

where the second trace term follows from Eq. (9.3-7). A block diagram of the continuous time LQG problem is presented in Fig. 9.3-1.

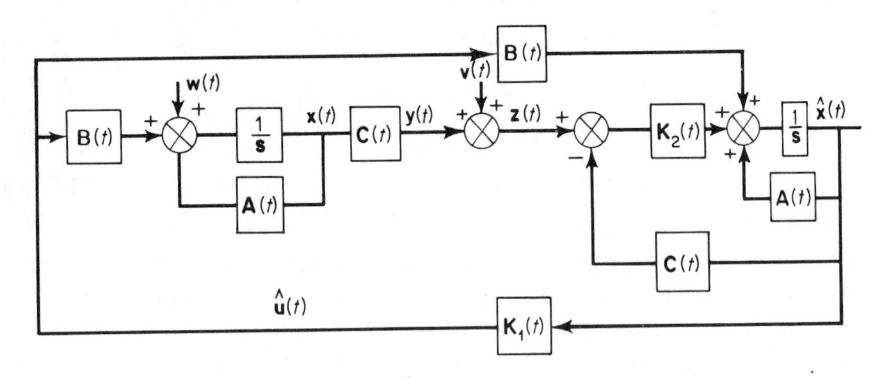

Fig. 9.3-1 Optimum linear combined estimation and control system.

The above results can be determined using the stochastic Hamilton-Jacobi-Bellman equation derived in Appendix C. From Eq. (9.3-11), we define

$$V[\hat{\mathbf{x}}(t), t] = \mathcal{E}\left\{\|\hat{\mathbf{x}}(t_f)\|^2_{\mathbf{S}} + \int_t^{t_f} [\|\hat{\mathbf{x}}(\tau)\|^2_{\mathbf{Q}_1(\tau)} + \|\hat{\mathbf{u}}(\tau)\|^2_{\mathbf{R}_1(\tau)}]\,d\tau \mid \hat{\mathbf{x}}(t)\right\} \tag{9.3-15}$$

where $\hat{\mathbf{u}}$ is assumed to be an optimal control. The stochastic Hamilton-Jacobi-Bellman equation then implies, in the context of the problem at hand, that

$$\frac{\partial V}{\partial t} + \|\hat{\mathbf{x}}(t)\|^2_{\mathbf{Q}_1(t)} + \|\hat{\mathbf{u}}(t)\|^2_{\mathbf{R}_1(t)} + \left(\frac{\partial V}{\partial \hat{\mathbf{x}}}\right)^T [\mathbf{A}(t)\hat{\mathbf{x}}(t) + \mathbf{B}(t)\hat{\mathbf{u}}(t)] + \frac{1}{2}\operatorname{tr}\left[\mathbf{Q}_2(t)\frac{\partial^2 V}{\partial \hat{\mathbf{x}}^2}\right] = 0 \tag{9.3-16}$$

with endpoint condition

$$V[\hat{\mathbf{x}}(t_f), t_f] = \|\hat{\mathbf{x}}(t_f)\|^2_{\mathbf{S}}. \tag{9.3-17}$$

Since we have assumed that $\hat{\mathbf{u}}$ achieves the minimum in the stochastic Hamilton-Jacobi-Bellman equation, we obtain from Eq. (9.3-16),

$$\hat{\mathbf{u}}(t) = -\frac{1}{2}\mathbf{R}_1^{-1}(t)\mathbf{B}^T(t)\left\{\frac{\partial V[\hat{\mathbf{x}}(t), t]}{\partial \hat{\mathbf{x}}(t)}\right\}, \tag{9.3-18}$$

such that Eq. (9.3-16) becomes with the optimal control

$$\frac{\partial V}{\partial t} + ||\hat{\mathbf{x}}(t)||^2_{\mathbf{Q}_1(t)} - \frac{1}{4}||\mathbf{B}^T(t)\frac{\partial V}{\partial \hat{\mathbf{x}}}||^2_{\mathbf{R}^{-1}(t)} + \left(\frac{\partial V}{\partial \hat{\mathbf{x}}}\right)^T \mathbf{A}(t)\hat{\mathbf{x}}(t) + \frac{1}{2}\operatorname{tr}\left[\mathbf{Q}_2(t)\frac{\partial^2 V}{\partial \hat{\mathbf{x}}^2}\right] = 0. \tag{9.3-19}$$

We now assume that a solution to this equation is of the form

$$V[\hat{\mathbf{x}}(t), t] = ||\hat{\mathbf{x}}(t)||^2_{\mathbf{P}_1(t)} + \eta(t). \tag{9.3-20}$$

Substituting Eq. (9.3-20) into Eq. (9.3-19) then implies that $\mathbf{P}_1(t)$ satisfies Eq. (9.3-13) and that in addition

$$\frac{d\eta(t)}{dt} = \operatorname{tr}[\mathbf{Q}_2(t)\mathbf{P}_1(t)], \qquad \boldsymbol{\eta}(t_f) = 0. \tag{9.3-21}$$

Determining $\partial V/\partial\hat{\mathbf{x}} = 2\mathbf{P}_1(t)\hat{\mathbf{x}}(t)$ from Eq. (9.3-20) and substituting the result into Eq. (9.3-18) then gives our former result, Eq. (9.3-12), as the optimal. Thus we have completed an alternate development of the separation theorem and certainty equivalence principle. Our earlier remarks for the discrete time case apply equally well here.

Example 9.3-1. Let us consider the minimization of

$$J = \mathcal{E}\left\{\tfrac{1}{2}x^2(t_f)s + \tfrac{1}{2}\int_0^{t_f}[2x^2(t) + u^2(t)]\,dt\right\}$$

for the system,

$$\dot{x} = -\tfrac{1}{2}x(t) + u(t) + w(t), \qquad x(0) = 10, \qquad z(t) = x(t) + v(t),$$

with $R_2 = \frac{1}{4}$, $Q_2 = 2$. We recall that we solved precisely the same two problems which result here when the separation theorem was used in Examples 5.1-1, and 8.2-1. Thus we need only complete the solution by illustrating the block diagram for the combined estimation and control as in Fig. 9.3-1. We note that the control gains have to be precomputed and stored, whereas the estimation gains may either be precomputed and stored or may be computed in real time. The control gain computation is independent of the estimation gain computation as a consequence of the separation theorem.

9.3-1 *The steady-state continuous-time LQG problem*

We now consider the time-invariant case; i.e., the matrices $\mathbf{A}$, $\mathbf{B}$, $\mathbf{C}$, $\mathbf{Q}_1$, $\mathbf{Q}_2$, $\mathbf{R}_1$, and $\mathbf{R}_2$ are constant, where $t_f = \infty$. The time invariance assumption implies that we can choose $t_o = 0$ without loss of generality. Following Tse

[15], we modify the cost function, Eq. (9.3-3), to the cost rate function

$$J_r = \lim_{T\to\infty} \frac{J_T}{T}, \tag{9.3-22}$$

where

$$J_T = \mathcal{E}\left\{\|\mathbf{x}(T)\|^2_{\mathbf{S}} + \int_0^T (\|\mathbf{x}(\tau)\|^2_{\mathbf{Q}_1} + \|\mathbf{u}(\tau)\|^2_{\mathbf{R}_1})\, d\tau\right\}. \tag{9.3-23}$$

For finite T, Eq. (9.3-12) implies that the optimal control is

$$\mathbf{u}^*(t) = -\mathbf{R}_1^{-1}\mathbf{B}^T\mathbf{P}_1(t)\hat{\mathbf{x}}(t), \tag{9.3-24}$$

where, from Eq. (9.3-13),

$$\dot{\mathbf{P}}_1(t) = -\mathbf{P}_1(t)\mathbf{A} - \mathbf{A}^T\mathbf{P}_1(t) + \mathbf{P}_1(t)\mathbf{B}\mathbf{R}_1^{-1}\mathbf{B}^T\mathbf{P}_1(t) - \mathbf{Q}_1, \qquad \mathbf{P}_1(T) = \mathbf{S} \tag{9.3-25}$$

and from Eqs. (9.3-8), (9.3-9), and (9.3-10),

$$\dot{\hat{\mathbf{x}}}(t) = \mathbf{A}\hat{\mathbf{x}}(t) + \mathbf{B}\mathbf{u}^*(t) + \mathbf{K}(t)[\mathbf{z}(t) - \mathbf{C}\hat{\mathbf{x}}(t)], \qquad \hat{\mathbf{x}}(0) = \bar{\mathbf{x}}(0) \tag{9.3-26}$$

$$\mathbf{K}(t) = \mathbf{P}_2(t)\mathbf{C}^T\mathbf{R}_2^{-1} \tag{9.3-27}$$

and

$$\dot{\mathbf{P}}_2(t) = \mathbf{A}\mathbf{P}_2 + \mathbf{P}_2\mathbf{A}^T - \mathbf{P}_2(t)\mathbf{C}^T\mathbf{R}_2^{-1}\mathbf{C}\mathbf{P}_2(t) + \mathbf{Q}_2, \qquad \mathbf{P}_2(0) = \mathbf{P}_{20}. \tag{9.3-28}$$

To be precise, we note that the solution of Eq. (9.3-25) is dependent on T and $\mathbf{S}$. Also the solution of Eq. (9.3-28) is dependent on the time 0 and $\mathbf{P}_2(0)$. Thus, we denote $\mathbf{P}_1(t, T, \mathbf{S}) = \mathbf{P}_1(t)$ and $\mathbf{P}_2[t, 0, \mathbf{P}_2(0)] = \mathbf{P}_2(t)$. Note also that as $T \longrightarrow \infty$, the limits

$$\lim_{T\to\infty} \mathbf{P}_1(t, T, \mathbf{S}) \quad \text{and} \quad \lim_{t\to\infty} \mathbf{P}_2[t, 0, \mathbf{P}_2(0)]$$

must be bounded so that $\mathbf{u}^*(t)$ will be bounded. Otherwise, if for some time the control $\mathbf{u}^*(t)$ is unbounded, $\mathbf{J}_r$ will be unbounded, and the control problem is meaningless. When $T = \infty$, we would also hope that as $t \longrightarrow \infty$, $\mathbf{P}_1(t)$ and $\mathbf{P}_2(t)$ would converge to the solutions of Eqs. (9.3-25) and (9.3-28) for the case where $\dot{\mathbf{P}}_1(t) = \dot{\mathbf{P}}_2(t) = 0$. These results hold under the assumptions of controllability and observability. Proofs can be found in [18, 19; see also 15]. That is, if the time-invariant system,

$$\dot{\mathbf{x}}(t) = \mathbf{A}\mathbf{x}(t) + \mathbf{B}\mathbf{u}(t) + \mathbf{w}(t)$$

$$\mathbf{z}(t) = \mathbf{C}\mathbf{x}(t) + \mathbf{v}(t)$$

is controllable and observable (see Chapter 7), then:

1. for any nonnegative definite $\mathbf{S}$,

$$\lim_{T\to\infty} \mathbf{P}_1(t, T, \mathbf{S}) = \mathbf{P}_1 \quad \text{satisfies}$$

$$\mathbf{0} = -\mathbf{P}_1\mathbf{A} - \mathbf{A}^T\mathbf{P}_1 + \mathbf{P}_1\mathbf{B}\mathbf{R}_1^{-1}\mathbf{B}^T\mathbf{P}_1 - \mathbf{Q}_1$$

2. for any nonnegative definite $\mathbf{P}_2(0)$,

$$\lim_{t_0 \to \infty} \mathbf{P}_2[t, -t_0, \mathbf{P}_2(0)] = \mathbf{P}_2 \quad \text{satisfies}$$

$$\mathbf{0} = \mathbf{A}\mathbf{P}_2 + \mathbf{P}_2\mathbf{A}^T - \mathbf{P}_2\mathbf{C}^T\mathbf{R}_2^{-1}\mathbf{C}\mathbf{P}_2 + \mathbf{Q}_2.$$

Thus, in the steady-state, the (time-invariant) optimal control is

$$\mathbf{u}^*(t) = -\mathbf{R}_1^{-1}\mathbf{B}^T\mathbf{P}_1\hat{\mathbf{x}}(t),$$

where

$$\dot{\hat{\mathbf{x}}}(t) = \mathbf{A}\hat{\mathbf{x}}(t) + \mathbf{B}\mathbf{u}^*(t) + \mathbf{K}[\mathbf{z}(t) - \mathbf{C}\hat{\mathbf{x}}(t)]$$

and

$$\mathbf{K} = \mathbf{P}_2\mathbf{C}^T\mathbf{R}_2^{-1}.$$

9.4 Extensions

There are many physical systems subject to control and noise in both systems dynamics and measurements that do not adequately satisfy the assumptions of the LQG problem. Several examples, presented in [7, 13, 14], have been mentioned. In these articles and throughout this chapter, however, the decision maker is allowed to have perfect memory of all past and present measurements, and if there are several decision makers, i.e., the control $\mathbf{u}(t)$ is a vector, then all decision makers share the same information. We now consider an example that violates the latter assumption and present its implications. We will restate the problem in an LQG problem formulation and note that the perfect memory assumption is then violated.

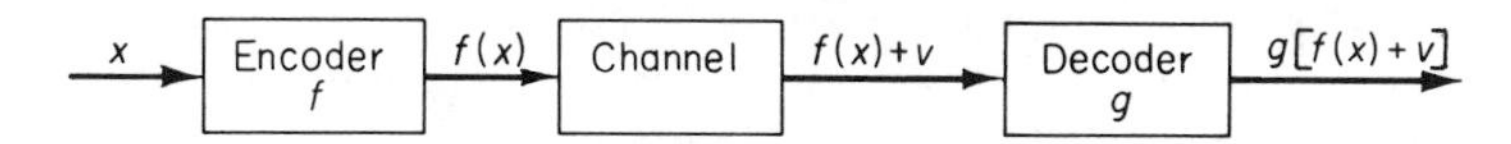

Fig. 9.4-1 Communications system.

Consider the communications system described in Fig. 9.4-1, following the problem considered in [18]. We wish to determine an encoder f and a decoder g such that the expression

$$J(f, g) = \mathcal{E}\{k^2[x - f(x)]^2 + [f(x) - g[f(x) + v]]^2\}, \tag{9.4-1}$$

is minimized, where $k^2 > 0$ and x_o and v are Gaussian. Without loss of generality, assume $\mathcal{E}\{x\} = \mathcal{E}\{v\} = 0$ and $\mathcal{E}\{v^2\} = 1$. Witsenhausen [19], in a famous counterexample, has shown that the best affine controller, i.e., modulator or coder, is not optimal over the set of all reasonably smooth pairs (f, g); however, an optimal control does exist and hence must be nonlinear. Note that the two controllers f and g do not share the same information.

Example 9.4-1. Let us restate the above communications problem in an LQG problem context and examine its implications. Denote $x(0) = x$, $\alpha(0, \cdot) = g$, and let $\alpha[1, x(0)] = f[x(0)] - x(0)$. Consider the state equations

$$x(1) = x(0) + u(0)$$
$$x(2) = x(1) - u(1)$$

and the measurement equations

$$z(0) = x(0)$$
$$z(1) = x(1) + v.$$

The cost function then becomes

$$J = \mathcal{E}\{k^2u^2(0) + x^2(2)\}, \qquad k^2 > 0,$$

where

$$u(0) = \alpha[0, z(0)]$$
$$u(1) = \alpha[1, z(1)].$$

The problem has now been recast as a time-varying LQG problem with the exception that the controller, upon making a decision at time $t = 1$, has "forgotten" $z(0)$. Thus, the violation of the perfect memory assumption implies that the controller is no longer linear in the best estimate of the state.

The above example clearly illustrates the effect of information pattern (i.e., what the controller knows and when the controller knows it) on optimal control structure and cost. Other examples of LQG problems with information patterns differing from the information pattern described in Sec. 9.1 (often referred to as decentralized control problems) can be found in [20] and its references. See also [21].

Even though the separation property is violated in some physical problems, it is still possible to invoke separation and then find the optimal control and optimum estimator in a "separated" fashion. There is no guarantee, of course, that results will be optimum if such an assumption is made.

9.5 Sensitivity analysis of combined estimation and control algorithms

In the past several years, combined estimation and control techniques have been applied to more and more types of systems. Typically, the control problem is solved as though all of the states were available for feedback; then the resulting control law is implemented using estimates of the states supplied by some type of estimator such as a Kalman filter. For nonlinear systems, this approach is often used for lack of a better alternative. For certain classes of systems, primarily the linear quadratic Gaussian (LQG) class, we have shown that this *separation principle* results in an overall optimum solution.

The algorithms for the LQG problem are just the regulator and filter algorithms of Chapters 5 and 8.

Optimum estimation and control, however, require perfect knowledge of system and noise parameters, an impossible requirement for any realistic process control problem. So, as a part of designing the best controller, the effects of parameter errors should be studied. We have considered the effects of parameter errors upon control and estimation in Chapters 7 and 8. Also needed for many systems control applications, however, is an approach which determines the effect of these parameter errors upon the cost for the combined estimation and control problems.

We will present an algorithm which determines the expected cost for a combined estimation and control problem for a linear system with quadratic cost function. The system may be subject to modeling errors, suboptimal gains, and incorrect prior statistics. The expected prior cost is shown to be a function of the system initial conditions and the solution to a matrix differential equation involving the parameter errors. For simplicity, we will consider only the continuous time case. Discrete time results are established in an analogous manner.

The problem statement for error analysis in combined estimation and control is as follows: A linear dynamic system is described by the equations

$$\dot{\mathbf{x}}(t) = \mathbf{A}(t)\mathbf{x}(t) + \mathbf{B}(t)\mathbf{u}(t) + \mathbf{w}(t) \tag{9.5-1}$$

$$\mathbf{z}(t) = \mathbf{H}(t)\mathbf{x}(t) + \mathbf{v}(t) \tag{9.5-2}$$

$$\mathcal{E}\{\mathbf{x}(t_o)\} = \boldsymbol{\mu}_o$$

$$\text{var}\{\mathbf{x}(t_o)\} = \mathbf{P}_2(t_o) = \mathbf{P}_{20},$$

where $\mathbf{w}(t)$ and $\mathbf{v}(t)$ are zero-mean, uncorrelated with themselves and with $\mathbf{x}(t_o)$, and

$$\text{cov}\{\mathbf{w}(t), \mathbf{w}(\tau)\} = \mathbf{Q}_2(t)\,\delta(t-\tau)$$

$$\text{cov}\{\mathbf{v}(t), \mathbf{v}(\tau)\} = \mathbf{R}_2(t)\,\delta(t-\tau).$$

The system is modeled by

$$\dot{\bar{\mathbf{x}}}(t) = \bar{\mathbf{A}}(t)\bar{\mathbf{x}}(t) + \bar{\mathbf{B}}(t)\mathbf{u}(t) + \bar{\mathbf{w}}(t) \tag{9.5-3}$$

$$\mathbf{z}(t) = \bar{\mathbf{H}}(t)\bar{\mathbf{x}}(t) + \bar{\mathbf{v}}(t) \tag{9.5-4}$$

$$\mathcal{E}\{\bar{\mathbf{x}}(t_o)\} = \bar{\boldsymbol{\mu}}_o$$

$$\text{var}\{\bar{\mathbf{x}}(t_o)\} = \bar{\mathbf{P}}_2(t_o) = \bar{\mathbf{P}}_{20}$$

with incorrect noise moments

$$\text{cov}\{\bar{\mathbf{w}}(t), \bar{\mathbf{w}}(\tau)\} = \bar{\mathbf{Q}}_2(t)\,\delta(t-\tau)$$

$$\text{cov}\{\bar{\mathbf{v}}(t), \bar{\mathbf{v}}(\tau)\} = \bar{\mathbf{R}}_2(t)\,\delta(t-\tau),$$

where the overbars indicate possibly incorrect parameters.

We consider a cost function

$$J = \frac{1}{2}\,\mathcal{E}\left\{\mathbf{x}^T(t_f)\mathbf{D}\mathbf{x}(t_f) + \int_{t_o}^{t_f} [\mathbf{x}^T(t)\mathbf{Q}_1(t)\mathbf{x}(t) + \mathbf{u}^T(t)\mathbf{R}_1(t)\mathbf{u}(t)]\right\}, \quad (9.5\text{-}5)$$

where $\mathcal{E}$ denotes the statistical expectation over all random variables throughout the time interval of interest, and a feedback controller

$$\mathbf{u}(t) = \bar{\mathbf{C}}(t)\hat{\mathbf{x}}(t) \quad (9.5\text{-}6)$$

with $\hat{\mathbf{x}}(t)$ with the estimate of $\mathbf{x}(t)$ given by

$$\dot{\hat{\mathbf{x}}}(t) = [\bar{\mathbf{A}}(t) + \bar{\mathbf{B}}(t)\bar{\mathbf{C}}(t)]\hat{\mathbf{x}}(t) + \bar{\mathbf{K}}(t)[\mathbf{z}(t) - \bar{\mathbf{H}}(t)\hat{\mathbf{x}}(t)], \qquad \hat{\mathbf{x}}(t_o) = \boldsymbol{\mu}_o(t), \quad (9.5\text{-}7)$$

where $\bar{\mathbf{C}}(t)$ and $\bar{\mathbf{K}}(t)$ may be suboptimal gains.

We desire to determine an expression for the expected cost which includes the effects of the modeling and gain errors. Our derivation of a cost algorithm for the stochastic controller proceeds as follows. For the controller given in Eq. (9.5-6), the cost function of Eq. (9.5-5) becomes

$$J = \frac{1}{2}\mathcal{E}\left\{\mathbf{x}^T(t_f)\mathbf{D}\mathbf{x}(t_f) + \int_{t_o}^{t_f} [\mathbf{x}^T(t)\mathbf{Q}_1(t)\mathbf{x}(t) + \hat{\mathbf{x}}^T(t)\bar{\mathbf{C}}^T(t)\bar{\mathbf{R}}_1(t)\bar{\mathbf{C}}(t)\hat{\mathbf{x}}(t)]\,dt\right\}. \quad (9.5\text{-}8)$$

Let

$$\mathbf{X}(t) = \mathcal{E}\left\{\begin{bmatrix}\mathbf{x}(t)\\ \tilde{\mathbf{x}}(t)\end{bmatrix}[\mathbf{x}^T(t), \tilde{\mathbf{x}}^T(t)]\right\} = \mathcal{E}\left\{\begin{bmatrix}\mathbf{x}(t)\mathbf{x}^T(t) & \mathbf{x}(t)\tilde{\mathbf{x}}^T(t)\\ \tilde{\mathbf{x}}(t)\mathbf{x}^T(t) & \tilde{\mathbf{x}}(t)\tilde{\mathbf{x}}^T(t)\end{bmatrix}\right\}, \quad (9.5\text{-}9)$$

where the error is defined, as in our previous efforts, as

$$\tilde{\mathbf{x}}(t) = \mathbf{x}(t) - \hat{\mathbf{x}}(t).$$

The cost function of Eq. (9.5-5) may be rewritten as

$$J = \frac{1}{2}\operatorname{tr}\left\{\begin{bmatrix}\mathbf{S} & \mathbf{0}\\ \mathbf{0} & \mathbf{0}\end{bmatrix}\mathbf{X}(t_f) + \int_{t_o}^{t_f}\left(\begin{bmatrix}\mathbf{Q}_1(t) & \mathbf{0}\\ \mathbf{0} & \mathbf{0}\end{bmatrix}\mathbf{X}(t) + \begin{bmatrix}\bar{\mathbf{C}}^T\mathbf{R}_1\bar{\mathbf{C}} & -\bar{\mathbf{C}}^T\mathbf{R}_1\bar{\mathbf{C}}\\ -\bar{\mathbf{C}}^T\mathbf{R}_1\bar{\mathbf{C}} & \bar{\mathbf{C}}^T\mathbf{R}_1\bar{\mathbf{C}}\end{bmatrix}\mathbf{X}(t)\right)dt\right\}, \quad (9.5\text{-}10)$$

where we will delete time arguments for notational simplicity. Now integration yields

$$\int_0^{t_f}\frac{d}{dt}[\mathbf{S}(t)\mathbf{X}(t)]\,dt = \mathbf{S}(t_f)\mathbf{X}(t_f) - \mathbf{S}(t_o)\mathbf{X}(t_o), \quad (9.5\text{-}11)$$

where $\mathbf{S}(t)$ is a symmetric $2n \times 2n$ matrix. Substituting Eq. (9.5-11) into Eq. (9.5-10) and letting

$$\mathbf{S}(t_f) = \begin{bmatrix}\mathbf{D} & \mathbf{0}\\ \mathbf{0} & \mathbf{0}\end{bmatrix}$$

yields

$$J = \tfrac{1}{2}\text{tr}\left\{\mathbf{S}(t_o)\mathbf{X}(t_o) + \int_{t_o}^{t_f}\left(\begin{bmatrix}\mathbf{Q}_1 & \mathbf{0}\\ \mathbf{0} & \mathbf{0}\end{bmatrix}\mathbf{X}(t) + \begin{bmatrix}\bar{\mathbf{C}}^T\mathbf{R}_1\bar{\mathbf{C}} & -\bar{\mathbf{C}}^T\mathbf{R}_1\bar{\mathbf{C}}\\ -\bar{\mathbf{C}}^T\mathbf{R}_1\bar{\mathbf{C}} & \bar{\mathbf{C}}^T\mathbf{R}_1\bar{\mathbf{C}}\end{bmatrix}\mathbf{X}(t) + \dot{\mathbf{S}}\mathbf{X} + \mathbf{S}\dot{\mathbf{X}}\right)dt\right\}. \tag{9.5-12}$$

From Eq. (9.5-9) we have, by taking time derivatives,

$$\dot{\mathbf{X}} = \boldsymbol{\varepsilon}\left\{\begin{bmatrix}\dot{\mathbf{x}}\mathbf{x}^T + \mathbf{x}\dot{\mathbf{x}}^T & \dot{\mathbf{x}}\tilde{\mathbf{x}}^T + \mathbf{x}\dot{\tilde{\mathbf{x}}}^T\\ \dot{\tilde{\mathbf{x}}}\mathbf{x}^T + \tilde{\mathbf{x}}\dot{\mathbf{x}}^T & \dot{\tilde{\mathbf{x}}}\tilde{\mathbf{x}}^T + \tilde{\mathbf{x}}\dot{\tilde{\mathbf{x}}}^T\end{bmatrix}\right\},$$

where $\dot{\mathbf{x}}$ is given by Eq. (9.5-1). Subtracting Eq. (9.5-7) and Eq. (9.5-1) and using Eq. (9.5-6) results in

$$\dot{\tilde{\mathbf{x}}} = \mathbf{A}\mathbf{x} - \bar{\mathbf{A}}\hat{\mathbf{x}} + \mathbf{B}\bar{\mathbf{C}}\hat{\mathbf{x}} - \overline{\mathbf{BC}}\hat{\mathbf{x}} + \mathbf{w} + \overline{\mathbf{KH}}\hat{\mathbf{x}} - \overline{\mathbf{KH}}\mathbf{x} - \bar{\mathbf{K}}\mathbf{v}$$

$$\tilde{\mathbf{x}}(t_o) = \boldsymbol{\mu}_o - \bar{\boldsymbol{\mu}}_o. \tag{9.5-13}$$

Now let modeling errors be defined by the symbology

$$\Delta\boldsymbol{\Xi} = \boldsymbol{\Xi} - \bar{\boldsymbol{\Xi}} \tag{9.5-14}$$

and substitute Eqs. (9.5-1), (9.5-13), and (9.5-14) into Eq. (9.5-12). This yields:

$$\dot{\mathbf{X}} = \mathbf{F}\mathbf{X} + \mathbf{X}\mathbf{F}^T + \mathbf{M} \tag{9.5-15}$$

with

$$\mathbf{X}(t_o) = \begin{bmatrix}\mathbf{P}_{20} + \Delta\mathbf{P}_{20} + \boldsymbol{\mu}_o\boldsymbol{\mu}_o^T & \mathbf{P}_{20} + \Delta\mathbf{P}_{20} + \boldsymbol{\mu}_o\Delta\boldsymbol{\mu}_o^T\\ \mathbf{P}_{20} + \Delta\mathbf{P}_{20} + \Delta\boldsymbol{\mu}_o\boldsymbol{\mu}_o^T & \mathbf{P}_{20} + \Delta\mathbf{P}_{20} + \Delta\boldsymbol{\mu}_o\boldsymbol{\mu}_o^T\end{bmatrix}, \tag{9.5-16}$$

where

$$F = \begin{bmatrix}\bar{\mathbf{A}} + \overline{\mathbf{BC}} + \Delta\mathbf{A} + \Delta\mathbf{B}\bar{\mathbf{C}} & -\overline{\mathbf{BC}} - \Delta\mathbf{B}\bar{\mathbf{C}}\\ \mathbf{A} + \Delta\mathbf{B}\bar{\mathbf{C}} - \bar{\mathbf{K}}\,\Delta\mathbf{H} & \bar{\mathbf{A}} - \Delta\mathbf{B}\bar{\mathbf{C}} - \overline{\mathbf{KH}}\end{bmatrix} \tag{9.5-17}$$

and

$$M = \begin{bmatrix}\mathbf{Q}_2 + \Delta\mathbf{Q}_2 & \mathbf{Q}_2 + \Delta\mathbf{Q}_2\\ \mathbf{Q}_2 + \Delta\mathbf{Q}_2 & \mathbf{Q}_2 + \Delta\mathbf{Q}_2 + \mathbf{K}\mathbf{R}_2\mathbf{K}^T + \mathbf{K}\,\Delta\mathbf{R}_2\mathbf{K}^T\end{bmatrix} \tag{9.5-18}$$

Thus it follows that

$$J = \tfrac{1}{2}\,\text{tr}\left\{\mathbf{S}(t_o)\mathbf{X}(t_o) + \int_{t_o}^{t_f}\left(\begin{bmatrix}\mathbf{Q}_1 & \mathbf{0}\\ \mathbf{0} & \mathbf{0}\end{bmatrix}\mathbf{X} + \begin{bmatrix}\mathbf{C}^T\mathbf{R}_1\mathbf{C} & -\mathbf{C}^T\mathbf{R}_1\mathbf{C}\\ -\mathbf{C}^T\mathbf{R}_1\mathbf{C} & \mathbf{C}^T\mathbf{R}_1\mathbf{C}\end{bmatrix}\mathbf{X} + \dot{\mathbf{S}}\mathbf{X} + \mathbf{S}\mathbf{F}\mathbf{X} + \mathbf{F}^T\mathbf{S}\mathbf{X} + \mathbf{S}\mathbf{M}\right)dt\right\}. \tag{9.5-19}$$

So if, for convenience, the coefficient of $\mathbf{X}$ is set to zero in Eq. (9.5-19), the resulting cost function is given by

$$J = \tfrac{1}{2}\text{tr}\left\{\mathbf{S}(t_o)\mathbf{X}(t_o) + \int_{t_o}^{t_f}(\mathbf{S}\mathbf{M})\,dt\right\} \tag{9.5-20}$$

$$\dot{\mathbf{S}} = -\mathbf{SF} - \mathbf{F}^T\mathbf{S} - \begin{bmatrix} \mathbf{Q}_1 & \mathbf{0} \\ \mathbf{0} & \mathbf{0} \end{bmatrix} - \begin{bmatrix} \bar{\mathbf{C}}^T\mathbf{R}_1\bar{\mathbf{C}} & -\bar{\mathbf{C}}^T\mathbf{R}_1\bar{\mathbf{C}} \\ -\bar{\mathbf{C}}^T\mathbf{R}_1\bar{\mathbf{C}} & \bar{\mathbf{C}}^T\mathbf{R}_1\bar{\mathbf{C}} \end{bmatrix} \tag{9.5-21}$$

$$\mathbf{S}(t_f) = \begin{bmatrix} \mathbf{D} & \mathbf{0} \\ \mathbf{0} & \mathbf{0} \end{bmatrix}.$$

Expanding Eq. (9.5-21) yields

$$\begin{aligned}
\dot{\mathbf{S}}_{11} = &-\mathbf{S}_{11}(\bar{\mathbf{A}} + \overline{\mathbf{BC}}) - (\bar{\mathbf{A}} + \overline{\mathbf{BC}})^T\mathbf{S}_{11} - \mathbf{S}_{11}(\Delta\mathbf{A} + \Delta\overline{\mathbf{BC}}) \\
&- (\Delta\mathbf{A} + \Delta\mathbf{B}\bar{\mathbf{C}})^T\mathbf{S}_{11} - \mathbf{S}_{12}(\Delta\mathbf{A} + \Delta\mathbf{B}\bar{\mathbf{C}} - \bar{\mathbf{K}}\,\Delta\mathbf{H}) \\
&- (\Delta\mathbf{A} + \Delta\mathbf{B}\bar{\mathbf{C}} - \bar{\mathbf{K}}\,\Delta\mathbf{H})^T\mathbf{S}_{12}^T - \mathbf{Q}_1 - \bar{\mathbf{C}}^T\mathbf{R}_1\bar{\mathbf{C}}
\end{aligned}$$

$$\mathbf{S}_{11}(t_f) = \mathbf{D} \tag{9.5-22}$$

$$\begin{aligned}
\dot{\mathbf{S}}_{12} = &\ \mathbf{S}_{11}\overline{\mathbf{BC}} + \mathbf{S}_{11}\,\Delta\overline{\mathbf{BC}} - \mathbf{S}_{12}(\mathbf{A} - \Delta\mathbf{B}\bar{\mathbf{C}} - \overline{\mathbf{KH}}) \\
&- (\bar{\mathbf{A}} + \overline{\mathbf{BC}})^T\mathbf{S}_{12} - (\Delta\mathbf{A} + \Delta\mathbf{B}\bar{\mathbf{C}})^T\mathbf{S}_{12} \\
&- (\Delta\mathbf{A} + \Delta\mathbf{B}\bar{\mathbf{C}} - \bar{\mathbf{K}}\,\Delta\mathbf{H})^T\mathbf{S}_{22} + \bar{\mathbf{C}}^T\mathbf{R}_1\bar{\mathbf{C}}
\end{aligned}$$

$$\mathbf{S}_{12}(t_f) = \mathbf{0}. \tag{9.5-23}$$

$$\begin{aligned}
\dot{\mathbf{S}}_{22} = &\ \mathbf{S}_{12}^T\overline{\mathbf{BC}} + (\overline{\mathbf{BC}})^T\mathbf{S}_{12} + \mathbf{S}_{12}^T\,\Delta\mathbf{B}\bar{\mathbf{C}} + (\Delta\mathbf{B}\bar{\mathbf{C}})^T\mathbf{S}_{12} \\
&- \mathbf{S}_{22}(\bar{\mathbf{A}} - \Delta\mathbf{B}\bar{\mathbf{C}} - \overline{\mathbf{KH}}) - (\bar{\mathbf{A}} - \Delta\mathbf{B}\bar{\mathbf{C}} - \overline{\mathbf{KH}})^T\mathbf{S}_{22} - \bar{\mathbf{C}}^T\mathbf{R}_1\mathbf{C}
\end{aligned}$$

$$\mathbf{S}_{22}(t_f) = \mathbf{0}. \tag{9.5-24}$$

These are the desired results. Equations (9.5-22) through (9.5-24) are solved backward in time from t_f to t_o, and Eq. (9.5-20) is used to compute the cost for the stochastic controller implemented with incorrect parameters, prior statistics and/or gain matrices. From this general result, several specialized cases follow. For example, if there are no parameter errors and the optimal values of $\mathbf{C}$ and $\mathbf{K}$ are used, Eq. (9.5-20) reduces to that previously derived, Eq. (9.3-14).

Example 9.5-1. We consider the first-order system

$$\dot{x} = -0.06x + 0.4u + w \qquad \mu_0 = \mathcal{E}\{x(t_0)\} = 5.2$$

$$z = 1.6x + v \qquad P_{20} = \text{var}\{x(t_0)\} = 4.5,$$

where w and v are zero-mean white Gaussian noises with $Q_2 = 4.2$, $R_2 = 3.7$. Suppose that the system model is

$$\dot{\bar{x}} = -0.1\bar{x} + u + \bar{w} \qquad \bar{\mu}_0 = 3.0$$

$$z = 2.0\bar{x} + \bar{v} \qquad \bar{P}_{20} = 2.0$$

and

$$Q_2 = 1.0 \qquad R_2 = 1.0$$

with cost function

$$J = \tfrac{1}{2}\mathcal{E}\{4x^2(t_f) + \int_{t_0}^{t_f} (x^2 + 10u^2)\,dt\}.$$

Equations (9.5-20) and (9.5-21) may now be used to determine the expected cost.

Four examples were considered, representing various modeling errors and gain approximations. Example A is the baseline case and contains no errors; that is, the model parameters are the same as the actual system parameters, and the optimal gains are used. The result is the optimum cost, and this is the system which would result from use of the separation theorem. Example C considers the effects of using the incorrect model. The appropriate equations are used to compute the control and filter gains but these equations contain the (incorrect) model parameters.

The control gain C must be computed backward in time from t_f and stored. If this is inconvenient, the control gain may be approximated by its constant steady state value. The filter gain K is computed forward in time so it may be conveniently computed on-line. This design approach is considered in Examples B and D for the correct and incorrect system models, respectively. The results of these four examples follow.

Example A:

Correct model parameters, optimal gains. This is the optimal, baseline case. Cost = 491.1.

Example B:

Correct model parameters, optimal filter gain, constant control gain (steady state value from Example A). Cost = 548.8.

Example C:

Incorrect model parameters; all gains computed with incorrect model parameters. Cost = 512.3.

Example D:

Incorrect model parameters; filter gain computed with incorrect model parameters; constant control gain (steady-state value from Example C). Cost = 519.0.

These examples demonstrate the use of the algorithm for several different situations. Examples C and D are typical of realistic process control problems —determining the added cost of modeling errors and gain approximations. Thus, an algorithm has been presented which determines the expected cost for a combined estimation and control problem with modeling errors. This algorithm is also useful in determining the cost of suboptimal gain approximations. If a realistic estimate of the error in some parameter yields an unacceptably high cost, then more effort can be directed toward precise identification of that system parameter. This work may be extended to the discrete case and to the case with correlated, nonzero-mean, colored noises in a straightforward but somewhat tedious manner.

In this section, we have examined sensitivity analysis concepts for continuous time linear stochastic control systems and have examined the use of the developed algorithms for a single example. Reference [22] extends these results to the case where a reduced order observer is used instead of the Kalman estimator. Although tedious to develop, sensitivity analysis algorithms of the type developed here can be of much utility in evaluating the effect of intentional suboptimizations and in determining the accuracy with which system parameter coefficients must be known.

References

1. KALMAN, R.E., ENGLAR, T.S., and BUCY, R.S., "Fundamental Study of Adaptive Control Systems." Wright-Patterson Air Force Base Tech. Rept., ASD-TDR-61-27, April 1962.
2. JOSEPH P.D., and TOU, J., "On Linear Control Theory." *Trans. AIEE*, 80, (1961), 193–96.
3. GUNCKEL, T.L., and FRANKLIN, G.F., "A General Solution for Linear, Sampled-data Control." *Trans. ASME*, Ser. D, 85, (1963), 197–203.
4. *Special Issue on Linear-Quadratic-Gaussian Problem*, *IEEE-AC*, **AC-16**, (December 1971).
5. WONHAM, W.M., "On the Separation Theorem of Stochastic Control." *SIAM J. Control*, **6**, (1968), 312–26.
6. ASTROM, K.J., *Introduction to Stochastic Control Theory*. Academic Press, New York, 1970.
7. BAR-SHALOM, Y., and TSE, E., "Dual Effect, Certainty Equivalence, and Separation in Stochastic Control." *IEEE-AC*, **AC-19**, (October 1974), 494–500.
8. DREYFUS, S.E., *Dynamic Programming and the Calculus of Variations*. Academic Press, New York, 1965.
9. FELDBAUM, A.A., *Optimal Control Systems*. Academic Press, New York, 1965.
10. STREIBEL, C., "Sufficient Statistics in the Optimal Control of Stochastic Systems." *J. Math. Anal. Appl.*, **12**, (1965), 576–92.
11. SAWARAGI, Y., and YOSHIKAWA, T., "Discrete-time Markovian Decision Processes with Incomplete State Information." *Annals Math. Statistics*, **41**, (1970), 78–86.
12. ATHANS, M., "The Role and Use of the Stochastic Linear-Quadratic-Gaussian Problem in Control System Design." *IEEE-AC*, **AC-16**, (December 1971), 529–51.
13. TSE, E., BAR-SHALOM, Y., and MEIER, L., III, "Wide-sense Adaptive Dual Control for Nonlinear Stochastic Systems." *IEEE-AC*, **AC-18**, (April 1973), 98–108.
14. SPEYER, J., DEYST, J., and JACOBSON, D., "Optimization of Stochastic Linear Systems With Additive Measurement and Process Noise Using Exponential Performance Criteria." *IEEE-AC*, **AC-19**, (August 1974), 358–66.
15. TSE, E., "On the Optimal Control of Stochastic Linear Systems." *IEEE-AC*, **AC-16**, (December 1971), 776–84.
16. MELSA, J.L., and SAGE, A.P., *An Introduction to Probability and Stochastic Processes*. Prentice-Hall, Inc., Englewood Cliffs, New Jersey, 1973.

17. SAGE, A.P., and MELSA, J.L., *Estimation Theory With Applications to Communications and Control.* McGraw-Hill Book Co., New York, 1971.

18. TSE, E., "On the Optimal Control of Linear Systems With Incomplete Information." Electronic Systems Laboratory, M.I.T., Cambridge, Massachusetts, ESL-R-412, January 1970.

19. WITSENHAUSEN, H.S., "A Counterexample in Stochastic Optimum Control." *SIAM J. Control*, **6**, (1968), 131–47.

20. CHONG, C.-Y., and ATHANS, M., "On the Periodic Coordination of Linear Stochastic Systems." *Automatica*, **12**, No. 3, (July 1976).

21. WITSENHAUSEN, H.S., "Separation of Estimation and Control, for Discrete-time Systems." *Proceedings IEEE*, **59**, (November 1971), 1557–66.

22. CARLOCK, G.W., and SAGE, A.P., "Sensitivity and Error Analysis Algorithms for Combined Estimation and Control Systems." *Int. J. Control*, 24, no. 3, (1975), 417–41.

Problems

1. Determine the optimal control for the system

$$\dot{x}_1 = x_2, \qquad \dot{x}_2 = u, \qquad z = x_1$$

$$J = \mathcal{E}\left\{\mathbf{x}^T(t_f)\begin{bmatrix} s_{11} & 0 \\ 0 & s_{22} \end{bmatrix}\mathbf{x}(t_f) + \int_{t_0}^{t_f} u^2(\tau)\,d\tau\right\},$$

where

$$r_2 = 1, \qquad \mathbf{Q}_2 = \begin{bmatrix} 0 & 0 \\ 0 & 1 \end{bmatrix}, \qquad \mathbf{P}_2(t_o) = \mathbf{0}, \qquad \mathbf{x}^T(t_o) = [1, 0].$$

Discuss the behavior of the resulting system as a function of **S** and t_f.

2. Derive and discuss in detail the requirements for the combined estimation and control of the system,

$$\dot{x}_1 = -x_1 + x_2 + w_1, \qquad \dot{x}_2 = u + w_2, \qquad z = x_2 + v$$

$$\bar{\mathbf{x}}(t_o) = \mathbf{x}_o, \qquad r_2 = 1, \qquad \mathbf{Q}_2 = \begin{bmatrix} 1 & 0 \\ 0 & 1 \end{bmatrix}, \qquad \mathbf{P}_2(t_o) = \begin{bmatrix} p_{11}(t_o) & 0 \\ 0 & p_{22}(t_o) \end{bmatrix},$$

so as to minimize

$$J = \mathcal{E}\left\{\int_{t_0}^{t_f}\left(\left\|\mathbf{x}^2(\tau)\right\|_{Q_1}^2 + u^2(\tau)\right)d\tau\right\},$$

where

$$\text{(a)} \quad \mathbf{Q}_1 = \begin{bmatrix} 1 & 0 \\ 0 & 1 \end{bmatrix}, \qquad \text{(b)} \quad \mathbf{Q}_1 = \begin{bmatrix} 0 & 0 \\ 0 & 1 \end{bmatrix}.$$

3. Determine the estimation and control algorithms for the system

$$\dot{\mathbf{x}}(t) = \mathbf{A}(t)\mathbf{x}(t) + \mathbf{B}(t)\mathbf{u}(t)$$
$$\mathbf{z}(t) = \mathbf{C}(t)\mathbf{x}(t) + \mathbf{D}(t)\mathbf{u}(t),$$

having cost function Eq. (9.3-3).

4. Show that for correlated plant and measurement noise $\mathcal{E}\{\mathbf{w}(t)\mathbf{v}^T(\tau)\} = \mathbf{S}(t)\delta(t-\tau)$, the Kalman filter equations become:

$$\dot{\hat{\mathbf{x}}}(t) = \mathbf{F}(t)\hat{\mathbf{x}}(t) + \mathbf{K}(t)[\mathbf{z}(t) - \mathbf{H}(t)\hat{\mathbf{x}}(t)]$$

$$\mathbf{K}(t) = [\mathbf{P}(t)\mathbf{H}^T(t) + \mathbf{G}(t)\mathbf{S}(t)]\mathbf{R}^{-1}(t)$$

$$\dot{\mathbf{P}}(t) = [\mathbf{F}(t) - \mathbf{G}(t)\mathbf{S}(t)\mathbf{R}^{-1}(t)\mathbf{H}(t)]\mathbf{P}(t) + \mathbf{P}(t)[\mathbf{F}(t) - \mathbf{G}(t)\mathbf{S}(t)\mathbf{R}^{-1}(t)\mathbf{H}(t)]^T$$
$$- \mathbf{P}(t)\mathbf{H}^T(t)\,\mathbf{R}^{-1}(t)\mathbf{H}(t)\mathbf{P}(t) + \mathbf{G}(t)[\mathbf{Q}(t) - \mathbf{S}(t)\mathbf{R}^{-1}(t)\mathbf{S}^T(t)]\mathbf{G}^T(t).$$

How is the combined estimation and control problem affected by the correlated plant and measurement noise?

5. Determine the combined estimation and control algorithms for both the discrete and continuous time case if:
   (a) there is no plant noise,
   (b) there is no measurement noise,
   (c) $\mathbf{P}(t_o) = \mathbf{P}(k_o) = \mathbf{0}$,
   (d) the system is completely unobserved; i.e., $\mathbf{C}(t) = \mathbf{C}(k) = \mathbf{0}$.

6. Determine the control and estimation algorithms for the discrete time problem when $\mathbf{u}(k) = \boldsymbol{\alpha}[k, \mathbf{Z}(k-m)]$, where m is a fixed integer.

7. In Example 9.4-1, let $\mathbf{x}(k) = \mathbf{x}(t_k)$, where $t_o = 0$, $t_{k+1} - t_k = 1$, and $t_f = t_{k_f}$. Approximate $\dot{\mathbf{x}}(t_k)$ by $x(t_{k+1}) - x(t_k)$, and solve for the associated discrete LQG problem. Compare your results to the continuous solution.

8. Verify Eq. (9.3-19).

9. Show that substitution of Eq. (9.3-20) into Eq. (9.3-19) implies that $\mathbf{P}_1(t)$ satisfies Eq. (9.3-13) and that $\eta(t)$ satisfies Eq. (9.3-21).

10. Show that Eq. (9.5-20) does in fact reduce to Eq. (9.3-14) for the case of no modeling errors. (Hint: Assume that $\mathbf{S}_{12} = 0$ and show that this does occur if $\mathbf{S}_{11} = \mathbf{P}_{10}$).

11. Conduct a sensitivity analysis study of Problem 1 similar to Example 9 5-1.

Computational methods in optimum systems control 10

In most of our work thus far we have used variational calculus, including the maximum principle, to reduce optimum systems control problems to two-point boundary-value differential or difference equations. This set of difference or differential equations normally may not be solved in a straightforward manner because of the occurrence of two-point boundary conditions rather than initial conditions. We have studied several problems which are notable exceptions to this statement in Chapter 5. In the majority of cases, however, solution of two-point boundary-value problems (TPBVP) is an extremely difficult task.

The difficulties in solving the two-point boundary-value proboems have led to a search for variational methods of a different kind, known as *direct methods*, which hopefully circumvent the problems associated with TPBVP solution by changing the variational problem into a problem of ordinary maxima and minima.

Perhaps the oldest techniques which are commonly used for this purpose are the Rayleigh-Ritz and the finite difference methods [1, 2, 3]. These methods are based on finding a sequence of functions which give successively smaller values to the functional to be minimized. In the Ritz method, for example, the trajectory and the control are expanded in terms of a weighted sum of a suitable set of functions, and a minimizing set of coefficients for these functions is found. These methods apparently have not been too popular due to the difficulties in finding a suitable set of basis functions and in determining the number of terms to use in the expansion.

Another direct method is that of *discrete dynamic programming* [4, 5, 6, 7]. Discrete dynamic programming is, in effect, the repeated, sequential, stage-by-stage application of the Hamilton-Jacobi equations or the optimality principle of Bellman. Discrete dynamic programming has an advantage of actually being simplified by the addition of control and state-space constraints. Its biggest disadvantage is the requirement for a truly fantastic amount of computer memory for any but very low order problems. We will give brief consideration to the polynomial approximation method of Bellman [7, 8] and the state increment method of Larson [9] which are two methods for the reduction of the memory requirements in the computational solution of optimal control problems by discrete dynamic programming.

A third and significantly different type of direct method is the *gradient method* or *method of steepest descent.* The gradient method consists of searching for an optimum by making each new iterate move in the direction of the gradient which points locally in the direction of the maximum decrease (or increase) of the cost function. The method was developed and applied to optimal control problems by Kelley [10, 11], and Bryson and Denham [12]. Various modifications to include penalty functions and other methods of treating equality and inequality constraints have been presented [13, 14, 15, 16, 17, 18, 19]. We shall study the approaches of Kelley, as well as Bryson and Denham, for continuous decision processes, as well as single-stage and multi-stage decision processes or discrete time systems.

These gradient methods have the ability to generate successively improved trajectories with very poor starting values. However, as we shall see, they tend to converge slowly as convergence is approached and, therefore, often require the selection and adjustment of several convergence parameters.

The gradient method is a first-order method since it is based on finding first-order effects of controls upon terminal constraints and the cost function. To improve the convergence of the gradient method near the optimal trajectory, second-order terms can be added. Due to the similarity of these terms to the terms present in the second variation, these methods are known as *methods based on second variations.* They do accelerate convergence of the gradient method, but as we shall see, they require good initial estimates of the initial control and trajectory for convergence.

Kelley [20], Breakwell, Speyer, and Bryson [21], Kelley, Kopp, and Moyer [22], and Merriam [23] have contributed to the development of this second variation method. A comparison of the second variation with other computational techniques is presented by Kopp and McGill, and Moyer and Pinkham [18, 19]. We shall summarize this method and point out its usefulness both as a means of control-law computation and as a means of generating a neighboring optimal closed-loop control.

The method of quasilinearization for the resolution of boundary-value problems arising in the solution of nonlinear differential equations has been widely developed and applied by Bellman and Kalaba, et al. in various works [24–27]. A formulation of the method for the solution of two-point boundary-value problems with inequality constraints and variable terminal time has been discussed by McGill and Kenneth [28] who refer to the method, perhaps more properly, as the generalized Newton-Raphson technique. Papers by Kopp and McGill [18], and Moyer and Pinkham [19], attempt to compare this approach with other iterative methods. Kumar and Sridhar [29], and Detchmendy and Sridhar [30], have considered the application of quasilinearization to the parameter identification problem in the noiseless case as well as in the case where measurement errors are present. Sage and Eisenberg [31] consider the estimation of time-varying parameters by means of quasilinearization associated with linear regression techniques as well as the use of the technique to model high-order systems with lower-order models. Eisenberg and Sage [32] use the method for the closed-loop optimization of fixed configuration systems.

Application of this technique to discrete systems appears to have been lightly treated. Henrici [33] discusses this approach in relation to a finite difference scheme for solution of a class of nonlinear boundary-value problems of second order, and he offers a proof for convergence of the proposed scheme. A similar approach to the solution of two-point boundary-value problems via finite difference techniques and use of quasilinearization has been presented by Sylvester and Meyer [34]. Sage and Burt [35], and Sage and Smith [36], introduced a formulation for the problem of parameter estimation and modeling in discrete systems.

There are, as might be expected, many other computational methods that have proven to be of value in optimum systems control. For example, the method of continuous time and discrete time-invariant imbedding has proven to be of considerable value in resolving certain problems involving stochastic nonlinear systems [37, 38]. Various mathematical programming algorithms [39–43] have also been useful in resolving various types of optimal control problems.

10.1 Discrete dynamic programming

Let us consider a very simple optimal control problem which we already know how to solve by a variety of methods—the Euler-Lagrange equations, the Pontryagin maximum principle, and the Hamilton-Jacobi-Bellman equations of continuous dynamic programming. We desire to minimize

$$J = \tfrac{1}{2}Sx^2(t_f) + \tfrac{1}{2}\int_{t_0}^{t_f} u^2(t)\,dt \tag{10.1-1}$$

for the system

$$\dot{x} = u(t), \qquad x(0) = x_o. \tag{10.1-2}$$

Let us call the initial state of the system $x(t_o) = c$. We may then define the functional

$$V(c, t_o) = \min_u \tfrac{1}{2} \int_{t_o}^{t_f} u^2(t)\,dt, \qquad V(c, t_f) = \tfrac{1}{2} S x^2(t_f), \tag{10.1-3}$$

where we see that $V(c, t_o)$, even though it is a functional of our control variable $u(t)$, is a function only of the initial state c and the time duration $t_f - t_o$ when the control $u(t)$ is optimal, $u(t) = \hat{u}(t)$, and where t_f is assumed fixed and known. As we have done before in our work with continuous dynamic programming, we rewrite $V(c, t_o)$ as

$$V(c, t_o) = \min_u \left[\tfrac{1}{2} \int_{t_o}^{t_o+\Delta t} u^2(t)\,dt + \tfrac{1}{2} \int_{t_o+\Delta t}^{t_f} u^2(t)\,dt \right]\cdot \tag{10.1-4}$$

By virtue of the definition of $V(c, t_o)$, and the optimality principle which we will state momentarily, this last equation becomes

$$V(c, t_o) = \min_u \left[\tfrac{1}{2} \int_{t_o}^{t_o+\Delta t} u^2(t)\,dt + V\left(c + \int_{t_o}^{t_o+\Delta t} \dot{x}(t)\,dt, t_o + \Delta t\right)\right]\cdot \tag{10.1-5}$$

In effect, we are asserting that we have decomposed our problem into two stages, the first stage lasting from t_o to $t_o + \Delta t$ sec, where the optimal control advances the system state c to $c + \int_{t_o}^{t_o+\Delta t} \dot{x}\,dt$, and where $t_f - \Delta t - t_o$ sec remain for the optimization. From $t_o + \Delta t$ sec until the terminal time, the system is controlled optimally with cost $V\left(c + \int_{t_o}^{t_o+\Delta t} \dot{x}(t)\,dt, t_o + \Delta t\right)\cdot$

We now consider that Δt is small enough so that the right-hand side of Eq. (10.1-5) may be expanded in a Taylor series about $\Delta t = 0$ which yields, for Eq. (10.1-5),

$$V(c, t_o) = \min_u \left[\frac{1}{2} u^2(t_o)\,\Delta t + V(c, t_o) + \frac{\partial V}{\partial c} \dot{x}(t_o)\,\Delta t + \frac{\partial V}{\partial t_o}\,\Delta t + \cdots \right]. \tag{10.1-6}$$

We now subtract $V(c, t_o)$ from both sides of this result, divide by Δt, and then take the limit as Δt becomes zero to obtain, since $\dot{x} = u$,

$$\frac{\partial V}{\partial t_o} + \min_u \left[\frac{\partial V}{\partial c} u(t_o) + \frac{1}{2} u^2(t_o) \right] = 0. \tag{10.1-7}$$

We immediately recognize this equation as the Hamilton-Jacobi-Bellman equation, Eq. (4.5-4) or (4.5-5). We can easily minimize this expression with respect to $u(t)$ to obtain $u(t_o) = -\partial V(c, t_o)/\partial c$ such that Eq. (10.1-7) becomes the Hamilton-Jacobi equation

$$\frac{\partial V(c, t_o)}{\partial t_o} - \frac{1}{2}\left[\frac{\partial V(c, t_o)}{\partial c}\right]^2 = 0, \qquad V(c, t_f) = \frac{1}{2} S x^2(t_f). \tag{10.1-8}$$

In general, an equation of this sort would be most difficult to solve. Because of the linearity of the system dynamics and the quadratic nature of the cost function, a solution for this particular case is quite simple. We assume $V(c, t_o) = \frac{1}{2}c^2p(t_o)$ and substitute the solution into Eq. (10.1-8), which immediately yields

$$\dot{p} = p^2(t), \qquad p(t_f) = S. \tag{10.1-9}$$

This equation is solved backward in time from $t = t_f$ to $t = t_o$, with the given condition at $t = t_f$, to obtain

$$p(t) = \frac{S}{1 + S(t_f - t)}. \tag{10.1-10}$$

We immediately know the cost function from the assumed solution, as well as the closed-loop control and optimum trajectory

$$V(x_o, t_o) = \frac{x^2(t_o)S}{2[1 + S(t_f - t_o)]}, \qquad \hat{u}(t) = -p(t)\hat{x}(t),$$
$$\hat{x}(t) = \left[\frac{1 + S(t_f - t)}{1 + S(t_f - t_o)}\right]x(t_o). \tag{10.1-11}$$

Thus we see that dynamic programming is indeed a direct method in that it attempts to directly minimize the given cost function without obtaining a two-point boundary-value problem. Unfortunately, the partial differential equation needed to do this is not often solved. A possible means of solution, which we discussed in Sec. 4.5 is to assume a series solution $V(c, t) = p_o(t) + cp_1(t) + c^2p_2(t) + \cdots$.

Another method which we wish to explore in this section consists essentially of digitizing Eq. (10.1-7) and is known as discrete dynamic programming. We will first consider discrete dynamic programming for the very simple problem we are considering here, and we will then make some general observations. We begin by digitizing our cost function and system dynamics, Eqs. (10.1-1) and (10.1-2), to obtain

$$J = \frac{1}{2}Sx^2(k_f) + \frac{T}{2}\sum_{k=k_o}^{k_f-1} u^2(k), \qquad x(k+1) = x(k) + Tu(k),$$
$$x(k_o) = x_o, \tag{10.1-12}$$

where $t_o = k_oT$ and $t_f = k_fT$ such that there are $k_f - k_o$ intervals of constant length T from t_o to t_f. This is a $(k_f - k_o)$-stage decision process since that many values of the control, $u(k_o)$, $u(k_o + 1) \cdots u(k_f - 1)$, are negotiable or subject to adjustment. Let us assume for convenience that we have a two-stage decision process such that $k_o = 0$, $k_f = 2$. The cost function, Eq. (10.1-12), becomes

$$J = \frac{1}{2}Sx^2(2) + \frac{T}{2}\sum_{h=0}^{1} u^2(k) = \frac{1}{2}Sx^2(2) + \frac{T}{2}u^2(0) + \frac{T}{2}u^2(1), \tag{10.1-13}$$

and we see that we have only two input control decisions to make.

Suppose that a choice of $u(0)$ has been made and that the corresponding system state after application of $u(0)$ is $x(1)$. Then there is only one control input $u(1)$ to adjust. There is no point, therefore, in including $u(0)$ in the cost function, and Eq. (10.1-13) becomes

$$J_1 = \frac{1}{2}Sx^2(2) + \frac{T}{2}u^2(1) = \frac{1}{2}S[x(1) + Tu(1)]^2 + \frac{T}{2}u^2(1). \quad (10.1\text{-}14)$$

We now desire to determine $u(1)$ to minimize Eq. (10.1-14), which is easily found as

$$\frac{\partial J_1}{\partial u(1)} = 0 = ST[x(1) + Tu(1)] + Tu(1), \qquad \hat{u}(1) = -\frac{Sx(1)}{1 + ST}. \quad (10.1\text{-}15)$$

We many consider this expression for the control $\hat{u}(1)$ as a fixed-decision rule defined for all possible states $x(1)$, and optimal for the last stage of operation. The corresponding minimum value of J_1 is then

$$J_{1\,\min} = \frac{S}{2}\frac{x^2(1)}{1 + ST}. \quad (10.1\text{-}16)$$

We now move back one stage and obtain the total cost for this case,

$$\begin{aligned} J &= \frac{T}{2}u^2(0) + J_{1\,\min} = \frac{T}{2}u^2(0) + \frac{S}{2}\frac{x^2(1)}{1 + ST} \\ &= \frac{T}{2}u^2(0) + \frac{S}{2}\frac{[x(0) + Tu(0)]^2}{1 + ST}. \end{aligned} \quad (10.1\text{-}17)$$

Our next step is to minimize Eq. (10.1-17) with respect to a choice of $u(0)$. This is easily accomplished, and we obtain

$$\hat{u}(0) = -\frac{Sx(0)}{1 + 2ST}, \qquad J_{\min} = \frac{Sx^2(0)}{2[1 + 2ST]} \quad (10.1\text{-}18)$$

for the optimum control and total minimum cost. The optimum trajectory is

$$x(0) = x(0), \qquad \hat{x}(1) = \frac{1 + ST}{1 + 2ST}x(0), \qquad \hat{x}(2) = \frac{\hat{x}(1)}{1 + ST} = \frac{x(0)}{1 + 2ST}. \quad (10.1\text{-}19)$$

Interestingly enough, the control and trajectory are the same for this dicrete version as they are for the original continuous problem. This is only because the optimum continuous control is a constant.

Each step in this optimuzation procedure involves minimization with respect to one variable only, whereas use of the discrete maximum principle of Chapter 6 would require, in effect, simultaneous minimization with respect to two variables. Let us now attempt to repeat this same procedure for a problem in which the Pontryagin maximum principle would lead to a nonlinear two-point boundary-value problem. Specifically, let us consider the cost function and system dynamics,

$$\begin{gathered} \dot{x} = -x^3 + u(t), \qquad x(0) = 1.0, \qquad x(1) = 0, \\ J = \tfrac{1}{2}\int_0^1 u^2(t)\,dt, \end{gathered} \quad (10.1\text{-}20)$$

which may be digitized into a four-stage decision process as

$$x(k+1) = x(k) - \frac{x^3(k)}{4} + \frac{u(k)}{4}, \qquad J = \frac{1}{8}\sum_{k=0}^{3} u^2(k). \tag{10.1-21}$$

Again we assume that we have applied $u(0)$, $u(1)$, $u(2)$, and now wish to determine the final stage control $u(3)$. The cost function becomes

$$J_3 = \tfrac{1}{8}u^2(3) = \tfrac{1}{8}[x^3(3) - 4x(3)]^2 \tag{10.1-22}$$

since we must require $x(4) = 0$. Thus the final stage control is actually fixed by the terminal equality constraint. We back up one stage and write for the cost from the second stage to the end,

$$J_2 = \tfrac{1}{8}u^2(2) + J_3 = \tfrac{1}{8}u^2(2) + \tfrac{1}{8}[x^3(3) - 4x(3)]^2. \tag{10.1-23}$$

But from the discrete system equation, this becomes

$$J_2 = \frac{1}{8}u^2(2) + \frac{1}{8}\left\{\left[x(2) - \frac{x^3(2)}{4} + \frac{u(2)}{4}\right]^3 - 4\left[x(2) - \frac{x^3(2)}{4} + \frac{u(2)}{4}\right]\right\}^2. \tag{10.1-24}$$

We minimize this with respect to a choice of $u(2)$, which results in an exceedingly difficult algebraic equation. We do see, though, that the best control $\hat{u}(2)$ will be a feedback control and will involve only $x(2)$. We then back up to stage 1, and then finally stage 0, to complete the problem. Clearly, even for this simple problem, a hand solution is not feasible. To get an answer close to the answer to the continuous problem, we may well have to consider 10 or perhaps even 100 stages. Thus machine solution appears to be the only possible approach.

We now turn to the general theory of discrete dynamic programming, as developed by Bellman, which is based on the following two principles which we already have used in our work in this section. The first of these is the *imbedding principle* which states that a problem with fixed initial states and a fixed interval of operation may always be viewed as a special case of a more general problem with a variable initial state and a variable operation time. Thus we may imbed the original problem into a whole class of problems whose solution inevitably provides the solution to our particular problem. A topic closely related to this, the invariant imbedding method, is discussed in [39, 40]. The other principle upon which dynamic programming is based is the *principle of optimality* which states that an optimal policy has the property that, whatever any initial states and decisions are, all remaining decisions must constitute an optimal policy with regard to the state which results from the first decision.

The general optimization problem consists of miminizing a cost function J:

$$J = \theta[\mathbf{x}(t_f), t_f] + \int_{t_r}^{t_f} \phi[\mathbf{x}(t), \mathbf{u}(t), t]\, dt \tag{10.1-25}$$

for a dynamic system

$$\dot{\mathbf{x}} = \mathbf{f}[\mathbf{x}(t), \mathbf{u}(t), t], \qquad \mathbf{x}(t_o) = \mathbf{x}_o. \tag{10.1-26}$$

We may discretize these to obtain

$$J = \theta[\mathbf{x}(k_f), k_f] + \sum_{k=k_o}^{k_f-1} \phi'[\mathbf{x}(k), \mathbf{u}(k), k] \tag{10.1-27}$$

$$\mathbf{x}(k+1) = \mathbf{f}'[\mathbf{x}(k), \mathbf{u}(k), k], \qquad \mathbf{x}(k_o) = \mathbf{x}_o. \tag{10.1-28}$$

The Hamilton-Jacobi-Bellman equations are:

$$\frac{\partial V[\mathbf{c}, t]}{\partial t} + \min_{\mathbf{u}} \left\{\phi(\mathbf{c}, \mathbf{u}, t) + \left[\frac{\partial V}{\partial \mathbf{c}}\right]^T \mathbf{f}(\mathbf{c}, \mathbf{u}, t)\right\} = 0, \qquad V[\mathbf{c}, t_f] = \theta[\mathbf{x}(t_f), t_f], \tag{10.1-29}$$

where V is the cost function or functional to be minimized and $\mathbf{f}$ is the associated "policy function" in dynamic programming terminology. By descretizing this equation or by employing the principle of optimality, we obtain the recurrence relation

$$V[\mathbf{c}, t] = \min_{\mathbf{u}} [\phi(\mathbf{c}, \mathbf{u}, t)\Delta t + V(\mathbf{c} + \mathbf{f}\Delta t, t + \Delta t)] \tag{10.1-30}$$

which may also be written as

$$V_k[\mathbf{c}] = \min_{\mathbf{u}} [\phi(\mathbf{c}, \mathbf{u}, k) + V_{k+1}(\mathbf{c} + \mathbf{f}\Delta t)]. \tag{10.1-31}$$

Equations (10.1-30) and (10.1-31) are known as the functional equations of dynamic programming. A special case of these equations is Eq. (10.1-5). Here V_i represents the cost at the ith stage, which includes the sum of the costs from that stage up to and including the cost of the final stage.

We previously alluded to the fact that the "curse of dimensionality" may seriously affect the prospects of our obtaining a computational solution to a discrete dynamic programming problem, even with the largest of modernday computers. For illustration of this, let us return to our simple first-order nonlinear optimization problem stated in Eqs. (10.1-20) and (10.1-21). We again consider that we have a four-stage decision process and divide the x axis from 0 to 1 into five values. We apply the principle of optimality and obtain the optimum trajectory. However, four stages are certainly not very many, nor are five discrete values for x. We may then try a nine–stage approximation with x discretized into ten values. Again, we may use the principle of optimality in order to determine the best trajectory. We note that the result is different from that obtained before, as well we might expect, since we are using a much finer grid structure. An even finer mesh will give a better approximation to the true answer to the continuous problem. In this first case, we have had to make four stages times five state values for a total of twenty grid points. In the second case, we have a total of nine times ten, or ninety grid points.

If we divide a scalar state variable into Γ discrete values, or n state variables into Γ different discrete values, and consider a k-stage decision process, we must have Γk grid points. For example, if we have a two-vector state variable and divide each component into 100 parts, then $\Gamma = 10^4$; if we consider a 100-stage approximation, we then have 10^6 grid points. Computers do not have sufficient high-speed memory to retain the value of the cost functions for this many transitions. Actually, not all of them need be stored in high-speed memory. We might store the "policy functions" $\mathbf{f}$ at time t in order to calculate the cost at time t from the cost at time $t + \Delta t$, and also the cost $V[\mathbf{c}, t]$ and $V[\mathbf{c} + \mathbf{f}\Delta t, t + \Delta t]$, such that high-speed memory need only be used to retain 4×10^4 points. This is the high-speed memory capacity of many a large digital computer, and thus we see that problems with more than two state variables may simply overtax the memory of the computer. For example, if we have four state variables each discretized into 100 points, we need at least $(100)^4$ words of memory for each of the four policy functions and each of the two cost functions, for a total of 6×10^8 words of memory, which is an absurdly large number. In addition, if we are considering a 100-stage process, and if each of the 6×10^8 words required one computation, we would need 6×10^{10} computations, counting the data shuffling, to replace the old cost and policy function values by new ones. Several schemes have been proposed to remedy this enormous computational burden. We will briefly consider two of them, polynomial approximation methods and state increment dynamic programming.

One of the major practical obstacles to the computational success of dynamic programming is the problem of representation of a function of one or more variables which is defined as a set of discrete values of its argument. The approach that we have taken thus far requires the storage of the cost and policy function for each discrete value of $\mathbf{c}$ and k and, as we have seen, will often require an excessive amount of computer storage space.

A possible solution to this problem is to represent the cost and policy functions as sums of polynomials, which has been called polynomial approximation. We will now consider two approaches, one of which is exceptionally simple conceptually and one which, in most cases, will require less computational effort. We will illustrate the first approach for policy function approximation and the second for cost function approximation, although either method may be used for cost or policy function approximation.

We will first consider the polynomial approximation of a scalar policy function for a known optimal control at the kth stage

$$f(x, \hat{u}, t_k) \simeq \sum_{m=0}^{M-1} a_m(t_k)x^m(t_k) = p[x(t_k)], \tag{10.1-32}$$

where M is a number which is hopefully much smaller than S, the number of discrete values of $x(t_k)$ at the kth stage. A method which we will call difference approximation in the next section can be used to effect a criterion for the determination of the M values of a_m. This criterion consists of finding the a_m such that, for the S values of $x(t_k)$, $x_i(t_k)$, $i \times 1, 2, \ldots, S$,

$$J = \tfrac{1}{2} \sum_{i=1}^{S} \{f(x_i, \hat{u}, t_k) - p[x_i(t_k)]\}^2 \tag{10.1-33}$$

is minimum. We easily determine the necessary requirements for a minimum by setting $\partial J / \partial a_h = 0$ to obtain

$$\frac{\partial J}{\partial a_h} = 0 = \sum_{i=1}^{S} x_i^h(t_k)\{p[x_i(t_k)] - f(x_i, \hat{u}_i, t_k)\}. \tag{10.1-34}$$

Thus we see that, in order to obtain the best polynomial coefficients by this criterion, we must solve, for each t_k, the M simultaneous linear algebraic equations represented by

$$\sum_{i=1}^{S} x_i^h(t_k)\left\{\sum_{m=0}^{M-1} a_m(t_k) x_i^m(t_k) - f(x_i, \hat{u}_i, t_k)\right\} = 0, \qquad h = 0. 1, 2, \ldots, M-1, \tag{10.1-35}$$

which may be quite a task for large S and M, particularly since we must make this calculation for every stage in the process. The cost function may be approximated by a similar process. We may extend this approach to polynomial approximation in a straightforward fashion to the case where $\mathbf{x}$ is an n-vector by approximating the policy function as

$$\mathbf{f}(\mathbf{x}, \hat{\mathbf{u}}, t_k) \simeq \sum_{m_1=0}^{M-1} \sum_{m_2=0}^{M-1} \cdots \sum_{m_n=0}^{M-1} a_{m_1 m_2 \ldots m_n} x_1^{m_1}(t_k) x_2^{m_2}(t_k) \cdots x_n^{m_n}(t_k). \tag{10.1-36}$$

In this case, a total of nM coefficients must be stored for the a's at each stage in the computation. For example, if there are four states in the policy function, and each state is discretized into 100 values, the direct approach to dynamic programming requires 10^6 words of memory just to store the policy function for a single stage. With the polynomial approximation method, if we use a fifth degree polynomial, such that $M = 6$, we need only twenty-four storage words to contain the a's for computation of the policy function at a single stage. The computation of the a's must be repeated for each stage in the process, and so we are trading computer solution time for high-speed storage requirements. Since this computation of the a's necessarily involves the inversion of an $M \times M$ matrix for each stage, we will now inquire into the use of a different version of polynomial approximation which will make use of the properties of orthogonal polynomials.

To present this approach to polynomial approximation in a simple form, we will first consider the case where we have a single state variable which, for convenience, we will assume to be restricted to the range $x(t) \in [-1, 1]$. Instead of storing the value of the cost and policy function at every discretized point in state-space, we choose to represent these functions by a finite series of orthogonal functions of the form

$$V[c, k] = \sum_{q=0}^{R-1} a_{kq} P_q(c). \tag{10.1-37}$$

If $P_q(c)$ is the qth Legendre polynomial, orthogonal on the interval $[-1, 1]$, we may then determine the coefficients a_{kq} from

$$a_{kq} = \frac{2q+1}{2} \int_{-1}^{1} V[c, k]P_q(c)\, dc. \tag{10.1-38}$$

This is an advantage to the Legendre polynomials since we can evaluate the coefficients from an integration rather than relying upon a differentiation or differencing process. The function $V[c, k]$ is now represented at all points in the interval $[-1, 1]$ by the R coefficients a_{kq}. Once we store these R values, we may then calculate $V[c, k]$ for any value of $c \in [-1, 1]$. This is an approximation, but by using a fifth- or sixth-order polynomial ($R = 6$ or 7), we would expect the approximation to be quite good.

Instead of calculating the a_{kq} from Eq. (10.1-38), Bellman [8] suggests the use of a Riemann approximation

$$\int_{-1}^{1} V[c, k]P_q(c)\, dc = \sum_{j=0}^{S} b_j V[c_j, k]P_q(c_j) \tag{10.1-39}$$

which is exact when $V[c, k]P_q(c)$ is a polynomial of degree $2S - 1$, when the c_j's are the zeros of the zeros of the Legendre polynomial of degree $S + 1$, and when the b_j's are the Christoffel numbers which are available to us up to quite large values of S [8].

For the case where $\mathbf{x}$ is an n-vector, these relations become

$$V[\mathbf{c}, k] = \sum_{q_1=0}^{R-1} \sum_{q_2=0}^{R-1} \cdots \sum_{q_n=0}^{R-1} a_{q_1 q_2 \ldots q_n} P_{q_1}(c_1)P_{q_2}(c_2) \ldots P_{q_n}(c_n) \tag{10.1-40}$$

$$a_{q_1 q_2 \ldots q_n} = \frac{(2q_1+1)(2q_2+1)\ldots(2q_n+1)}{2^n} \times$$

$$\int_{-1}^{1}\int_{-1}^{1} \cdots \int_{-1}^{1} V[\mathbf{c}, k]P_{q_1}(c_1) \ldots P_{q_n}(c_n)\, dc_1, dc_2 \ldots dc_n. \tag{10.1-41}$$

The integral on the right-hand side of Eq. (10.1-41) may be evaluated by any convenient method. Since the spacing of the grid points is arbitrary, we find it convenient to space the grid points in each dimension at the zeros of a Legendre polynomial and evaluate the multiple integral of Eq. (10.1-41) by an n-dimensional extension of Gaussian quadrature.

To see how this simplifies the cost function calculation, we may start with Eq. (10.1-31), where we may assume $\theta = 0$ such that $V_{k_f} = 0$ and

$$V_{k_f-1}[c] = \min_u \phi[c, u, k_f]. \tag{10.1-42}$$

We then convert $V_{k_f-1}[c]$ into the sequence $a_{11}, a_{12}, \ldots, a_{1,R-1}$ by use of Eqs. (10.1-38) and (10.1-39). Since the b_j and the $P_q(c_j)$ are fixed constants determined from Eqs. (10.1-38) and (10.1-39), the product may be stored and called w_{jq} such that

$$a_{1q} = \sum_{j=0}^{S} w_{jq} V[c_j, k]. \tag{10.1-43}$$

Next we determine the cost at the next stage accoing to Eq. (10.1-30) or (10.1-31):

$$V_{k_f-2}[c_j] = \min_u \{\phi[c_j, u, k_f - 1] + V_{k_f-1}[c_j + f\,\Delta t]\}, \qquad j = 1, 2, \ldots, S. \tag{10.1-44}$$

We only need to compute the S values of $V_{k_f-2}[c_j]$ since the value of V_{k_f-1} can be obtained from Eqs. (10.1-43) and (10.1-37) as

$$V_{k_f-1}[c_j + f\,\Delta t] = \sum_{q=0}^{R-1} a_{1q} p_q(c_j + f\,\Delta t). \tag{10.1-45}$$

We then determine a_{2k} and obtain V_{k_f-3} from V_{k_f-2} by a procedure similar to the foregoing. The only storage needed at the kth stage is storage of the sequence a_{kq} (R of them) and the constants W_{kq}($S + 1$ of them). This storage requirement may turn out to be quite negligible. Unfortunately for the n-dimensional case, calculation of the required coefficients $a_{kq_1q_2\ldots q_n}$ may still be a very time-consuming process if we are to approximate the cost function by a polynomial of order five or more. Another attempt to overcome this dimensionality barrier is the method of state increment dynamic programming introduced by Larson [9]. This we will now consider.

The essential difference between state increment dynamic programming and the conventional method is the choice of the time interval over which a given control is applied. In conventional dynamic programming, the sampling interval Δt is fixed, whereas in state increment dynamic programming, the time increment δt is the minimum time required for at least one of the state variables to change by one increment. Thus the next state for a given control is known to lie within some small neighborhood of the point at which we apply control. By use of this procedure, we compute the optimal control in blocks which may cover a long time interval but only a small distance along each state variable. In a real sense, we are incorporating sampling interval sensitivity with the discrete dynamic programming method.

We again assume the system equations and cost function

$$\dot{\mathbf{x}} = \mathbf{f}(\mathbf{x}, \mathbf{u}, t), \qquad \mathbf{x}(t_o) = \mathbf{x}_o, \qquad J = \theta[\mathbf{x}(t), t] + \int_{t_o}^{t_f} \phi[\mathbf{x}, \mathbf{u}, t]\,dt \tag{10.1-46}$$

which we discretize as

$$\mathbf{x}(t + \delta t) = \mathbf{x}(t) + \mathbf{f}[\mathbf{x}(t), \mathbf{u}(t), t]\,\delta t, \qquad \mathbf{x}(t_o) = \mathbf{x}_o, \tag{10.1-47}$$

and we use the principle of optimality,

$$V[\mathbf{c}, t] = \min_u \{\phi(\mathbf{c}, \mathbf{u}, t)\,\delta t + V[\mathbf{c} + \mathbf{f}(\mathbf{c}, u, t)\,\delta t, t + \delta t]\}, \tag{10.1-48}$$

as the computational algorithm in order to obtain the optimal control and trajectory. Each state variable is constrained such that $\beta_i^-(t) \leq x_i(t) \leq \beta_i^+(t)$, and within each range β, each state variable is quantized in uniform increments Δx_i. In state increment dynamic programming, we determine the time interval δt such that at least one of the state variables changes by one increment.

$$\delta t = \min \left\{ \frac{|\Delta x_i|}{|f_i(\mathbf{x}, \mathbf{u}, t)|} \right\}, \qquad i = 1, 2, \ldots, n. \tag{10.1-49}$$

We see that the next state always lies within some small neighborhood of the present state, on the surface of an n-dimensional cube centered at the present state and with length $2x_i$ along each x_i axis.

The first step in preparing a problem for computation is to partition the x-t space into blocks, each of which covers w_i increments along the x_i axis and ΔT sec along the t axis. We consider the boundary between adjacent blocks as being in both blocks. The next step is to constrain the next state resulting from a control to be in the same block as the system state when the control is applied. We reduce high-speed memory requirements because only a few values of the cost function per state in the block are needed in the high-speed memory, whereas one value of the cost function for each state in the entire space is needed in conventional dynamic programming. To permit trajectories to transfer from block to block, values of the optimal cost function are computed along the boundary between blocks and stored in high-speed memory; they are used in order to evaluate the cost of controls applied at states within a single increment of the boundary in an adjacent block which has not yet been computationally processed.

Larson gives the following comparison of state increment dynamic programming with conventional dynamic programming [9]. In the conventional procedure, if there are n state variables and N_i quantization levels for the ith state variable, the high-speed memory requirement is

$$N = 4 \prod_{i=1}^{n} N_i \tag{10.1-50}$$

words, as we have previously indicated. In state increment dynamic programming, if a block covers $w_i \Delta x_i$ units along its ith axis such that there are $w_i + 1$ different quantization levels in each block, and if a linear interpolation in time is used, the high-speed memory requirement is

$$N' = 2 \prod_{i=1}^{n} (w_i + 1). \tag{10.1-51}$$

If, in addition, points on the boundary of a previously computed block are stored, and if there are S optimal points in an interval Δt where $S \Delta t = \Delta T$, the memory requirement is

$$N'' = 2 \prod_{i=1}^{n} w_i + S \left[\prod_{i=1}^{n} (w_i + 1) - \prod_{i=1}^{n} w_i \right]. \tag{10.1-52}$$

For w_i on the order of 3 to 5, and S in the vicinity of 5 to 10, a great reduction in the high-speed memory requirements is possible with no increase in computing time compared to conventional dynamic programming. The actual computation proceeds in much the same way as that for conventional dynamic programming. Blocks are selected and oriented so that most interblock transitions take place in the direction of most likely movement of the system trajectory. The block at $t_f - \Delta T$ is processed first. Computation at blocks in the time interval $(m-1)\Delta T \leq t - t_o \leq m\Delta T$ are made on the basis of the data which we obtain at $t = m\Delta T$ by the computations at the previous block in the time interval $m\Delta T \leq t - t_o \leq (m+1)\Delta T$. To start the computation, one value of the optimum cost function must be known for each state variable. Generally, but not necessarily, we know or precompute these values at the final time t_f.

In this section we introduced the subject of discrete dynamic programming. The dimensionality barrier is perhaps the biggest deterrent to the success of discrete dynamic programming. There are many advantages to dynamic programming, not the least of which is that it is a useful alternate viewpoint. In addition, there are some problems, such as those involving control and/or state variable inequality constraints, in which dynamic programming may be simpler to apply than other approaches. In fact these inequality constraints may actually aid in the solution of a discrete dynamic programming problem in that they narrow down the region of state and control space into which we must look for a possible solution to our problem. Also, as we have seen in Chapter 9, a stochastic version of continuous dynamic programming can be formulated.

10.2 Gradient techniques

A rather straightforward direct numerical method which has been in use for a number of years is the steepest descent or gradient method. The method was originally conceived to treat single-stage decision processes or nonlinear programming problems. Much recent interest has centered upon the use of the method for multistage decision processes and continuous optimal control problems.

10.2-1 Gradient techniques for single-stage decision processes

In this subsection we will first consider the minimization of the cost function

$$J = \theta(\mathbf{x}, \mathbf{u}) \tag{10.2-1}$$

by appropriate choice of $\mathbf{u}$, subject to the equality constraint

$$\mathbf{f}(\mathbf{x}, \mathbf{u}) = 0, \tag{10.2-2}$$

where $\mathbf{x}$ and $\mathbf{f}$ are n-vectors, $\mathbf{u}$ is an m-vector, and θ is a scalar. Later we will consider this same problem with inequality constraints present. We have thus formulated a nonlinear programming or single-stage decision problem. As we know from Chapter 2, we may define a Hamiltonian

$$H(\mathbf{x}, \boldsymbol{\lambda}, \mathbf{u}) = \theta(\mathbf{x}, \mathbf{u}) + \boldsymbol{\lambda}^T\mathbf{f}(\mathbf{x}, \mathbf{u}) \tag{10.2-3}$$

and then set the gradient of the Hamiltonian with respect to $\mathbf{u}$ equal to the null vector,

$$\frac{\partial H}{\partial \mathbf{u}} = \frac{\partial \theta(\mathbf{x}, \mathbf{u})}{\partial \mathbf{u}} + \left[\frac{\partial \mathbf{f}^T(\mathbf{x}, \mathbf{u})}{\partial \mathbf{u}}\right]\boldsymbol{\lambda} = 0, \tag{10.2-4}$$

and the gradient of the Hamiltonian with respect to $\mathbf{x}$ also equal to the null vector,

$$\frac{\partial H}{\partial \mathbf{x}} = \frac{\partial \theta(\mathbf{x}, \mathbf{u})}{\partial \mathbf{x}} + \frac{\partial \mathbf{f}^T(\mathbf{x}, \mathbf{u})}{\partial \mathbf{x}}\boldsymbol{\lambda} = 0. \tag{10.2-5}$$

These provide the first necessary condition for a minimum. In addition, the matrix

$$\begin{bmatrix} \dfrac{\partial}{\partial \mathbf{x}}\dfrac{\partial H}{\partial \mathbf{x}} & \dfrac{\partial}{\partial \mathbf{u}}\dfrac{\partial H}{\partial \mathbf{x}} \\ \left[\dfrac{\partial}{\partial \mathbf{u}}\dfrac{\partial H}{\partial \mathbf{x}}\right]^T & \dfrac{\partial}{\partial \mathbf{u}}\dfrac{\partial H}{\partial \mathbf{u}} \end{bmatrix} \tag{10.2-6}$$

must be nonnegative definite along the "trajectory" $\mathbf{f}(\mathbf{x}, \mathbf{u}) = \mathbf{0}$. In general, we will not be able to solve the nonlinear algebraic equations resulting from Eqs. (10.2-4) and (10.2-5). Since along $\mathbf{f}(\mathbf{x}, \mathbf{u}) = \mathbf{0}$ we have $J = H$, we see that, approximately, if we set $\partial H/\partial \mathbf{x} = \mathbf{0}$,

$$\Delta J = \left[\frac{\partial H}{\partial \mathbf{u}}\right]^T \Delta\mathbf{u}. \tag{10.2-7}$$

If we wish to make the greatest or steepest change in J, ΔJ, we could perhaps calculate the gradient $\partial H/\partial \mathbf{u}$ and make a change in J by making $\Delta\mathbf{u}$ directed opposite to the gradient

$$\Delta\mathbf{u} = -K\left[\frac{\partial H}{\partial \mathbf{u}}\right]. \tag{10.2-8}$$

Thus we see that the change in J is such as to make J smaller, which we desire since we want to minimize J

$$\Delta J = -K\left[\frac{\partial H}{\partial \mathbf{u}}\right]^T\left[\frac{\partial H}{\partial \mathbf{u}}\right]. \tag{10.2-9}$$

To start the procedure, we will assume a knowledge of a nonoptimum control vector $\mathbf{u}^N$. From this we determine the value of $\mathbf{x}$, $\mathbf{x}^N$, to satisfy the equality constraint, Eq. (10.2-2),

$$\mathbf{f}(\mathbf{x}^N, \mathbf{u}^N) = \mathbf{0}, \tag{10.2-10}$$

and then we find the value of the adjoint to insure $\partial J/\partial \mathbf{x}^N = \mathbf{0}$, from Eq. (10.2-5),

$$\boldsymbol{\lambda}^N = -\left[\frac{\partial \mathbf{f}^T(\mathbf{x}^N, \mathbf{u}^N)}{\partial \mathbf{x}^N}\right]^{-1}\left[\frac{\partial \theta(\mathbf{x}^N, \mathbf{u}^N)}{\partial \mathbf{x}^N}\right]. \tag{10.2-11}$$

We then compute the gradient vector, from Eq. (10.2-3),

$$\frac{\partial H}{\partial \mathbf{u}^N} = \frac{\partial \theta(\mathbf{x}^N, \mathbf{u}^N)}{\partial \mathbf{u}^N} + \left[\frac{\partial \mathbf{f}^T(\mathbf{x}^N, \mathbf{u}^N)}{\partial \mathbf{u}^N}\right]\boldsymbol{\lambda}^N. \tag{10.2-12}$$

This vector will be null only when we actually satisfy the condition for a minimum. We use the gradient vector to compute the steepest descent such that we obtain a change un $\mathbf{u}$, from Eq. (10.2-8),

$$\Delta \mathbf{u}^N = -K\left[\frac{\partial H(\mathbf{x}^N, \boldsymbol{\lambda}^N, \mathbf{u}^N)}{\partial \mathbf{u}^N}\right]. \tag{10.2-13}$$

We then use the new value of $\mathbf{u}$,

$$\mathbf{u}^{N+1} = \mathbf{u}^N + \Delta \mathbf{u}^N, \tag{10.2-14}$$

to compute $\mathbf{x}^{N+1}$, $\boldsymbol{\lambda}^{N+1}$ and repeat the iterative process until the change in the cost function,

$$\Delta J^N = -K\left[\frac{\partial H(\mathbf{x}^N, \boldsymbol{\lambda}^N, \mathbf{u}^N)}{\partial \mathbf{u}^N}\right]^T\left[\frac{\partial H(\mathbf{x}^N, \boldsymbol{\lambda}^N, \mathbf{u}^N)}{\partial \mathbf{u}^N}\right], \tag{10.2-15}$$

between iterations is below some particular small value.

Example 10.2-1. To get a physical feeling for this elementary gradient method, let us consider minimization of $J = x^2 + u^2$ subject to the equality constraint $xu = 1$. This is a very easy problem to solve analytically since we may write the Hamiltonian as $H = x^2 + u^2 + \lambda(xu - 1)$ and set $\partial H/\partial u = \partial H/\partial x = 0$ to obtain $u = \pm 1$, $x = \pm 1$, $\lambda = -2$, $J_{\min} = 2$. Let us suppose that we do not know the correct solution and assume a trial value $u = u^0$. For this particular example, the pertinent equations for the gradient technique, Eqs. (10.2-10) through (10.2-15), become:

$$\mathbf{f}(\mathbf{x}^N, \mathbf{u}^N) = \mathbf{0} = x^N u^N - 1$$

$$\boldsymbol{\lambda}^N = -\left[\frac{\partial \mathbf{f}^T}{\partial \mathbf{x}^N}\right]^{-1}\frac{\partial \theta}{\partial \mathbf{x}^N} = -\frac{2x^N}{u^N}$$

$$\frac{\partial H}{\partial \mathbf{u}^N} = \frac{\partial \theta}{\partial \mathbf{u}^N} + \left[\frac{\partial \mathbf{f}^T}{\partial \mathbf{u}^N}\right]\boldsymbol{\lambda}^N = 2u^N + x^N\lambda^N$$

$$\Delta \mathbf{u}^N = -K\frac{\partial H}{\partial \mathbf{u}^N} = -K(2u^N + x^N\lambda^N) = -2K\left[u^N - \frac{1}{(u^N)^3}\right]$$

$$\Delta J^N = -K\left[\frac{\partial H}{\partial \mathbf{u}^N}\right]^T\left[\frac{\partial H}{\partial \mathbf{u}^N}\right] = -K(2u^N + x^N\lambda^N)^2$$

$$u^{N+1} = u^N + \Delta u^N.$$

Thus for this particularly simple problem, we determine successive values of the control by solving the difference equation

$$u^{N+1} = u^N - 2K\left[u^N - \frac{1}{(u^N)^3}\right].$$

We see that the equilibrium values for the difference equation are ± 1. Typically, we would pick K such that the change in the cost function is 5 percent or 10 percent at each iteration. For example, if we pick $u^0 = 2$, then $x^0 = \frac{1}{2}$ from $f(x, u) = 0$ and $J^0 = 2^2 + (0.5)^2 = 4.25$. Thus if ΔJ is to be about 10 percent of J, we then use $K = 0.1$. For this particular case, the sequence of control values which we obtain is $\{u\} = \{2.000, 1.625, 1.347, 1.159, 1.056, 1.010\}$. The convergence of the process is fairly rapid in this case. The rate of convergence may be increased, up to a point, by increasing K. There is considerable risk in doing this, as we can see by letting $K = 0.5$, such that $u^{N+1} = 1/(u^N)^3$. Now if we start with $u^0 = 2$, we obtain the sequence $\{u\} = \{2.000, 0.125, 51.100, \ldots\}$, and we see that the sequence is diverging.

By use of the second method of Lyapunov, we determine the convergence region of the sequence as a function of K and the initial value u^0. If we let $a^N = u^N - 1$, which transfers the equilibrium point at 1 to the origin, we obtain the difference equation

$$a^{N+1} = a^N - 2K\left[a^N + 1 - \frac{1}{(a^N + 1)^3}\right].$$

Then if we use for a Lyapunov V^N function $(a^N)^2$ and obtain $\Delta V = V^{N+1} - V^N$, we have

$$\Delta V = -4K(a^N)^2\left\{1 + \frac{1}{a^N} - \frac{1}{a^N(a^N + 1)^3} - K\left[1 + \frac{1}{a^N} - \frac{1}{a^N(a^N + 1)^3}\right]^2\right\}.$$

For the sequence to be stable, ΔV must be negative. Thus the stability boundary for the equilibrium value of $u = 1$ is determined by

$$K < \left[1 + \frac{1}{u^N - 1} - \frac{1}{(u^N - 1)(u^N)^3}\right]^{-1}, \qquad u^N > 0.$$

For values of u^0 less than zero, we attempt to converge to the equilibrium value of -1. We may easily show that the stability boundary in that case is determined by

$$K < \left[1 + \frac{1}{u^N + 1} - \frac{1}{(u^N + 1)(u^N)^3}\right]^{-1}, \qquad u^N < 0.$$

For the particular case where the initial value of u is 2 and $K = \frac{1}{2}$, we obtain $u = 2.000, 0.125, \ldots,$ such that the stability inequality yields $K < \frac{8}{15}, \frac{7}{4096}, \ldots$. Thus we see that our solution for $K = 0.5$ could not possibly converge.

The steps in this "first-order" gradient procedure which we have just derived are:

1. Determine $\mathbf{u}^i$
2. Obtain $\mathbf{x}^i$ from $\mathbf{f}(\mathbf{x}^i, \mathbf{u}^i) = \mathbf{0}$
3. Evaluate $\boldsymbol{\lambda}^i$ from

$$\boldsymbol{\lambda}^i = -\left[\frac{\partial \mathbf{f}^T(\mathbf{x}^i, \mathbf{u}^i)}{\partial \mathbf{x}^i}\right]^{-1} \frac{\partial \theta(\mathbf{x}^i, \mathbf{u}^i)}{\partial \mathbf{x}^i}$$

4. Determine

$$\frac{\partial H(\mathbf{x}^i, \mathbf{u}^i, \boldsymbol{\lambda}^i)}{\partial \mathbf{u}^i} = \frac{\partial \theta(\mathbf{x}^i, \mathbf{u}^i)}{\partial \mathbf{u}^i} + \frac{\partial \mathbf{f}^T(\mathbf{x}^i, \mathbf{u}^i)}{\partial \mathbf{x}^i}\boldsymbol{\lambda}^i$$

5. Compute

$$\mathbf{u}^{i+1} = \mathbf{u}^i - K^i\left[\frac{\partial H(\mathbf{x}^i, \mathbf{u}^i, \boldsymbol{\lambda}^i)}{\partial \mathbf{u}^i}\right]$$

6. Repeat the procedure until there is no change in control from iteration to iteration.

Example 10.2-2. Let us suppose that we desire to solve the set of linear algebraic equations represented by the equation $\mathbf{Ax} = \mathbf{b}$. Clearly, the solution is $\mathbf{x} = \mathbf{A}^{-1}\mathbf{b}$. We wish to inquire into the possibility of a steepest descent solution to this set of equations, which might possibly enable us to avoid taking the inverse of the $\mathbf{A}$ matrix. We may define an error vector as $\mathbf{e} = \mathbf{Ax} - \mathbf{b}$ and then attempt to find the value of $\mathbf{x}$ which minimizes the cost function,

$$J = \mathbf{e}^T\mathbf{R}\mathbf{e} = (\mathbf{u}^T\mathbf{A}^T - \mathbf{b}^T)\mathbf{R}(\mathbf{Au} - \mathbf{b}),$$

where we are using $\mathbf{u}$ rather than $\mathbf{x}$ since, in the recast version of the problem, $\mathbf{u}$ is a *control* rather than a *state*. From Eqs. (10.2-13) and (10.2-14) we obtain

$$\Delta\mathbf{u}^N = -2K\mathbf{A}^T\mathbf{R}(\mathbf{Au}^N - \mathbf{b}), \qquad \mathbf{u}^{N+1} = \mathbf{u}^N - 2K\mathbf{A}^T\mathbf{R}(\mathbf{Au}^N - \mathbf{b}),$$

and we see that we can avoid the use of matrix inversion by the gradient technique for this problem. We do not obtain the solution for $\mathbf{A}^{-1}$ but only the solution for $\mathbf{u}$ (or $\mathbf{x}$). It is possible, however, to obtain the matrix $\mathbf{A}^{-1}$ by making all components of the $\mathbf{b}$-vector zero except the ith component which is made equal to 1. The value of the vector $\mathbf{x}$ so determined may be called $\mathbf{x}^i$. We may find the $\mathbf{A}^{-1}$ matrix by letting i take on the values $1, 2, \ldots, n$, where $\mathbf{x}$ is an n-vector, and then using $\mathbf{A}^{-1} = [\mathbf{x}^1 \vdots \mathbf{x}^2 \vdots \cdots \mathbf{x}^n]$.

We will now offer a slightly different approach to the problem of this subsection where we wish to minimize Eq. (10.2-1) subject to the equality constraint of Eq. (10.2-2). In some few cases, it is possible to explicitly solve Eq. (10.2-2) for the n values of $\mathbf{x}$ in terms of the m values of $\mathbf{u}$ and substitute. these relations in Eq. (10.2-1), obtaining a scalar performance index which is a function of only the m control variables. In this case, no Lagrange multipliers are necessary. However, this approach is often not feasible.

Now we will consider a technique known as the gradient projection technique. In the neighborhood of some nominal values $\bar{\mathbf{x}}$ and $\bar{\mathbf{u}}$, we may write as a good approximation for Eq. (10.2-2),

$$\mathbf{f}(\mathbf{x}, \mathbf{u}) = \mathbf{f}(\bar{\mathbf{x}}, \bar{\mathbf{u}}) + \left[\frac{\partial \mathbf{f}(\bar{\mathbf{x}}, \bar{\mathbf{u}})}{\partial \bar{\mathbf{x}}}\right][\mathbf{x} - \bar{\mathbf{x}}] + \left[\frac{\partial \mathbf{f}(\bar{\mathbf{x}}, \bar{\mathbf{u}})}{\partial \bar{\mathbf{u}}}\right][\mathbf{u} - \bar{\mathbf{u}}] = \mathbf{0}. \quad (10.2\text{-}16)$$

Now if we assume that Eq. (10.2-2) is satisfied at $\mathbf{x} = \bar{\mathbf{x}}$ and $\mathbf{u} = \bar{\mathbf{u}}$, we then have a system of linear equations in $\Delta\bar{\mathbf{x}}$ and $\Delta\bar{\mathbf{u}}$ for the vanishing of $\mathbf{f}(\bar{\mathbf{x}}, \bar{\mathbf{u}})$

to first order where $\Delta\bar{\mathbf{x}} = \mathbf{x} - \bar{\mathbf{x}}$, $\Delta\bar{\mathbf{u}} = \mathbf{u} - \bar{\mathbf{u}}$,

$$\left[\frac{\partial \mathbf{f}(\bar{\mathbf{x}}, \bar{\mathbf{u}})}{\partial \bar{\mathbf{x}}}\right]\Delta\bar{\mathbf{x}} + \left[\frac{\partial \mathbf{f}(\bar{\mathbf{x}}, \bar{\mathbf{u}})}{\partial \bar{\mathbf{u}}}\right]\Delta\bar{\mathbf{u}} = \mathbf{0}. \tag{10.2-17}$$

We can now solve for the n values of the increment vector $\Delta\bar{\mathbf{x}}$ since $\partial\mathbf{f}/\partial\bar{\mathbf{x}}$ must have an inverse. We may substitute these values in Eq. (10.2-1) and determine the gradient of θ or J subject to these determined constraints. In a geometrical sense, this is a projection of the free gradient vector upon the m-dimensional control space determined by the intersection of the n hyperplanes of Eq. (10.2-17). We now consider a first-order change in J,

$$\Delta J = \left[\frac{\partial \theta(\bar{\mathbf{x}}, \bar{\mathbf{u}})}{\partial \bar{\mathbf{x}}}\right]^T \Delta\bar{\mathbf{x}} + \left[\frac{\partial \theta(\bar{\mathbf{x}}, \bar{\mathbf{u}})}{\partial \mathbf{u}}\right]^T \Delta\bar{\mathbf{u}}. \tag{10.2-18}$$

From Eq. (10.2-17), this expression becomes

$$\Delta J = \left\{-\left[\frac{\partial \theta(\bar{\mathbf{x}}, \bar{\mathbf{u}})}{\partial \bar{\mathbf{x}}}\right]^T \left[\frac{\partial \mathbf{f}(\bar{\mathbf{x}}, \bar{\mathbf{u}})}{\partial \bar{\mathbf{x}}}\right]^{-1} \left[\frac{\partial \mathbf{f}(\bar{\mathbf{x}}, \bar{\mathbf{u}})}{\partial \bar{\mathbf{u}}}\right] + \left[\frac{\partial \theta(\bar{\mathbf{x}}, \bar{\mathbf{u}})}{\partial \bar{\mathbf{u}}}\right]^T\right\} \Delta\bar{\mathbf{u}}. \tag{10.2-19}$$

Clearly, the terms within the btackets of Eq. (10.2-19) are the same as those obtained from Eqs. (10.2-4) and (10.2-5). Thus we have demonstrated the equivalence of the two approaches.

We may naturally question whether there is any improvement to be obtained by incorporating second-order terms in development of these gradient algorithms. If we retain terms of first and second order in our expansion of the cost function, we obtain, since we have $J = H$, along $\mathbf{f}(\mathbf{x}, \mathbf{u}) = \mathbf{0}$,

$$\Delta J^N = \left[\frac{\partial H(\mathbf{x}^N, \mathbf{u}^N, \boldsymbol{\lambda}^N)}{\partial \mathbf{x}^N}\right]^{\mathrm{T}} \Delta\mathbf{x}^N + \left[\frac{\partial H(\mathbf{x}^N, \mathbf{u}^N, \boldsymbol{\lambda}^N)}{\partial \mathbf{u}^N}\right]^{\mathrm{T}} \Delta\mathbf{u}^N + \tfrac{1}{2}[\Delta\mathbf{x}^{N\mathrm{T}}\, \Delta\mathbf{u}^{N\mathrm{T}}] \times$$
$$\begin{bmatrix} \dfrac{\partial^2 H(\mathbf{x}^N, \mathbf{u}^N, \boldsymbol{\lambda}^N)}{(\partial \mathbf{x}^N)^2} & \dfrac{\partial}{\partial \mathbf{u}^N}\dfrac{\partial H(\mathbf{x}^N, \mathbf{u}^N, \boldsymbol{\lambda}^N)}{\partial \mathbf{x}^N} \\ \left[\dfrac{\partial}{\partial \mathbf{u}^N}\dfrac{\partial H(\mathbf{x}^N, \mathbf{u}^N, \boldsymbol{\lambda}^N)}{\partial \mathbf{x}^N}\right]^{\mathrm{T}} & \dfrac{\partial^2 H(\mathbf{x}^N, \mathbf{u}^N, \boldsymbol{\lambda}^N)}{(\partial \mathbf{u}^N)^2} \end{bmatrix} \begin{bmatrix} \Delta\mathbf{x}^N \\ \Delta\mathbf{u}^N \end{bmatrix}. \tag{10.2-20}$$

We will adjust $\boldsymbol{\lambda}^N$ such that Eq. (10.2-5) is satisfied and $\partial H/\partial \mathbf{x} = \mathbf{0}$. Also we will require that Eq. (10.2-2) hold for first-order changes; thus we will again require Eq. (10.2-17) to be valid. By substituting Eqs. (10.2-5) and (10.2-17) in the foregoing, we obtain:

$$\Delta J^N = \left[\frac{\partial H(\mathbf{x}^N, \mathbf{u}^N, \boldsymbol{\lambda}^N)}{\partial \mathbf{u}^N}\right]^{\mathrm{T}} \Delta\mathbf{u}^N + \frac{1}{2}\,\Delta\mathbf{u}^{N\mathrm{T}}\left[-\left[\frac{\partial}{\partial \mathbf{u}}\frac{\partial H}{\partial \boldsymbol{\lambda}}\right]^{\mathrm{T}}\left[\frac{\partial}{\partial \mathbf{x}}\frac{\partial H}{\partial \boldsymbol{\lambda}}\right]^{\mathrm{T}} \mathbf{I}\right] \times$$
$$\begin{bmatrix} \dfrac{\partial^2 H}{\partial \mathbf{x}^2} & \dfrac{\partial}{\partial \mathbf{u}}\dfrac{\partial H}{\partial \mathbf{x}} \\ \left[\dfrac{\partial}{\partial \mathbf{u}}\dfrac{\partial H}{\partial \mathbf{x}}\right]^{\mathrm{T}} & \dfrac{\partial^2 H}{\partial \mathbf{u}^2} \end{bmatrix} \begin{bmatrix} -\left[\dfrac{\partial}{\partial \mathbf{x}}\dfrac{\partial H}{\partial \boldsymbol{\lambda}}\right]\left[\dfrac{\partial}{\partial \mathbf{u}}\dfrac{\partial H}{\partial \boldsymbol{\lambda}}\right] \\ \mathbf{I} \end{bmatrix} \Delta\mathbf{u}^N, \tag{10.2-21}$$

where H, $\mathbf{x}$, $\mathbf{u}$, and $\boldsymbol{\lambda}$ are evaluated on the nth iteration. We see that we now have the possibility of minimizing ΔJ with respect to choice of $\Delta \mathbf{u}^N$; this is a standard problem in matrix calculus. We obtain:

$$\Delta \mathbf{u}^N = -\left\{\left[\frac{\partial}{\partial \mathbf{u}}\frac{\partial H}{\partial \boldsymbol{\lambda}}\right]^{\mathrm{T}}\left[\frac{\partial}{\partial \mathbf{x}}\frac{\partial H}{\partial \boldsymbol{\lambda}}\right]^{\mathrm{T}}\left[\frac{\partial^2 H}{\partial \mathbf{x}^2}\right]\left[\frac{\partial}{\partial \mathbf{x}}\frac{\partial H}{\partial \boldsymbol{\lambda}}\right]\left[\frac{\partial}{\partial \mathbf{u}}\frac{\partial H}{\partial \boldsymbol{\lambda}}\right] + \left[\frac{\partial^2 H}{\partial \mathbf{u}^2}\right] - \left[\frac{\partial}{\partial \mathbf{u}}\frac{\partial H}{\partial \boldsymbol{\lambda}}\right]^{\mathrm{T}}\left[\frac{\partial}{\partial \mathbf{x}}\frac{\partial H}{\partial \boldsymbol{\lambda}}\right]^{\mathrm{T}}\left[\frac{\partial}{\partial \mathbf{u}}\frac{\partial H}{\partial \mathbf{x}}\right] - \left[\frac{\partial}{\partial \mathbf{u}}\frac{\partial H}{\partial \mathbf{x}}\right]\left[\frac{\partial}{\partial \mathbf{x}}\frac{\partial H}{\partial \boldsymbol{\lambda}}\right]\left[\frac{\partial}{\partial \mathbf{u}}\frac{\partial H}{\partial \boldsymbol{\lambda}}\right]\right\}^{-1}\frac{\partial H}{\partial \mathbf{u}} \tag{10.2-22}$$

and see that we do indeed have a procedure which specifies the change in control from iteration to iteration. Since we have minimized ΔJ to quadratic terms in $\Delta \mathbf{x}$ and $\Delta \mathbf{u}$, we expect faster convergence in this second-order gradient or second variation method than we obtained in the first-order gradient method. However, this is obtained at the expense of greater computational complexity and increased convergence difficulties. The steps in the computation are relatively straightforward and are quite similar to those of the first-order gradient method. They can be conveniently summarized as:

1. Select a $\mathbf{u}^i$
2. Determine $\mathbf{x}^i$ from $\mathbf{f}(\mathbf{x}^i, \mathbf{u}^i) = \mathbf{0}$
3. Determine
$$H(\mathbf{x}^i, \mathbf{u}^i, \boldsymbol{\lambda}^i) = \theta(\mathbf{x}^i, \mathbf{u}^i) + \boldsymbol{\lambda}^{i\mathrm{T}}\mathbf{f}(\mathbf{x}^i, \mathbf{u}^i)$$
4. Evaluate $\boldsymbol{\lambda}^i$ from
$$\boldsymbol{\lambda}^i = -\left[\frac{\partial \mathbf{f}(\mathbf{x}^i, \mathbf{u}^i)}{\partial \mathbf{x}^i}\right]^{-1}\frac{\partial \theta(\mathbf{x}^i, \mathbf{u}^i)}{\partial \mathbf{x}^i} \quad \text{or} \quad \frac{\partial H(\mathbf{x}^i, \mathbf{u}^i, \boldsymbol{\lambda}^i)}{\partial \mathbf{x}^i} = 0$$
5. Obtain necessary derivatives so as to compute $\Delta \mathbf{u}^i$ from Eq. (10.2-22)
6. Determine $\qquad \mathbf{u}^{i+1} = \mathbf{u}^i + \Delta \hat{\mathbf{u}}^i$
7. Repeat the computation until $\Delta \mathbf{u}^i$ changes little from iteration to iteration.

Example 10.2-3. Again we consider the problem of solving $\mathbf{Au} = \mathbf{b}$ by minimizing the cost function

$$J = \theta(\mathbf{u}) = \tfrac{1}{2}(\mathbf{Au} - \mathbf{b})^{\mathrm{T}}\mathbf{R}(\mathbf{Au} - \mathbf{b}).$$

We obtain, from Eq. (10.2-22),

$$\mathbf{u}^{i+1} = \mathbf{u}^i - [\mathbf{A}^{\mathrm{T}}\mathbf{RA}]^{-1}\mathbf{A}^{\mathrm{T}}\mathbf{R}(\mathbf{Au}^i - \mathbf{b}) = \mathbf{A}^{-1}\mathbf{b}.$$

We see that convergence now is obtained in one step regardless of $\mathbf{u}^i$. We obtain $\mathbf{u}^{i+1} = \mathbf{A}^{-1}\mathbf{b}$, which is the known solution. This feature of one-step convergence is inherent in linear problems. However, the computational algorithm involves $\mathbf{A}^{-1}$ and, in this simple case, the reason for using an iterative technique is to avoid problems inherent in taking the matrix inverse. In any case, the greatest use of the gradient algorithm is to solve sets of nonlinear algebraic equations, and for this the second variation method sometimes has great advantages.

We have optimized the function $J = \theta(\mathbf{u})$ by a first-order gradient method and a second-order gradient or second variation method. For the first order gradient method we obtained

$$\Delta\mathbf{u}^N = -K^N \frac{d\theta(\mathbf{u}^N)}{d\mathbf{u}^N}, \tag{10.2-23}$$

whereas for the second variation method we obtained

$$\Delta u^N = -\left[\frac{d^2\theta(\mathbf{u}^N)}{(d\mathbf{u}^N)^2}\right]^{-1} \frac{d\theta(\mathbf{u}^N)}{d\mathbf{u}^N}. \tag{10.2-24}$$

The first-order method possesses the advantage of computational simplicity and the disadvantage of slow convergence near the optimum unless the optimum first-order gradient method is employed, which is generally very difficult to accomplish. The second variation method possesses the advantage of rapid convergence near the optimum solution and the disadvantages of needing a closer to optimum value of the initual $\mathbf{u}$ for convergence and considerably greater computational complexity.

The conjugate gradient method [38, 40, 41] of Fletcher and Powell attempts to combine the best features of both methods while lessening their disadvantages. Rather than attempting to compute $[d^2\theta(\mathbf{u})/(d\mathbf{u})^2]^{-1}$, a sequence of direction vectors $\boldsymbol{\gamma}^1, \boldsymbol{\gamma}^2, \ldots$ is generated that is conjugate or orthogonal with respect to $d^2\theta(\mathbf{u})/(d\mathbf{u})^2$ in that

$$\boldsymbol{\gamma}^{iT}\left[\frac{d^2\theta(\mathbf{u}^i)}{(d\mathbf{u}^i)^2}\right]\boldsymbol{\gamma}^j = 0, \qquad i \neq j. \tag{10.2-25}$$

Then a series of searches is made along each of the conjugate direction vectors $\boldsymbol{\gamma}^N$ to find the optimum length in that direction in which to proceed. Thus we use

$$\mathbf{u}^{N+1} = \mathbf{u}^N - K^N\boldsymbol{\gamma}^N, \tag{10.2-26}$$

where K^N is a positive scalar chosen in an optimum way such that

$$K^N = \min_{K_N} \theta(\mathbf{u}^N - K_N\boldsymbol{\gamma}^N), \tag{10.2-27}$$

and thus we implement the optimum gradient method.

It is straightforward to show that the algorithm

$$\boldsymbol{\gamma}^i = -\frac{d\theta(\mathbf{u}^i)}{d\mathbf{u}^i} + \frac{\| d\theta(\mathbf{u}^i)/d\mathbf{u}^i \|^2}{\| d\theta(\mathbf{u}^{i-1})/d\mathbf{u}^{i-1} \|^2}\boldsymbol{\gamma}^{i-1} \tag{10.2-28}$$

generates a set of conjugate direction vectors as defined by Eq. (10.2-25). We start by assuming that $\boldsymbol{\gamma}^1 = d\theta(\mathbf{u}^1)/d\mathbf{u}^1$. In practice, the determination of an optimum K^N is all but impossible. It is generally straightforward to try several candidate values of K^N in close proximity to the value used on the previous iteration and select the one which yields a minimum of Eq. (10.2-27). The method proceeds as follows:

1. Choose $\mathbf{u}^i$
2. Determine $\boldsymbol{\gamma}^i = \dfrac{d\theta(\mathbf{u}^i)}{d\mathbf{u}^i}$
3. Determine K^i so as to minimize $\theta(\mathbf{u}^i - K^i\boldsymbol{\gamma}^i)$
4. Compute $\mathbf{u}^{i+1} = \mathbf{u}^i - K^i\boldsymbol{\gamma}^i$
5. Determine

$$\gamma^{i+1} = -\frac{d\theta(\mathbf{u}^{i+1})}{d\mathbf{u}^{i+1}} + \frac{\|d\theta(\mathbf{u}^{i+1})/d\mathbf{u}^{i+1}\|^2}{\|d\theta(\mathbf{u}^i)/d\mathbf{u}^i\|^2}\gamma^i$$

6. Repeat computation for a new iterate starting at step 3. Continue until $\mathbf{u}^i$ does not change appreciably from iteration to iteration.

Example 10.2-4. We consider the determination of $\mathbf{u}$ from the relation $\mathbf{Au} = \mathbf{b}$. The quadratic cost function, $J = \theta(\mathbf{u}) = \frac{1}{2}(\mathbf{Au} - \mathbf{b})^T\mathbf{R}(\mathbf{Au} - \mathbf{b})$, is used. We assume a solution $\mathbf{u}^1$ and obtain $\boldsymbol{\gamma}^i = \mathbf{A}^T\mathbf{R}(\mathbf{Au}^1 - \mathbf{b})$. Then we note that the computation proceeds as $\mathbf{u}^2 = \mathbf{u}^1 - k^1\mathbf{A}^T\mathbf{R}(\mathbf{Au}^1 - \mathbf{b})$, where k^1 is determined so as to minimize

$$\begin{aligned} J^2 = \theta(\mathbf{u}^2) &= \tfrac{1}{2}(\mathbf{Au}^1 - \mathbf{AA}^T\mathbf{RAu}^1k^1 + \mathbf{AA}^T\mathbf{Rb}k^1 - \mathbf{b})^T \\ &\quad \times \mathbf{R}(\mathbf{Au}^1 - \mathbf{AA}^T\mathbf{RAu}^1k^1 + \mathbf{AA}^T\mathbf{Rb}k^1 - \mathbf{b}) \\ &= [(\mathbf{I} - \mathbf{AA}^T\mathbf{R}k^1)(\mathbf{Au}^1 - \mathbf{b})]^T\mathbf{R}[(\mathbf{I} - \mathbf{AA}^T\mathbf{R}k^1)(\mathbf{Au}^1 - \mathbf{b})]. \end{aligned}$$

This value of k is easily obtained as

$$k^1 = \frac{(\mathbf{Au}^1 - \mathbf{b})^T\mathbf{RAA}^T\mathbf{R}(\mathbf{Au}^1 - \mathbf{b})}{(\mathbf{Au}^1 - \mathbf{b})^T\mathbf{RAA}^T\mathbf{RAA}^T\mathbf{R}(\mathbf{Au}^1 - \mathbf{b})}.$$

Thus u^2 is determined. From this, k^2 is found, and then u^3 .... It is possible to show that this problem will converge to the correct solution in M steps, where $\mathbf{u}$ is an M-vector. This statement applies to any quadratic cost linear constraint problem when solved with the conjugate gradient method. Unfortunately, it does not apply to nonquadratic cost or nonlinear equality constraint problems.

It is a relatively simple matter for us to show that the minimization problem for the cost function and equality constraint $J = \theta(\mathbf{x}, \mathbf{u})$ and $\mathbf{f}(\mathbf{x}, \mathbf{u}) = \mathbf{0}$ proceeds in a fasion analogous to the procedure just outlined for the noconstraint case. All that is necessary is to substitute $H(\mathbf{x}^N, \mathbf{u}^N, \boldsymbol{\lambda}^N)$ for $H(\mathbf{u}^N)$ in the aforementioned steps of the conjugate gradient procedure. Of course, it is also necessary to solve the additional relations

$$\frac{\partial H}{\partial \boldsymbol{\lambda}^N} = \frac{\partial H}{\partial \mathbf{x}^N} = \mathbf{0} \tag{10.2-29}$$

at each iteration stage. We have introduced the static gradient or single-stage decision process here primarily because of the simplicity of the presentation rather than because of any direct use of these procedures in optimum systems control. However, these methods carry over to the multistage control or decision process or continuous control or decision process problems with only minor modifications.

In this subsection, we need only consider state variable equality constraints since, for a single-stage decision process, an inequality constraint is either an equality constraint or no constraint at all. Naturally, we will devote considerably more time to inequality constraints when we consider continuous optimization problems with the gradient technique. We first wish to consider the minimization of Eq. (10.2-1) subject to the constraint of Eq. (10.2-2) and the q-vector equality constraint $\mathbf{g}(\mathbf{x}) = \mathbf{0}$ where $q \leq n - 1$. By use of the Lagrange multiplier, we write the Hamiltonian as

$$H(\mathbf{x}, \boldsymbol{\lambda}, \boldsymbol{\Gamma}, \mathbf{u}) = \theta(\mathbf{x}, \mathbf{u}) + \boldsymbol{\lambda}^T\mathbf{f}(\mathbf{x}, \mathbf{u}) + \boldsymbol{\Gamma}^T\mathbf{g}(\mathbf{x})$$

and then set $\partial H/\partial \mathbf{u} = \mathbf{0}$ and $\partial H/\partial \mathbf{x} = \mathbf{0}$ to give the set of four nonlinear algebraic vector equations which establishes the necessary conditions for a minimum

$$\frac{\partial \theta(\mathbf{x}, \mathbf{u})}{\partial \mathbf{u}} + \left[\frac{\partial \mathbf{f}^T(\mathbf{x}, \mathbf{u})}{\partial \mathbf{u}}\right]\boldsymbol{\lambda} = \mathbf{0}, \qquad \mathbf{f}(\mathbf{x}, \mathbf{u}) = \mathbf{0}$$

$$\frac{\partial \theta(\mathbf{x}, \mathbf{u})}{\partial \mathbf{x}} + \left[\frac{\partial \mathbf{f}^T(\mathbf{x}, \mathbf{u})}{\partial \mathbf{x}}\right]\boldsymbol{\lambda} + \left[\frac{\partial \mathbf{g}^T(\mathbf{x})}{\partial \mathbf{x}}\right]\boldsymbol{\Gamma} = \mathbf{0}, \qquad \mathbf{g}(\mathbf{x}) = \mathbf{0}.$$

Unfortunately, we may not follow the procedure which we used initially in this section to obtain a successful gradient procedure. The reason is that we can adjust $\mathbf{x}$ such that $\mathbf{f}(\mathbf{x}, \mathbf{u}) = \mathbf{0}$ for arbitrary $\mathbf{u}$, but this value of $\mathbf{x}$ will almost certainly not satisfy $\mathbf{g}(\mathbf{x}) = \mathbf{0}$. A procedure due to Bryson and Denham circumvents this difficulty [12]. Next we will examine a penalty function method in order to accomplish the same goal.

In this penalty function method due to Kelley [11], we eliminate the state variable equality constraint by adding to the cost function a penalty for violation of the constraint. Thus the problem we have posed of minimizing

$$J = \theta(\mathbf{x}, \mathbf{u}) \tag{10.2-30}$$

subject to the equality constraints

$$\mathbf{f}(\mathbf{x}, \mathbf{u}) = \mathbf{0}, \qquad \mathbf{g}(\mathbf{x}) = \mathbf{0}, \tag{10.2-31}$$

becomes, using the penalty function approach, one of minimizing

$$J = \theta(\mathbf{x}, \mathbf{u}) + \mathbf{g}^T(\mathbf{x})\mathbf{N}\mathbf{g}(\mathbf{x}) \tag{10.2-32}$$

subject to

$$\mathbf{f}(\mathbf{x}, \mathbf{u}) = \mathbf{0},$$

where $\mathbf{N}$ is a positive definite diagonal weighting matrix. For increasingly large $\mathbf{N}$, it is reasonable to expect that the violations of the constraint $\mathbf{g}(\mathbf{x}) = \mathbf{0}$ will become smaller. Of course, the equality constraint $\mathbf{f}(\mathbf{x}, \mathbf{u}) = \mathbf{0}$ could also be included in the penalty function, and we would then have the problem of minimizing the free problem where **N** and **M** are positive definite diagonal weighting matrices

$$J = \theta(\mathbf{x}, \mathbf{u}) + \mathbf{g}^T(\mathbf{x})\mathbf{N}\mathbf{g}(\mathbf{x}) + \mathbf{f}^T(\mathbf{x}, \mathbf{u})\mathbf{M}\mathbf{f}(\mathbf{x}, \mathbf{u}). \qquad (10.2\text{-}33)$$

We have previously mentioned inequality constraints for single-stage decision processes. As we alluded, we may first assume that the inequality constraint is never violated; hence we ignore the inequality constraint, start the gradient solution, and proceed until the minimum of $J = \theta(\mathbf{x}, \mathbf{u})$ is reached or until one or more of the inequality constraints are violated. The violated constraint is then treated as if it were an equality constraint, and the computation is resumed. We may also apply the penalty function approach to inequality constraints by converting the inequality constraint $\mathbf{g}(\mathbf{x}) \geq \mathbf{0}$ into a penalty function such that the cost function becomes

$$J = \theta(\mathbf{x}, \mathbf{u}) + \mathbf{g}^T(\mathbf{x})\mathbf{N}\mathbf{H}(\mathbf{g})\mathbf{g}(\mathbf{x}), \qquad (10.2\text{-}34)$$

where **N** is a positive definite diagonal weighting matrix and where $\mathbf{H}(\mathbf{g})$ is a Heaviside diagonal matrix such that $\mathrm{H}_{ii}(\mathbf{g}) = 1$ if $g_i(x) \leq 0$, and $\mathrm{H}_{ii}(\mathbf{g}) = 0$ if $g_i(x) \geq 0$. Again the penalty function is a measure of our violation of the constraint, and other than quadratic losss penalty functions may be used.

10.2-2
The gradient technique for continuous decision processes—the gradient in function space

In this subsection we wish to extend the gradient procedure so that it is capable of solving some typical optimal control problems. We will first assume that there are no inequality constraints, that the initial and terminal times are fixed, and that the initial state is fixed and the terminal state is unspecified. Thus we wish to minimize

$$J = \theta[\mathbf{x}(t_f), t_f] + \int_{t_0}^{t_f} \phi[\mathbf{x}(t), \mathbf{u}(t), t]\, dt \qquad (10.2\text{-}35)$$

for the system

$$\dot{\mathbf{x}} = \mathbf{f}[\mathbf{x}(t), \mathbf{u}(t), t], \qquad \mathbf{x}(t_o) = \mathbf{x}_o. \qquad (10.2\text{-}36)$$

We will later remove many of the foregoing restrictions. From our past study of optimization, we can immediately proceed with the solution to this problem.

We define the Hamiltonian

$$H[\mathbf{x}(t), \mathbf{u}(t), \boldsymbol{\lambda}(t), t] = \phi[\mathbf{x}(t), \mathbf{u}(t), t] + \boldsymbol{\lambda}^T(t)\mathbf{f}[\mathbf{x}(t), \mathbf{u}(t), t] \quad (10.2\text{-}37)$$

and then set

$$\frac{\partial H}{\partial \mathbf{x}} = -\dot{\boldsymbol{\lambda}} = \frac{\partial \phi(\mathbf{x}, \mathbf{u}, t)}{\partial \mathbf{x}} + \left[\frac{\partial \mathbf{f}^T(\mathbf{x}, \mathbf{u}, t)}{\partial \mathbf{x}}\right]\boldsymbol{\lambda}(t) \qquad (10.2\text{-}38)$$

with the terminal condition

$$\boldsymbol{\lambda}(t_f) = \frac{\partial \theta[\mathbf{x}(t_f), t_f]}{\partial \mathbf{x}(t_f)}. \tag{10.2-39}$$

Also, we minimize the Hamiltonian with respect to a choice of **u**. For the case where any control is admissible, we use

$$\frac{\partial H}{\partial \mathbf{u}} = \mathbf{0} = \frac{\partial \phi(\mathbf{x}, \mathbf{u}, t)}{\partial \mathbf{u}} + \frac{\partial \mathbf{f}^T(\mathbf{x}, \mathbf{u}, t)}{\partial \mathbf{u}} \boldsymbol{\lambda}(t). \tag{10.2-40}$$

We shall now guess a solution for the optimal control, and therefore, we will not obtain $\partial H/\partial \mathbf{u} = \mathbf{0}$. We will solve the differential system equality constraint, Eq. (10.2-36), with the assumed value of **u** and also will solve the adjoint Eq. (10.2-38) backward in time from t_f to t_o with the terminal conditions of Eq. (10.2-39). Thus the incremental first-order change in the cost function, Eq. (10.2-35), becomes, for a control differing by an amount $\Delta\mathbf{u}(t)$ from $\mathbf{u}(t)$,

$$\Delta J = \int_{t_o}^{t_f} \left\{ \frac{\partial H[\mathbf{x}(t), \mathbf{u}(t), \boldsymbol{\lambda}(t), t]}{\partial \mathbf{u}(t)} \right\}^T \Delta\mathbf{u}(t)\, dt, \tag{10.2-41}$$

as can easily be demonstrated from Eq. (10.2-38). If we wish to make the largest change in ΔJ, we would calculate the gradient $\partial H/\partial \mathbf{u}$ and then make $\Delta\mathbf{u}$ directed opposite to the gradient

$$\Delta\mathbf{u}(t) = -K(t) \frac{\partial H[\mathbf{x}(t), \mathbf{u}(t), \boldsymbol{\lambda}(t), t]}{\partial \mathbf{u}(t)}. \tag{10.2-42}$$

Again, we see that the change is such as to make J smaller if $K(t) > 0$, which is desirable since we wish to minimize J. To start the procedure, we assume that we have some nonoptimal control $\mathbf{u}^N(t)$. We determine $\mathbf{x}^N(t)$ by solving Eq. (10.2-36) and $\boldsymbol{\lambda}^N(t)$ by solving Eq. (10.2-37) with the terminal condition, Eq. (10.2-39). We then determine $\partial H/\partial \mathbf{u}^N(t)$ from Eq. (10.2-40) which becomes

$$\frac{\partial H}{\partial \mathbf{u}^N} = \frac{\partial \phi(\mathbf{x}^N, \mathbf{u}^N, t)}{\partial \mathbf{u}^N} + \frac{\partial \mathbf{f}^T(\mathbf{x}^N, \mathbf{u}^N, t)}{\partial \mathbf{u}^N} \boldsymbol{\lambda}^N \tag{10.2-43}$$

and then compute $\Delta\mathbf{u}(t)$ from Eq. (10.2-42), where $K(t)$ is a nonnegative time function. The predicted change in J, ΔJ^N, may, if desired, be calculated from Eq. (10.2-41). A new trial value of **u**

$$\mathbf{u}^{N+1}(t) = \mathbf{u}^N(t) + \Delta\mathbf{u}^N(t) \tag{10.2-44}$$

is then found and the procedure repeated until either the control or the cost function does not change significantly from iteration to iteration.

We may summarize these steps in the first-order continuous time gradient method as follows:

1. Determine the Hamiltonian

$$H = \phi[\mathbf{x}(t), \mathbf{u}(t), t] + \boldsymbol{\lambda}^T(t)\mathbf{f}[\mathbf{x}(t), \mathbf{u}(t), t]$$

2. Guess initial values of $\mathbf{u}(t)$ and $\mathbf{x}(t_o)$

3. With these $\mathbf{u}^N(t)$ and $\mathbf{x}^N(t_o)$, solve Eq. (10.2-36) for $\mathbf{x}^N(t)$
4. Solve the adjoint equations

$$\dot{\boldsymbol{\lambda}}^N = -\frac{\partial H}{\partial \mathbf{x}^N(t)}, \qquad \boldsymbol{\lambda}^N(t_f) = \frac{\partial \theta[\mathbf{x}^N(t_f)]}{\partial \mathbf{x}^N(t_f)}$$

   backward in time
5. Determine the incremental change

$$\Delta \mathbf{u}^N(t) = -K^N \frac{\partial H}{\partial \mathbf{u}^N(t)}$$

6. Compute new trial value

$$\mathbf{u}^{N+1}(t) = \mathbf{u}^N(t) + \Delta \mathbf{u}^N(t)$$

7. Repeat the computation starting at step 3. Continue until the change is "slight" from iteration to iteration.

Example 10.2-5. Here we will examine the gradient solution of the simple problem of minimizing $J = \frac{1}{2}Sx^2(1) + \frac{1}{2}\int_0^1 u^2(t)\,dt$ for the system $\dot{x} = u(t)$, $x(0) = 1$. The adjoint equation for this problem is easily shown to be $\dot{\lambda} = 0$ with the terminal condition $\lambda(1) = Sx(1)$. Let us guess the initial solution $u(t) = u^0$, a constant. Then we generate the state for this control which is $x^0(t) = 1 + u^0 t$ and the adjoint which is $\lambda^0(t) = S(1 + u^0)$. Finally, we determine $\partial H/\partial u^N$ which is $u^N(t) + \lambda^N(t) = u^0 + S(1 + u^0)$. The next control is then $u^1(t) = u^0 - K(t)\,\partial H/\partial u^0 = u^0 - K(t)[u^0 + S(1 + u^0)]$. Thus we see that the iterations for this simple problem, if we assume a time-invariant K, are proceeding as

$$u^{N+1} = u^N(1 - K - KS) - KS.$$

The equilibrium point, where $u^{N+1} = u^N$ is easily determined to be $u^N = -S/(1 + S)$, which we recognize as being the correct solution for this problem. Just as in the simple problem which we considered in the previous subsection, we can discern stability requirements as a function of K and the initial value of u, u^0. As we see from the equation for u^{N+1}, convergence is slower as the correct value of u is approached. This is characteristic of the gradient method, and we can circumvent it by increasing K as convergence is approached.

For this very simple example, we may obtain an optimum gradient-adjustment procedure. We would like to change K such that we get as close as possible to the minimum cost for each iteration. In other words, we wish to minimize $\Delta J = J^{N+1} - J^N$. For this problem

$$J^N = \tfrac{1}{2}Sx^N(1)^2 + \tfrac{1}{2}\int_0^1 [u^N(t)]^2\,dt = \tfrac{1}{2}S[1 + u^N]^2 + \tfrac{1}{2}[u^N]^2$$

$$J^{N+1} = \tfrac{1}{2}S[1 + u^N(1 - K - KS) - KS]^2 + \tfrac{1}{2}[u^N(1 - K - KS) - KS]^2$$

and, as we might expect, minimizing J^{N+1} with respect to a choice of K yields $K_{\text{optimum}} = 1/(S + 1)$. If we examine this further, we see that, for this value of K, the next value of u will be always $u^{N+1} = -S/(1 + S)$, which is precisely the correct value. Use of the optimum gradient technique in this case ensures convergence in only one step. This will not occur, in general, even for linear problems.

For higher-order linear problems or for nonlinear problems of any order, the determination of the optimum K for the gradient procedure is a very difficult task. There is still some merit in adjusting K, and a practically efficient method consists of using the past value of K, K^N, one-half the past value, and two and ten times the past value of K^N to determine Δu^N, which in turn determines $u^{N+1} = u^N + \Delta u^N$. The resulting four values of u^{N+1} are then used to determine x^{N+1} and J^{N+1}. The value of $K^{N+1}(\frac{1}{2}K^N, K^N, 2K^N, \text{or } 10K^N)$ which produces the smallest J^{N+1} is then used for the $(N+1)$th iteration by the gradient procedure.

Example 10.2-6. As a second, more realistic example of the gradient in function space technique for the solution of optimal control problems, let us consider the minimization of $J = \frac{1}{2}\int_0^1 (x^2 + u^2)\,dt$ for the system $\dot{x} = -x^2 + u$, $x(0) = 10.0$. We first need to determine the adjoint and the gradient equations. For this problem, the Hamiltonian is $H = \frac{1}{2}x^2 + \frac{1}{2}u^2 - \lambda x^2 + \lambda u$, and thus the adjoint equation is $\dot{\lambda} = -x + 2\lambda x$ with the terminal condition $\lambda(1) = 0$. The control gradient is $\partial H/\partial u = u + \lambda$. Suppose that we guess the initial control $u^0(t) = 0$, which is not too unreasonable, since the final value of the control, $u(1)$, should be zero, and use $K = 1$. To implement the gradient method, the steps we must take are:

1. Determine $x^N(t)$ from $u^N(t)$ for $t \in [0, 1]$ by

$$\dot{x}^N = -[x^N(t)]^2 + u^N(t), \qquad x^N(0) = 10$$

2. Determine $\lambda^N(t)$ from $x^N(t)$ by

$$\dot{\lambda}^N = -x^N(t) + 2\lambda^N(t)x^N(t), \qquad \lambda^N(1) = 0$$

3. Determine $\partial H/\partial u^N$ from

$$\frac{\partial H}{\partial u^N} = u^N(t) + \lambda^N(t)$$

4. Determine $\Delta u^N(t)$ and ΔJ^N from

$$\Delta u^N(t) = -K\frac{\partial H}{\partial u^N} = -Ku^N(t) - K\lambda^N(t)$$

$$\Delta J^N(t) = -K\int_0^1 \left[\frac{\partial H}{\partial u^N}\right]^2 dt = -K\int_0^1 [u^N(t) + \lambda^N(t)]^2\,dt$$

5. Compute the control for the next iteration

$$u^{N+1}(t) = u^N(t) + \Delta u^N(t)$$

6. Shuffle data and repeat the procedure, starting at 1, until $\Delta u^N(t)$ or $\Delta J^N(t)$ changes very little from iteration to iteration.

For the particular initial control assumed here, we obtain for the various steps in the procedure:

1. $u^0(t) = 0, \qquad x^0(t) = 10/(1 + 10t)$
2. $\lambda^0(t) = \frac{1}{2}[1 - (1 + 10t)^2/121]$
3. $\partial H/\partial u^0 = \frac{1}{2}[1 - (1 + 10t)^2/121]$
4. $\Delta u^0(t) = -\frac{1}{2}[1 - (1 + 10t)^2/121]$

$$\Delta J^0(t) = -\frac{1}{4}\int_0^1 [1 - (1 + 10t)^2/121]^2\, dt = -(0.0458)$$

5. $u^1(t) = u^0(t) + \Delta u^0(t) = -\frac{1}{2}[1 - (1 + 10t)^2/121].$

Figures 10.2-1 and 10.2-2 illustrate the *zero*th and first iteration of the control and trajectory as well as the final iteration or correct value. It is interesting

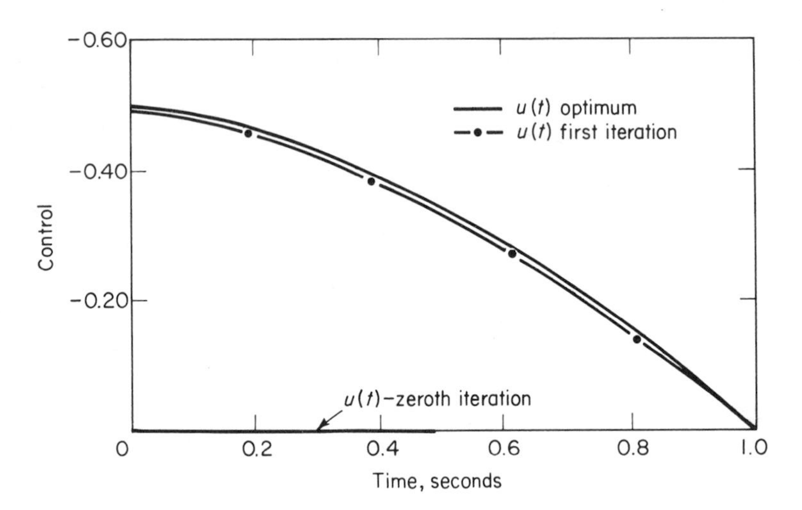

Fig. 10.2-1 Control versus time for Example 10.2-6.

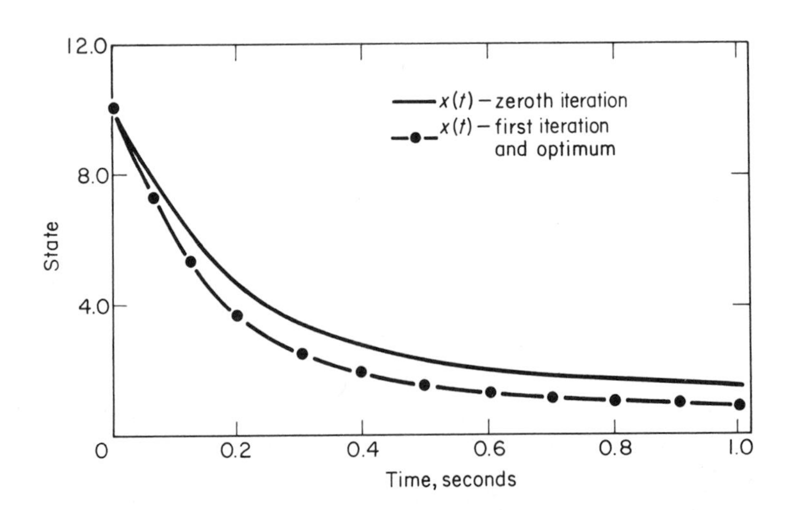

Fig. 10.2-2 State versus time for Example 10.2-6.

to note that, for this particular choice of K, the trajectories have almost converged to the correct value after just one iteration.

We now wish to extend our efforts to cover the case where some components of the state are specified at the fixed terminal time. We will consider the minimization of

$$J = \theta[\mathbf{x}(t_f), t_f] + \int_{t_o}^{t_f} \phi[\mathbf{x}(t), \mathbf{u}(t), t]\, dt \tag{10.2-45}$$

for the system where

$$\dot{\mathbf{x}} = \mathbf{f}[\mathbf{x}(t), \mathbf{u}(t), t], \qquad \mathbf{x}(t_o) = \mathbf{x}_o. \tag{10.2-46}$$

Also, some terminal components of the state vector are specified by the q-vector equality constraint

$$\mathbf{N}[\mathbf{x}(t_f), t_f] = \mathbf{0}. \tag{10.2-47}$$

As in the previous subsection, we can formulate the maximum principle for this problem and obtain the gradient procedure in a straightforward fashion. The difficulty with this, as we have seen, is that satisfying Eq. (10.2-46) by finding $\mathbf{x}^N(t)$ for a given $\mathbf{u}^N(t)$ will not satisfy Eq. (10.2-47). Thus we need to incorporate Eq. (10.2-47) directly into the gradient procedure.

The penalty function approach of Kelley [11] may be used to obtain the solution to the fixed endpoint problem. In this case we add a penalty to the cost function, Eq. (10.2-45), for violation of the terminal manifold equality constraint; thus we minimize

$$J = \mathbf{N}^T[\mathbf{x}(t_f), t_f]\mathbf{G}\mathbf{N}[\mathbf{x}(t_f), t_f] + \theta[\mathbf{x}(t_f), t_f] + \int_{t_o}^{t_f} \phi[\mathbf{x}(t), \mathbf{u}(t), t]\, dt \tag{10.2-48}$$

with the equality constraint

$$\dot{\mathbf{x}} = \mathbf{f}[\mathbf{x}(t), \mathbf{u}(t), t], \qquad \mathbf{x}(t_o) = \mathbf{x}_o, \tag{10.2-49}$$

where $\mathbf{G}$ is a positive definite diagonal weighting matrix. For increasingly large $\mathbf{G}$, we expect that the violations of the constraint $\mathbf{N} = \mathbf{0}$ will become smaller and smaller. A partially unresolved question is: How large should $\mathbf{G}$ be? The answer can generally be obtained only by numerical experimentation.

Example 10.2-7. To illustrate the application of this method for the fixed right-hand end problem, let us consider the minimization of $J = \frac{1}{2}\int_0^1 u^2\, dt$ for the system $\dot{x} = u(t)$, where $x(0) = 1$, $x(1) = 0$. We assume an initial value for the control, a constant u^0 such that $x^0 = 1 + u^0 t$.

We implement the penalty function approach to this problem by considering the cost function as $J = Gx^2(1) + \frac{1}{2}\int_0^1 u^2(t)\, dt$ with the equality constraint $\dot{x} = u(t)$, $x(0) = 1$. This is just Example (10.2-5) with G replaced by S, so there is no need to repeat the solution. We again comment that a potential source of difficulty lies in the fact that we get the correct answer only for infinite G, although a value of $G = 100$ yields only 1 percent error, which may well be acceptable, and a value $G = 10^4$ yields a 0.01 percent error.

Example 10.2-8. Let us now consider the minimization of

$$J = \frac{1}{2}\int_0^1 \rho^2(t)\,dt$$

for the nuclear reactor system described by a single group of delayed neutrons

$$\dot{n} = \frac{\rho - \beta}{\Lambda} n(t) + \lambda c(t)$$

$$\dot{c} = \frac{\beta}{\Lambda} n(t) - \lambda c(t),$$

where

n = neutron flux density $\quad \beta = 0.0064$ = fractions of precursors formed

c = precursor density $\quad \Lambda = 0.001$ sec = neutron lifetime

ρ = reactivity $\quad \lambda = 0.1\ \text{sec}^{-1}$ = precursor decay constant.

We desire to find the control which will drive the neutron flux density to a desired value while minimizing the integral of control squared.

$$n(0) = 0.5\ \text{K.W.}, \qquad n(1) = 5.0$$

$$c(0) = 32.0\ \text{K.W.}, \qquad c(1) = \text{free}.$$

In using the penalty function approach, we rewrite the cost function as

$$J = \tfrac{1}{2}S[n(1) - 5]^2 + \tfrac{1}{2}\int_0^1 \rho^2(t)\,dt$$

with the same kinetics equations as before.

Since the number of steps necessary for convergence is so dependent upon the initial reactivity $\rho^0(t)$, K, ΔN_o, $(\Delta p_o)^2$ and any strategy used to change these values during the computation, it is doubtful that comparisons of the

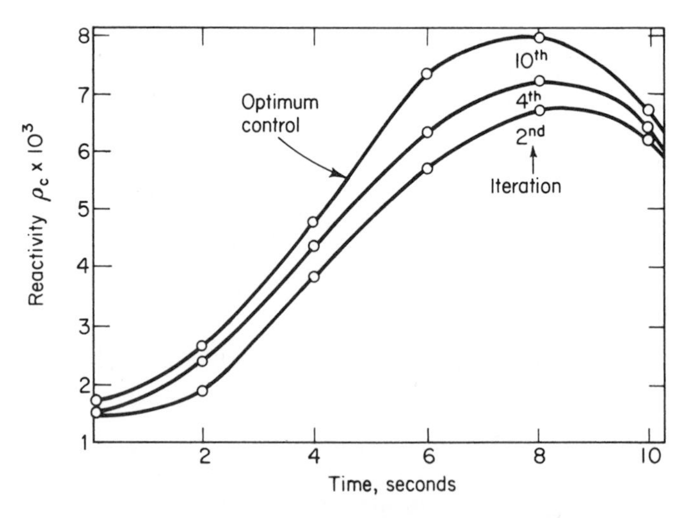

Fig. 10.2-3 Reactivity versus time for Example 10.2-8.

details of the computations have any definite meaning. Figures 10.2-3 and 10.2-4 illustrate typical iterations using the penalty function approach for this problem.

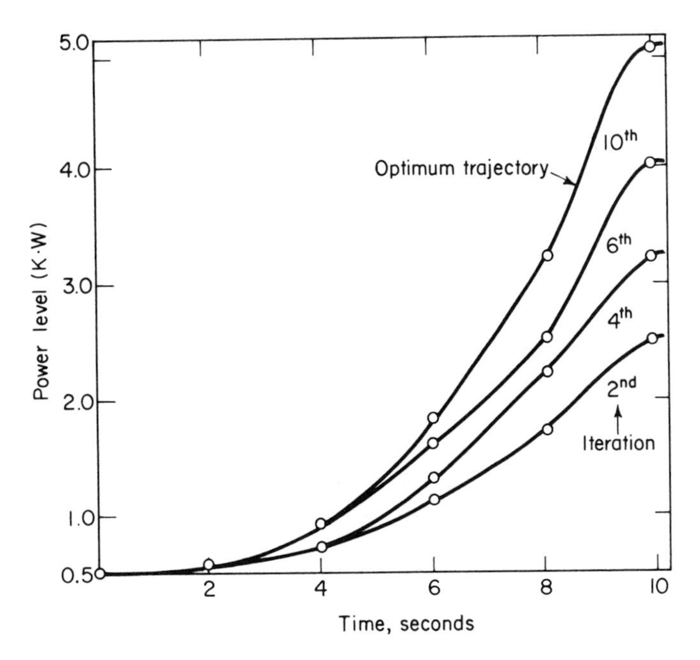

Fig. 10.2-4 Flux density versus time for Example 10.2-8.

We will conclude our study of the gradient technique for continuous decision processes by considering nonfixed terminal time problems and problems with state and/or control variable inequality constraints. We will illustrate the use of the gradient procedure only with the penalty function approach, although it is certainly possible to use the Bryson-Denham procedure for variable terminal time problems [12], as well as for problems with state and control variable inequality constraints [13, 14].

As we have seen in Chapter 4, Sec. 4.2-2, the maximum principle equations for a variable terminal-time problem involve the following relations at the unknown terminal time

$$H[\mathbf{x}(t_f), \mathbf{u}(t_f), \boldsymbol{\lambda}(t_f), t_f] + \frac{\partial \theta[\mathbf{x}(t_f), t_f]}{\partial t_f} + \frac{\partial \mathbf{N}^T[\mathbf{x}(t_f), t_f]}{\partial t_f}\mathbf{v} = 0 \quad (10.2\text{-}50)$$

$$\boldsymbol{\lambda}(t_f) = \frac{\partial \theta[\mathbf{x}(t_f), t_f]}{\partial \mathbf{x}(t_f)} + \frac{\partial \mathbf{N}^T[x(t_f), t_f]}{\partial \mathbf{x}(t_f)}\mathbf{v} = \mathbf{0} \quad (10.2\text{-}51)$$

$$\mathbf{N}[\mathbf{x}(t_f), t_f] = \mathbf{0}. \quad (10.2\text{-}52)$$

We have seen in this subsection that we cannot directly force $\mathbf{N}[\mathbf{x}(t_f), t_f]$

$= \mathbf{0}$ at each and every iteration in the gradient procedure even for fixed t_f. We demonstrated that we could include this terminal state equality constraint as a penalty and assume, in the reformulated problem, that there is no terminal equality constraint. Similarly, we cannot use Eqs. (10.2-50) and (10.2-51) directly in the gradient procedure since we must integrate the adjoint equations backward in time from t_f to determine H, and we do not know t_f until we have solved Eq. (10.2-50). By use of perturbation techniques, Bryson and Denham [12] are able to determine a "predicted" terminal time and effectively use these equations. An alternate method due to Kelley [11] consists of determining and computing dJ/dt on the Nth iteration, with a knowledge of $\mathbf{x}^N(t)$ and $\mathbf{u}^N(t)$. Since the terminal time is free, we choose a terminal time such that the cost function is a minimum with respect to this terminal time or, in other words, $dJ/dt = 0$. It is possible that the time determined from $dJ/dt = 0$ may not be the terminal time, since it could well be the time which maximizes rather than minimizes the cost function. If there is any question of this, we can examine the sign of d^2J/dt^2 when $dJ/dt = 0$.

Thus for a problem posed as one of minimizing

$$J = \theta[\mathbf{x}(t_f), t_f] + \int_{t_o}^{t_f} \phi[\mathbf{x}(t), \mathbf{u}(t), t]\, dt \tag{10.2-53}$$

for the system

$$\dot{\mathbf{x}} = \mathbf{f}[\mathbf{x}(t), \mathbf{u}(t), t], \qquad \mathbf{x}(t_o) = \mathbf{x}_o \tag{10.2-54}$$

with the terminal manifold

$$\mathbf{N}[\mathbf{x}(t_f), t_f] = \mathbf{0}, \tag{10.2-55}$$

where t_f is not fixed, we reformulate the problem as one of minimizing

$$J' = \tfrac{1}{2}\|\mathbf{N}[\mathbf{x}(t_f), t_f]\|_{\mathbf{G}}^2 + \theta[\mathbf{x}(t_f), t_f] + \int_{t_o}^{t_f} \phi[\mathbf{x}(t), \mathbf{u}(t), t]\, dt \tag{10.2-56}$$

for the system described by Eq. (10.2-54), where t_f is determined by

$$\frac{dJ'}{dt_f} = 0 = \left\{\frac{\partial \mathbf{N}^T}{\partial t_f} + \left[\frac{\partial \mathbf{N}}{\partial \mathbf{x}(t_f)}\right]^T \dot{\mathbf{x}}(t_f)\right\} \mathbf{G}\mathbf{N} + \frac{\partial \theta}{\partial t_f} + \left[\frac{\partial \theta}{\partial \mathbf{x}(t_f)}\right]^T \dot{\mathbf{x}}(t_f) + \phi[\mathbf{x}(t_f), \mathbf{u}(t_f), t_f]. \tag{10.2-57}$$

The computation then proceeds in the normal way for the gradient in function space procedure as we have previously outlined.

We are also interested in incorporating control inequality constraints

$$\mathbf{g}[\mathbf{x}(t), \mathbf{u}(t), t] \geq \mathbf{0} \tag{10.2-58}$$

and state variable inequality constraints

$$\mathbf{h}[\mathbf{x}(t), t] \geq \mathbf{0} \tag{10.2-59}$$

into our gradient procedure. Here again we will use only the penalty function method. We considered control and state variable equality constraints at some length in Sec. 4.3 and found that a possible way of incorporating a control inequality constraint was to convert it into an equivalent equality constraint. Thus Eq. (10.2-58) would be replaced by

$$(y_i)^2 = g_i[\mathbf{x}(t), \mathbf{u}(t), t], \qquad i = 1, 2, \dots, r \tag{10.2-60}$$

which will force g_i to be greater than or equal to zero, since $(y_i)^2$ must be greater than or equal to zero. y_i is treated as if it were another control variable, and the problem is solved in the usual way. We will use this approach in the next section to solve a constrained optimization problem via the quasilinearization approach. Here, we cannot use this equation directly since we cannot force g_i to be greater than zero. We have a $\mathbf{u}^N$ and solve $\dot{\mathbf{x}}^N = \mathbf{f}^N$ to obtain $\mathbf{x}^N$ which fixes $\mathbf{g}$. However, we can include a penalty for exceeding the inequality. Thus, we replace Eq. (10.2-60) by a penalty term inside the cost function such that, in part, we have

$$J = \dots + \int_{t_o}^{t_f} \sum_{i=1}^{r} |g_i[\mathbf{x}(t), \mathbf{u}(t), t]|^p \mathrm{H}(g_i)\, dt + \dots, \tag{10.2-61}$$

where p is any positive power which we choose and $\mathrm{H}(g_i)$ is a step function defined by

$$\mathrm{H}(g_i) = \begin{cases} 0, & g_i \geq 0 \\ K_i, & g_i < 0. \end{cases} \tag{10.2-62}$$

We are able to control the depth of penetration of the control inequality constraint by adjusting p and K_i.

The state variable inequality constraint requires a somewhat different treatment, as we know from Chapter 4. We replace the constraint Eq. (10.2-61) by the differential equation

$$\dot{x}_{n+1} = f_{n+1} = |h_1(\mathbf{x}, t)|^{p_1}\mathrm{H}(h_1) + |h_2(\mathbf{x}, t)|^{p_2}\mathrm{H}(h_2) + \dots + |h_s(\mathbf{x}, t)|^{p_s}\mathrm{H}(h_s),$$
$$x_{n+1}(t_o) = 0, \qquad x_{n+1}(t_f) = 0, \tag{10.2-63}$$

or the differential equations

$$\begin{aligned} \dot{x}_{n+1} &= |h_1(x, t)|^{p_1}\mathrm{H}(h_1), & x_{n+1}(t_o) &= 0, & x_{n+1}(t_f) &= 0 \\ \dot{x}_{n+2} &= |h_2(x, t)|^{p_2}\mathrm{H}(h_2), & x_{n+2}(t_o) &= 0, & x_{n+2}(t_f) &= 0 \\ &\vdots \\ \dot{x}_{n+s} &= |h_s(x, t)|^{p_s}\mathrm{H}(h_s), & x_{n+s}(t_o) &= 0, & x_{n+s}(t_f) &= 0. \end{aligned} \tag{10.2-64}$$

Again we cannot require Eqs. (10.2-63) and (10.2-64) in the gradient procedure because there is no way to ensure that the constraint will not be violated. So we add the final values of the x's as penalty functions such that the part of the reformulated cost function becomes either of the following:

$$J = \dots + |x_{n+1}(t_f)|^p K + \dots, \qquad x_{n+1}(t_o) = 0 \tag{10.2-65}$$

$$J = \dots + \sum_{j=1}^{s} |x_{n+i}(t_f)|^{p_i} K_i + \dots, \qquad x_{n+i}(t_o) = 0, \qquad i = 1, 2, \dots, s. \tag{10.2-66}$$

For convenience, the p's are quite often selected as square terms, although this is not necessary. The gradient computation then proceeds in the usual way. The key features of this computation are indicated in the following examples.

Example 10.2-9. A problem which has been successfully solved by the gradient method as well as by the second variation approach [19] and the quasilinearization method (which we will present in later sections) is that of the minimum time transfer from the orbit of Earth to the orbit of Mars. These orbits are assumed to be circular and coplanar, and gravitational attractions of the planets on the rocket are neglected. The normalized dynamics and boundary conditions are assumed to be given by:

$$\dot{r} = w \qquad \text{(radial velocity)}$$

$$w = \frac{v^2}{r} - \frac{\mu}{r^2} + \frac{T \sin\theta}{m_o + \dot{m}t} \qquad \text{(radial acceleration)}$$

$$\dot{v} = \frac{-wv}{r} + \frac{T\cos\theta}{m_o + \dot{m}t} \qquad \text{(circumferential acceleration)}$$

$u(t) = \theta(t) =$ thrust steering angle measured up from local horizontal
$x_1(t_o) = r(0) = 1.0, \qquad x_2(t_f) = 1.525$
$x_2(t_o) = w(0) = 0.0, \qquad x_2(t_f) = 0.0$
$x_3(t_o) = v(0) = 1.0, \qquad x_3(t_f) = .8098$
$\mu =$ gravitational constant $= 1.0, \qquad T =$ thrust magnitude $= 0.1405$
$m_o =$ initial mass $= 1.0, \qquad \dot{m} =$ mass consumption rate $= 0.0749.$

A possible penalty function augmented cost function for the problem is

$$J = t_f + \tfrac{1}{2}s_{11}[x_1(t_f) - 1.525]^2 + \tfrac{1}{2}s_{22}[x_2(t_f)]^2 + \tfrac{1}{2}s_{33}[x_3(t_f) - 0.8098]^2,$$

and thus the criterion for determining the final time is

$$\frac{dJ}{dt_f} = 0 = 1 + s_{11}[x_1(t_f) - 1.525]\dot{x}_1(t_f) + s_{22}x_2(t_f)\dot{x}_2(t_f) + s_{33}[x_3(t_f) - 0.8098]\dot{x}_3(t_f),$$

where we have previously given the various derivatives.

We start the solution to the problem by assuming an initial $\theta^0(t)$ from which we determine the initial trajectory $\mathbf{x}^0(t)$ and the cost function derivative dJ/dt_f. When this derivative goes through zero (and if the second derivative is positive), we have determined the first iteration final time. With this final time and the value of $\mathbf{x}^0(t)$, we solve the adjoint equations backward from t_f to t_o with the final conditions

$$[\boldsymbol{\lambda}^0(t_f)]^T = [s_{11}\{x_1(t_f) - 1.525\}, \quad s_{22}x_2(t_f), \quad s_{33}\{x_3(t_f) - 0.8098\}].$$

We then have enough data to compute the gradient $\partial H/\partial u^0$ and determine the control increment $\Delta u^0(t) = -K[\partial H/\partial u^0]$ and the next trial control $u^1(t) = u^0(t) + \Delta u^0(t)$. The process is repeated until convergence is obtained.

One difficulty with the gradient scheme is that of determining just when the solution has converged, since the convergence may be very slow as the correct control and trajectory are approached. We have previously mentioned the possibility of determining an optimum value of K, and we indicated that this was not feasible for anything other than low-order linear problems. A non-optimum but computationally feasible method of trying K^{N+1} equal to $\frac{1}{2}K^N$, K^N, $2K^N$, $10K^N$, and using the value which produced lowest cost was suggested. If we are using penalty functions, we may also attempt to increase their size. In this problem we could double the **S** matrix after apparent convergence had been reached to see if this significantly refines the solution. Another method [22] consists of shifting the terminal values such that, after a convergence has been reached with values $\mathbf{x}^{C_1}(t_f)$ and an error $\Delta\mathbf{x}^{C_1}(t_f) = \bar{\mathbf{x}}(t_f) - \mathbf{x}^{C_1}(t_f)$, where $\bar{\mathbf{x}}(t_f)$ represents the desired final state, the problem is restarted with the penalty function augmented cost function

$$J' = t_f + \tfrac{1}{2}\|\mathbf{x}(t_f) - \bar{\mathbf{x}}(t_f) + \Delta\mathbf{x}^{C_1}(t_f)\|_{\mathbf{S}}^2$$

which simply biases the desired final value by the error in achieving the desired final value from a previous group of iterations. This method of shifting terminal values may be repeated several times if necessary to refine the end conditions. Clearly this method is also applicable to cases where only certain components of $\mathbf{x}(t_f)$ are specified.

A possible initial control and the final optimal control thrust angle are as shown in Fig. 10.2-5. The number of iterations required for convergence is again highly dependent upon the particular computational strategy used, such as the gain K and the penalty function values, as well as any shifted terminal values.

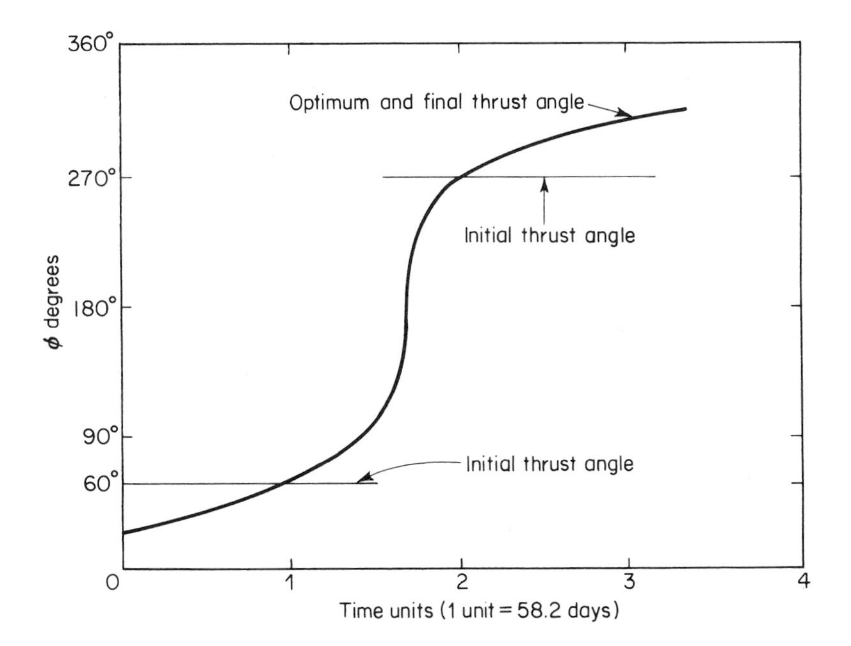

Fig. 10.2-5 Thrust steering angle versus time for Example 10.2-9.

Example 10.2-10. Let us consider the penalty function approach to the gradient in function space method for the computation of the control and trajectory to transfer the linear system $\dot{x}_1 = x_2(t)$, $\dot{x}_2 = u(t)$, $|u(t)| \leq 1$, from the state $x_1(0) = 10$, $x_2(0) = 0$, to the origin $x_1(t_f) = x_2(t_f) = 0$ in minimum time.

We reformulate the problem as one of minimizing

$$J = \tfrac{1}{2}s_{11}x_1^2(t_f) + \tfrac{1}{2}s_{12}x_2^2(t_f) + t_f + \tfrac{1}{2}\int_0^{t_f} Q[u^2(t) - 1]\mathrm{H}[1 - u^2(t)]\, dt$$

for the system

$$\dot{x}_1 = x_2(t), \qquad \dot{x}_2 = u(t), \qquad x_1(0) = 10, \qquad x_2(0) = 10.$$

The pertinent equations needed for the gradient method are the state and adjoint equations and the $\partial H/\partial u$ equation for updating the control. We see that these equations are

$$\dot{\mathbf{x}}^N = \begin{bmatrix} 0 & 1 \\ 0 & 0 \end{bmatrix} \mathbf{x}^N(t) + \begin{bmatrix} 0 \\ 1 \end{bmatrix} u^N(t), \qquad \mathbf{x}^N(0) = \begin{bmatrix} 10 \\ 0 \end{bmatrix}$$

$$\dot{\boldsymbol{\lambda}}^N = -\begin{bmatrix} 0 & 0 \\ 1 & 0 \end{bmatrix} \boldsymbol{\lambda}^N(t), \qquad \boldsymbol{\lambda}^N(t_f) = \begin{bmatrix} s_{11}x_1^N(t_f) \\ s_{22}x_2^N(t_f) \end{bmatrix}$$

$$\frac{\partial H}{\partial u^N} = \lambda_2^N(t) - Qu^N(t)\mathrm{H}(1 - [u^N(t)]^2).$$

In addition, since the terminal time is unspecified, we have the relation $H + \partial\theta/\partial t_f = 0$ at the terminal time. Thus we have

$$\lambda_1^N(t_f)x_2^N(t_f) + \lambda_2^N(t_f)u(t_f) + \tfrac{1}{2}Q\{[u^N(t_f)]^2 - 1\}\mathrm{H}\{1 - [u^N(t_f)]^2\} = 0.$$

This would indeed be a difficult condition to apply since we must start integrating the adjoint backward in time from the unknown terminal time. We therefore see the merit in using $dJ/dt_f = 0$ as a criterion for stopping the iterations. From the reformulated cost function, we have

$$\frac{dJ^N}{dt_f} = s_{11}x_1^N(t_f)\dot{x}_1^N(t_f) + s_{22}x_2^N(t_f)\dot{x}_2^N(t_f) + 1 + \tfrac{1}{2}Q\{[u^N(t_f)]^2 - 1\}\mathrm{H}\{1 - [u^N(t_f)]^2\}.$$

The procedure for calculating the optimal control is as follows. We assume an initial control $u^0(t)$, then calculate the system state and dJ^0/dt_f. When $dJ^0/dt_f = 0$, we have the first iteration final time. With this final time we solve the adjoint equation backward. We then have all the information we require to calculate $\Delta u^N(t) = -K[\partial H/\partial u^N]$ and the next control iterate $u^{N+1}(t) = u^N(t) + \Delta u^N(t)$. The iterations proceed until the change in the control Δu^N from iteration to iteration is very small.

We now assume that the control selected initially is $u^0 = -1$. We solve the system equations to obtain

$$u^0(t) = -1, \qquad x_1^0(t) = 10 - \frac{t^2}{2}, \qquad x_2^0(t) = -t, \qquad 0 < t < \sqrt{10}.$$

Suppose that we apply this control for $\sqrt{10}$ sec and then switch to the $+1$ control. The equations for the control and state become

$$u^0(t) = 1, \qquad x_1^0(t) = \frac{(t - 2\sqrt{10})^2}{2}, \qquad x_2^0(t) = t - 2\sqrt{10}, \qquad t > 10.$$

The time derivatives of the cost function becomes

$$\frac{dJ^0}{dt} = s_{11}\left(\frac{t^3}{2} - 10t\right) + s_{22}(t) + 1, \qquad 0 < t < 10$$

$$\frac{dJ^0}{dt} = \frac{s_{11}(t - 2\sqrt{10})^3}{2} + s_{22}(t - 2\sqrt{10}) + 1, \qquad t > 10.$$

When this time derivative is zero, we have a candidate for the final time. If the second derivative is positive at the candidate final time, we use that value of the candidate final time as the final time for the next iteration. It turns out that we have actually used the optimal control as the initial control for this case. However, we see that the derivative will not go to zero at $t = 2\sqrt{10}$ to give the true final time. Thus, as we expect, we are going to obtain errors in the solution of this problem directly dependent upon the sizes of the penalty functions. From the expression for dJ^0/dt, $t > \sqrt{10}$, we can show that a valid approximation, for s_{11} and s_{22} reasonably large, is

$$t_f \approx 2\sqrt{10} - \frac{1}{s_{22}}$$

and, therefore, conclude that it is easy in this case to ensure very small error with only moderately large penalty functions.

In a more realistic situation, we would not be able to guess the correct control initially. In this case, the value of the state vector at the determined final time is some nonzero $[\mathbf{x}^0(t_f)]^T = [x_1^0(t_f)x_2^0(t_f)]$, and we then solve the adjoint equations

$$\dot{\lambda}_1^0 = 0, \qquad \dot{\lambda}_2^0 = -\lambda_1^0(t), \qquad \lambda_1^0(t_f) = s_{11}x_1^0(t_f), \qquad \lambda_2^0(t_f) = s_{22}x_2^0(t_f)$$

backward in time from t_f to 0. We are then able to determine a numerical value for the gradient $\partial H/\partial u^0$ and determine the next trial control,

$$u^1(t) = u^0(t) + \Delta^0 u(t) = u^0(t) - K\left[\frac{\partial H}{\partial u^0(t)}\right],$$

and thus repeat the iterations until convergence is obtained.

Example 10.2-11. As a final example of optimal control computation using the gradient procedure, let us consider a brachistochrone problem with an inequality constraint on the admissible state-space. Specifically, we consider a particle falling in a constant gravitational field g for a fixed time t_f with a given initial speed $x_3(0)$. We wish to adjust the path of the particle such that final value of the horizontal coordinate $x_1(t_f)$ is maximized. The value of the vertical coordinate $x_2(t)$ is unspecified at the final time. We wish to find the optimum path with the state variable inequality constraint that $x_2 - 0.4x_1 - 0.2 \leq 0$ for the system described by

$$\begin{aligned} \dot{x}_1 &= x_3 \cos u, & x_1(0) &= 0 \\ \dot{x}_2 &= x_3 \sin u, & x_2(0) &= 0 \\ \dot{x}_3 &= g \sin u, & x_3(0) &= 0.07195 \\ g &= 1, & t_f &= 1.720. \end{aligned}$$

This problem has an analytical solution as obtained by Dreyfus [44]. It has previously been solved by the quasilinearization computational technique by McGill [45] and by a gradient method by Jazwinski [46] for a minimum time formulation of the problem. In Chapter 4 we examined this problem and used the maximum principle to formulate the two-point boundary value problem whose solution would yield the optimal trajectory and control. For a detailed account of this TPBVP we refer to Example 4.4-3.

Here we convert the state variable inequality constrain to a differential equation by letting

$$\dot{x}_4 = [x_2(t) - 0.4x_1(t) - 0.20]^2 \mathrm{H}(0.20 + 0.4x_1 - x_2), \qquad x_4(0) = 0,$$

where $\mathrm{H}(g) = 0$ if $g > 0$ and 1 if $g < 0$. We cannot require that $x_4(t_f) = 0$, so we place a penalty for $x_4(t_f)$ being larger than zero and write the modified cost function as

$$J = -x_1(t_f) + \tfrac{1}{2}Sx_4^2(t_f), \qquad t_f = 1.720.$$

The system equations are:

$$\begin{aligned}
\dot{x}_1 &= x_3 \cos u, & x_1(0) &= 0 \\
\dot{x}_2 &= x_3 \sin u, & x_2(0) &= 0 \\
\dot{x}_3 &= \sin u, & x_3(0) &= 0.07195 \\
\dot{x}_4 &= (x_2 - 0.4x_1 - 0.20)^2 \mathrm{H}(0.2 + 0.4x_1 - x_2), & x_4(0) &= 0.
\end{aligned}$$

The computation by the gradient technique proceeds in the usual way. We assume a $u^0(t)$ and calculate $\mathbf{x}^0(t)$ by integrating $\dot{\mathbf{x}}^0 = \mathbf{f}^0$ forward in time from t_o to t_f. We then integrate the adjoint equations backward in time from t_f to t_o with the terminal conditions $\boldsymbol{\lambda}^T(t_f) = [-1, 0, 0, Sx_4(t_f)]$. This enables us to compute $\partial H/\partial u^0$ and thus obtain the control for the next iteration

$$u^1(t) = u^0(t) + \Delta u^0(t) = u^0(t) - K\frac{\partial H}{\partial u^0}.$$

If we assume an initial control $u^0(t) = \pi/6$, we obtain for the *zero*th trajectory

$$x_1^0(t) = 0.062t + 0.216t^2, \qquad x_2^0(t) = 0.036t + 0.125t^2$$
$$x_3^0(t) = 0.07195 + 0.5t, \qquad x_4^0(t) = 0.$$

The adjoint equations, when solved, yield

$$\lambda_1^0(t) = -1, \qquad \lambda_2^0(t) = 0, \qquad \lambda_3^0(t) = 0.866(t - 1.72), \qquad \lambda_4^0(t) = 0.$$

Thus the next control becomes

$$u^1(t) = u^0(t) - K\frac{\partial H}{\partial u^0} = \frac{\pi}{6} + K(1.264 - t).$$

We now choose some suitable value of K and continue the iterations. The initial control did not cause the trajectory to enter the forbidden region, although the final value is fairly close. A value of K of 0.1 will cause a slight penetration of the forbidden region. Thus we need to specify an S in order to

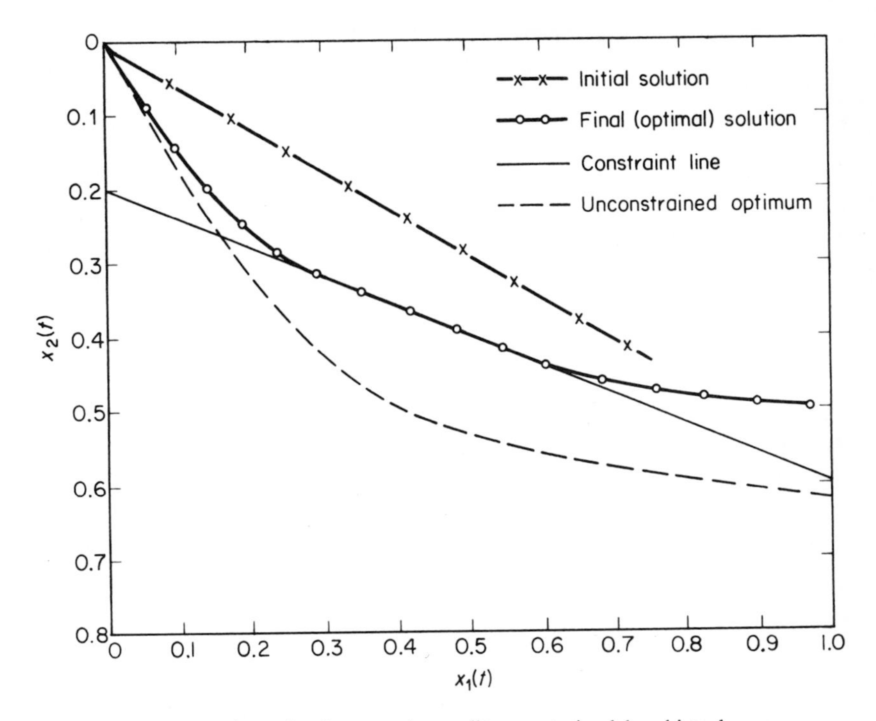

Fig. 10.2-6 Trajectories for state inequality constrained brachistochrone for Example 10.2-11.

continue the computation. If we use $S = 0$, we have no inequality constraint, and the solution converges upon the correct zero constraint solution in eight iterations. For $S = 10^3$ or greater, the final converged solution is indistinguishable from a plot of the true solution. Figure 10.2-6 gives some of the results of the computation.

10.2-3 *The gradient method for multistage decision processes*

We may easily extend the work of the previous subsections to include discrete processes. We wish to minimize

$$J = \theta[\mathbf{x}(k_f), k_f] + \sum_{k=k_o}^{k_f-1} \phi[\mathbf{x}(k), \mathbf{u}(k), k], \tag{10.2-67}$$

where we assume that θ and ϕ include any penalty functions necessary for the consideration of control and state variable inequalities, state variable inequalities, and terminal manifold equality constraints. The difference system is

$$\mathbf{x}(k+1) = \mathbf{f}[\mathbf{x}(k), \mathbf{u}(k), k], \qquad \mathbf{x}(k_o) = \mathbf{x}_o. \tag{10.2-68}$$

The Hamiltonian for this problem is, from Chapter 6,

$$H = \phi[\mathbf{x}(k), \mathbf{u}(k), k] + \boldsymbol{\lambda}^T(k+1)\mathbf{f}[\mathbf{x}(k), \mathbf{u}(k), k]. \tag{10.2-69}$$

We set $\partial H/\partial \mathbf{x}(k) = \boldsymbol{\lambda}(k)$ to obtain the adjoint equation

$$\boldsymbol{\lambda}(k) = \frac{\partial H}{\partial \mathbf{x}(k)} = \frac{\partial \phi}{\partial \mathbf{x}(k)} + \left[\frac{\partial \mathbf{f}^T}{\partial \mathbf{x}(k)}\right]\boldsymbol{\lambda}(k+1), \tag{10.2-70}$$

where the terminal condition on the adjoint equation is

$$\boldsymbol{\lambda}(k_f) = \frac{\partial \theta[\mathbf{x}(k_f), k_f]}{\partial \mathbf{x}(k_f)}. \tag{10.2-71}$$

The computation proceeds as follows. We guess or determine a $\mathbf{u}^N(k)$, $k = k_o, \ldots, k_f$, and then determine $\mathbf{x}^N(k)$ from Eq. (10.2-68). This allows us to solve the adjoint equation, Eq. (10.2-70), backward from stage k_f with the terminal condition of Eq. (10.2-71) to stage k_o. Then we calculate $\partial H/\partial \mathbf{u}^N$,

$$\frac{\partial H}{\partial \mathbf{u}^N(k)} = \frac{\partial \phi[\mathbf{x}^N(k), \mathbf{u}^N(k), k]}{\partial \mathbf{u}^N(k)} + \left[\frac{\partial \mathbf{f}^T[\mathbf{x}(k), \mathbf{u}(k), k]}{\partial \mathbf{u}(k)}\right]\boldsymbol{\lambda}(k+1) \tag{10.2-72}$$

which will not normally be zero. The perturbation in the cost function with perturbations in $\mathbf{x}(k)$ and $\mathbf{u}(k)$ may be obtained by rewriting Eq. (10.2-67) as

$$\begin{aligned} J &= \theta[\mathbf{x}(k_f), k_f] + \sum_{k=k_o}^{k_f-1} [H - \boldsymbol{\lambda}^T(k+1)\mathbf{x}(k+1)] \\ &= \boldsymbol{\lambda}^T(k_o)\mathbf{x}(k_o) + \theta[\mathbf{x}(k_f), k_f] - \boldsymbol{\lambda}^T(k_f)\mathbf{x}(k_f) + \sum_{k=k_o}^{k_f-1} [H - \boldsymbol{\lambda}^T(k)\mathbf{x}(k)]. \end{aligned} \tag{10.2-73}$$

Thus we obtain:

$$\Delta J = +\boldsymbol{\lambda}^T(k_o)\,\Delta\mathbf{x}(k_o) + \left[\frac{\partial \theta}{\partial \mathbf{x}(k_f)} - \boldsymbol{\lambda}(k_f)\right]^T \Delta\mathbf{x}(k_f) + \sum_{k=k_o}^{k_f-1} \left\{\left[\frac{\partial H}{\partial \mathbf{x}(k)} - \boldsymbol{\lambda}(k)\right]^T \Delta\mathbf{x}(k) + \left[\frac{\partial H}{\partial \mathbf{u}(k)}\right]^T \Delta\mathbf{u}(k)\right\}\cdot \tag{10.2-74}$$

From Eqs. (10.2-70) and (10.2-71), and the fact that since $\mathbf{x}(k_o)$ is given $\Delta\mathbf{x}(k_o) = \mathbf{0}$, we have

$$\Delta J^N = \sum_{k=k_o}^{k_f-1} \left\{\frac{\partial H[\mathbf{x}^N(k), \mathbf{u}^N(k), \boldsymbol{\lambda}^N(k+1), k]}{\partial \mathbf{u}^N(k)}\right\}^T \Delta\mathbf{u}^N(k). \tag{10.2-75}$$

Therefore, to minimize ΔJ^N, we set

$$\Delta\mathbf{u}^N(k) = -K(k)\frac{\partial H}{\partial \mathbf{u}^N(k)} \tag{10.2-76}$$

and compute a new control with which to repeat the procedure from

$$\mathbf{u}^{N+1}(k) = \mathbf{u}^N(k) + \Delta\mathbf{u}^N(k). \tag{10.2-77}$$

We continue the iterations until there is little change from iteration to iteration. Since the procedure is so similar to that for continuous processes, there is little need for us to elaborate further for the discrete case.

10.3
Optimization based on the second variation

A natural extension of the gradient method of the previous section consists of adding one more term, the second variation, in the expansion of J. As we have seen, the gradient method is essentially a first-order method, since it is based on finding the first-order effects of the control on the cost function to be minimized and possible terminal constraints. By adding the second variation to our expansion of ΔJ, we obtain a direct second-order method which, as we shall see, achieves the goal of an improved convergence rate near optimum control and trajectory (as $\partial H/\partial \mathbf{u}$ becomes $\mathbf{0}$), but at the expense of some of the desirable features of the gradient procedure. The principal difficulty is the necessity of initiating the second-variation procedure with a rather good approximation to the optimal control, and the greatly increased complexity of the computational algorithms. Of course, the gradient computation may serve to generate initial iterations for the second variation, or the two procedures may be combined. Our initial description of the second variation method is necessarily detailed. The Riccati transformation approach, discussion of which concludes this section, minimizes much of this detail.

As before, we desire to minimize

$$J = \theta[\mathbf{x}(t_f), t_f] + \int_{t_o}^{t_f} \phi[\mathbf{x}(t), \mathbf{u}(t), t]\, dt, \tag{10.3-1}$$

where θ and ϕ may include penalty functions of various types. We wish to minimize Eq. (10.3-1) subject to the equality constraint for the differential system

$$\dot{\mathbf{x}} = \mathbf{f}[\mathbf{x}(t), \mathbf{u}(t), t], \qquad \mathbf{x}(t_o) = \mathbf{x}_o \tag{10.3-2}$$

with the terminal manifold equality constraint

$$\mathbf{N}[\mathbf{x}(t_f), t_f] = \mathbf{0}. \tag{10.3-3}$$

We adjoin Eqs. (10.3-2) and (10.3-3) to Eq. (10.3-1) in the usual way and define the Hamiltonian,

$$H[\mathbf{x}(t), \mathbf{u}(t), \boldsymbol{\lambda}(t), t] = \phi[\mathbf{x}(t), \mathbf{u}(t), t] + \boldsymbol{\lambda}^T(t)\mathbf{f}[\mathbf{x}(t), \mathbf{u}(t), t], \tag{10.3-4}$$

such that the adjoined cost function, Eq. (10.3-1), becomes

$$J = \Theta[\mathbf{x}(t_f), \mathbf{v}, t_f] + \int_{t_o}^{t_f} \{H[\mathbf{x}(t), \mathbf{u}(t), \mathbf{k}(t), t] - \boldsymbol{\lambda}^T(t)\dot{\mathbf{x}}(t)\}\, dt, \tag{10.3-5}$$

where

$$\Theta[\mathbf{x}(t), \mathbf{v}, t_f] = \theta[\mathbf{x}(t_f), t_f] + \mathbf{v}^T\mathbf{N}[\mathbf{x}(t_f), t_f]. \tag{10.3-6}$$

We now expand J to obtain ΔJ just as we did in Chapters 3 and 4. We will regard t_f as being unspecified so that we easily obtain, after some simplification,

$$\Delta J = J_1 + J_2, \tag{10.3-7}$$

where

$$J_1 = \Delta t_f\left\{H[\mathbf{x}(t_f), \mathbf{u}(t_f), \boldsymbol{\lambda}(t_f), t_f] + \frac{\partial\Theta[\mathbf{x}(t_f), \mathbf{v}, t_f]}{\partial t_f}\right\} +$$
$$\Delta_t\mathbf{x}^T(t_f)\left\{\frac{\partial\Theta[\mathbf{x}(t_f), \mathbf{v}, t_f]}{\partial\mathbf{x}(t_f)} - \boldsymbol{\lambda}(t_f)\right\} +$$
$$\Delta_t\mathbf{x}^T(t_o)\boldsymbol{\lambda}(t_o) - \Delta t_o H[\mathbf{x}(t_o), \mathbf{u}(t_o), \boldsymbol{\lambda}(t_o), t_o] +$$
$$\int_{t_o}^{t_f} \Delta\mathbf{x}^T(t)\left\{\frac{\partial H[\mathbf{x}(t), \mathbf{u}(t), \boldsymbol{\lambda}(t), t]}{\partial\mathbf{x}(t)} + \dot{\boldsymbol{\lambda}}\right\} dt +$$
$$\int_{t_o}^{t_f} \Delta\mathbf{u}^T(t)\left\{\frac{\partial H[\mathbf{x}(t), \mathbf{u}(t), \boldsymbol{\lambda}(t), t]}{\partial\mathbf{u}(t)}\right\} dt \tag{10.3-8}$$

$$J_2 = \frac{1}{2}\Delta_t\mathbf{x}^T(t_f)\frac{\partial^2\Theta[\mathbf{x}(t_f), \mathbf{v}, t_f]}{\partial\mathbf{x}^2(t_f)}\Delta_t\mathbf{x}(t_f) + \Delta t_f\Delta_t\mathbf{x}^T(t_f)\frac{\partial^2\Theta[\mathbf{x}(t_f), \mathbf{v}, t_f]}{\partial\mathbf{x}(t_f)\,\partial t_f} +$$
$$\frac{1}{2}(\Delta t_f)^2\frac{\partial^2\Theta[\mathbf{x}(t_f), \mathbf{v}, t_f]}{\partial t_f^2} + \frac{1}{2}\left\{\Delta t\,\Delta_t\mathbf{x}^T(t)\dot{\boldsymbol{\lambda}} + (\Delta t)^2\frac{\partial H[\mathbf{x}(t), \mathbf{u}(t), \boldsymbol{\lambda}(t), t]}{\partial t} +\right.$$
$$\left.\Delta t\,\Delta\mathbf{x}^T(t)\frac{\partial H[\mathbf{x}(t), \mathbf{u}(t), \boldsymbol{\lambda}(t), t]}{\partial\mathbf{x}(t)} + 2\Delta t\,\Delta\mathbf{u}^T(t)\frac{\partial H[\mathbf{x}(t), \mathbf{u}(t), \boldsymbol{\lambda}(t), t]}{\partial\mathbf{u}(t)}\right\}\Bigg|_{t=t_o}^{t=t_f} +$$
$$\frac{1}{2}\int_{t_o}^{t_f}[\Delta\mathbf{x}^T(t)\;\Delta\mathbf{u}^T(t)]\begin{bmatrix}\dfrac{\partial^2 H}{\partial\mathbf{x}^2} & \left[\dfrac{\partial}{\partial\mathbf{u}}\dfrac{\partial H}{\partial\mathbf{x}}\right] \\ \left[\dfrac{\partial}{\partial\mathbf{u}}\dfrac{\partial H}{\partial\mathbf{x}}\right]^T & \dfrac{\partial^2 H}{\partial\mathbf{u}^2}\end{bmatrix}\begin{bmatrix}\Delta\mathbf{x}(t)\\ \Delta\mathbf{u}(t)\end{bmatrix}dt. \tag{10.3-9}$$

In these equations, J_1 becomes the first variation if the change $\Delta_t\mathbf{x}$ and $\Delta\mathbf{u} = \Delta_t\mathbf{u}$ are the variations $\boldsymbol{\delta}\mathbf{x}$ and $\boldsymbol{\delta}\mathbf{u}$. Likewise, J_2 becomes the second variation $\delta^2 J$ if $\Delta_t\mathbf{x}$ and $\Delta\mathbf{u}$ become $\boldsymbol{\delta}\mathbf{x}$ and $\boldsymbol{\delta}\mathbf{u}$. As in Chapter 4, $\mathbf{h}(t) = \Delta\mathbf{x}(t)$ is the change in $\mathbf{x}(t)$ only and does not include a variation in the terminal time, which yields the total change $\Delta_t\mathbf{x}(t_f) = \Delta\mathbf{x}(t_f) + \dot{\mathbf{x}}(t_f)\,\Delta t_f$.

Instead of minimizing J, we shall be content to require Eq. (10.3-7) to be as small as possible. Also, we require Eq. (10.3-2) to hold for small perturbations from the nominal. The latter requirement yields

$$\Delta\dot{\mathbf{x}}(t) = \frac{\partial\mathbf{f}[\mathbf{x}(t), \mathbf{u}(t), t]}{\partial\mathbf{x}(t)}\Delta\mathbf{x}(t) + \frac{\partial\mathbf{f}[\mathbf{x}(t), \mathbf{u}(t), t]}{\partial\mathbf{u}(t)}\Delta\mathbf{u}(t), \qquad \Delta\mathbf{x}(t_o) = \Delta\mathbf{x}_o. \tag{10.3-10}$$

In terms of the variational Hamiltonian, this can be written as

$$\Delta\dot{\mathbf{x}}(t) = \left[\frac{\partial}{\partial\mathbf{x}}\frac{\partial H}{\partial\boldsymbol{\lambda}}\right]\Delta\mathbf{x}(t) + \left[\frac{\partial}{\partial\mathbf{u}}\frac{\partial H}{\partial\boldsymbol{\lambda}}\right]\Delta\mathbf{u}(t), \qquad \Delta\mathbf{x}(t_o) = \Delta\mathbf{x}_o. \tag{10.3-11}$$

We now wish to minimize ΔJ subject to the equality constraint on the perturbation of $\mathbf{x}$ given by Eq. (10.3-11). We choose $\boldsymbol{\lambda}(t)$ so that

$$-\dot{\boldsymbol{\lambda}} = \frac{\partial H}{\partial\mathbf{x}} = \frac{\partial\phi}{\partial\mathbf{x}} + \frac{\partial\mathbf{f}^T}{\partial\mathbf{x}}\boldsymbol{\lambda}(t), \qquad \boldsymbol{\lambda}(t_f) = \frac{\partial\Theta[\mathbf{x}(t_f), \mathbf{v}, t_f]}{\partial\mathbf{x}(t_f)} \tag{10.3-12}$$

and t_f such that

$$H[\mathbf{x}(t_f), \mathbf{u}(t_f), \boldsymbol{\lambda}(t_f), t_f] + \frac{\partial \Theta[\mathbf{x}(t_f), \mathbf{v}, t_f]}{\partial t_f} = 0. \tag{10.3-13}$$

Furthermore, for a first-order extremum, we choose $\mathbf{u}(t)$ such that

$$\frac{\partial H}{\partial \mathbf{u}} = \mathbf{0} = \frac{\partial \phi}{\partial \mathbf{u}} + \frac{\partial \mathbf{f}^T}{\partial \mathbf{u}} \boldsymbol{\lambda}(t). \tag{10.3-14}$$

We wish to consider the foregoing perturbations and to require that

$$\Delta \mathbf{N}[\mathbf{x}(t_f), t_f] = \left[\frac{\partial \mathbf{N}^T[\mathbf{x}(t_f), t_f]}{\partial \mathbf{x}(t_f)}\right]^T \Delta_t \mathbf{x}(t_f) = \Delta \mathbf{N}_f. \tag{10.3-15}$$

The expression for ΔJ in Eq. (10.3-7) becomes, with the equalities of Eqs. (10.3-12), (10.3-13), (10.3-14), and (10.3-15), the terminal manifold requirement $\Delta \mathbf{N}[\mathbf{x}(t_f), t_f] = \Delta \mathbf{N}_f$, and the constraint that the perturbation equation (10.3-10) in $\Delta \mathbf{x}(t)$ is satisfied,

$$\Delta J = \Delta_t^T \mathbf{x}(t_o) \boldsymbol{\lambda}(t_o) - \Delta t_o H[\mathbf{x}(t_o), \mathbf{u}(t_o), \boldsymbol{\lambda}(t_o), t_o] + \Delta \mathbf{v}^T \{\Delta \mathbf{N}[\mathbf{x}(t_f), t_f] - \mathbf{N}_f\} +$$

$$\frac{1}{2}[\Delta_t \mathbf{x}^T(t_f)\ \Delta t_f] \begin{bmatrix} \dfrac{\partial^2 \Theta[\mathbf{x}(t_f), \mathbf{v}, t_f]}{\partial \mathbf{x}^2(t_f)} & \dfrac{\partial^2 \Theta[\mathbf{x}(t_f), \mathbf{v}, t_f]}{\partial \mathbf{x}(t_f)\, \partial t_f} \\ \dfrac{\partial^2 \Theta[\mathbf{x}(t_f), \mathbf{v}, t_f]}{\partial \mathbf{x}(t_f)\, \partial t_f} & \dfrac{\partial^2 \Theta[\mathbf{x}(t_f), \mathbf{v}, t_f]}{\partial t_f^2} \end{bmatrix} \begin{bmatrix} \Delta_t \mathbf{x}(t_f) \\ \Delta t_f \end{bmatrix} +$$

$$\frac{1}{2}\Bigg\{\Delta t\, \Delta_t \mathbf{x}^T(t) \dot{\boldsymbol{\lambda}} + (\Delta t)^2 \frac{\partial H[\mathbf{x}(t), \mathbf{u}(t), \boldsymbol{\lambda}(t), t]}{\partial t} +$$

$$\Delta t\, \Delta \mathbf{x}^T(t) \frac{\partial H[\mathbf{x}(t), \mathbf{u}(t), \boldsymbol{\lambda}(t), t]}{\partial t} + 2\Delta t\, \Delta \mathbf{u}^T(t) \frac{\partial H[\mathbf{x}(t), \mathbf{u}(t), \boldsymbol{\lambda}(t), t]}{\partial \mathbf{u}(t)}\Bigg\}\Bigg|_{t=t_o}^{t=t_f} +$$

$$\frac{1}{2} \int_{t_o}^{t_f} \Bigg\{ [\Delta \mathbf{x}^T(t)\ \Delta \mathbf{u}^T(t)] \begin{bmatrix} \dfrac{\partial^2 H}{\partial \mathbf{x}^2} & \left[\dfrac{\partial}{\partial \mathbf{u}} \dfrac{\partial H}{\partial \mathbf{x}}\right] \\ \left[\dfrac{\partial}{\partial \mathbf{u}} \dfrac{\partial H}{\partial \mathbf{x}}\right]^T & \dfrac{\partial^2 H}{\partial \mathbf{u}^2} \end{bmatrix} \begin{bmatrix} \Delta \mathbf{x}(t) \\ \Delta \mathbf{u}(t) \end{bmatrix} +$$

$$2\Delta \boldsymbol{\lambda}^T(t) \left[\left\{\frac{\partial}{\partial \mathbf{x}} \frac{\partial H}{\partial \boldsymbol{\lambda}}\right\} \Delta \mathbf{x}(t) + \left\{\frac{\partial}{\partial \mathbf{u}} \frac{\partial H}{\partial \boldsymbol{\lambda}}\right\} \Delta \mathbf{u}(t) - \Delta \dot{\mathbf{x}} \right] + 2\Delta \mathbf{u}^T(t) \frac{\partial H}{\partial \mathbf{u}} \Bigg\} dt. \tag{10.3-16}$$

We now apply the calculus of variations to this cost function. We let $\Delta_t \mathbf{x}(t_f)$ take on a variation $\boldsymbol{\delta} \Delta_t \mathbf{x}(t_f)$; $\Delta \mathbf{x}(t)$ take on a variation $\boldsymbol{\delta} \Delta \mathbf{x}(t)$; $\Delta \mathbf{u}(t)$ take on the variation $\boldsymbol{\delta} \Delta \mathbf{u}(t)$; and Δt_f take on a variation $\delta \Delta t_f$. We then set $\delta \Delta J = 0$ to obtain the necessary requirement for a minimum. Alternatively, we may define another Hamiltonian $\mathcal{H}$ as

$$\mathcal{H}[\Delta \mathbf{x}(t), \Delta \mathbf{u}(t), \Delta \boldsymbol{\lambda}(t), t] = \boldsymbol{\Phi} + \Delta \boldsymbol{\lambda}^T(t) \mathbf{F} + \Delta \mathbf{u}^T(t) \frac{\partial H}{\partial \mathbf{u}}, \tag{10.3-17}$$

where

$$\Phi = \tfrac{1}{2}[\Delta\mathbf{x}^T(t)\,\Delta\mathbf{u}^T(t)]\begin{bmatrix} \dfrac{\partial^2 H}{\partial \mathbf{x}^2} & \left[\dfrac{\partial}{\partial \mathbf{u}}\dfrac{\partial H}{\partial \mathbf{x}}\right] \\ \left[\dfrac{\partial}{\partial \mathbf{u}}\dfrac{\partial H}{\partial \mathbf{x}}\right]^T & \dfrac{\partial^2 H}{\partial \mathbf{u}^2} \end{bmatrix}\begin{bmatrix}\Delta\mathbf{x}(t) \\ \Delta\mathbf{u}(t)\end{bmatrix} \tag{10.3-18}$$

$$\mathbf{F} = \left[\frac{\partial}{\partial \mathbf{x}}\frac{\partial H}{\partial \boldsymbol{\lambda}}\right]\Delta\mathbf{x}(t) + \left[\frac{\partial}{\partial \mathbf{u}}\frac{\partial H}{\partial \boldsymbol{\lambda}}\right]\Delta\mathbf{u}(t) \tag{10.3-19}$$

and apply the Pontryagin maximum principle for the case where t_f is free. As we noticed in Chapter 4, the details of satisfying the transversality conditions and determining the final time become a bit lengthy. Obtaining the adjoint and control equations is relatively easy. By either of the approaches suggested before, we obtain the system equation, the adjoint equation, and the control equation:

$$\Delta\dot{\mathbf{x}} = \left[\frac{\partial}{\partial \mathbf{x}}\frac{\partial H}{\partial \boldsymbol{\lambda}}\right]\Delta\mathbf{x}(t) + \left[\frac{\partial}{\partial \mathbf{u}}\frac{\partial H}{\partial \boldsymbol{\lambda}}\right]\Delta\mathbf{u}(t), \qquad \Delta\mathbf{x}(t_o) = \Delta\mathbf{x}_o \tag{10.3-20}$$

$$\Delta\dot{\boldsymbol{\lambda}} = -\frac{\partial \mathcal{H}}{\partial \mathbf{x}} = -\frac{\partial^2 H}{\partial \mathbf{x}^2}\Delta\mathbf{x}(t) - \left[\frac{\partial}{\partial \mathbf{u}}\frac{\partial H}{\partial \mathbf{x}}\right]\Delta\mathbf{u}(t) - \left[\frac{\partial}{\partial \boldsymbol{\lambda}}\frac{\partial H}{\partial \mathbf{x}}\right]\Delta\boldsymbol{\lambda}(t) \tag{10.3-21}$$

$$\mathbf{0} = \frac{\partial \mathcal{H}}{\partial \mathbf{u}} = \frac{\partial^2 H}{\partial \mathbf{u}^2}\Delta\mathbf{u}(t) + \left[\frac{\partial}{\partial \mathbf{u}}\frac{\partial H}{\partial \mathbf{x}}\right]^T\Delta\mathbf{x}(t) + \left[\frac{\partial}{\partial \mathbf{u}}\frac{\partial H}{\partial \boldsymbol{\lambda}}\right]^T\Delta\boldsymbol{\lambda}(t) + \frac{\partial H}{\partial \mathbf{u}}, \tag{10.3-22}$$

where, as we can easily show,

$$\frac{\partial}{\partial \mathbf{x}}\frac{\partial H}{\partial \boldsymbol{\lambda}} = \frac{\partial \mathbf{f}}{\partial \mathbf{x}}, \qquad \frac{\partial}{\partial \boldsymbol{\lambda}}\frac{\partial H}{\partial \mathbf{x}} = \left[\frac{\partial \mathbf{f}}{\partial \mathbf{x}}\right]^T, \qquad \frac{\partial}{\partial \mathbf{u}}\frac{\partial H}{\partial \boldsymbol{\lambda}} = \frac{\partial \mathbf{f}}{\partial \mathbf{u}}. \tag{10.3-23}$$

From Chapter 4, we see that the terminal requirements may be expressed as those given by Speyer [47] and modified for our problem:

$$\Delta\mathbf{x}^T(t_f)\left\{\frac{\partial \mathbf{N}^T[\mathbf{x}(t_f), t_f]}{\partial \mathbf{x}(t_f)}\right\} + \Delta t_f\left\{\frac{\partial \mathbf{N}[\mathbf{x}(t_f), t_f]}{\partial t_f} + \left[\frac{\partial \mathbf{N}^T[\mathbf{x}(t_f), t_f]}{\partial \mathbf{x}(t_f)}\right]^T \mathbf{f}[\mathbf{x}(t_f), \mathbf{u}(t_f), t_f]\right\} = \Delta\mathbf{N}_f \tag{10.3-24}$$

$$\Delta\boldsymbol{\lambda}(t_f) + \Delta t_f\dot{\boldsymbol{\lambda}}(t_f) - \frac{\partial^2\Theta[\mathbf{x}(t_f), \mathbf{v}, t_f]}{\partial \mathbf{x}^2(t_f)}\Delta\mathbf{x}(t_f) - \Delta t_f\frac{\partial^2\Theta[\mathbf{x}(t_f), \mathbf{v}, t_f]}{\partial \mathbf{x}(t_f)\,\partial t_f} - \Delta t_f\frac{\partial \phi[\mathbf{x}(t_f), \mathbf{u}(t_f), t_f]}{\partial \mathbf{x}(t_f)} - \Delta t_f\frac{\partial^2\Theta[\mathbf{x}(t_f), \mathbf{v}, t_f]}{\partial \mathbf{x}^2(t_f)}\mathbf{f}[\mathbf{x}(t_f), \mathbf{u}(t_f), t_f] - \frac{\partial \mathbf{N}^T[\mathbf{x}(t_f), t_f]}{\partial \mathbf{x}(t_f)}\Delta\mathbf{v} = \mathbf{0}. \tag{10.3-25}$$

$$\boldsymbol{\lambda}^T(t_g)\frac{\partial \mathbf{f}[\mathbf{x}(t_f), \mathbf{u}(t_f), t_f]}{\partial \mathbf{x}(t_f)}\Delta\mathbf{x}(t_f) + \boldsymbol{\lambda}^T(t_f)\frac{\partial \mathbf{f}[\mathbf{x}(t_f), \mathbf{u}(t_f), t_f]}{\partial t_f}\Delta t_f + \mathbf{f}^T[\mathbf{x}(t_f), \mathbf{u}(t_f), t_f]\Delta\boldsymbol{\lambda}(t_f) + \left\{\frac{\partial^2\Theta[\mathbf{x}(t_f), \mathbf{v}, t_f]}{\partial \mathbf{x}(t_f)\,\partial t_f}\right\}^T\Delta\mathbf{x}(t_f) +$$

$$\left\{\frac{\partial\phi[\mathbf{x}(t_f), \mathbf{u}(t_f), t_f]}{\partial\mathbf{x}(t_f)}\right\}^T \Delta\mathbf{x}(t_f) + \frac{\partial^2\Theta[\mathbf{x}(t_f), \mathbf{v}, t_f]}{\partial t_f^2}\,\Delta t_f +$$

$$\left\{\frac{\partial^2\Theta[\mathbf{x}(t_f), \mathbf{v}, t_f]}{\partial\mathbf{x}(t_f)\,\partial t_f}\right\}^T \mathbf{f}[\mathbf{x}(t_f), \mathbf{u}(t_f), t_f]\,\Delta t_f + \frac{\partial\phi[\mathbf{x}(t_f), \mathbf{u}(t_f), t_f]}{\partial t_f}\,\Delta t_f +$$

$$\left\{\frac{\partial\phi[\mathbf{x}(t_f), \mathbf{v}, t_f]}{\partial\mathbf{x}(t_f)}\right\}^T \mathbf{f}[\mathbf{x}(t_f), \mathbf{u}(t_f), t_f]\,\Delta t_f + \left\{\frac{\partial\mathbf{N}[\mathbf{x}(t_f), t_f]}{\partial t_f}\right\}^T \Delta\mathbf{v} = 0. \tag{10.3-26}$$

In many cases the terminal time will be specified, and the terminal conditions simplify considerably to become

$$\Delta\mathbf{x}^T(t_f)\left\{\frac{\partial\mathbf{N}^T[\mathbf{x}(t_f), t_f]}{\partial\mathbf{x}(t_f)}\right\} = \Delta\mathbf{N}_f \tag{10.3-27}$$

$$\Delta\boldsymbol{\lambda}(t_f) = \frac{\partial^2\Theta[\mathbf{x}(t_f), \mathbf{v}, t_f]}{\partial\mathbf{x}^2(t_f)}\,\Delta\mathbf{x}(t_f) + \frac{\partial\mathbf{N}^T[\mathbf{x}(t_f), t_f]}{\partial\mathbf{x}(t_f)}\,\Delta\mathbf{v}. \tag{10.3-28}$$

Of particular interest and significance is the fact that we can solve Eq. (10.3-22) for the incremental control. We obtain for $\partial H/\partial\mathbf{u} = 0$,

$$\Delta\mathbf{u}(t) = -\left[\frac{\partial^2 H}{\partial\mathbf{u}^2}\right]^{-1}\left[\left\{\frac{\partial}{\partial\mathbf{u}}\frac{\partial H}{\partial\mathbf{x}}\right\}^T \Delta\mathbf{x}(t) + \left\{\frac{\partial}{\partial\mathbf{u}}\frac{\partial H}{\partial\boldsymbol{\lambda}}\right\}^T \Delta\boldsymbol{\lambda}(t)\right] \tag{10.3-29}$$

in all cases except where $[\partial^2 H/\partial\mathbf{u}^2]$ is singular. If $[\partial^2 H/\partial\mathbf{u}^2]$ is singular at points or on an interval of time, we say that the problem has a singular extremal subarc (such as discussed in Chapter 5), and the second-variation method, as we present it here, will fail. It is convenient to rewrite Eq. (10.3-29) in the form

$$\Delta\mathbf{u}(t) = \mathbf{A}(t)\Delta\mathbf{x}(t) + \mathbf{B}(t)\Delta\boldsymbol{\lambda}(t),$$

$$\mathbf{A}(t) = -\left[\frac{\partial^2 H}{\partial\mathbf{u}^2}\right]^{-1}\left[\frac{\partial}{\partial\mathbf{u}}\frac{\partial H}{\partial\mathbf{x}}\right]^T, \qquad \mathbf{B}(t) = -\left[\frac{\partial^2 H}{\partial\mathbf{u}^2}\right]^{-1}\left[\frac{\partial}{\partial\mathbf{u}}\frac{\partial H}{\partial\boldsymbol{\lambda}}\right]^T. \tag{10.3-30}$$

The perturbed state and adjoint equations, Eqs. (10.3-21) and (10.3-22), can be rewritten as

$$\begin{bmatrix}\Delta\dot{\mathbf{x}}\\ \Delta\dot{\boldsymbol{\lambda}}\end{bmatrix} = \begin{bmatrix}\mathbf{C}_{11}(t) & \mathbf{C}_{12}(t)\\ \mathbf{C}_{21}(t) & \mathbf{C}_{22}(t)\end{bmatrix}\begin{bmatrix}\Delta\mathbf{x}(t)\\ \Delta\boldsymbol{\lambda}(t)\end{bmatrix}, \tag{10.3-31}$$

where

$$\mathbf{C}_{11}(t) = \left[\frac{\partial}{\partial\mathbf{x}}\frac{\partial H}{\partial\boldsymbol{\lambda}}\right] + \left[\frac{\partial}{\partial\mathbf{u}}\frac{\partial H}{\partial\boldsymbol{\lambda}}\right]\mathbf{A}(t) = \frac{\partial\mathbf{f}}{\partial\mathbf{x}} + \frac{\partial\mathbf{f}}{\partial\mathbf{u}}\mathbf{A}(t) \tag{10.3-32}$$

$$\mathbf{C}_{12}(t) = \left[\frac{\partial}{\partial\mathbf{u}}\frac{\partial H}{\partial\boldsymbol{\lambda}}\right]\mathbf{B}(t) = \frac{\partial\mathbf{f}}{\partial\mathbf{u}}\mathbf{B}(t) \tag{10.3-33}$$

$$\mathbf{C}_{21}(t) = -\frac{\partial^2 H}{\partial\mathbf{x}^2} - \left[\frac{\partial}{\partial\mathbf{u}}\frac{\partial H}{\partial\mathbf{x}}\right]\mathbf{A}(t) \tag{10.3-34}$$

$$\mathbf{C}_{22}(t) = -\mathbf{C}_{11}^T(t) = -\frac{\partial}{\partial\boldsymbol{\lambda}}\frac{\partial H}{\partial\mathbf{x}} - \left[\frac{\partial}{\partial\mathbf{u}}\frac{\partial H}{\partial\mathbf{x}}\right]\mathbf{B}(t). \tag{10.3-35}$$

Let us first consider perturbations in the initial state variables $\Delta\mathbf{x}(t_o)$, with no perturbation in the terminal condition $\Delta\mathbf{N}[\mathbf{x}(t_f), t_f] = \mathbf{0}$, as well as no perturbation in the final time. There are then n independent quantities in Eqs. (10.3-24), (10.3-25), and (10.3-26) which we will call $\Delta\mathbf{\Gamma}(t)$. We will now find n solutions to Eq. (10.3-31) which result from setting each one of the $\Delta\mathbf{\Gamma}(t_f)$ equal to unity, with the other $n - 1$ equal to zero. Equation (10.3-31) is then solved backward n times to yield

$$\begin{bmatrix} \Delta\mathbf{x}(t) \\ \Delta\boldsymbol{\lambda}(t) \end{bmatrix} = \begin{bmatrix} \boldsymbol{\varphi}_{x\Gamma}(t, t_f) \\ \boldsymbol{\varphi}_{\lambda\Gamma}(t, t_f) \end{bmatrix} \Delta\mathbf{\Gamma}(t_f). \tag{10.3-36}$$

We can thus write the terminal variation in terms of initial perturbations as

$$\Delta\mathbf{\Gamma}(t_f) = \boldsymbol{\varphi}_{x1}^{-1}(t_o, t_f)\,\Delta\mathbf{x}(t_o), \tag{10.3-37}$$

where $\boldsymbol{\varphi}_{x\Gamma}$ is nonsingular if the variations in the initial conditions are linearly independent. $\Delta\mathbf{\Gamma}(t_f)$ is related to the state and adjoint variables by

$$\begin{bmatrix} \Delta\mathbf{x}(t_f) \\ \Delta\boldsymbol{\lambda}(t_f) \end{bmatrix} = \mathbf{G}(t_f)\,\Delta\mathbf{\Gamma}(t_f), \tag{10.3-38}$$

where $\mathbf{G}(t_f)$ is a $2n \times n$ matrix of known terminal values. We can also obtain the perturbation in the state and adjoint variables due to initial perturbations $\Delta\mathbf{x}(t_o)$ from Eqs. (10.3-36) and (10.3-37) as

$$\begin{bmatrix} \Delta\mathbf{x}(t) \\ \Delta\boldsymbol{\lambda}(t) \end{bmatrix} = \begin{bmatrix} \boldsymbol{\varphi}_{x\Gamma}(t, t_f) \\ \boldsymbol{\varphi}_{\lambda\Gamma}(t, t_f) \end{bmatrix} \boldsymbol{\varphi}_{x\Gamma}^{-1}(t_o, t_f)\,\Delta\mathbf{x}(t_o). \tag{10.3-39}$$

The perturbation time t_o is not particularly significant, so we may repla e t_o in Eq. (10.3-39) be any time t_k. From Eq. (10.3-30) we can determine the closed-loop control law:

$$\begin{aligned} \Delta\mathbf{u}(t) &= [\mathbf{A}(t)\boldsymbol{\varphi}_{x\Gamma}(t, t_f) + \mathbf{B}(t)\boldsymbol{\varphi}_{\lambda\Gamma}(t, t_f)]\boldsymbol{\varphi}_{x\Gamma}^{-1}(t_o, t_f)\,\Delta\mathbf{x}(t_o) \\ &= [\mathbf{A}(t)\boldsymbol{\varphi}_{x\Gamma}(t, t_f) + \mathbf{B}(t)\boldsymbol{\varphi}_{\lambda\Gamma}(t, t_f)]\boldsymbol{\varphi}_{x\Gamma}^{-1}(t, t_f)\,\Delta\mathbf{x}(t). \end{aligned} \tag{10.3-40}$$

We note the very close resemblance between this method of control and that presented earlier in the regulator problem section of Chapter 5. We now see that we have a definitive method for determining the various weighting matrices **R**, **Q**, and **S** for the regulator problem in terms of the original cost function. It is a simple matter to show that this regulator problem becomes one of minimizing

$$\begin{aligned} J = \tfrac{1}{2}\Delta\mathbf{x}^T(t_f)S\,\Delta\mathbf{x}(t_f) + \tfrac{1}{2}\int_{t_o}^{t_f} \{&\Delta\mathbf{x}^T(t)Q(t)\,\Delta\mathbf{x}(t) + \Delta\mathbf{u}^T(t)M^T(t)\,\Delta\mathbf{x}(t) + \\ &\Delta\mathbf{x}^T(t)M(t)\,\Delta\mathbf{u}(t) + \Delta\mathbf{u}^T(t)R(t)\,\Delta\mathbf{u}(t)\}\,dt \end{aligned} \tag{10.3-41}$$

for the system

$$\Delta\dot{\mathbf{x}} = A(t)\,\Delta\mathbf{x}(t) + B(t)\,\Delta\mathbf{u}(t), \qquad \Delta\mathbf{x}(t_o) = \Delta\mathbf{x}_o, \tag{10.3-42}$$

where

$$A(t) = \frac{\partial\mathbf{f}[\hat{\mathbf{x}}(t), \hat{\mathbf{u}}(t), t]}{\partial\hat{\mathbf{x}}(t)}, \qquad B(t) = \frac{\partial\mathbf{f}[\hat{\mathbf{x}}(t), \hat{\mathbf{u}}(t), t]}{\partial\hat{\mathbf{u}}(t)}, \tag{10.3-43}$$

and where $\hat{\mathbf{u}}(t)$ and $\hat{\mathbf{x}}(t)$ are nominal (optimal) controls. We determine the

$\boldsymbol{R}$, $\boldsymbol{Q}$, $\boldsymbol{S}$, and $\boldsymbol{M}$ matrices by evaluating various derivatives of the Hamiltonian about the optimal trajectory

$$Q(t) = \frac{\partial^2 H}{\partial \hat{\mathbf{x}}^2(t)}, \qquad R(t) = \frac{\partial^2 H}{\partial \hat{\mathbf{u}}^2(t)}$$
$$S = \frac{\partial^2 \Theta}{\partial \hat{\mathbf{x}}^2(t_f)}, \qquad M(t) = \frac{\partial}{\partial \hat{\mathbf{u}}(t)} \frac{\partial H}{\partial \hat{\mathbf{x}}(t)}. \tag{10.3-44}$$

We may show that the optimum closed-loop control for this problem is

$$\Delta\mathbf{u}(t) = -R^{-1}(t)[M^T(t) + B^T(t)P(t)]\,\Delta\mathbf{x}(t) = K(t)\,\Delta\mathbf{x}(t), \tag{10.3-45}$$

where

$$\dot{\mathbf{P}} = [-A^T(t) + M(t)R^{-1}(t)B^T(t)]P(t) + P(t)[-A(t) + B(t)R^{-1}(t)M^T(t)] + P(t)B(t)R^{-1}(t)B^T(t)P(t) - Q(t) + M(t)R^{-1}(t)M^T(t) \tag{10.3-46}$$

with the terminal condition $P(t_f) = S$. Thus both systems can be represented by the block diagram of Fig. 10.3-1. Problem 27 demonstrates that this

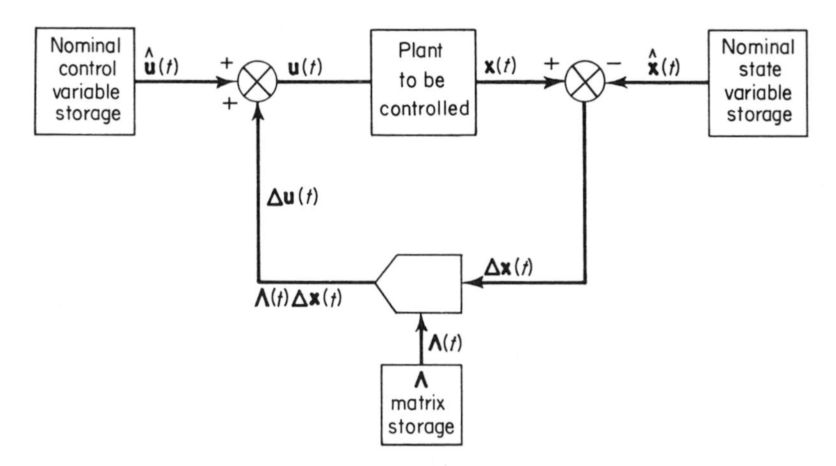

Fig. 10.3-1 "Lambda matrix" control system.

method of control can also be used for a change in terminal manifold $\Delta\mathbf{N}_f$ and that the controls computed for each deviation may be added as shown in Fig. 10.3-2. These gain matrices are called *lambda matrices* by Breakwell, Speyer, and Bryson [21] in their original work on this problem.

We will now find $n + q + 1$ solutions to Eq. (10.3-31) by setting, in succession, each of the n components of $\Delta\mathbf{x}(t_f)$, each of the q components of $\Delta\mathbf{N}[\mathbf{x}(t_f), t_f]$, and the term Δt_f equal to unity, with all other terms but the one whose solution we are seeking set equal to zero. We thus obtain

$$\begin{bmatrix} \Delta\mathbf{x}(t) \\ \Delta\boldsymbol{\lambda}(t) \end{bmatrix} = \begin{bmatrix} \boldsymbol{\varphi}_{xx_f}(t, t_f) & \boldsymbol{\varphi}_{x\nu}(t, t_f) & \boldsymbol{\varphi}_{xt_f}(t, t_f) \\ \boldsymbol{\varphi}_{\lambda x_f}(t, t_f) & \boldsymbol{\varphi}_{\lambda\nu}(t, t_f) & \boldsymbol{\varphi}_{\lambda t_f}(t, t_f) \end{bmatrix} \begin{bmatrix} \Delta\mathbf{x}(t_f) \\ \Delta\boldsymbol{\nu} \\ \Delta t_f \end{bmatrix}, \tag{10.3-47}$$

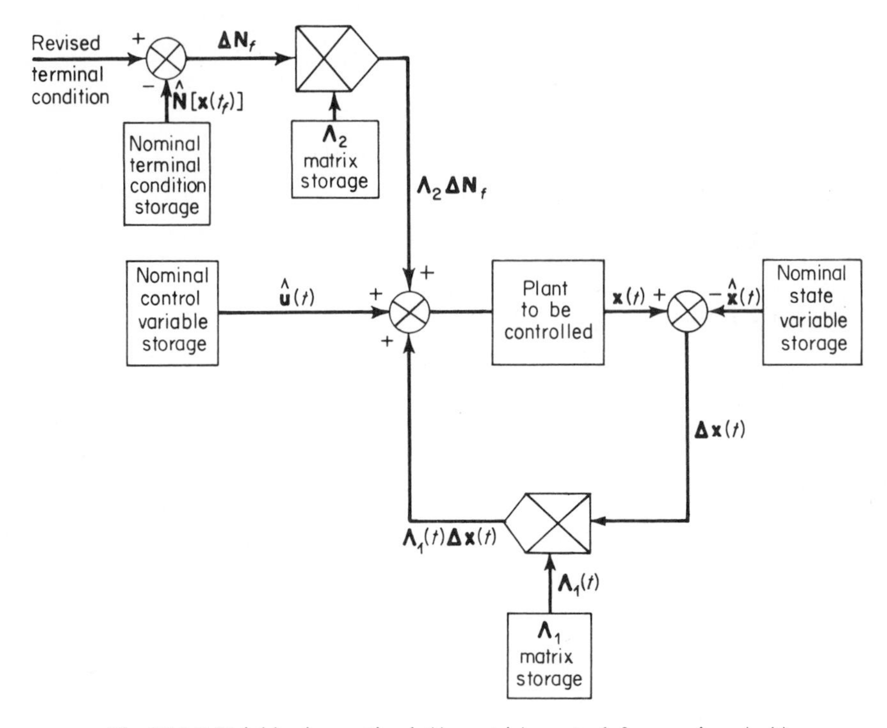

Fig. 10.3-2 Neighboring optimal (Λ matrix) control for varying $\Delta\mathbf{x}(t)$ and $\Delta\mathbf{N}_f$.

where the various $\boldsymbol{\varphi}$ matrices in Eq. (10.3-47) are not state transition matrices but are simply matrices of the unit solutions for $\Delta\mathbf{x}$, $\Delta\mathbf{N}$, and Δt; they are not even square matrices. If we evaluate Eq. (10.3-47) at time t_o and use the terminal conditions of Eqs. (10.3-24), (10.3-25), and (10.3-26) it follows that we have sufficient information to form

$$\begin{bmatrix} \Delta\mathbf{x}(t_o) \\ \Delta\mathbf{N}_f \\ \Delta\tau \end{bmatrix} = \boldsymbol{\Xi} \begin{bmatrix} \Delta x(t_f) \\ \Delta\mathbf{v} \\ \Delta t_f \end{bmatrix}. \tag{10.3-48}$$

If the $(n + q + 1)$ square matrix $\boldsymbol{\Xi}$ is singular, we have what is called a *conjugate point*, and we cannot construct an optimum control system using the second variation in a neighborhood of the nominal trajectory. Various aspects of the conjugate point problem have been investigated by Lee [48]. If there is no conjugate point, we may solve Eq. (10.3-48) for $[\Delta\mathbf{x}(t_f)\Delta\mathbf{v}\Delta t_f]^T$, insert this solution in Eq. (10.3-47), and then use Eq. (10.3-30) to obtain the closed-loop control

$$\Delta\mathbf{u}(t) = [\mathbf{A}(t), \mathbf{B}(t)]\begin{bmatrix} \boldsymbol{\varphi}_{xx_f}(t, t_f) & \boldsymbol{\varphi}_{x\nu}(t, t_f) & \boldsymbol{\varphi}_{xt_f}(t, t_f) \\ \boldsymbol{\varphi}_{\lambda x}(t, t_f) & \boldsymbol{\varphi}_{x\nu}(t, t_f) & \boldsymbol{\varphi}_{\lambda t_f}(t, t_f) \end{bmatrix} \boldsymbol{\Xi}^{-1} \begin{bmatrix} \Delta\mathbf{x}(t_o) \\ \Delta\mathbf{N}_f \\ \Delta\tau \end{bmatrix}. \tag{10.3-49}$$

For the optimal solution, the multiplier $\Delta\tau$ is zero, and we see that we may in fact superimpose the controls due to varying $\Delta\mathbf{x}(t_o)$ and $\Delta\mathbf{N}_f$. Again, there is nothing particularly important here about the time t_o, and we may use $\Delta\mathbf{x}(t)$ in Eq. (10.3-49) if we compute $\boldsymbol{\Xi}^{-1}$ as a function of t rather than t_o.

Quite clearly, this scheme may be used for either continuous or sampled data. It is desirable to develop a discrete version of the second-variation procedure. Since the calculations, either for optimization or control, are to be accomplished on a digital computer, it is conceivable that this discrete version would be computationally more efficient as well as more accurate. We have developed a discrete version of the gradient procedure in the preceding section. In the following sections, we will consider a discrete as well as a continuous version of quasilinearization. Sage and Melsa [38] present a discussion of a discrete second-variation method, but since the development would of necessity by quite lengthy, we will not present it here.

We have seen that the second-variation method may be used for both optimization and control and have given a block diagram for the use of this procedure in control. We will now consider a typical computational procedure for the determination of an optimal control.

Just as in the gradient method, we start the computation by selecting a control $\mathbf{u}^N(t)$ and determining the resulting state vector $\mathbf{x}^N(t)$ by solution of Eq. (10.3-2) for $t \in [t_o, t_f]$. If t_f is not specified, we may determine the time when $dJ^N/dt_f = 0$ and use it as t_f^N. Here J is the original cost function plus penalty functions as in the gradient procedure. We integrate the adjoint equations (10.3-12) backward for $t \in [t_f^N, t_o]$. We then determine the linearized equations for the state, adjoint, and control, Eqs. (10.3-20), (10.3-21), and (10.3-22). These equations are then solved for the transition matrix with $\Delta\mathbf{x}(t_o) = \mathbf{0}$ and the terminal conditions of Eqs. (10.3-24), (10.3-25), and (10.3-26). This determines $\Delta\mathbf{x}^N(t)$ and $\Delta\boldsymbol{\lambda}^N(t)$ for $t \in [t_o, t_f]$ and allows us to determine numerical values for $\Delta\mathbf{u}^N(t)$ from Eq. (10.3-30). A new control is formed from $\mathbf{u}^{N+1}(t) = \mathbf{u}^N(t) + \Delta\mathbf{u}^N(t)$, and the calculation is repeated. An advantage to this procedure is that it is second order, and no arbitrary gain K is needed to add $\mathbf{u}^N(t)$ to $\Delta\mathbf{u}^N(t)$. We gain this advantage at the expense of a considerably more complicated set of computational algorithms. Also, as we should expect and shall soon see demonstrated, this method converges in one step for variational problems where the associated two-point boundary-value problem is linear. A disadvantage to this method is that the region of convergence is normally less than that for the gradient method. Convergence is, however, much more rapid than for the gradient method when good starting values are used. As we shall see, after a brief diversion for some examples, we may introduce a Riccati transformation to considerably simplify the computation of optimal controls via a combined gradient second-variation technique. This will also allow us to control $\partial H/\partial \mathbf{u}^N$.

Example 10.3-1. We will now consider a very simple example which nevertheless illustrates the key features of the computation of an optimal control by means of second variations. We consider the minimization of

$$J = \tfrac{1}{2}x^2(1) + \tfrac{1}{2}\int_0^{t_f} u^2(t)\,dt$$

with fixed $t_f = 1$ for the system

$$\dot{x} = -x(t) + u(t), \qquad x(0) = 1.$$

Just as in the steepest descent or gradient method, we need to obtain and integrate the adjoint equation backward in time. This equation is obtained from the Hamiltonian

$$H(x, u, \lambda, t) = \tfrac{1}{2}u^2 - \lambda x + \lambda u,$$

and the transversality condition is obtained from $\lambda(t_f) = \partial\theta/\partial x(t_f)$ such that we have

$$\dot{\lambda} = \lambda(t), \qquad \lambda(1) = x(1).$$

We now need to determine the "linearized" equations for the state, adjoint, and control, for what is sometimes called the accessory problem. From Eqs. (10.3-20), (10.3-21), and (10.3-22), with the associated transversality conditions of Eqs. (10.3-27) and (10.3-28), these become:

$$\Delta\dot{x} = -\Delta x(t) + \Delta u(t), \qquad \Delta x(0) = \Delta x_0$$
$$\Delta\dot{\lambda} = \Delta\lambda(t), \qquad \Delta\lambda(1) = \Delta x(1)$$
$$0 = \Delta u(t) + \Delta\lambda(t) + u(t) + \lambda(t).$$

Next we obtain the accessory homogeneous Hamilton canonic equations

$$\begin{bmatrix} \Delta\dot{x} \\ \Delta\dot{\lambda} \end{bmatrix} = \begin{bmatrix} -1 & -1 \\ 0 & 1 \end{bmatrix}\begin{bmatrix} \Delta x(t) \\ \Delta\lambda(t) \end{bmatrix}, \qquad \begin{bmatrix} \Delta x(0) \\ \Delta\lambda(1) \end{bmatrix} = \begin{bmatrix} \Delta x_o \\ \Delta x(1) \end{bmatrix}.$$

We obtain the transition matrix and thus have the homogeneous solution,

$$\begin{bmatrix} \Delta x(t) \\ \Delta\lambda(t) \end{bmatrix} = \boldsymbol{\varphi}(t, t_f)\begin{bmatrix} \Delta x(t_f) \\ \Delta\lambda(t_f) \end{bmatrix} = \begin{bmatrix} e^{-(t-t_f)} & \tfrac{1}{2}[e^{-(t-t_f)} - e^{(t-t_f)}] \\ 0 & e^{t-t_f} \end{bmatrix}\begin{bmatrix} \Delta x(t_f) \\ \Delta\lambda(t_f) \end{bmatrix}.$$

Since the problem we are considering here results in a linear two-point boundary value problem, the accessory problem is effectively decoupled from the original problem, and we see that the last equation represents just the Hamiltonian canonic equations for the original problem which we may easily solve. For a nonlinear problem, this separation does not occur, and it is instructive in this simple case to go through the details which we might actually encounter in a nonlinear problem.

If we use a first estimate for the control $u^0(t) = 0$, we obtain from the state equation,

$$x^0(t) = e^{-t},$$

and from the adjoint equation,

$$\lambda^0(t) = e^{(t-1)}x^0(1) = e^{(t-2)}.$$

The initial value of the adjoint is, therefore, $\lambda^0(0) = e^{-2}$. Next we solve the accessory problem, Eq. (10.3-31), with zero initial $\Delta x(t)$, to obtain

$$\Delta x(t) = \tfrac{1}{2}[e^{-t} - e^t][\Delta\lambda(t_o) + e^{-2}], \qquad \Delta\lambda(t) = e^t \Delta\lambda(t_o).$$

We wish to find $\Delta\lambda(t_o)$ such that the terminal conditions on $\lambda(t_f)$ are satisfied. The pertinent equations are solved with $\Delta x(0) = 0$, $u^0(t) = 0$ and $\lambda^0(t) = e^{(t-2)}$. They are:

$$\begin{bmatrix} \Delta x(t) \\ \Delta\lambda(t) \end{bmatrix} = \boldsymbol{\varphi}(t, t_o)\begin{bmatrix} \Delta x(0) \\ \Delta\lambda(0) \end{bmatrix} + \int_0^t \boldsymbol{\varphi}(t, \tau)\begin{bmatrix} -1 \\ 0 \end{bmatrix}\lambda^0(\tau)\, d\tau.$$

To meet the boundary condition, we must have $\Delta\lambda(1) = \Delta x(1)$; we, therefore, have for $\Delta\lambda^0(0)$,

$$\Delta\lambda^0(0) = \frac{e^{-3} - e^{-1}}{3e^1 - e^{-1}}.$$

Hence, the next control iterate is the old iterate plus the incremental value just obtained,

$$u^1(t) = u^0(t) + \Delta u^0(t) = u^0(t) - \Delta\lambda(t) - \lambda^0(t) = \frac{-2e^{t-1}}{3e^1 - e^{-1}}.$$

This is the optimal control, and so we see that convergence has been obtained in one iteration. This occurs for any case in which the two-point boundary value problem is linear. The steps in the procedure for obtaining this control are fairly lengthy. A Riccati transformation can be used to advantage on the problem of computing an optimal control, as we shall see after we examine a simple case with a conjugate point, and a neighboring optimal control problem. We shall formulate this Riccati transformation method in such a fashion that we can control the change in $\partial H/\partial \mathbf{u}$ from iteration to iteration.

Example 10.3-2. We will now examine the neighboring optimal control system for startup of a very simple zero-group, delayed neutron reactor model given by

$$\dot{n} = \rho(t)n(t), \qquad n(0) = n_o = 1,$$

where the cost function is

$$J = \tfrac{1}{2}\alpha[n(1) - 100]^2 + \tfrac{1}{2}\int_0^1 \rho^2(t)\, dt.$$

As we can easily determine, the Hamiltonian and canonic equations for this problem are:

$$\begin{aligned} H &= \tfrac{1}{2}\rho^2(t) + \lambda(t)\rho(t)n(t) \\ \dot{n} &= \rho(t)n(t), \qquad n(0) = 1 \\ \dot{\lambda} &= -\rho(t)\lambda(t), \qquad \lambda(1) = \alpha[n(1) - 100] \\ 0 &= \rho(t) + \lambda(t)n(t). \end{aligned}$$

We are fortunate in that we may obtain an analytical solution to this problem as

$$\hat{n}(t) = e^{4.6t}, \qquad \hat{p}(t) = 4.6, \qquad \hat{\lambda}(t) = -4.6e^{-4.6t}$$

for the case where $\alpha = 0.85$. This yields $\hat{n}(1) = 98$. To implement the closed-loop control, we need the matrices of Eqs. (10.3-43) and (10.3-44) which are, for this example,

$$\mathbf{A}(t) = \hat{p}(t), \qquad \mathbf{B}(t) = \hat{n}(t)$$

$$\mathbf{Q}(t) = 0, \qquad \mathbf{R}(t) = 1, \qquad \mathbf{S} = \alpha, \qquad \mathbf{M}(t) = \hat{\lambda}(t).$$

The optimum closed-loop controller for this example is then determined from Eqs. (10.3-45) and (10.3-46) as

$$\Delta p(t) = [\hat{\lambda}(t) + \hat{n}(t)p(t)]\,\Delta x(t) = [-4.6e^{-4.6t} + e^{4.6t}p(t)]\,\Delta x(t)$$

$$\dot{p} = -18.4p(t) + e^{9.2t}p^2(t) + 21.6e^{-9.2t}, \qquad p(1) = 0.85.$$

The system is implemented as illustrated in Fig. 10.3-3. We have not explored all of the merits of this type of closed-loop control. Among other things, it can be shown to reduce errors between the nominal (optimal) state variables and the actual state variables due to such things as initial errors in knowledge of $n(t_o)$. The scheme is also effective with sufficiently small errors in knowledge of the plant equation $\dot{n} = pn$ (i.e., the scheme works reasonably well if implemented on a system described by $\dot{n} = 1.05pn$), and noise introduced at various points in the loop.

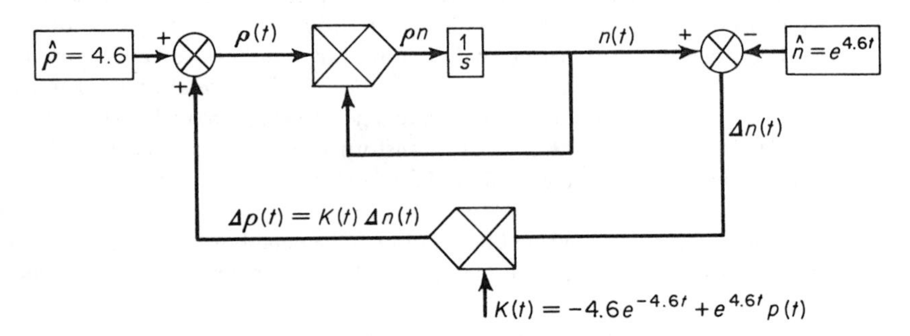

Fig. 10.3-3 Neighboring optimal control of a simple reactor for Example 10.3-2.

In Example 10.3-2 we considered a much more realistic and complicated version of startup of a nuclear reactor. At that time we indicated that the best values of **Q**, **S**, and **R** could be determined by experimentation. We now see that the second variation has furnished us with a precise way of determining weighting matrices for that example.

Example 10.3-3. Let us now consider the problem of determining a neighboring optimum closed-loop control system for a problem first considered by Lee [48] in which we desire to minimize

$$J = \tfrac{1}{2}x^2(t_f) + \tfrac{1}{2}\int_0^{t_f} [x^2(t) + u^2(t)]\,dt,$$

where t_f is fixed, for the nonlinear system

$$\dot{x} = -\sqrt{12 - 2x^2(t)} + u(t), \qquad x(0) = -1.$$

We can easily determine the canonic equations for this problem as

$$\dot{x} = -\sqrt{12 - 2x^2(t)} + u(t), \qquad x(0) = -1$$

$$\dot{\lambda} = -x(t) - \frac{2x(t)\lambda(t)}{\sqrt{12 - 2x^2(t)}}, \qquad \lambda(t_f) = x(t_f)$$

$$u(t) = -\lambda(t).$$

Let us assume that the final time is $5\pi/3$ sec. We may then show that the solution to the canonic equations is

$$x(t) = -2\cos\left(t + \frac{\pi}{3}\right)$$

$$u(t) = 2\left[\sin\left(t + \frac{\pi}{3}\right) + \sqrt{3 - 2\cos^2\left(t + \frac{\pi}{3}\right)}\right].$$

We wish to again determine the neighboring optimal control. We may easily show that

$$A(t) = \frac{2\hat{x}(t)}{\sqrt{12 - 2\hat{x}^2(t)}}, \qquad B(t) = 1$$

$$Q(t) = 1 - \frac{24\hat{u}(t)}{[12 - 2\hat{x}^2(t)]^{3/2}}, \qquad R(t) = 1, \qquad S = 1, \qquad M = 0$$

such that the closed-loop control is obtained from

$$\Delta u(t) = -p(t)\,\Delta x(t),$$

where

$$\dot{p} = -2A(t)p(t) + p^2(t) - Q(t), \qquad p(t_f) = 1.$$

It is possible to solve this by the methods presented in Chapter 8, with the result,

$$p(t) = A(t) - \tan[t_f - \tan^{-1} 3 - t].$$

This $p(t)$ goes to infinity at $t = 7\pi/6 - \tan^{-1} 3 = 2.415$. Since this time is almost in the middle of the optimization interval, we see that it is quite impossible to implement the closed-loop control $\Delta u(t) = -p(t)\,\Delta x(t)$. The reason for this is that our problem possesses a conjugate point. We may show this in another way by showing that the Ξ matrix of Eqs. (10.3-31) and (10.3-48) is singular. Yet another way of illustrating is to show that, for this particular case, the Q matrix for the accessory problem is negative at some times in the interval $[t_o, t_f]$. We know from our previous studies in Chapters 3, 4, 5, and in particular 8, that the Q matrix must be at least nonnegative definite to indeed establish a minimum for the accessory problem.

We will conclude our discussions in this section by describing a Riccati transformation of the accessory problem for the second-variation method.

In addition to including these second-order terms, we will also include first-order effects such that the computational method to be discussed represents a unification and extension of the second-variation and gradient techniques. This procedure was first described by McReynolds and Bryson [17]. To give a unified presentation of the method and to further our understanding of the material we have just presented on the second variation, it is convenient to repose our original problem.

We desire to minimize

$$J = \theta[\mathbf{x}(t_f), t_f] + \int_{t_o}^{t_f} \phi[\mathbf{x}(t), \mathbf{u}(t), t]\, dt \tag{10.3-50}$$

for the differential system

$$\dot{\mathbf{x}} = \mathbf{f}[\mathbf{x}(t), \mathbf{u}(t), t], \qquad \mathbf{x}(t_o) = \mathbf{x}_o \tag{10.3-51}$$

with the terminal manifold equality constraint

$$\mathbf{N}[\mathbf{x}(t_f), t_f] = \mathbf{0}. \tag{10.3-52}$$

The cost function may include penalty functions for various inequality constraints. We will assume that the final time is fixed in our development, although it is certainly possible to consider the case where the final time is unspecified [17]. As before, we define

$$\Theta[\mathbf{x}(t_f), \mathbf{v}, t_f] = \theta[\mathbf{x}(t_f), t_f] + \mathbf{v}^T\mathbf{N}[\mathbf{x}(t_f), t_f] \tag{10.3-53}$$

$$H[\mathbf{x}(t), \mathbf{u}(t), \boldsymbol{\lambda}(t), t] = \phi[\mathbf{x}(t), \mathbf{u}(t), t] + \boldsymbol{\lambda}^T(t)\mathbf{f}[\mathbf{x}(t), \mathbf{u}(t), t] \tag{10.3-54}$$

such that we may write the adjoined cost function, Eq. (10.3-50), as

$$J = \Theta[\mathbf{x}(t_f), \mathbf{v}, t_f] + \int_{t_o}^{t_f} \{H[\mathbf{x}(t), \mathbf{u}(t), \boldsymbol{\lambda}(t), t] - \boldsymbol{\lambda}^T(t)\dot{\mathbf{x}}\}\, dt. \tag{10.3-55}$$

Necessary conditions for an extremum are:

$$\dot{\mathbf{x}} = \frac{\partial H}{\partial \boldsymbol{\lambda}}, \qquad \mathbf{x}(t_o) = \mathbf{x}_o \tag{10.3-56}$$

$$\dot{\boldsymbol{\lambda}} = -\frac{\partial H}{\partial \mathbf{x}}, \qquad \boldsymbol{\lambda}(t_f) = \frac{\partial \Theta}{\partial \mathbf{x}(t_f)} \tag{10.3-57}$$

$$\frac{\partial H}{\partial \mathbf{u}} = \mathbf{0}, \qquad \mathbf{N}[\mathbf{x}(t_f), t_f] = \mathbf{0}. \tag{10.3-58}$$

We will start our computational procedure, just as in the gradient method, by selecting an initial control $\mathbf{u}^0(t)$ and integrating the state equation (10.3-51) or (10.3-56) forward in time $\forall t \in [t_o, t_f]$. Next, the adjoint equations are solved backward in time with the terminal conditions of Eq. (10.3-57). Unless we have actually used the optimal control as $\mathbf{u}^0(t)$, we will not satisfy either of the relationships in Eq. (10.3-58). Now we consider a linear perturbation in the requirements for an extremum, Eqs. (10.3-56), (10.3-57), and (10.3-58), about the given $\mathbf{u}(t)$ and the computed $\mathbf{x}(t)$ and $\boldsymbol{\lambda}(t)$. We easily obtain:

$$\Delta\dot{\mathbf{x}} = \left[\frac{\partial}{\partial \mathbf{x}}\frac{\partial H}{\partial \boldsymbol{\lambda}}\right]\Delta\mathbf{x}(t) + \left[\frac{\partial}{\partial \mathbf{u}}\frac{\partial H}{\partial \boldsymbol{\lambda}}\right]\Delta\mathbf{u}(t) = \frac{\partial \mathbf{f}}{\partial \mathbf{x}}\Delta\mathbf{x}(t) + \frac{\partial \mathbf{f}}{\partial \mathbf{u}}\Delta\mathbf{u}(t),$$

$$\Delta\mathbf{x}(t_o) = \Delta\mathbf{x}_o \tag{10.3-59}$$

$$\Delta\dot{\boldsymbol{\lambda}} = -\frac{\partial^2 H}{\partial \mathbf{x}^2}\Delta\mathbf{x}(t) - \left[\frac{\partial}{\partial \boldsymbol{\lambda}}\frac{\partial H}{\partial \mathbf{x}}\right]\Delta\boldsymbol{\lambda}(t) - \left[\frac{\partial}{\partial \mathbf{u}}\frac{\partial H}{\partial \mathbf{x}}\right]\Delta\mathbf{u}(t) \tag{10.3-60}$$

$$\Delta\boldsymbol{\lambda}(t_f) = \frac{\partial^2 \Theta}{\partial \mathbf{x}^2(t_f)}\Delta\mathbf{x}(t_f) + \left[\frac{\partial}{\partial \mathbf{v}}\frac{\partial \Theta}{\partial \mathbf{x}(t_f)}\right]\Delta\mathbf{v} \tag{10.3-61}$$

$$\Delta\left(\frac{\partial H}{\partial \mathbf{u}}\right) = \left[\frac{\partial}{\partial \mathbf{u}}\frac{\partial H}{\partial \mathbf{x}}\right]^T\Delta\mathbf{x}(t) + \frac{\partial^2 H}{\partial \mathbf{u}^2}\Delta\mathbf{u}(t) + \left[\frac{\partial}{\partial \mathbf{u}}\frac{\partial H}{\partial \boldsymbol{\lambda}}\right]^T\Delta\boldsymbol{\lambda}(t),$$

$$\Delta\mathbf{N}_f = \frac{\partial \mathbf{N}}{\partial \mathbf{x}}\Delta\mathbf{x}(t_f) \tag{10.3-62}$$

where the partial derivatives are, of course, evaluated about the known state, control, and adjoint for a particular iteration.

Equations (10.3-59) through (10.3-62) represent a linear two-point boundary value problem in the perturbed variables in terms of a specified perturbed initial condition $\Delta\mathbf{x}(t_o)$, a specified perturbed terminal manifold $\Delta\mathbf{N}_f$, and a specified perturbation in the control gradient $\Delta(\partial H/\partial\mathbf{u})$. When we considered the second-variation method, we assumed that we were close enough to the optimum so that $\partial H/\partial\mathbf{u}$ could be considered as if it were $\mathbf{0}$. In the gradient procedure, the change in control from one iteration to the next was directly proportional to this $\partial H/\partial\mathbf{u}$ term. Here, we will consider both the first- and second-order terms and assume that $\partial^2 H/\partial\mathbf{u}^2$ is nonsingular (i.e., no singular solutions or singular subarcs exist) such that we obtain, from Eq. (10.3-62),

$$\Delta\mathbf{u}(t) = -\left[\frac{\partial^2 H}{\partial \mathbf{u}^2}\right]^{-1}\left\{-\Delta\left(\frac{\partial H}{\partial \mathbf{u}}\right) + \left[\frac{\partial}{\partial \mathbf{u}}\frac{\partial H}{\partial \mathbf{x}}\right]^T\Delta\mathbf{x}(t) + \left[\frac{\partial}{\partial \mathbf{u}}\frac{\partial H}{\partial \boldsymbol{\lambda}}\right]^T\Delta\boldsymbol{\lambda}(t)\right\}. \tag{10.3-63}$$

By substituting this relation into the perturbed state and adjoint equations (10.3-59) and (10.3-60), we find that

$$\Delta\dot{\mathbf{x}} = \mathbf{C}_{11}(t)\,\Delta\mathbf{x}(t) + \mathbf{C}_{12}(t)\,\Delta\boldsymbol{\lambda}(t) + \mathbf{v}(t) \tag{10.3-64}$$

$$\Delta\dot{\boldsymbol{\lambda}} = \mathbf{C}_{21}(t)\,\Delta\mathbf{x}(t) - \mathbf{C}_{11}^T(t)\,\Delta\boldsymbol{\lambda}(t) + \mathbf{w}(t), \tag{10.3-65}$$

where $\mathbf{C}_{11}(t)$, $\mathbf{C}_{12}(t)$, $\mathbf{C}_{21}(t)$, $\mathbf{v}(t)$, and $\mathbf{w}(t)$ are evaluated about the known iterates as:

$$\mathbf{C}_{11}(t) = \frac{\partial}{\partial \mathbf{x}}\frac{\partial H}{\partial \boldsymbol{\lambda}} - \left[\frac{\partial}{\partial \mathbf{u}}\frac{\partial H}{\partial \boldsymbol{\lambda}}\right]\left[\frac{\partial^2 H}{\partial \mathbf{u}^2}\right]^{-1}\left[\frac{\partial}{\partial \mathbf{u}}\frac{\partial H}{\partial \mathbf{x}}\right]^T \tag{10.3-66}$$

$$\mathbf{C}_{12}(t) = -\left[\frac{\partial}{\partial \mathbf{u}}\frac{\partial H}{\partial \boldsymbol{\lambda}}\right]\left[\frac{\partial^2 H}{\partial \mathbf{u}^2}\right]^{-1}\left[\frac{\partial}{\partial \mathbf{u}}\frac{\partial H}{\partial \boldsymbol{\lambda}}\right]^T \tag{10.3-67}$$

$$\mathbf{C}_{21}(t) = -\frac{\partial^2 H}{\partial \mathbf{x}^2} + \left[\frac{\partial}{\partial \mathbf{u}}\frac{\partial H}{\partial \mathbf{x}}\right]\left[\frac{\partial^2 H}{\partial \mathbf{u}^2}\right]^{-1}\left[\frac{\partial}{\partial \mathbf{u}}\frac{\partial H}{\partial \mathbf{x}}\right]^T \tag{10.3-68}$$

$$\mathbf{v}(t) = \left[\frac{\partial}{\partial \mathbf{u}} \frac{\partial H}{\partial \boldsymbol{\lambda}}\right]\left[\frac{\partial^2 H}{\partial \mathbf{u}^2}\right]^{-1} \Delta\left(\frac{\partial H}{\partial \mathbf{u}}\right) \tag{10.3-69}$$

$$\mathbf{w}(t) = -\left[\frac{\partial}{\partial \mathbf{u}} \frac{\partial H}{\partial \mathbf{x}}\right]\left[\frac{\partial^2 H}{\partial \mathbf{u}^2}\right]^{-1} \Delta\left(\frac{\partial H}{\partial \mathbf{u}}\right). \tag{10.3-70}$$

These equations are different from those of Eq. (10.3-31) only due to the forcing terms $\mathbf{w}(t)$ and $\mathbf{v}(t)$.

We now introduce what is called an *inhomogenous* Riccati transformation [3] which differs but slightly from the homogenous Riccati transformation used in Chapters 5 and 7. We let

$$\Delta\boldsymbol{\lambda}(t) = \mathbf{P}(t)\,\Delta\mathbf{x}(t) + \mathbf{R}(t)\,\Delta\mathbf{v} + \mathbf{h}(t) \tag{10.3-71}$$

$$\Delta\mathbf{N}_f = \mathbf{R}^T(t)\,\Delta\mathbf{x}(t) + \mathbf{Q}(t)\,\Delta\mathbf{v} + \mathbf{g}(t). \tag{10.3-72}$$

By differentiating these two vector equations with respect to time, we obtain

$$\Delta\dot{\boldsymbol{\lambda}} = \mathbf{P}\,\Delta\dot{\mathbf{x}} + \dot{\mathbf{P}}\,\Delta\mathbf{x}(t) + \dot{\mathbf{R}}\,\Delta\mathbf{v} + \dot{\mathbf{h}} \tag{10.3-73}$$

$$\mathbf{0} = \mathbf{R}^T\,\Delta\dot{\mathbf{x}} + \dot{\mathbf{R}}^T\,\Delta\mathbf{x} + \dot{\mathbf{Q}}\,\Delta\mathbf{v} + \dot{\mathbf{g}}. \tag{10.3-74}$$

From Eqs. (10.3-71) and (10.3-64) we have

$$\Delta\dot{\mathbf{x}} = [\mathbf{C}_{11}(t) + \mathbf{C}_{12}(t)\mathbf{P}(t)]\,\Delta\mathbf{x}(t) + \mathbf{C}_{12}(t)\mathbf{R}(t)\,\Delta\mathbf{v} + \mathbf{C}_{12}(t)\mathbf{h}(t) + \mathbf{v}(t). \tag{10.3-75}$$

We next equate Eqs. (10.3-65) and (10.3-73) and use Eqs. (10.3-71) and (10.3-75) to eliminate $\Delta\boldsymbol{\lambda}(t)$ and $\Delta\dot{\mathbf{x}}$. We finally obtain:

$$\begin{aligned}&[\mathbf{C}_{21}(t) - \mathbf{C}_{11}^T(t)\mathbf{P}(t) - \mathbf{P}(t)\mathbf{C}_{11}(t) - \mathbf{P}(t)\mathbf{C}_{12}(t)\mathbf{P}(t) - \dot{\mathbf{P}}]\,\Delta\mathbf{x}(t) - \\ &[\mathbf{C}_{11}^T(t)\mathbf{R}(t) + \mathbf{P}(t)\mathbf{C}_{12}(t)\mathbf{R}(t) + \dot{\mathbf{R}}]\,\Delta\mathbf{v} - [\mathbf{C}_{11}(t)\mathbf{h}(t) + \mathbf{P}(t)\mathbf{C}_{12}(t)\mathbf{h}(t) + \\ &\mathbf{P}(t)\mathbf{v}(t) - \mathbf{w}(t) + \dot{\mathbf{h}}] = \mathbf{0}.\end{aligned} \tag{10.3-76}$$

In an entirely similar way, we also obtain:

$$\begin{aligned}&[\dot{\mathbf{R}}^T + \mathbf{R}^T(t)\mathbf{C}_{11}(t) + \mathbf{R}^T(t)\mathbf{C}_{12}(t)\mathbf{P}(t)]\,\Delta\mathbf{x}(t) + [\mathbf{R}^T(t)\mathbf{C}_{12}(t)\mathbf{R}(t) + \dot{\mathbf{Q}}]\,\Delta\mathbf{v} + \\ &[\mathbf{R}^T(t)\mathbf{C}_{12}(t)\mathbf{h}(t) + \mathbf{R}^T(t)\mathbf{v}(t) + \dot{\mathbf{g}}] = \mathbf{0}.\end{aligned} \tag{10.3-77}$$

These two equations, (10.3-76) and (10.3-77), must be true for arbitrary values of $\Delta\mathbf{x}(t)$ and $\Delta\mathbf{v}$, and thus the coefficients of $\Delta\mathbf{x}(t)$ and $\Delta\mathbf{v}$ must vanish. Also, by comparing the terminal conditions of Eqs. (10.3-61) and (10.3-62) with the Riccati transformation equations, (10.3-71) and (10.3-72), and again realizing that these equations hold for arbitrary $\Delta\mathbf{x}(t)$ and $\Delta\mathbf{v}$, we have the terminal conditions necessary to solve these equations. The "Riccati type" equations are:

$$\dot{\mathbf{P}} = -\mathbf{C}_{11}^T(t)\mathbf{P}(t) - \mathbf{P}(t)\mathbf{C}_{11}(t) - \mathbf{P}(t)\mathbf{C}_{12}(t)\mathbf{P}(t) + \mathbf{C}_{21}(t), \quad \mathbf{P}(t_f) = \frac{\partial^2 \Theta}{\partial \mathbf{x}^2(t_f)} \tag{10.3-78}$$

$$\dot{\mathbf{R}} = -[\mathbf{C}_{11}^T(t) + \mathbf{P}(t)\mathbf{C}_{12}(t)]\mathbf{R}(t), \qquad \mathbf{R}(t_f) = \frac{\partial \mathbf{N}}{\partial \mathbf{x}(t_f)} \tag{10.3-79}$$

$$\dot{\mathbf{Q}} = -\mathbf{R}^T(t)\mathbf{C}_{12}(t)\mathbf{R}(t), \qquad \mathbf{Q}(t_f) = \mathbf{0} \tag{10.3-80}$$

$$\dot{\mathbf{h}} = -[\mathbf{C}_{11}^T(t) + \mathbf{P}(t)\mathbf{C}_{12}(t)]\mathbf{h}(t) - \mathbf{P}(t)\mathbf{v}(t) + \mathbf{w}(t), \qquad \mathbf{h}(t_f) = \mathbf{0} \tag{10.3-81}$$

$$\dot{\mathbf{g}} = -\mathbf{R}^T(t)[\mathbf{C}_{12}(t)\mathbf{h}(t) + \mathbf{v}(t)], \qquad \mathbf{g}(t_f) = \mathbf{0}. \tag{10.3-82}$$

By integrating these last five equations backward in time, together with the perturbed adjoint Eq. (10.3-60), we are able to compute the required change in the perturbed manifold multiplier $\Delta\mathbf{v}$ from Eq. (10.3-72) as

$$\Delta\mathbf{v} = \mathbf{Q}^{-1}(t_o)[\Delta\mathbf{N}_f - \mathbf{g}(t_o) - \mathbf{R}^T(t_o)\Delta\mathbf{x}(t_o)] \tag{10.3-83}$$

in terms of the desired small change in $\Delta(\partial H/\partial\mathbf{u})$. We must specify this in order to compute $\mathbf{w}(t)$ and $\mathbf{v}(t)$ and the desired small changes in $\Delta\mathbf{N}_f$ and $\Delta\mathbf{x}(t_o)$. Once we know $\Delta\mathbf{v}$, we may then solve Eq. (10.3-75) forward in time and use Eqs. (10.3-63) and (10.3-71) to determine $\Delta\mathbf{u}(t)$, $\Delta\boldsymbol{\lambda}(t)$, and $\Delta\mathbf{x}_o(t)$, which are used to update the old control, adjoint, and state for a new iteration.

With the method as developed here, we really need only the perturbed control $\Delta\mathbf{u}(t)$ to start the new iteration. However, there are several rather obvious modifications of the basic procedure in which we may need or desire $\Delta\mathbf{x}(t)$ and $\Delta\boldsymbol{\lambda}(t)$. The method may also be implemented as a neighboring optimum control scheme as illustrated in Figs. 10.3-1 and 10.3-2. To show this we substitute Eq. (10.3-71) into Eq. (10.3-63) to obtain

$$\Delta\mathbf{u}(t) = -\left[\frac{\partial^2 H}{\partial\mathbf{u}^2}\right]^{-1}\left\{-\Delta\left(\frac{\partial H}{\partial\mathbf{u}}\right) + \left[\frac{\partial}{\partial\mathbf{u}}\frac{\partial H}{\partial\mathbf{x}}\right]^T\Delta\mathbf{x} + \left[\frac{\partial}{\partial\mathbf{u}}\frac{\partial H}{\partial\boldsymbol{\lambda}}\right]^T\mathbf{P}(t)\,\Delta\mathbf{x}(t) + \mathbf{R}(t)\,\Delta\mathbf{v} + \mathbf{h}(t)\right\}. \tag{10.3-84}$$

An advantage to this McReynolds and Bryson method over the gradient procedure is that we have rather precise control over the desired changes $\Delta(\partial H/\partial\mathbf{u})$ and $\Delta\mathbf{N}_f$ from iteration to iteration. The method possesses the advantage over the basic second-variation method in that the use of the first-order terms, as well as the second-order terms, ensures convergence over a much wider range of initial controls. Both of the second-variation methods can compute closed-loop controls, whereas the gradient method computes an open-loop control. A method very similar to this has been considered by Bullock [49], and Schley and Lee [50], with generally excellent results. This method is also somewhat similar to the quasilinearization method in the following section. An auxiliary advantage of this method is that it is possible, in many cases, to fix the number of iterations before the computation is run. For example, if N iterations are to be used, we may then pick the change in the control gradient and the terminal manifold from step to step as

$$\Delta^r\left(\frac{\partial H}{\partial\mathbf{u}}\right) = -\xi^r\frac{\partial H^{r-1}}{\partial\mathbf{u}} \tag{10.3-85}$$

$$\Delta^r\mathbf{N}_f = -\xi^r\mathbf{N}^{r-1}[\mathbf{x}(t_f), t_f], \tag{10.3-86}$$

where $\xi^r = r/N$ and $r = 1, 2, \ldots, N$. In a fashion similar to that of Example 10.3-3, we may show that no conjugate points exist if the matrix $\mathbf{P}(t) - \mathbf{R}(t)\mathbf{Q}^{-1}(t)\mathbf{R}^T(t)$ is finite in the interval $[t_o, t_f]$.

Example 10.3-4. To illustrate the computation of an optimal control by this method, which has been called a successive sweep method, let us consider a very simple problem with a free right-hand end. We desire to minimize

$$J = \frac{1}{2}\int_0^{t_f} (x_1^2 + x_2^2 + u^2)\, dt$$

for the driven sinusoidal oscillator

$$\dot{x}_1 = x_2(t), \qquad x_1(0) = x_{10}$$
$$\dot{x}_2 = -x_1(t) + u(t), \qquad x_2(0) = x_{20}.$$

Just as in the steepest descent or gradient method, we need to obtain and integrate the adjoint equations backward in time. The adjoints and their terminal conditions are obtained from the Hamiltonian,

$$H[\mathbf{x}, u, \boldsymbol{\lambda}, t] = \lambda_1(t)x_2(t) + \lambda_2(t)[-x_1(t) + u(t)] + \tfrac{1}{2}[x_1^2(t) + x_2^2(t) + u^2(t)],$$

and the transversality conditions, in the usual way, as

$$\dot{\lambda}_1 = \lambda_2(t) - x_1(t), \qquad \lambda_1(t_f) = 0$$
$$\dot{\lambda}_2 = -\lambda_1(t)x_2(t), \qquad \lambda_2(t_f) = 0.$$

We now need to determine the "linearized" equation for the state, adjoint, and control for the "accessory" problem from Eqs. (10.3-59), (10.3-60), (10.3-61), and (10.3-62). These become:

$$\Delta\dot{\mathbf{x}} = \begin{bmatrix}\Delta x_1\\ \Delta x_2\end{bmatrix} = \begin{bmatrix}0 & 1\\ -1 & 0\end{bmatrix}\Delta\mathbf{x}(t) + \begin{bmatrix}0\\ 1\end{bmatrix}\Delta u(t), \qquad \Delta\mathbf{x}(0) = \Delta\mathbf{x}_o$$

$$\Delta\dot{\boldsymbol{\lambda}} = \begin{bmatrix}\Delta\dot{\lambda}_1\\ \Delta\dot{\lambda}_2\end{bmatrix} = -\begin{bmatrix}1 & 0\\ 0 & 1\end{bmatrix}\Delta\mathbf{x}(t) - \begin{bmatrix}0 & -1\\ 1 & 0\end{bmatrix}\Delta\boldsymbol{\lambda}(t), \qquad \boldsymbol{\lambda}(t_f) = \mathbf{0}$$

$$\Delta\left(\frac{\partial H}{\partial u}\right) = \Delta u(t) + \Delta\lambda_2(t).$$

The Riccati equations which we need to solve are from Eqs. (10.3-78) through (10.3-82):

$$\dot{p}_{11} = 2p_{12}(t) + p_{12}^2(t) - 1, \qquad p_{11}(t_f) = 0$$
$$\dot{p}_{12} = p_{22}(t) - p_{11}(t) + p_{12}(t)p_{22}(t), \qquad p_{12}(t_f) = 0$$
$$\dot{p}_{22} = -p_{12}(t) - 1 + p_{22}^2(t), \qquad p_{22}(t_f) = 0$$
$$\dot{h}_1 = h_2(t) + p_{12}(t)\left[h_2(t) - \Delta\left(\frac{\partial H}{\partial u}\right)\right], \qquad h_1(t_f) = 0$$
$$\dot{h}_2 = -h_1(t) + p_{22}\left[h_2(t) - \Delta\left(\frac{\partial H}{\partial u}\right)\right], \qquad h_2(t_f) = 0.$$

These equations are solved backward in time, and then Eq. (10.3-84), which is here

$$\Delta u(t) = \Delta\left(\frac{\partial H}{\partial u}\right) - p_{21}(t)\,\Delta x_1(t) - p_{22}(t)\,\Delta x_2(t) - h_2(t),$$

is used to update the control. Since we are now using a combination gradient, second-variation procedure, the computation will not normally converge in one iteration unless we let, from Eq. (10.3-85) with $r = N = 1$,

$$\Delta^1\left(\frac{\partial H}{\partial u}\right) = \frac{\partial H^0}{\partial u},$$

in which case convergence is obtained in a single iteration.

Example 10.3-5. We now desire to minimize the cost function

$$J = \tfrac{1}{2}x^2(t_f) + \tfrac{1}{2}\int_0^{t_f} u^2(t)\,dt$$

for the nonlinear system described by

$$\dot{x} = -x^2(t) + u(t), \qquad x(0) = 10.$$

The Hamiltonian and adjoint equation for our problem are

$$H[x(t), u(t), \lambda(t), t] = \tfrac{1}{2}u^2(t) + \lambda(t)[-x^2(t) + u(t)]$$
$$\dot{\lambda} = 2x(t)\lambda(t), \qquad \lambda(t_f) = x(t_f).$$

The linearized equations for the perturbed system are from Eqs. (10.3-59), (10.3-60), (10.3-61), and (10.3-62), where we will now use symbols to indicate iteration number

$$\Delta\dot{x}^N = -2x^{N-1}(t)\,\Delta x^N(t) + \Delta u^N(t), \qquad \Delta x^N(0) = \Delta x_0^N$$
$$\Delta\dot{\lambda}^N = 2x^{N-1}(t)\,\Delta\lambda^N(t) + 2\lambda^{N-1}(t)\,\Delta x^N(t), \qquad \Delta\lambda^N(t_f) = \Delta x^N(t_f)$$
$$\Delta\left(\frac{\partial H}{\partial u}\right) = \Delta u(t) + \Delta\lambda(t).$$

The various terms in Eqs. (10.3-66) through (10.3-70) are

$$C_{11}^N(t) = -2x^{N-1}(t), \qquad C_{12}^N(t) = -1, \qquad C_{21}^N(t) = 2\lambda^{N-1}(t)$$
$$v^N(t) = \Delta^N\left(\frac{\partial H}{\partial u}\right), \qquad w(t) = 0$$

We therefore need to solve the Riccati-type equations from Eqs. (10.3-79) through (10.3-82),

$$\dot{p}^N = 4x^{N-1}(t)p^N(t) + [p^N(t)]^2 + 2\lambda^{N-1}(t), \qquad p^N(t_f) = 1$$
$$\dot{h}^N = [2x^{N-1}(t) + p^N(t)]h^N(t) - p^N(t)\,\Delta^N\left(\frac{\partial H}{\partial u}\right), \qquad h^N(t_f) = 0.$$

These are solved backward in time as always. The next control is then computed from $u^{N+1}(t) = u^N(t) + \Delta u^N(t)$, where $\Delta u^N(t)$ is obtained from Eq. (10.3-84) as

$$\Delta u^N(t) = -\left\{-\Delta^N\left(\frac{\partial H}{\partial u}\right) + p^N\,\Delta x^N(t) + h^N(t)\right\}\cdot$$

A computer solution of these equations converges to the correct solution in four iterations, starting with an initial guess for the control of zero. A potential source of trouble with this method is the appearance of conjugate points at some stage in the computation. Even if the true optimal solution does not possess a conjugate point, there is no guarantee that the computation, on some iteration, will not pass through a conjugate point for that iteration. Bullock [49] notes the occurrence of this in attempting to solve a forced van der Pol equation optimization problem by this method.

We might also remark that we may consider approximately optimum linearized stochastic control problems for nonlinear systems by use of the second variation method as represented in the *lambda matrix* control system of Figs. 10.3-1 and 10.3-2. This is achieved by use of a linearized Kalman filter such as derived in Chapter 8 to generate estimates of system states. These estimated states are used as the true states for purposes of closed-loop control determination. Of course the separation theorem does not apply for nonlinear systems. However, we may consider the linearized cost function and system of Eqs. (10.3-41) through (10.3-46), put expectation signs in front of the cost function to the system equation, and add an observation equation which can be linearized about an operating point. The separation theorem, under the assumptions stated in Chapter 9, is valid for this linearized system.

10.4 Quasilinearization

Now that we have considered several direct methods for the solution of optimum systems control problems, we turn our attention to the solution of these problems using the indirect method of quasilinearization.

10.4-1 Continuous-time quasilinearization

We will consider a differential system of the form

$$\dot{\mathbf{x}} = \mathbf{f}(\mathbf{x}, t). \tag{10.4-1}$$

This vector differential equation is assumed to have a unique solution over the interval $t \in [t_o, t_f]$ and is subject to the multipoint boundary conditions

$$\langle \mathbf{C}(t_i), \mathbf{x}(t_i) \rangle = b_i, \tag{10.4-2}$$

where $\langle \ , \ \rangle$ indicates inner product and where $t_i \in t$, $i \in j$, $j = 1, 2, \ldots, n$.

In the foregoing equations, $\mathbf{x}(t)$ is an n-dimensional state vector, $\mathbf{f}$ is an n-dimensional vector function, $\mathbf{C}(t_i)$ is an n-dimensional coefficient vector, and the b_i are a set of n scalars corresponding to the n boundary conditions.

We determine the trajectories of the state equations by solving Eq. (10.4-1) subject to the boundary conditions imposed by Eq. (10.4-2). Solving this nonlinear, multipoint, boundary value problem is generally a formidable undertaking.

Continuous quasilinearization is a technique whereby a nonlinear, multipoint, boundary value problem is transformed into a more readily solvable linear, nonstationary boundary value problem. This technique involves the study of a sequence of vectors, $\{\mathbf{x}^N(t)\}$, which can be made to approximate the true solution of the nonlinear system. These sequences of vectors obey

$$\dot{\mathbf{x}}^{N+1} = \mathbf{f}(\mathbf{x}^N, t) + [J_{\mathbf{x}^N}\mathbf{f}(\mathbf{x}^N, t)][\mathbf{x}^{N+1}(t) - \mathbf{x}^N(t)], \quad N = 0, 1, 2, \ldots \qquad (10.4\text{-}3)$$

with boundary conditions given by

$$\langle \mathbf{C}(t_i), \mathbf{x}^{N+1}(t_i) \rangle = b_i, \qquad i = 1, 2, \ldots, n. \qquad (10.4\text{-}4)$$

The Jacobian matrix $J_{\mathbf{x}^N}\mathbf{f}$ is defined by†

$$[J_{\mathbf{x}^N}\mathbf{f}(\mathbf{x}^N, t)] = \frac{\partial \mathbf{f}(\mathbf{x}^N)}{\partial \mathbf{x}^N} = \frac{\partial(f_1, f_2, \ldots, f_n)}{\partial(x_1, x_2, \ldots, x_n)} \qquad (10.4\text{-}5)$$

which has the ijth element

$$J_{\mathbf{x}^N}\mathbf{f}(\mathbf{x}^N, t)\Big]_{ij} = \frac{\partial f_i(\mathbf{x})}{\partial x_j}\bigg|_{\mathbf{x}=\mathbf{x}^N(t)} \cdot \qquad (10.4\text{-}6)$$

It is apparent that Eq. (10.4-3) is just the first two terms of the Taylor series expansion of Eq. (10.4-1) on the $(N+1)$th iteration,

$$\dot{\mathbf{x}}^{N+1} = \mathbf{f}(\mathbf{x}^{N+1}, t), \qquad (10.4\text{-}7)$$

expanded about the Nth iteration.

$$\mathbf{f}(\mathbf{x}^{N+1}, t) = \mathbf{f}(\mathbf{x}^N, t) + \left[\frac{\partial \mathbf{f}(\mathbf{x}, t)}{\partial \mathbf{x}}\right]\bigg|_{\mathbf{x}=\mathbf{x}^N(t)} [\mathbf{x}^{N+1}(t) - \mathbf{x}^N(t)]. \qquad (10.4\text{-}8)$$

It is necessary for us to somehow obtain, approximate, or guess at the value of the vector $\mathbf{x}^0(t)$, the initial estimate of $\mathbf{x}(t)$. It is not necessary that $\mathbf{x}^0(t)$ satisfy the boundary conditions. The recursive scheme for the $(N+1)$th approximation of $\mathbf{x}(t)$ is then determined by a solution of the linear, nonstationary, differential equation represented by Eq. (10.4-3) and subject to the boundary conditions given by Eq. (10.4-4). The general solution of Eq. (10.4-3) is

$$\mathbf{x}^{N+1}(t) = \boldsymbol{\varphi}^{N+1}(t, t_o)\mathbf{x}^{N+1}(t_o) + \mathbf{p}^{N+1}(t), \qquad (10.4\text{-}9)$$

where $\boldsymbol{\varphi}(t, t_o)$ is the fundamental solution matrix of

$$\frac{\partial \boldsymbol{\varphi}^{N+1}(t, t_o)}{\partial t} = [J_{\mathbf{x}^N}\mathbf{f}]\boldsymbol{\varphi}^{N+1}(t, t_o) \qquad (10.4\text{-}10)$$

$$\boldsymbol{\varphi}^{N+1}(t_o, t_o) = \mathbf{I}; \qquad (10.4\text{-}11)$$

†A cost function J never has a variable for a subscript, as does the Jacobian $J_{\mathbf{x}}$.

$\mathbf{p}(t)$, which is the particular solution vector for Eq. (10.4-9), obeys the differential equation,

$$\dot{\mathbf{p}}^{N+1} = \mathbf{f}(\mathbf{x}^N, t) - [J_{\mathbf{x}^N}\mathbf{f}]\mathbf{x}^N + [J_{\mathbf{x}^N}\mathbf{f}]\mathbf{p}^{N+1}(t), \tag{10.4-12}$$

with the boundary conditions

$$\mathbf{p}^{N+1}(t_o) = \mathbf{0}. \tag{10.4-13}$$

The initial condition vector $\mathbf{x}^{N+1}(t_o)$, is determined so as to match the n boundary conditions given by

$$\langle \mathbf{C}(t_i), \boldsymbol{\varphi}^{N+1}(t_i, t_o)\mathbf{x}^{N+1}(t_o) + \mathbf{p}^{N+1}(t_i)\rangle = b_i, \qquad i = 1, 2, \ldots, n. \tag{10.4-14}$$

Once $\mathbf{x}^{N+1}(t_o)$ is determined, then $\mathbf{x}^{N+1}(t)$ is immediately available to us from Eq. (10.4-9).

The sequence of vectors $\{\mathbf{x}^{N+1}(t)\}$ which converges to the true solution $\mathbf{x}(t)$ can be shown [27] to obey the relationship

$$\|\mathbf{x}^{N+1}(t) - \mathbf{x}^N(t)\| \leq k\,\|\mathbf{x}^N(t) - \mathbf{x}^{N-1}(t)\|, \tag{10.4-15}$$

where $k \neq f(N)$, thus exhibiting a quadratic convergence. In addition, for a large class of problems, the sequence has a monotone convergence as well [27]. These properties are of great significance in the practical application of quasilinearization.

The employment of quasilinearization in the optimal control and identification of systems is of particular interest to systems control engineers. For an interesting example where an analytic solution to the quasilinearized equations may be found, we refer the reader to Example 10.4-1. In the procedure just described, $\mathbf{x}^{N+1}(t_o)$ was selected so as to match a series of n boundary conditions. A unique solution for those two-point boundary value problems characteristic of optimal control almost always exists from a knowledge of n boundary conditions. However, in the identification problem, a series of n measurements need not uniquely determine the initial condition vector [31].

To alleviate these uniqueness difficulties, and also difficulties which arise from noisy measurements and poor modeling, an additional number of measurements, hence boundary conditions, are added. Linear regression techniques can then be used to provide an estimate of the initial condition vector. For noisy data, linear regression provides an unbiased estimate of the true initial-condition vector. For noise-free data, linear regression solves the problems posed by uniqueness difficulties and poor modeling.

In order to employ linear regression techniques, a reformulation of Eq. (10.4-14) is desirable. Let $\mathbf{q}_D$ be an m-dimensional set of measured points corresponding to our b_i in Eq. (10.4-14). Let $\mathbf{q}^{N+1}$ be the corresponding set of points obtained from the trajectories of $\mathbf{x}^{N+1}(t)$. We may express $\mathbf{q}^{N+1}$ as

$$\mathbf{q}^{N+1} = \boldsymbol{\Psi}\mathbf{x}^{N+1}(t_o) + \boldsymbol{\gamma}^{N+1}. \tag{10.4-16}$$

The matrix $\mathbf{\Psi}$ and the vector $\boldsymbol{\gamma}$ are composed of elements from the fundamental solution matrix and the particular solution vector which we evaluate at the appropriate sampling times. An estimate of the initial-condition vector is to be determined which minimizes the difference between $\mathbf{q}_D$ and $\mathbf{q}^{N+1}$ in a least-squares sense. Mathematically, we desire to minimize

$$J[\mathbf{x}^{N+1}(t_o), \mathbf{q}_d] = \|\mathbf{q}_d - \mathbf{q}^{N+1}\|^2_{\mathbf{R}^{N+1}} \tag{10.4-17}$$

which is, from Eq. (10.4-16),

$$J[\mathbf{x}^{N+1}(t_o), \mathbf{q}_d] = \|\mathbf{q}_d - \boldsymbol{\gamma}^{N+1} - \mathbf{\Psi}\mathbf{x}^{N+1}(t_o)\|^2_{\mathbf{R}^{N+1}} \tag{10.4-18}$$

with respect to $\mathbf{x}^{N+1}(t_o)$. In these equations, $\mathbf{R}^{N+1}$ is a nonnegative definite, symmetric matrix. The estimate of the initial condition vector is determined if we equate the gradient of J with respect to $\mathbf{x}^{N+1}(t_o)$ to zero and solve for $\mathbf{x}^{N+1}(t_o)$. The result is

$$\mathbf{x}^{N+1}(t_o) = [\mathbf{\Psi}^T\mathbf{R}^{N+1}\mathbf{\Psi}]^{-1}\mathbf{\Psi}^T\mathbf{R}^{N+1}[\mathbf{q}_d - \boldsymbol{\gamma}^{N+1}]. \tag{10.4-19}$$

Quite often, $\mathbf{R}^{N+1} = \mathbf{I}$, and Eq. (10.4-19) reduces to

$$\mathbf{x}^{N+1}(t_o) = [\mathbf{\Psi}^T\mathbf{\Psi}]^{-1}\mathbf{\Psi}^T(\mathbf{q}_d - \boldsymbol{\gamma}^{N+1}). \tag{10.4-20}$$

When the number of measurements is equal to the order of the system, Eq. (10.4-20) becomes

$$\mathbf{x}^{N+1}(t_o) = \mathbf{\Psi}^{-1}(\mathbf{q}_d - \boldsymbol{\gamma}^{N+1}) \tag{10.4-21}$$

which corresponds to the boundary conditions of Eq. (10.4-14). It is natural and desirable for us to inquire into the computer storage requirements for this least-squares curve fitting. The $\mathbf{\Psi}$ matrix may be accumulated in the computer for each time at which there is a data point, thus making it unnecessary to store the state transition matrix at each data point. It is easy to show that an equivalent expression for Eq. (10.4-19) is

$$\mathbf{x}^{N+1}(t_o) = \mathbf{M}^{-1}\mathbf{S}, \tag{10.4-22}$$

where we must evaluate on the $(N + 1)$th iteration,

$$\mathbf{M} = \sum_i \boldsymbol{\varphi}^T(t_i, t_o)\mathbf{C}(t_i)\mathbf{R}_i\mathbf{C}^T(t_i)\boldsymbol{\varphi}(t_i, t_o) \tag{10.4-23}$$

$$\mathbf{S} = \sum_i \boldsymbol{\varphi}^T(t_i, t_o)\mathbf{C}(t_i)\mathbf{R}_i[b_i - \mathbf{C}^T(t_i)\mathbf{p}(t_i)] \tag{10.4-24}$$

with past definitions of $\mathbf{R}$, $\mathbf{C}$, b, $\mathbf{p}$, and $\boldsymbol{\varphi}$ where $\sum_i$ includes all data points.

An extension of the foregoing method has been presented [31] which allows us to add new data points at any iteration in the procedure, or to subtract data points, as well as to determine the optimum spacings for a fixed number of data points. This extension involves the use of the matrix inversion lemma, presented in Appendix D, to avoid inverting a large $\mathbf{M}$ or $\mathbf{\Psi}$ matrix where the number of data points is large.

An alternate method of treating the minimization of the cost function,

Eq. (10.4-18), is to use the discrete maximum principle of Chapter 6. This results in a nonlinear two-point boundary value difference equation whose solution can be resolved via the technique of discrete quasilinearization.

10.4-2
Discrete time quasilinearization

Our development of discrete quasilinearization follows the development of continuous quasilinearization. We consider a system of nonlinear difference equations

$$\mathbf{x}(k+1) = \mathbf{f}[\mathbf{x}(k), k], \qquad k \in [k_o, k_f] \tag{10.4-25}$$

with two-point boundary conditions

$$\langle \mathbf{C}_i(j), \mathbf{x}(j) \rangle = b_i(j), \qquad j = k_o, k_f, \qquad i = 1, 2, \ldots, n/2, \tag{10.4-26}$$

where $\mathbf{x}$ and $\mathbf{c}$ are n-vectors, $\langle \ , \ \rangle$ denotes the inner product, and the period of observation of the solution vector is $t = k_oT$ to $t = k_fT$. Utilizing the method of discrete quasilinearization, an iterative procedure is established for successively approximating the solution to Eq. (10.4-25) by solutions of the system of linear difference equations

$$\mathbf{x}^{N+1}(k+1) = \mathbf{f}[\mathbf{x}^N(k), k] + \{J_{\mathbf{x}^N}\mathbf{f}[\mathbf{x}^N(k), k]\}[\mathbf{x}^{N+1}(k) - \mathbf{x}^N(k)], \tag{10.4-27}$$

where $J_{\mathbf{x}^N}\mathbf{f}$ is the Jacobian matrix of partial derivatives having as its ijth element the partial derivative

$$\left.\frac{\partial f_i}{\partial x_j}\right|_{\mathbf{x}=\mathbf{x}^N(k)},$$

and $\mathbf{x}^N$ indicates the solution at the Nth iteration. Equation (10.4-27) is obtained by a Taylor series expansion just as in the continuous case. We may rewrite Eq. (10.4-27) as

$$\mathbf{x}^{N+1}(k+1) = \mathbf{A}(k)\mathbf{x}^{N+1}(k) + \mathbf{u}(k), \tag{10.4-28}$$

where

$$\mathbf{A}(k) = J_{\mathbf{x}^N}\mathbf{f}[\mathbf{x}^N(k), k], \qquad \mathbf{u}(k) = \mathbf{f}[\mathbf{x}^N(k), k] - \{J_{\mathbf{x}^N}\mathbf{f}[\mathbf{x}^N(k), k]\}\mathbf{x}^N(k). \tag{10.4-29}$$

Since Eq. (10.4-28) is linear in the $(N+1)$th approximation, we may determine a solution by generating the homogeneous and particular solutions and imposing the boundary conditions of Eq. (10.4-26). Let $\boldsymbol{\varphi}^{N+1}(k, l)$ be the fundamental solution matrix of

$$\boldsymbol{\varphi}^{N+1}(k+1, l) = \mathbf{A}(k)\boldsymbol{\varphi}^{N+1}(k, l), \qquad \boldsymbol{\varphi}^{N+1}(l, l) = \mathbf{I}. \tag{10.4-30}$$

The particular solution $\mathbf{p}^{N+1}(k)$ is generated by the n-vector equation

$$\mathbf{p}^{N+1}(k+1) = \mathbf{A}(k)\mathbf{p}^{N+1}(k) + \mathbf{u}(k), \qquad \mathbf{p}^{N+1}(k_o) = \mathbf{0}. \tag{10.4-31}$$

The solution to Eq. (10.4-28) is thus given by

$$\mathbf{x}^{N+1}(k) = \boldsymbol{\varphi}^{N+1}(k, l)\mathbf{x}^{N+1}(l) + \mathbf{p}^{N+1}(k), \tag{10.4-32}$$

where we obtain the constant vecor $\mathbf{x}^{N+1}(l)$ from the boundary conditions by solving the equations,

$$\langle \mathbf{C}_i(j), \boldsymbol{\varphi}^{N+1}(j, k_o)\mathbf{x}^{N+1}(k_o) + \mathbf{p}^{N+1}(j)\rangle = b_i(j),$$
$$j = k_o, k_f, \qquad i = 1, 2, \ldots, \frac{n}{2}\cdot \tag{10.4-33}$$

An initial or *zero*th trajectory $\mathbf{x}^0$ may be generated if we select initial conditions for the unknown elements of $\mathbf{x}(k_o)$ and solve Eq. (10.4-28) forward in time. In practice, we will seldom use Eq. (10.4-32) to obtain the (N + 1)th trajectory. Such an approach requires retention of the complete homogeneous and particular solution trajectories with a corresponding requirement for considerable computer memory storage. An approach having reduced memory storage requirements consists of retaining only $\mathbf{p}^{N+1}(k_o)$, $\boldsymbol{\varphi}^{N+1}(k_f, k_o)$, and $\mathbf{p}^{N+1}(k_f)$ in memory until evaluation of $\mathbf{x}^{N+1}(k_o)$ by means of Eq. (10.4-33). The $(N + 1)$th trajectory $\mathbf{x}^{N+1}(k)$ is then generated from Eq. (10.4-28) where we obtain the requisite initial-condition vector elements from $\mathbf{x}^{N+1}(k_o)$. This trajectory is stored in computer memory as the final solution or, for evaluation of $\mathbf{A}(k)$ and $\mathbf{u}(k)$, for the next iteration.

In this section, we developed a discrete version of quasilinearization. This could have been developed for the general case of a multipoint boundary value problem in a fashion almost exactly the same as presented here. Our choice of the two-point boundary value problem development has been made to complement the more general continuous-time presentation.

10.4-3
Solution of two-point boundary value problems of optimal control by quasilinearization

In all of our work thus far, we have seen that the equations whose solutions resolve an optimum control problem are two-point boundary value differential or difference equations. For linear systems with quadratic cost functions, the resulting TPBVP is linear and can be solved by use of the superposition principle or by conversion to a nonlinear matrix Riccati equation which has conditions prescribed at the final time only. For nonlinear systems, or linear systems with nonquadratic cost functions, the resulting

TPBVP are nonlinear. Iterative techniques must generally be used to solve such problems. The last section presented various gradient and second-variation methods for overcoming nonlinear two-point boundary value problems. Here we wish to examine the quasilinearization approach for the solution of these problems.

SOLUTION OF CONTINUOUS OPTIMAL CONTROL PROBLEMS WITH FIXED TERMINAL TIME: By application of the maximum principle, or the Euler-Lagrange equations, to an optimum control problem, we inevitably find that we need to solve

$$\dot{\mathbf{z}} = \mathbf{h}(\mathbf{z}, t) \tag{10.4-34}$$

in order to establish the optimum trajectory and, therefore, the optimum control vector. In Eq. (10.4-34), half the initial conditions are specified at the initial time t_o and half at the final time t_f. Thus the two-point boundary value problem is of a type where the quasilinearization approach developed here is directly applicable. The only exception occurs when the initial or final time is not fixed, since the boundary conditions are then specified at unknown times. As we will soon see, a minor modification to the procedure will allow us to treat unspecified initial and/or terminal time problems.

Example 10.4-1. Consider the TPBVP resulting from minimization of

$$J = \frac{1}{2}\int_0^1 (x^2 + u^2)\,dt$$

for the system

$$\dot{x} = -x^2 + u, \qquad x(0) = 10.0.$$

The TPBVP resulting from the application of the Euler-Lagrange equations or the maximum principle is

$$\dot{x} = -x^2 - \lambda, \qquad x(0) = 10$$
$$\dot{\lambda} = -x + 2\lambda x, \qquad \lambda(1) = 0.$$

We quasilinearize this about the Nth iteration and obtain the quasilinearized equations:

$$\dot{x}^{N-1} = -(x^N)^2 - 2x^N(x^{N+1} - x^N) - \lambda^{N+1}, \qquad x^{N+1}(0) = 10$$
$$\dot{\lambda}^{N+1} = -x^{N+1} + 2\lambda^N x^N + 2x^N(\lambda^{N+1} - \lambda^N) +$$
$$2\lambda^N(x^{N+1} - x^N), \qquad \lambda^{N+1}(1) = 0.$$

We may write these equations in the form

$$\dot{\mathbf{y}} = \mathbf{A}(t)\mathbf{y}(t) + \mathbf{u}(t),$$

where

$$\mathbf{y}^T(t) = [x^{N+1}(t), \lambda^{N+1}(t)]$$

$$\mathbf{A}(t) = \begin{bmatrix} -2x^N(t) & -1 \\ -1 + 2\lambda^N(t) & 2x^N(t) \end{bmatrix}, \qquad \mathbf{u}(t) = \begin{bmatrix} [x^N(t)]^2 \\ -2\lambda^N(t)x^N(t) \end{bmatrix}.$$

This resulting linear time-varying vector differential equation may now be solved in any convenient way. We have discussed two methods in Chapter 5. One method is to use the Riccati transformation of Sec. 10.3, which decouples the boundary value problem into two initial-condition problems; another method is to use the superposition principle. The iterative solution is continued until two successive iterates differ by some small prescribed amount.

Example 10.4-2. We desire to change the flux density in a nuclear reactor from some value n_o to a value n_f such that the energy required to drive the control rods is minimized. We therefore wish to minimize the cost function which approximates this desire

$$J = \frac{1}{2}\int_{t_o}^{t_f} [\rho^2 + \alpha(n - n_f)^2]\, dt.$$

We will assume that the reactor can be adequately represented by the point-kinetics equations with one group of delayed neutrons

$$\dot{n} = \frac{\rho - \beta}{\Lambda} n + \lambda c, \qquad \dot{c} = \frac{\beta n}{\Lambda} - \lambda c.$$

The following parameter values are assumed: $\beta = 0.0068$, $\Lambda = 10^{-3}$ sec, $\lambda = 0.1\ \text{sec}^{-1}$, $n_o = 5$ K.W., $c_o = 64 n_o$, $\alpha = 5 \times 10^{-6}$, $t_o = 0$, $t_f = 1$ sec. The TPBVP is obtained as

$$\dot{n} = \frac{-\beta}{\Lambda} n + \lambda c - \Gamma_1 \frac{n^2}{\Lambda^2}, \qquad \dot{c} = \frac{\beta n}{\Lambda} - \lambda c$$

$$\dot{\Gamma}_1 = -\alpha(n - n_f) + \Gamma_1\left[\frac{\Gamma_1 n}{\Lambda^2} + \frac{\beta}{\Lambda}\right] - \Gamma_2 \frac{\beta}{\Lambda}, \qquad \dot{\Gamma}_2 = -\lambda(\Gamma_1 - \Gamma_2)$$

with the boundary conditions $n(t_o) = 5$, $c(t_o) = 64 n_o = 320$ K.W., and $\Gamma_1(t_f) = \Gamma_2(t_f) = 0$.

The quasilinearized equations take the form $\dot{\mathbf{y}} = \mathbf{A}\mathbf{y} + \mathbf{u}$, where the initial and final conditions are $y_1(0) = 5$, $y_2(0) = 320$, $y_3(1) = y_4(1) = 0$, and where

$$\mathbf{A}(t) = \begin{bmatrix} \frac{-\beta}{\Lambda} - \frac{2\Gamma_1^N(t) n^N(t)}{\Lambda^2} & \lambda & -\left[\frac{n^N(t)}{\Lambda}\right]^2 & 0 \\ \frac{\beta}{\Lambda} & -\lambda & 0 & 0 \\ -\alpha + \left[\frac{\Gamma_1^N(t)}{\Lambda}\right]^2 & 0 & \frac{\beta}{\Lambda} + \frac{2\Gamma_1^N(t) n^N(t)}{\Lambda^2} & \frac{-\beta}{\Lambda} \\ 0 & 0 & -\lambda & +\lambda \end{bmatrix}$$

$$\mathbf{u}^T(t) = \left[2\Gamma_1^N(t)[n^N(t)]^2,\ 0,\ \alpha n_f - \frac{2n^N(t)[\Gamma_1^N(t)]^2}{\Lambda^2},\ 0\right].$$

We start the quasilinearization solution by assuming that $c(t) = c_o$ and $n(t) = n_o + (n_f - n_o)t$ for the *zero*th iteration. Figure 10.4-1 indicates the quasilinearized flux density solution at each stage in the iteration.

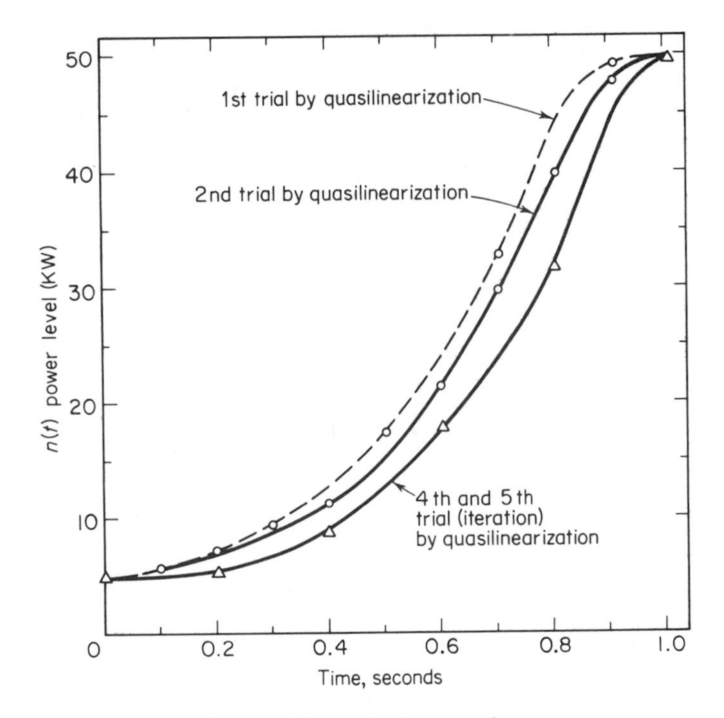

Fig. 10.4-1 Power level versus time.

SOLUTION OF SPECIFIC OPTIMAL CONTROL PROBLEMS BY QUASILINEARIZATION: In a nonlinear control system, it is not generally possible to express the optimal control law as a product of the state vector and a time-varying gain. In fact, a solution of the nonlinear, partial differential equations relating the optimal control to the optimum trajectory, $\hat{\mathbf{u}}(t)$ to $\hat{\mathbf{x}}(t)$, is normally not feasible. Furthermore, the optimal control is highly dependent, in a nonlinear manner, on the initial state vector $\mathbf{x}(t_o)$. This means that, for most nonlinear control systems, only open-loop control laws are available even though closed-loop control laws are more desirable. The desirability of closed-loop control laws has led to the development of *specific optimal control.*

The specific optimal control (S.O.C.) problem is defined in the following manner. We are given a plant with a state equation of the form

$$\dot{\mathbf{x}} = \mathbf{f}(\mathbf{x}, \mathbf{u}, t), \qquad \mathbf{x}(t_o) = \mathbf{x}_o. \tag{10.4-35}$$

We desire to determine the unknown parameters in a control law of the form

$$\mathbf{u} = \mathbf{h}(\mathbf{y}, \mathbf{b}), \tag{10.4-36}$$

where $\mathbf{y}$ is a p-dimensional vector which is a known function of the state, and $\mathbf{b}$ is a q-dimensional constant vector to be determined such that an index of performance of the form

$$J = \int_{t_0}^{t_f} \phi[\mathbf{x}, \mathbf{u}, t]\, dt \tag{10.4-37}$$

is minimized.

The specific optimal control problem can be solved by a simple reformulation of the above problem. The reformulation is carried out by substitution of Eq. (10.4-36) into Eqs. (10.4-35) and (10.4-37), which gives us

$$\dot{\mathbf{x}} = \mathbf{f}[\mathbf{x}, \mathbf{h}(\mathbf{y}, \mathbf{b}), t], \qquad J(\mathbf{b}) = \int_{t_0}^{t_f} \phi[\mathbf{x}, \mathbf{h}(\mathbf{y}, \mathbf{b}), t]\, dt. \tag{10.4-38}$$

The fact that $\mathbf{y}$ is a known function of $\mathbf{x}$ results in

$$\dot{\mathbf{x}} = f(\mathbf{x}, \mathbf{b}, t), \qquad J(\mathbf{b}) = \int_{t_0}^{t_f} \phi(\mathbf{x}, \mathbf{b}, t)\, dt. \tag{10.4-39}$$

Since $\mathbf{b}$ is a constant vector, we adjoin the vector equation $\dot{\mathbf{b}} = \mathbf{0}$ to the differential system equations of Eq. (10.4-39). Thus the S.O.C. problem has been imbedded in the problem of finding the vector $\mathbf{b}$, which minimizes the cost function and associated differential constraints. It is now in the form for application of optimal control theory. Using Pontryagin's maximum principle, we may solve the S.O.C. problem by solving a two-point boundary value problem.

The employment of S.O.C. in lieu of the true optimal control leads to a degradation in the performance of the control system. The form of the control law, Eq. (10.4-36), is very instrumental in determining how well the performance of the S.O.C. compares to that of the true optimal control. In practice, for ease of implementation, it is sometimes desirable to select a control law of quadratic or linear form. We shall now illustrate the salient features of specific optimum control by three examples.

Example 10.4-3. We consider the system described by

$$\dot{x} = -x^2 + u, \qquad x(0) = 10.0.$$

We desire to employ an S.O.C. of the form $u(t) = ax(t)$, where the parameter a is to be chosen so as to minimize

$$J(a) = \tfrac{1}{2} \int_0^{1.0} (x^2 + u^2)\, dt.$$

We reformulate this problem by adjoining the equation $\dot{a} = 0$ to the system dynamics. Thus we have the system and cost function

$$\dot{x} = -x^2 + ax, \qquad x(0) = 10.0, \qquad \dot{a} = 0$$

$$J(a) = \tfrac{1}{2} \int_0^{1.0} (x^2 + a^2x^2)\, dt.$$

The S.O.C. problem now becomes a problem of obtaining the value of a which minimizes the cost function subject to the associated differential constraints. Thus we have formulated a problem of parameter optimization and have a single-stage decision process. In theory, the methods of Chapter 2 could be used to resolve the problem. However, treatment of the problem as one of dynamic optimization will be considered here. Often, the resulting computational algorithms for this type of problem are easier to implement than if static optimization methods are used. The canonic equations and boundary conditions are easily obtained as

$$\begin{aligned} \dot{x} &= -x^2 + ax & x_1(0) &= 10.0 \\ \dot{a} &= 0 & \lambda_2(0) &= 0.0 \\ \dot{\lambda}_1 &= -a\lambda_1 + 2\lambda_1 x - x - a^2 x & \lambda_1(1) &= 0.0 \\ \dot{\lambda}_2 &= -\lambda_1 x - a^2 x & \lambda_2(1) &= 0.0. \end{aligned}$$

The solution of the canonic equations, subject to the boundary conditions, yields the parameter a and, hence, the specific optimal control law. This TPBVP may be solved by quasilinearization. The value of a is found to be $a = -0.1038$. The optimal value of the cost function is calculated as $J_{\text{opt}} = 4.5108$. The value of the cost function when S.O.C. is used is $J_{\text{SOC}} = 4.5483$. A comparison of the optimal and specific optimal controls and trajectories are shown in Figs. 10.4-2 and 10.4-3. It should be noted that this optimal control problem was the subject of Example 10.4-1.

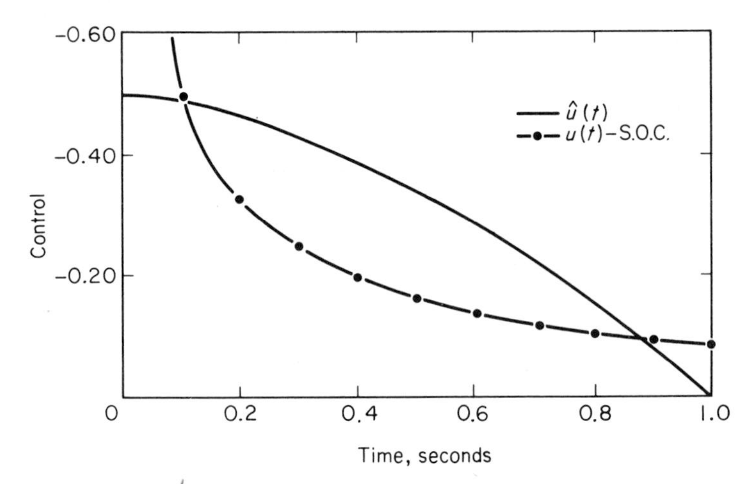

Fig. 10.4-2 Optimal and specific optimal control laws associated with Example 10.4-3.

Example 10.4-4. We consider the second-order nonlinear system that is represented by the state equations

$$\begin{aligned} \dot{x}_1 &= x_2 + 0.01x_2^3, & x_1(0) &= 2.0 \\ \dot{x}_2 &= -4x_1 - 5x_2 + 4u, & x_2(0) &= 0.0. \end{aligned}$$

We desire to employ an S.O.C. of the form $u = bx_1$. The parameter b is to be chosen so as to minimize the cost function

$$J(b) = \frac{1}{2}\int_0^{1.0} [x_1^2 + (bx_1)^2]\,dt.$$

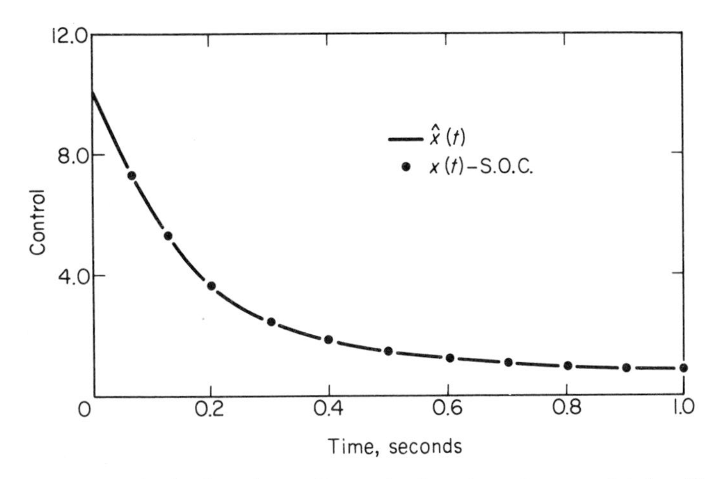

Fig. 10.4-3 Optimal and specific optimal trajectories associated with Example 10.4-3.

The canonic equations resulting from optimization theory are

$$\dot{x}_1 = x_2 + 0.01x_2^3 \qquad x_1(0) = 2.0$$

$$\dot{x}_2 = -4x_1 - 5x_2 + 4bx_1 \qquad x_2(0) = 0.0$$

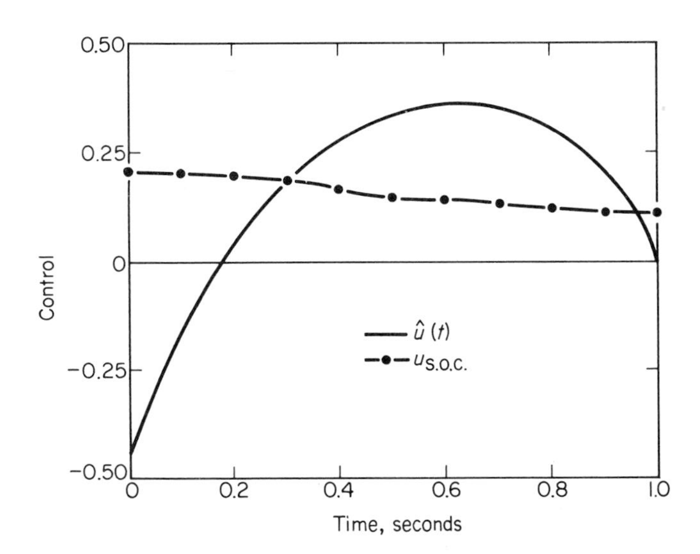

Fig. 10.4-4 Optimal and specific optimal control for the system discussed in Example 10.4-4.

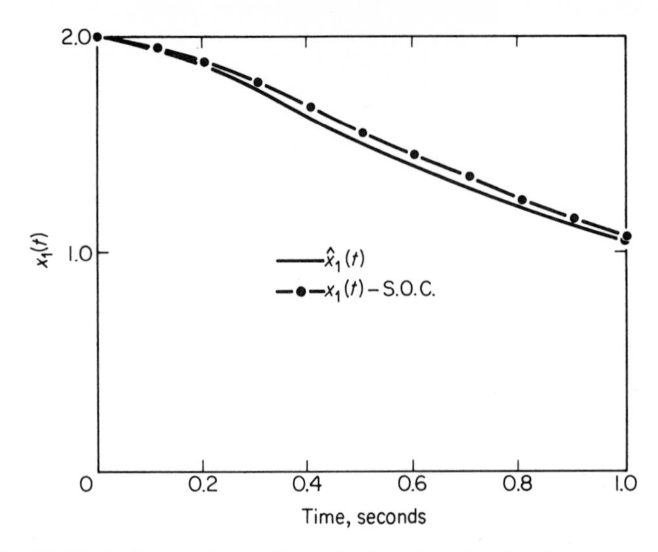

Fig. 10.4-5 Optimal and specific optimal trajectories for the system of Example 10.4-4—$x_1(t)$ versus time.

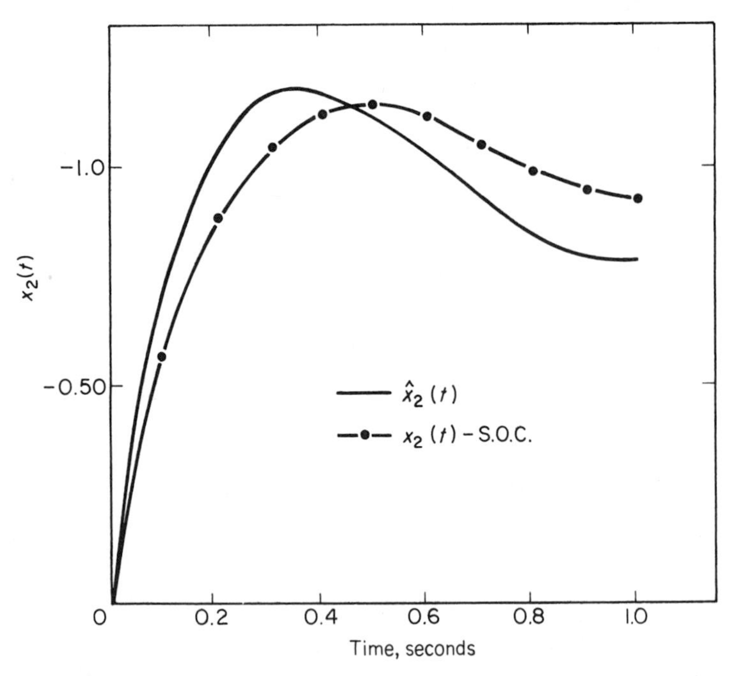

Fig. 10.4-6 Optimal and specific optimal trajectories for the system of Example 10.4-4—$x_2(t)$ versus time.

$$\dot{b} = 0 \qquad \lambda_3(0) = 0.0$$
$$\dot{\lambda}_1 = 4\lambda_2 - 4b\lambda_2 - x_1 - b^2x_1 \qquad \lambda_1(1) = 0.0$$
$$\dot{\lambda}_2 = 5\lambda_2 - \lambda_1 - 0.03x_2^2\lambda_1 \qquad \lambda_2(1) = 0.0$$
$$\dot{\lambda}_3 = -4\lambda_2 x_1 - bx_1^2 \qquad \lambda_3(1) = 0.0.$$

They constitute a sixth-order **TPBVP** which we may solve using quasilinearization. The value of the parameter is found to be $b = +0.0970$. The minimum value of the cost function is calculated to be $J_{\min} = 1.7174$. The value of the cost function when the specific optimal control is employed is $J_{\text{SOC}} = 1.7527$. A comparison of the specific optimal and the optimal control and trajectories are shown in Figs. 10.4-4, 10.4-5, and 10.4-6.

Example 10.4-5. We now desire to find a specific optimum control for the reactor system of Example 10.4-2. Since it has been shown that proportional control of the desired reactivity cannot be satisfactory, we attempt to

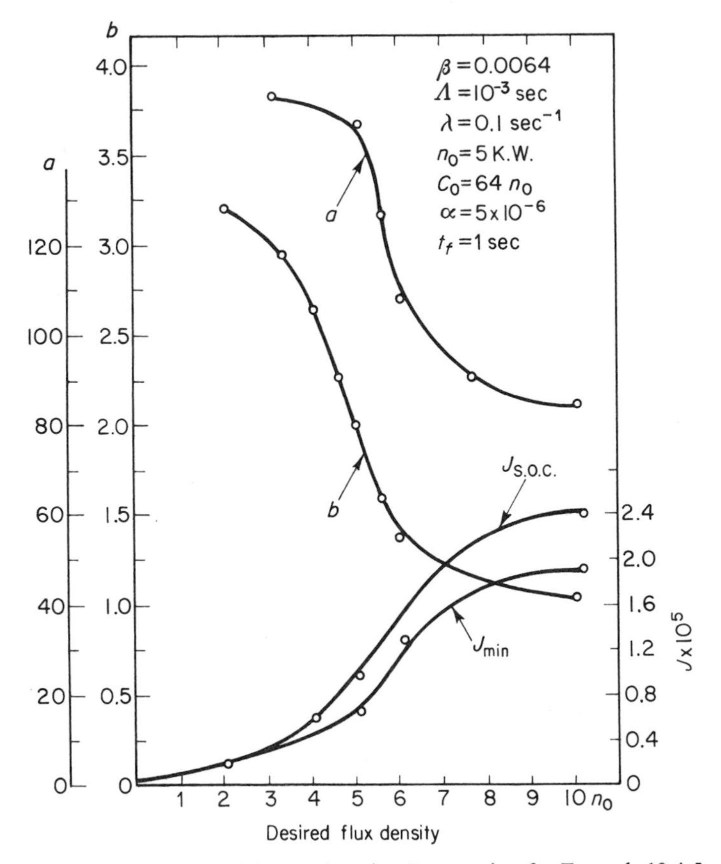

Fig. 10.4-7 a and b and the cost function J versus time for Example 10.4-5.

find a proportional plus integral control of the form

$$\rho = a(n - n_f) + b \int_{t_0}^{t_f} (n - n_f)\, dt.$$

The final two-point boundary value problem that we obtain has four state and four adjoint variables; hence, the **A** matrix of the quasilinearized system is an 8×8 matrix. Considerable computational simplification is achieved when we realize that solution to many of the equations is trivial. Solution by the quasilinearization procedure indicates that the best values of the coefficients a and b in the control law are highly dependent upon the final time t_f as well as on the initial and final desired states. Thus it is desirable to vary the closed-loop controller parameters as function of the neutron density and the allowed startup time if true optimum performance is to be achieved. This suggests the desirability of an optimum adaptive controller as a means of further improving system performance. Salient features of the results of the quasilinearization procedure are illustrated in Fig. 10.4-7. Figure 10.4-8 illustrates the superiority of the specific optimal control closed-loop system as the neutron lifetime is varied.

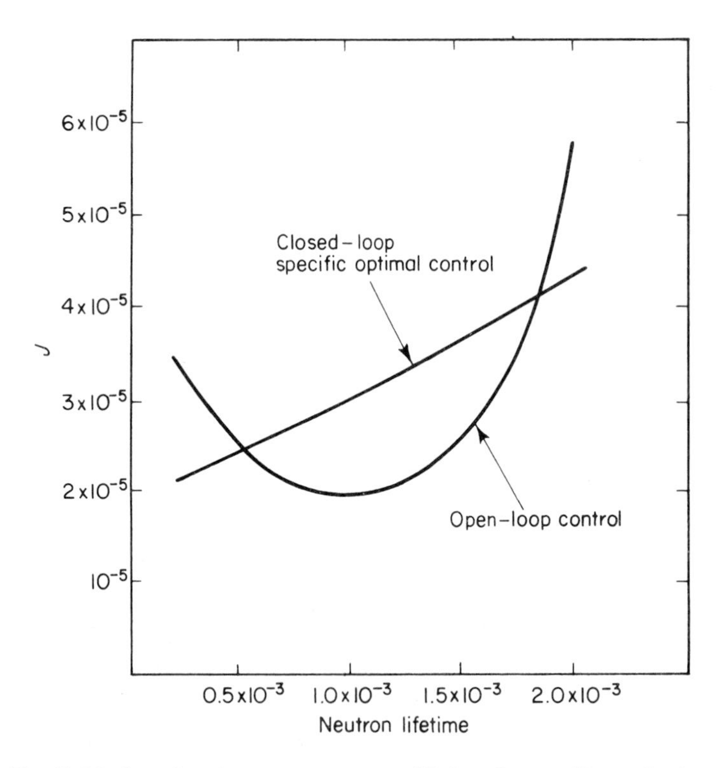

Fig. 10.4-8 Cost function versus neutron lifetime for specific optimal control and optimum open-loop control for Example 10.4-5.

SOLUTION OF OPTIMAL CONTROL PROBLEMS WITH UNDETERMINED TERMINAL TIME: We have thus far successfully applied the technique of quasilinearization to two-point boundary values resulting from optimization theory with fixed initial and terminal times. Equality and inequality constraints can be written as equality constraints and handled by the Lagrange multiplier method [51, 52]. Alternately, penalty functions may be used.

We have not, thus far, considered any problems with variable terminal time. Two ways to incorporate this method into the quasilinearization approach have been suggested. McGill and Kenneth [28] and Kenneth and Taylor [53] consider problems with inequality constraints as well as those with variable terminal times. The former can be converted into equality constraints and pose no problem other than perhaps increased convergence difficulties, i.e., a narrower region of initial trajectories which finally converge to the true solution. They treat the variable terminal time problem in the following manner. The TPBVP is written as $\dot{\mathbf{x}} = \mathbf{f}(\mathbf{x}, t)$ with half the boundary conditions given at the initial time and half at the terminal time, which is not known. This problem presents difficulties since the terminal time is not known. In order to resolve this difficulty one (or more) of the states, whose terminal condition is specified, is specified at the initial time. An initial guess at the final time is made, and the iteration proceeds until changes between iterations are below some level. Then the final time is adjusted by a finite difference or discrete quasilinearization method until the changes in final time between iterations are below some level.

Long [54] suggests an alternate approach which is more systematic in that the boundary conditions are satisfied at each iteration, whereas in the McGill and Kenneth method, the initial condition on a state which should have a prescribed final condition is satisfied only after the optimum terminal time is found. The alternate approach proceeds as follows.

The original cost function is assumed to be

$$J = \int_{t_o}^{t_f} \phi(\mathbf{x}, \mathbf{u}, t)\, dt, \tag{10.4-40}$$

where t_f is not specified. We make a change of variable by letting $t = a\tau$, where τ is the new independent variable and a is a constant to be determined by the optimization procedure. The beginning and terminal times may be taken as $\tau_o = 0$ and $\tau_f = 1$ such that, when a is determined, we can compute the terminal time $t_f = a\tau_f$. Naturally, this change of time scale must be incorporated into the dynamic system equations and in any other equality constraints as well as in any inequality constraints in the optimization problem.

Example 10.4-6. We now desire to find a minimum time controller for a second-order system. The problem is to minimize

$$J = \int_0^{t_f} dt$$

for the system described by

$$\dot{x}_1 = x_2, \qquad x_1(0) = x_{10}$$

$$\dot{x}_2 = u, \qquad x_2(0) = x_{20}$$

with the inequality constraint $|u(t)| \leq 1$. At the unknown terminal time, we desire $x_1(t_f) = x_2(t_f) = 0$. The solution proceeds as follows:

We convert the inequality constraint to an equality constraint by writing

$$[1 - u(t)][u(t) + 1] = \alpha^2(t), \qquad \alpha^2(t) + u^2(t) - 1 = 0.$$

The Hamiltonian then becomes

$$H = \lambda_1(t)x_2(t) + \lambda_2(t)u(t) + \lambda_3(t)[\alpha^2(t) + u^2(t) - 1] + 1.$$

The canonic equations, solution of which yields the optimum control and state variable, are obtained in the usual fashion:

$$\frac{dx_1(t)}{dt} = x_2(t), \qquad x_1(0) = x_{10}$$

$$\frac{dx_2(t)}{dt} = u(t), \qquad x_2(0) = x_{20}$$

$$\frac{d\lambda_1(t)}{dt} = 0, \qquad x_1(t_f) = 0$$

$$\frac{d\lambda_2(t)}{dt} = -\lambda_1(t), \qquad x_2(t_f) = 0$$

with the coupling relations

$$\lambda_1(t)x_2(t) + \lambda_2(t)u(t) + 1 = 0$$

$$\lambda_2(t) + 2\lambda_3(t)u(t) = 0$$

$$\alpha^2(t) + u^2(t) - 1 = 0$$

$$\alpha(t)\lambda_3(t) = 0.$$

In these equations, the terminal time is not known. Thus we assume $t = a\tau$, we take $\tau_o = 0$, $\tau_f = 1$, and then quasilinearize to obtain:

$$\left[\frac{dx_1(\tau)}{d\tau}\right]^{N+1} = a^N x_2^{N+1}(\tau) + x_2^N(\tau)a^{N+1} - x_2^N(\tau)a^N, \quad x_1^{N+1}(0) = x_{10}$$

$$\left[\frac{dx_2(\tau)}{d\tau}\right]^{N+1} = a^N u^{N+1}(\tau) + u^N(\tau)a^{N+1} - a^N u^N(\tau), \qquad x_2^{N+1}(0) = x_{20}$$

$$\left[\frac{d\lambda_1(\tau)}{d\tau}\right]^{N+1} = 0, \qquad x_1^{N+1}(\tau_f) = 0$$

$$\left[\frac{d\lambda_2(\tau)}{d\tau}\right]^{N+1} = -a^N\lambda_1^{N+1}(\tau) - \lambda_1^N(\tau)a^{N+1} + \lambda_1^N(\tau)a^N, \quad x_2^{N+1}(\tau_f) = 0$$

$$\lambda_1^{N+1}(\tau)x_2^N(\tau) + \lambda_1^N(\tau)x_2^{N+1}(\tau) - \lambda_1^N(\tau)x_2^N(\tau) +$$
$$\lambda_2^{N+1}(\tau)u^N(\tau) + \lambda_2^N(\tau)u^{N+1}(\tau) - \lambda_2^N(\tau)u^N(\tau) + 1 = 0$$

$$\lambda_2^{N+1}(\tau) + 2\lambda_3^N(\tau)u^{N+1}(\tau) + 2u^N(\tau)\lambda_3^{N+1}(\tau) - 2u^N(\tau)\lambda_3^N(\tau) = 0$$
$$-[\alpha^N(\tau)]^2 + 2\alpha^N(\tau)\alpha^{N+1}(\tau) - [u^N(\tau)]^2 + 2u^N(\tau)u^{N+1}(\tau) - 1 = 0$$
$$-\alpha^N(\tau)\lambda_3^N(\tau) + \alpha^N(\tau)\lambda_3^{N+1}(\tau) + \lambda_3^N(\tau)\alpha^{N+1}(\tau) = 0.$$

These quasilinearized equations may be written in the standard vector-matrix notation. An initial trajectory is determined; a^{N+1} is forced to be constant and adjusted so that the two-point boundary conditions are satisfied. In this particular case, computer experiment demonstrates to us that the solution converges to the proper solution in from three to six iterations, depending upon the values assumed for x_{10} and x_{20}. Since this minimum time problem was rather thoroughly solved in Chapter 5, we will not repeat the solution here.

This method can be extended to multipoint boundary value problems where several terminal times are unknown. If, for example, an initial time $t_o = 0$ is known, but not two additional times t_1 and t_2, then we may make the time-scale change $t = a\tau$ for $a < \tau < 1$, and $t = b\tau$ for $t = a + b(\tau - 1)$ and $1 < \tau < 2$. Interesting computational results for an example of this type are provided by Long [54].

In the last three subsections, we developed the quasilinearization computational algorithms. We illustrated some of the applications of quasilinearization to the determination of optimal controls for discrete time systems. Numerous other applications are also available. Applications to identification are discussed in [31, 36, 38]. Further applications to closed-loop systems with random forcing functions are considered in [55, 56], which combine the quasilinearization approach with sensitivity analysis methods. Use of quasilinearization to determine coordination variables for estimation of states and parameters in large-scale systems is considered in [57, 58, 59, 60].

References

1. COURANT, R., and HILBERT, D., *Methods of Mathematical Physics.* Vol. 1. Interscience Publishers, New York, 1953.

2. KANTOROVICH, L.V., and KRYLOV, V.I., *Approximate Methods of Higher Analysis.* Interscience Publishers, New York, and P. Noordhoff Ltd., Groningen, The Netherlands, 1958.

3. GELFAND, I.M., and FOMIN, S.V., *Calculus of Variations.* Prentice-Hall, Inc., Englewood Cliffs, New Jersey, 1963.

4. BELLMAN, R., *Dynamic Programming.* Princeton University Press, Princeton, New Jersey, 1957.

5. BELLMAN, R., *Adaptive Control Processes, A Guided Tour.* Princeton University Press, Princeton, New Jersey, 1961.

6. DREYFUS, S.E., *Dynamic Programming and the Calculus of Variations*. Academic Press, New York, 1965.
7. BELLMAN, R., "On the Determination of Optimal Trajectories Via Dynamic Programming," G. Leitman, ed. *Optimization Techniques*. Academic Press, New York, 1962.
8. BELLMAN, R., KALABA, R., and KOTKIN, B., "Polynomial Approximation—A New Computational Technique in Dynamic Programming—I, Allocation Processes." *Mathematics of Computation*, **17**, 1963, 155–61.
9. LARSON, R.E., "Dynamic Programming with Reduced Computational Requirements." *IEEE Trans. Autom. Control*, AC-10, (1965), 135–43.
10. KELLEY, H.J., "Gradient Theory of Optimal Flight Paths." *ARS J.*, 30, (1960), 947–54.
11. KELLEY, H.J., "Method of Gradients," G. Leitman, ed. *Optimization Techniques*. Academic Press, New York, 1962, Chapter 6.
12. BRYSON, A.E., and DENHAM, W.F., "A Steepest Ascent Method for Solving Optimum Programming Problems." *J. Applied Mechanics*, **29**, (1962), 247–57.
13. BRYSON, A.E., et al., "Optimal Programming Problems with Inequality Constraints I: Necessary Conditions for Extremal Solutions." *AIAA J.*, **1**, (1963), 2544–50.
14. DENHAM, W.F., and BRYSON, A.E., "Optimal Programming Problems with Inequality Constraints II: Solution by Steepest Ascent." *AIAA J.*, **2**, (1964), 25–34.
15. STANCIL, R.T., "A New Approach to Steepest Ascent Trajectory Optimization." *AIAA J.*, **2**, (1964), 1365–70.
16. GOLDSTEIN, A.A., "On Steepest Descent." *SIAM J. Control*, Ser. A, **3**, (1965), 147–51.
17. MCREYNOLDS, S.R., and BRYSON, A.E., "A Successive Sweep Method for Solving Optimal Programming Problems." *Proceedings Joint Autom. Control Conf.*, (August 1965), pp. 551–55.
18. KOPP, R.E., and MCGILL, R., "Several Trajectory Optimization Techniques; Part I: Discussion," A.V. Balakrishnan and L.W. Neustadt, eds. *Computing Methods in Optimization Problems*. Academic Press, New York, 1964, pp. 65–89.
19. MOYER, H.G., and PINKHAM, G., "Several Trajectory Optimization Techniques; Part II: Application." [18, pp. 91–105].
20. KELLEY, H.J., "Guidance Theory and Extremal Fields," *IRE Trans. Autom. Control*, **AC-7**, (1962), 75–81.
21. BREAKWELL, J.V., SPEYER, J.L., and BRYSON, A.E., "Optimization and Control of Nonlinear Systems using the Second Variation." *SIAM J. Control*, Ser. A, **1**, (1963), 193–223.
22. KELLEY, H.J., KOPP, R.E., and MOYER, H.G., "A Trajectory Optimization Technique Based upon the Theory of the Second Variation." *Progress in Astronautics and Aeronautics*, **14**, (1964), 559–82.

23. MERRIAM, C.W., *Optimization Theory and the Design of Feedback Control Systems*. McGraw-Hill Book Co., New York, 1964.

24. BELLMAN, R., and KALABA, R., *Quasilinearization and Nonlinear Boundary-Value Problems*. Elsevier Press, New York, 1965.

25. BELLMAN, R.E., KAGIWADA, H.H., and KALABA, R. "Quasilinearization, Boundary Value Problems and Linear Programming." The Rand Corporation, Memo RM-4284-PR, September 1964.

26. BELLMAN, R., KALABA, R., and SRIDHAR, R., "Adaptive Control via Quasilinearization and Differential Approximation." The Rand Corporation, Memo. RM-3928-PR, November 1963.

27. KALABA, R., "On Nonlinear Differential Equations, The Maximum Operation, and Monotone Convergence." *J. Math. and Mechanics*, **8**, (1959), 519–74.

28. MCGILL, R., and KENNETH, P., "Solution of Variational Problems by Means of a Generalized Newton-Raphson Operator." *AIAA J.*, **2**, (1964), 1761–66.

29. KUMAR, K.S.P., and SRIDHAR, R., "On the Identification of Control Systems by the Quasilinearization Method." *IEEE Tran. Autom. Control*, **AC-9**, **2**, (April 1964), 151–54.

30. DETCHMENDY, D.M., and SRIDHAR, R., "On the Experimental Determination of the Dynamical Characteristics of Physical Systems." *Proceedings of the Natl. Electronics Conf.*, **21**, (1965), 575–80.

31. SAGE, A.P., and EISENBERG, B.R., "Experiments in Nonlinear and Nonstationary System Identification via Quasilinearization and Differential Approximation." *Proceedings Joint Autom. Control Conf.*, (1965), pp. 522–30.

32. EISENBERG, B.R., and SAGE, A.P., "Closed-Loop Optimization of Fixed Configuration Systems." *Int. J. Control*, **3**, **2**, (1966), 183–94.

33. HENRICI, P., *Discrete Variable Methods in Ordinary Differential Equations*. Wiley, New York, 1962.

34. SYLVESTER, R.J., and MEYER, F., "Two Point Boundary Value Problems by Quasilinearization." *J. Soc. Ind. Appl. Math.*, **13**, (June 1965), 586–602.

35. SAGE, A.P., and BURT, R.W., "Optimum Design and Error Analysis of Digital Integrators for Discrete System Simulation." *American Fed. of Infor. Processing Societies, Proceedings F.J.C.C.*, **27**, Part *I*, (1965), 903–14.

36. SAGE, A.P., and SMITH, S.L., "Real-Time Digital Simulation for Systems Control." *I.E.E.E. Proceedings*, **54**, **12**, (December 1966), 1802–12.

37. SAGE, A.P., and MELSA, J.L., *Estimation Theory with Applications to Communications and Control*. McGraw-Hill Book Co., New York, 1971.

38. SAGE, A.P., and MELSA, J.L., *System Identification*. Academic Press, New York, 1971.

39. BRACKEN, J., and MCCORMICK, G.P., *Selected Applications of Nonlinear Programming*. Wiley, New York, 1968.

40. FIACCO, A.B., and MCCORMICK, G.P., *Nonlinear Programming: Sequential Unconstrained Minimization Techniques.* Wiley, New York, 1968.

41. TABAK, D., and KUO, B.C., *Optimal Control by Mathematical Programming.* Prentice-Hall, Inc., Englewood Cliffs, New Jersey, 1971.

42. DYER, P., and MCREYNOLDS, S., *The Computation and Theory of Optimal Controls.* Academic Press, New York, 1970.

43. POLAK, E., *Computational Methods in Optimization: A Unified Approach.* Academic Press, New York, 1971.

44. DREYFUS, S., "Variational Problems with State Variable Inequality Constraints." Rand Corporation, Paper P-2605, 1962, pp. 72–85.

45. MCGILL, R., "Optimal Control, Inequality State Constraints and the Generalized Newton-Raphson Algorithm." *SIAM J. Control*, Ser. A, **3**, **2**, (1965), 291–98.

46. JAZWINSKI, A.H., "Optimal Trajectories and Linear Control of Nonlinear Systems." *AIAA J.*, 2, no. 8, (August 1964), 1371–79.

47. SPEYER, J.L., "Optimization and Control Using Perturbation Theory to Find Neighboring Optimum Paths." Raytheon Co. Tech. Rept., BR-2142, December 1962.

48. LEE, I., "Optimal Trajectory, Guidance, and Conjugate Points." *Information and Control*, **8**, (1965), 589–606.

49. BULLOCK, T., "Computational of Optimal Controls by a Method Based on Second Variations." SUDAER Rept. 297, Stanford University, January 1967.

50. SCHLEY, C.H. JR., and LEE, I., "Optimal Control Computation by the Newton-Raphson Method and the Riccati Transformation." *IEEE Trans. Autom. Control*, **AC-12**, **2**, (April 1967), 139–44.

51. MCGILL, R., "Optimal Control, Inequality State Constraints, and the Generalized Newton-Raphson Algorithm." *SIAM J. Control*, Ser. A, **3**, (1965), 291–98.

52. KENNETH, P., and MCGILL, R., "Two Point Boundary Value Problem Techniques," C.T. Leondes, ed., *Advances in Control Systems*, Vol. 3. Academic Press, New York, 1966, pp. 69–110.

53. KENNETH, P., and TAYLOR, G.E., "Solution of Variational Problems With Bounded Control Variables by Means of the Generalized Newton-Raphson Method," A. Lavi, ed. *Recent Advances in Optimization Techniques.* Wiley, New York, 1966, pp. 471–87.

54. LONG, R.S., "Newton-Raphson Operator; Problems With Undetermined End Points." *A.I.A.A. J.*, **3**, *7*, (July 1965), 1351–52.

55. SBAITI, A.A., and SAGE, A.P., "System Optimization Using Quasilinearization and Sensitivity Analysis." *Int. J. Control*, **16**, *2*, (1972), 343–52.

56. SBAITI, A.A., and SAGE, A.P., "Suboptimal Closed-Loop Control Using Sensitivity Techniques." *Proceedings Milwaukee Symposium on Autom. Controls and Computers*, March 1975, pp. 437–50.

57. GUINZY, N.J., and SAGE, A.P., "System Identification in Large Scale Systems with Hierarchical Structures." *Computers and Electrical Eng.*, **1**, *1*, (1973), 23–42.

58. SMITH, N.J., and SAGE, A.P., "Hierarchical Structuring for Systems Identification." Infor. Science, **7**, *1*, (January 1974), 49–72.

Problems

1. Illustrate the continuous- and discrete-time dynamic programming methods for the minimization of the quadratic cost function for a first-order linear system

$$J = \tfrac{1}{2}\int_0^1 (x^2 + u^2)\,dt, \qquad \dot{x} = u, \qquad x(0) = 1.$$

2. Extend the results of Problem 1 to include the general linear regulator problem. Compare this solution technique with that presented in Chapter 5.

3. Obtain numerical results for Problem 1 as a function of the number of time and amplitude quantization levels used.

4. Obtain the functional equation whose solution minimizes

$$J = \tfrac{1}{2}\int_{t_o}^{t_f} \{\rho^2(t) + \alpha[n(t) - N_d]^2\}\,dt$$

for the system

$$\dot{n} = \rho(t)n(t), \qquad n(0) = n_o.$$

Obtain the solution by discrete dynamic programming where a three-stage approximation is used. Repeat the problem for a ten-stage approximation. Compare the effort involved, computer storage and computation time, and accuracy for the two solutions.

5. Outline the state increment dynamic programming approach to Problem 4.

6. Obtain the discrete dynamic programming solution to minimize

$$J = \tfrac{1}{2}\int_0^1 u^2(t)\,dt$$

for the system

$$\dot{x} = -x(t) + u(t), \qquad x(0) = 1, \qquad x(1) = 0$$

as a function of the number of stages used in the approximation. Compare this with the true solution to the continuous problem.

7. Obtain the functional equation to minimize

$$J = \tfrac{1}{2}\int_{t_o}^{t_f} (x_1^2 + x_2^2)\,dt$$

for the system

$$\dot{x}_1 = x_2, \qquad \dot{x}_2 = u, \qquad x_1(t_o) = x_{10}, \qquad x_2(t_o) = x_{20}, \qquad |u| \leq 1.$$

8. Discuss and illustrate the use of cost and policy function polynomial approximation for the system and cost function

$$J = \tfrac{1}{2}\int_0^1 (x^2 + u^2)\,dt, \qquad \dot{x} = -x^3 + u, \qquad x(0) = 10.$$

9. Illustrate the use of discrete dynamic programming to solve the minimum time problem where $x(t_f) = \mathbf{0}$ and

$$J = \int_{t_o}^{t_f} dt, \quad \dot{x}_1 = x_2, \quad \dot{x}_2 = u, \quad x_1(t_o) = x_{10}, \quad x_2(t_o) = x_{20}, \quad |u| \leq 1.$$

10. Find the solution to the equation $x^2 = 12$ by means of the steepest descent procedure.

11. Find the values of x, y, and z which yield the minimum of $J = x^2 + y^2 + z^2$ with the equality constraint $xyz = 1$.

12. Find by the gradient method the control vector $\mathbf{u}$ which minimizes $J = \mathbf{x}^T\mathbf{Q}\mathbf{x} + \mathbf{u}^T\mathbf{R}\mathbf{u}$ for the system $\mathbf{x} + \mathbf{B}\mathbf{u} + \mathbf{c} = \mathbf{0}$, with the equality state constraint $\mathbf{D}\mathbf{x} + \mathbf{e} = \mathbf{0}$.

13. Solve the algebraic equation $f(x) = x^3 + 4x^2 + 5x + 2 = 0$ for its roots by considering the minimization of $J = f^2(x)$ by the gradient technique.

14. Solve the linear programming problem of maximizing $J = x_1 + x_2 + x_3$, subject to the inequality constraints $x_1 + x_2 \leq 2$ and $x_2 + 2x_3 \leq 4$, by the gradient technique.

15. Illustrate the determination of the damping ratio ζ of the system $C(s)/R(s) = 1/(1 + 2\zeta s + s^2)$ such as to minimize $J = \int_0^\infty [r(t) - c(t)]^2 \, dt$, where $r(t) = 1$, $t \geq 0$, $c(0) = \dot{c}(0) = 0$. Sketch a hybrid computer realization of the gradient technique for this problem.

16. Verify Eq. (10.2-41).

17. Determine the convergence region for Example 10.2-5.

18. Contrast the Denham-Bryson [14] method with the penalty function approach used in this text.

19. Use the gradient procedure to find the control and trajectory which minimize $J = \int_0^1 (x^2 + u^2) \, dt$ for the system $\dot{x} = u$, $x(0) = 1$.

20. Use the gradient procedure to find the control and trajectory which minimize $J = \int_0^1 u^2 \, dt$ for the system $\dot{x} = -x + u$, $x(0) = 1$, where we specify $x(1) = 0$. Use the penalty function technique.

21. Use the gradient method to find the control and trajectory which minimize $J = \int_0^{t_f} (1 + u^2) \, dt$ for the system $\dot{x} = u$, $x(0) = 1$, $x(t_f) = 0$, where t_f is unspecified.

22. Use the gradient method to find the control and trajectory which minimize $J = \alpha x^2(t_f) + \int_0^{t_f} (1 + u^2) \, dt$ for the system $\dot{x} = u$, $x(0) = 1$, where t_f is unspecified.

23. It is possible to consider the adjoint variable as an "influence" coefficient. Show that the term $\boldsymbol{\lambda}(t_o)$, obtained by integrating the adjoint equations backward, represents the influence of a change in $\mathbf{x}(t_o)$, $\Delta\mathbf{x}(t_o)$, upon the cost function. Also show that $\partial H/\partial \mathbf{u}$ represents the influence of an impulse change in $\mathbf{u}(t)$ upon the cost function. From this, show that $\boldsymbol{\lambda}^T(t_o)\,\Delta\mathbf{x}(t_o) = \boldsymbol{\lambda}^T(t_f)\,\Delta\mathbf{x}(t_f)$, a very important relation which relates injection errors to final errors.

24. Determine the requisite algorithms for the gradient procedure by extending to the continuous case the gradient projection method of Sec. 10.2-1 for single-stage decision processes.

25. Illustrate the gradient method for the determination of the control which minimizes

$$J = 0.005 \sum_{k=o}^{99} [x^2(k) + u^2(k)]$$

for the system $x(k+1) = x(k) - 0.01x^3(k) + u(k)$, $x(0) = 10$. Compare with the results of Problem 8.

26. Determine the minimum stage controller $J = k_f$ for the system $x_1(k+1) = x_1(k) + x_2(k)$, $x_2(k+1) = x_2(k) + u(k)$, $x_1(0) = 10$, $x_2(0) = 0$, where we require $x_1(k_f) = x_2(k_f) = 0$ and restrict $u(k)$ such that $|u| \leq 1$.

27. For the case where $\Delta\mathbf{x}(t_o) = 0$ but where we desire to change the terminal manifold ΔN_f, show by a procedure analogous to that used to obtain Eq. (10.3-40) that the control may be expressed by

$$\Delta\mathbf{u}(t) = [\mathbf{A}(t)\boldsymbol{\varphi}_{xv}(t, t_o) + \mathbf{B}(t)\boldsymbol{\varphi}_{\lambda v}(t, t_o)]\boldsymbol{\varphi}_{vv}^{-1}(t_f, t_o)\,\Delta\mathbf{N}_f$$

which may be linearly added to the control for $\Delta\mathbf{x}(t)$, as shown in Eq. (10.3-49). See Fig. 10.3-2. What are the $\boldsymbol{\varphi}$'s?

28. Demonstrate the validity of Eqs. (10.3-45) and (10.3-46).

29. Show the equivalence of Eqs. (10.3-45) and (10.3-40).

30. Illustrate the use of the second variation to obtain the control to minimize $J = \int_0^1 (x^2 + u^2)\,dt$ for the linear system $\dot{x} = -x(t) + u(t), x(0) = 1, x(1) = 0$. Use the method both to obtain the optimum control $u(t)$ and to determine a closed-loop system.

31. Repeat Problem 30 for the cost function $J = \frac{1}{2}\int_0^{t_f} (1 + u^2)\,dt$ and system $\dot{x} = -x^3 + u$, $x(0) = 10$, $x(t_f) = 0$.

32. Consider the minimization of $J = \frac{1}{2}x^3(3\pi) + \frac{1}{2}\int_0^{3\pi} (u^2 - 2x^2)\,dt$ for the system $\dot{x} = x + u$. Show that $x(t) = \cos t$ and that a conjugate point exists in the neighboring optimal control problem where we wish to minimize $\Delta J = J_1 + \frac{1}{2}\Delta x^2(t_f) + \frac{1}{2}\int_0^{3\pi} [\Delta u^2(t) - 2\Delta x^2(t)]\,dt$ for the system $\Delta\dot{x} = \Delta x + \Delta u$, $\Delta x(t_o) = \Delta x_o$ where J_1 is defined as in Eq. (10.3-8).

33. Obtain the computational algorithms to obtain the optimal control and trajectory for the cost function and system

$$J = \frac{1}{2}\int_0^1 [x^2(t) + u^2(t)]\,dt, \qquad x = -x^3(t) + u(t), \qquad x(0) = 10.$$

Use the gradient method, the second-variation method, and the successive sweep method. Compare the effort used in obtaining the algorithms; also compare the solution effort.

34. Repeat Problem 33 with the additional constraint that $x(1) = 0$.

35. Discuss the approximate optimal closed-loop linearized control of the system

$$\dot{x} = -x^3(t) + u(t) + w(t)$$
$$z(t) = x(t) + v(t),$$

where $w(t)$ and $v(t)$ are zero-mean white Gaussian random noise terms with variance coefficients $q(t) = 3$, $r(t) = 2.0$. The expected initial condition is $\mathcal{E}\{x(0)\} = 10$. The cost function to be minimized is

$$J = \tfrac{1}{2}\mathcal{E}\left\{\int_0^1 [x^2(t) + u^2(t)]\right\} dt$$

Compare the solution to this problem with that for Problem 33.

36. Extend your solution of Problem 35 to the general case.

37. For the system $\dot{x} = -ax$ with constant parameter a, find a by the quasilinearization procedure such that $x(0) = 1$, $x(1) = 0.5$. For the *zero*th iteration, assume that $x^0(t) = 1 - 0.5t$, $a^0(t) = 0.5$.

38. For the system $\dot{x} = -ax$ with constant parameter a, find the value of a such that the solution to the system equation best fits (in the least-square-error sense) the points $x(0) = 1$, $x(0.5) = 0.75$, $x(1) = 0.5$, $x(2) = 0.25$. Set up the quasilinearization computational algorithms in detail first. Assume a *zero*th iterate of $x^0(t) = 1 - 0.5t$ and $a^0 = 0.5$.

39. It is desired to minimize the cost function $J = \frac{1}{2}\int_0^2 u^2\,dt$ for the system $\dot{x} = -x^3 + u$ with $x(0) = 10$, $x(2) = 0$. Find the quasilinearized equations and the optimum control and trajectory.

40. Find the minimum time control for

$$\dot{x}_1 = x_2 \qquad x_1(0) = 10$$
$$\dot{x}_2 = -2x_2 - x_1 + u \qquad x_2(0) = 1$$

such that the cost function $J = \int_0^{t_f} dt$ is minimized, where $x_1(t_f) = x_2(t_f) = 0$ and t_f is, of course, unspecified. The control inequality constraint is $|u| \leq 1$.

41. Find the constant parameters a and b for a specific optimal problem for the system

$$\dot{x}_1 = x_2 \qquad x_1(0) = 10$$
$$\dot{x}_2 = -2x_2 - x_1 + u \qquad x_2(0) = 0$$

with

$$J = \tfrac{1}{2}\int_0^1 (x_1^2 + x_2^2 + 0.1u^2)\,dt,$$

where $u = ax_1 + bx_2$. If sufficient computation time is available, obtain data such as to determine the best feedback parameters a and b as a function of various initial states $x_1(0)$ and $x_2(0)$. Replace $\dot{x}_1 = x_2$ by $\dot{x}_1 = \alpha x_2$ and investigate the effect of varying α from a nominal value of 1.

42. Consider the system $\dot{x} = -ax^3 + u + w$, with constant but unknown parameter a, such that $\dot{a} = 0$, and with known input u and unknown input w. The output x is corrupted with noise v before observation. Thus the observed quantity is

$$z = x(t) + v(t).$$

Find the quasilinearized equations which give the best least-squares curve fit. Find the quasilinearized equations for the dynamic equations

$$\dot{\hat{x}} = -\hat{a}\hat{x}^3 + u + \hat{w}$$

$$\dot{\hat{a}} = 0$$

$$z = \hat{x}(t) + \hat{v}(t)$$

with cost function

$$J = \frac{1}{2}\int_{t_o}^{t_f} \{\hat{v}^2(t) + \alpha\hat{w}^2(t)\}\, dt.$$

43. Find the discrete quasilinearization equations for the two-point boundary value problem which minimizes

$$J = \frac{1}{2}\sum_{k=0}^{K-1} u^2(k)$$

for the system

$$x(k+1) = x(k) - Tx^3(k) + u(k)$$

with $x(k = 0) = 10$, $x(k = K) = 0$. Compare the solution of this problem for values of sample period T such that $KT = 2$. Compare these results with the numerical results of Problem 39 and comment.

44. Develop a continuous quasilinearization procedure to solve linear system problems with fixed time of control and quadratic error criteria where the terminal manifold equations are nonlinear.

45. Apply the procedure from Problem 44 to the particular case where

$$\dot{x}_1 = x_2, \qquad \dot{x}_2 = u, \qquad x_1(0) = x_2(0) = 0, \qquad J = \frac{1}{2}\int_0^1 u^2\, dt,$$

$$x_1^2(1) + x_2^2(1) = 1.$$

46. Develop the discrete version of Problems 44 and 45.

47. Illustrate how we may quasilinearize $\mathbf{N}$ and θ so as to minimize $J = \theta[\mathbf{x}(t_f)] + \int_{t_o}^{t_f} \phi\, dt$ for the system $\dot{\mathbf{x}} = \mathbf{f}(\mathbf{x}, \mathbf{u}, t)$ with $\mathbf{x}(t_o) = \mathbf{x}_o$, $\mathbf{N}[\mathbf{x}(t_f)] = \mathbf{0}$, and t_o and t_f fixed.

48. Find the quasilinearized equations which minimize $J = \frac{1}{2}[x_2^2(t_f) + x_3^2(t_f)]^{1/2} + \frac{1}{2}\int_0^{t_f} \alpha u^2\, dt$ for the system $\dot{x}_1 = x_2, \dot{x}_2 = x_3, \dot{x}_3 = u, x_1(0) = 10, x_2(0) = x_3(0) = 0, x_1(t_f = 1) = 0$.

49. Verify Eqs. (10.4-22) through (10.4-24). What do the corresponding equations become for discrete quasilinearization?

50. Determine the quasilinearization algorithms necessary to resolve the optimal control problem of minimization of

$$J = \int_0^{10} [10x^2(t) + u^2(t)] \, dt$$

for the system

$$\dot{x} = -x(t) - 0.2x^3(t) + u(t)$$

$$x(0) = 2.0.$$

Now replace the term 0.2 in the above by the parameter ϵ. Use sensitivity analysis and quasilinearization to determine the optimum open–loop control. Contrast and compare solution complexity with that resulting from using the basic quasilinearization algorithms.

51. Let $\mu(t) = ax(t)$ where a is a linear constant function of time $\dot{a} = 0$, and determine the optimum a for Problem 50.

Appendices

The algebra, calculus, and differential equations of vectors and matrices

A

This appendix is intended to delineate the various vector-matrix operations encountered in this text. The symbols for vectors are boldface lower case Arabic or upper or lower case Greek letters, e.g., $\mathbf{x}$ or $\boldsymbol{\alpha}$. The symbols for matrices are boldface upper case Arabic or upper or lower case Greek letters, e.g., $\mathbf{F}$ or $\boldsymbol{\varphi}$. Brackets [ ] are used around matrices only where clarity is improved. The transpose of a vector or matrix is indicated by a superscript following the symbol, e.g., $\mathbf{x}^T$ or $\mathbf{F}^T$. All vector symbols not followed by the transpose superscript are column vectors. The inverse of a nonsingular matrix is indicated by the superscript (-1) following the matrix symbol, e.g., $\mathbf{F}^{-1}$. The inverse transpose of a matrix is indicated by the superscript $(-T)$ following the matrix symbol, e.g., $\mathbf{F}^{-T}$. The basic symbols, definitions, and operations which we use are as follows.

A.1 Matrix algebra

A.1-1 Matrix

We will call a set of mn quantities, real or complex, arranged in a rectangular array of n columns and m rows a matrix of order (m, n) or $m \times n$. If $m = n$, we will call the matrix a square matrix. A matrix is denoted in the following manner:

$$\mathbf{F} = \begin{bmatrix} f_{11} & f_{12} & \cdots & f_{1n} \\ f_{21} & f_{22} & \cdots & f_{2n} \\ \cdot & & & \\ \cdot & & & \\ \cdot & & & \\ f_{m1} & f_{m2} & \cdots & f_{mn} \end{bmatrix}.$$

A.1-2
Row vector

A set of n quantities arranged in a row is a matrix of order $(1, n)$. We will call it a row vector:

$$\mathbf{x}^T = [x_1, x_2, x_3, \ldots, x_n].$$

A.1-3
Column vector

We will call a set of n quantities arranged in a column a matrix of order $(n, 1)$ or a column vector,

$$\mathbf{x} = \begin{bmatrix} x_1 \\ x_2 \\ x_3 \\ \cdot \\ \cdot \\ \cdot \\ x_n \end{bmatrix}.$$

A.1-4
Scalar

A matrix of order $(1, 1)$ will be called a scalar. A vector with only a single component is likewise a scalar.

A.1-5
Diagonal matrix

We will define a diagonal matrix as a square matrix in which all elements off the main diagonal are zero,

$$\mathbf{F} = \begin{bmatrix} f_{11} & 0 & 0 \\ 0 & f_{22} & 0 \\ 0 & 0 & f_{33} \end{bmatrix}.$$

A.1-6
Identity matrix

We will define the identity matrix as a diagonal matrix having unit diagonal elements,

$$\mathbf{I} = \begin{bmatrix} 1 & 0 & 0 \\ 0 & 1 & 0 \\ 0 & 0 & 1 \end{bmatrix}.$$

A.1-7
Null or zero matrix

The matrix [**0**], which has all zero elements, will be called a null or zero matrix.

A.1-8
Equality of two matrices

We will say that two matrices **F** and **G** are equal if, and only if, they are of the same order and all their corresponding elements are equal

$$[\mathbf{F}] - [\mathbf{G}] = [\mathbf{0}].$$

A.1-9
Singular and nonsingular matrices

We will call a square matrix [**F**], for which the determinant of its elements, det [**F**], is zero, a singular matrix. If det [**F**] $\neq$ **0**, then we say that [**F**] is a nonsingular matrix.

A.1-10
Matrix transpose

We form the transpose of a matrix [**F**] by the interchange of rows and columns. Thus the transpose of the matrix in Sec. 1.1 is

$$\mathbf{F}^T = \begin{bmatrix} f_{11} & f_{21} & f_{31} & \cdots & f_{n1} \\ f_{12} & & & & \\ \cdot & & & & \\ \cdot & & & & \\ \cdot & & & & \\ f_{1n} & & & & f_{nm} \end{bmatrix}.$$

A.1-11
Orthogonal matrix

We will call a matrix [**F**] such that $\mathbf{F}^T\mathbf{F} = \mathbf{F}\mathbf{F}^T = \mathbf{I}$ an orthogonal matrix.

A.1-12
Symmetric and skew-symmetric matrices

We will define a symmetric matrix as one that is square and symmetrical about the main diagonal, $\mathbf{F}^T = \mathbf{F}$. $\mathbf{F}^T = -\mathbf{F}$ is the requirement for a skew-symmetric matrix.

A.1-13
Cofactor

We will define the cofactor of an element in the ith row and jth column of a matrix as the value of the determinant formed by writing the elements of the matrix as a determinant, then deleting the ith row and jth column, and giving to it the sign $(-1)^{i+j}$.

A.1-14
Adjoint matrix

We may obtain the adjoint matrix of a square matrix $\mathbf{F}$ by replacing each element by its cofactor and transposing. We will denote the adjoint by $[\text{adj } \mathbf{F}]$. We note that the adjoint matrix differs from the matrix of an adjoint system.

A.1-15
Minor of a matrix or determinant

The matrix or determinant obtained by suppressing m rows and m columns of the matrix $\mathbf{F}$ or the det $[\mathbf{F}]$ of the matrix $\mathbf{F}$ of order (n, n), will be called a minor of order $(n - m)$.

A.1-16
Characteristic matrix, characteristic equations, and eigenvalues

The characteristic matrix of a square matrix $\mathbf{F}$ of order (n, n) and with constant elements is

$$[\mathbf{C}] = [\lambda \mathbf{I} - \mathbf{F}].$$

We will call the equation, det $[\mathbf{C}] = \det [\lambda \mathbf{I} - \mathbf{F}] = 0$, the characteristic equation of the matrix $\mathbf{F}$. The n roots of the characteristic equation $\det [\lambda \mathbf{I} - \mathbf{F}] = 0$, which will be of degree n in λ, will be called the eigenvalues of the matrix $\mathbf{F}$.

A.1-17
Addition of matrices

We will perform the addition of two matrices $\mathbf{F}$ and $\mathbf{G}$ of the same order by adding their corresponding elements and by writing the sums as the corresponding elements of the resultant matrix $\mathbf{F} + (\mathbf{G} + \mathbf{H}) = (\mathbf{F} + \mathbf{G}) + \mathbf{H}$.

A.1-18
Multiplication of matrices

SCALAR MULTIPLICATION: A matrix $\mathbf{F}$ is multiplied by a scalar k if we multiply all the elements of $\mathbf{F}$ by k. We may easily show that scalar multiplication is commutative in that $k\mathbf{F} = \mathbf{F}k$. Thus we assume that the matrix is defined over a commutative field.

MULTIPLICATION OF TWO VECTORS: We will obtain multiplication of a row vector $\mathbf{x}^T$ by a column vector $\mathbf{y}$ by multiplying corresponding elements in each vector and then adding the products. The product is a scalar and is denoted by

$$\mathbf{x}^T\mathbf{y} = [x_1, x_2, x_3, \ldots, x_n]\begin{bmatrix} y_1 \\ y_2 \\ y_3 \\ \cdot \\ \cdot \\ \cdot \\ y_n \end{bmatrix} = (x_1y_1 + x_2y_2 + x_3y_3 + \cdots + x_ny_n).$$

The scalar $\mathbf{x}^T\mathbf{y}$ is often called the inner product and given the symbol $\langle \mathbf{x}, \mathbf{y} \rangle$. This inner product can also be written with respect to a matrix as

$$\langle \mathbf{x}, \mathbf{Py} \rangle = \mathbf{x}^T\mathbf{Py}$$
$$\langle \mathbf{P}^T\mathbf{x}, \mathbf{y} \rangle = \mathbf{x}^T\mathbf{Py}.$$

When $\mathbf{x}$ and $\mathbf{y}$ are the same vector, the inner product is equivalent to the square of the Euclidean norm

$$\mathbf{x}^T\mathbf{x} = ||\mathbf{x}||^2 = \langle \mathbf{x}, \mathbf{x} \rangle$$
$$\mathbf{x}^T\mathbf{Px} = ||\mathbf{x}||_{\mathbf{P}}^2 = \langle \mathbf{x}, \mathbf{Px} \rangle.$$

An interesting property of the norm is that $||\mathbf{x} + \mathbf{y}|| \leq ||\mathbf{x}|| + ||\mathbf{y}||$.

In a similar way, the outer product can be defined. In this case, it is not necessary that the $\mathbf{x}$ and $\mathbf{y}$ vectors have the same dimension. The outer product is defined by

$$\mathbf{x} > < \mathbf{y} = \mathbf{x}\mathbf{y}^T =$$

$$\begin{bmatrix} x_1 \\ x_2 \\ \cdot \\ \cdot \\ \cdot \\ x_n \end{bmatrix}[y_1, y_2, \ldots, y_m] = \begin{bmatrix} (x_1y_1) & (x_1y_2) & (x_1y_3) & \cdots & (x_1y_m) \\ (x_2y_1) & (x_2y_2) & \cdots & & \\ \cdot & & & & \\ \cdot & & & & \\ \cdot & & & & \\ (x_ny_1) & (x_ny_2) & \cdots & & (x_ny_m) \end{bmatrix}.$$

MULTIPLICATION OF TWO MATRICES: Our definition of multiplication will be such that two matrices can be multiplied together only when the number of columns of the first is equal to the number of rows of the second. The product of two matrices will be defined by the equation

$$\mathbf{H}_{ij} = \mathbf{FG}_{ij} = \sum_{k=1}^{k=p} f_{ik}g_{kj},$$

where the order of the matrices **F**, **G**, and **H** are (m, p), (p, n), and (m, n), respectively. We may express this symbolically in the form, $(m, p)\cdot(p, n) = (m, n)$.

A.1-19
The inverse of a matrix

We may show that $[\mathbf{F}]\,[\text{adj}\,\mathbf{F}] = [\text{adj}\,\mathbf{F}]\mathbf{F} = \mathbf{I}\det[\mathbf{F}]$, where **I** is the nth-order identity matrix and **F** is an nth-order square matrix. If **F** is a nonsingular matrix, $\det[\mathbf{F}] \neq 0$, and there exists a matrix **G** such that,

$$\mathbf{FG} = \mathbf{GF} = \mathbf{I}, \qquad \mathbf{G} = \frac{[\text{adj}\,\mathbf{F}]}{\det[\mathbf{F}]} = \frac{[\text{adj}\,\mathbf{F}]}{|\mathbf{F}|} = \mathbf{F}^{-1}.$$

The matrix **G** will be defined as the inverse matrix of **F**. We will denote it by $\mathbf{F}^{-1}$. The inverse of a product matrix is the product of the inverses of the separate matrices in the reverse order. Thus we may write

$$(\mathbf{AB})^{-1} = \mathbf{B}^{-1}\mathbf{A}^{-1}.$$

A.1-20
Transpose of a product of matrices

If **F** and **G** are conformable matrices such that we may define the product **FG**, then we may obtain the transpose of this product by

$$(\mathbf{FG})^T = \mathbf{G}^T\mathbf{T}^T.$$

A.1-21
The Cayley-Hamilton theorem

If **F** is a square matrix, then we write

$$\Delta(\lambda) = \det[\lambda\mathbf{I} - \mathbf{F}] = 0$$

as its characteristic equation. We may show that

$$\Delta([\mathbf{F}]) = 0.$$

This is a consequence of the Cayley-Hamilton theorem, which states that every square matrix satisfies its own characteristic equation in a matrix sense.

A.1-22 *Transformation of a square matrix to diagonal form*

We are given the square matrix **F**. The equation $\mathbf{y} = \mathbf{Fx}$ expresses a linear transformation between the column vectors **x** and **y**. If we require that $\mathbf{y} = \lambda\mathbf{x}$, then $\lambda\mathbf{x} = \mathbf{Fx}$. The possible values of λ are $\lambda_1, \lambda_2, \ldots, \lambda_n$, which are the eigenvalues of **F**. In what is to follow, we shall assume that the eigenvalues are distinct.

To each eigenvalue of **F** there corresponds an eigenvector $\mathbf{x}^i$ which satisfies the relation,

$$\lambda_i \mathbf{x}_i = F\mathbf{x}^i, \qquad i = 1, 2, 3, \ldots, n.$$

If we construct the square matrix,

$$[\mathbf{X}] = [(\mathbf{x})^1, (\mathbf{x})^2, (\mathbf{x})^3, \ldots, (\mathbf{x})^n].$$

We can write

$$\mathbf{XG} = \mathbf{FX}, \qquad \mathbf{G} = \begin{bmatrix} \lambda_1 & 0 & \cdots & 0 \\ 0 & \lambda_2 & \cdots & 0 \\ \cdot & & & \cdot \\ \cdot & & & \cdot \\ \cdot & & & \cdot \\ 0 & 0 & \cdots & \lambda_n \end{bmatrix} = \text{diag}\,[\boldsymbol{\lambda}]$$

and show that

$$\mathbf{F} = \mathbf{XGX}^{-1}$$

which is the required transformation which converts **F** to diagonal form. This tells us that if $\mathbf{y} = \mathbf{Fx}$, $\mathbf{x} = \mathbf{Xu}$, and $\mathbf{y} = \mathbf{Xv}$, then

$$\mathbf{v} = \mathbf{X}^{-1}\mathbf{y} = \mathbf{X}^{-1}\mathbf{Fx} = \mathbf{X}^{-1}\mathbf{FXu}$$
$$\mathbf{v} = \mathbf{Gu}.$$

A.1-23 *Quadratic forms*

We will define a quadratic form in **x** as

$$Q = \mathbf{x}^T\mathbf{Fx} = \sum_{i=1}^{n}\sum_{j=1}^{n} f_{ij}x_ix_j,$$

where **F** is a square symmetric matrix. A quadratic form is said to be positive definite if $Q > 0$ for $\mathbf{x} \neq \mathbf{0}$. A square symmetric matrix is said to be positive definite if its corresponding quadratic form is positive definite. A necessary and sufficient test for a positive definite matrix **F** is that the following inequalities be satisfied:

$$|f_{11}| > 0, \qquad \begin{vmatrix} f_{11} & f_{12} \\ f_{21} & f_{22} \end{vmatrix} > 0, \qquad \begin{vmatrix} f_{11} & f_{12} & f_{13} \\ f_{21} & f_{22} & f_{23} \\ f_{31} & f_{32} & f_{33} \end{vmatrix} > 0, \qquad \text{etc.}$$

If the matrix is positive semidefinite, the greater than signs in the foregoing are replaced by greater than or equal to signs. Negative definite and negative semidefinite matrices are defined such that their negatives are positive definite and positive semidefinite respectively.

A.2 Differentiation of matrices and vectors

A.2-1 Differentiation with respect to a scalar (time)

We define differentiation of a vector or matrix with respect to a scalar in the following way:

$$\mathbf{z} = \mathbf{z}(t), \qquad \frac{d\mathbf{z}}{dt} = \left[\frac{dz_1}{dt}, \frac{dz_2}{dt}, \ldots, \frac{dz_m}{dt}\right]^T$$

$$\mathbf{F} = \mathbf{F}(t), \qquad \frac{d\mathbf{F}}{dt} = \begin{bmatrix} \frac{df_{11}}{dt} & \frac{df_{12}}{dt} & \cdots & \frac{df_{1n}}{dt} \\ \frac{df_{21}}{dt} & \frac{df_{22}}{dt} & \cdots & \frac{df_{2n}}{dt} \\ \vdots & \vdots & & \vdots \\ \frac{df_{m1}}{dt} & \frac{df_{m2}}{dt} & \cdots & \frac{df_{mn}}{dt} \end{bmatrix}.$$

A.2-2 Differentiation with respect to a vector

Differentiation of vectors with respect to vectors are defined as follows:

$$f = f(\mathbf{x}) = \text{scalar}, \qquad \frac{df}{d\mathbf{x}} = \left[\frac{df}{dx_1}, \frac{df}{dx_2}, \ldots, \frac{df}{dx_n}\right]^T.$$

This is often called a gradient and written $\nabla_{\mathbf{x}} f = \frac{df}{d\mathbf{x}}$.

$$\mathbf{z} = \mathbf{z}(\mathbf{x}), \qquad \frac{d\mathbf{z}^T}{d\mathbf{x}} = \begin{bmatrix} \frac{dz_1}{dx_1} & \frac{dz_2}{dx_1} & \cdots & \frac{dz_m}{dx_1} \\ \frac{dz_1}{dx_2} & \frac{dz_2}{dx_2} & \cdots & \frac{dz_m}{dx_2} \\ \vdots & \vdots & & \vdots \\ \frac{dz_1}{dx_n} & \frac{dz_2}{dx_n} & \cdots & \frac{dz_m}{dx_n} \end{bmatrix}.$$

This is sometimes called the Jacobian matrix and written $J_{\mathbf{x}}\mathbf{z}(\mathbf{x})$, where

$$\left[\frac{d\mathbf{z}}{d\mathbf{x}}\right]_{ij} = [J_{\mathbf{x}}\{\mathbf{z}(\mathbf{x})\}]_{ij} = \left[\frac{\partial z_i}{\partial x_j}\right].$$

A.2-3
Operations involving partial derivatives

Consider the functions: $f = f(\mathbf{y}, \mathbf{x}, t)$, $\mathbf{y} = \mathbf{y}(\mathbf{x}, t)$, and $\mathbf{x} = \mathbf{x}(t)$. We define the following operations:

$$\frac{df}{d\mathbf{x}} = \left[\frac{\partial \mathbf{y}^T}{\partial \mathbf{x}}\right]\frac{\partial f}{\partial \mathbf{y}} + \frac{\partial f}{\partial \mathbf{x}}$$

$$\frac{df}{dt} = \left[\frac{\partial f}{\partial \mathbf{x}} + \left\{\frac{\partial \mathbf{y}^T}{\partial \mathbf{x}}\right\}\left(\frac{\partial f}{\partial \mathbf{y}}\right)\right]^T \frac{d\mathbf{x}}{dt} + \left[\frac{\partial f}{\partial \mathbf{y}}\right]^T \frac{\partial \mathbf{y}}{\partial t} + \frac{\partial f}{\partial t}.$$

Similar operations on the vector function are defined by

$$\mathbf{z} = \mathbf{z}(\mathbf{y}, \mathbf{x}, t), \qquad \mathbf{y} = \mathbf{y}(\mathbf{x}, t), \qquad \mathbf{x} = \mathbf{x}(t)$$

$$\frac{d\mathbf{z}}{d\mathbf{x}} = \left[\frac{\partial \mathbf{z}^T}{\partial \mathbf{y}}\right]^T \left[\frac{\partial \mathbf{y}^T}{\partial \mathbf{x}}\right]^T + \frac{\partial \mathbf{z}}{\partial \mathbf{x}}$$

$$\frac{d\mathbf{z}}{dt} = \left[\frac{\partial \mathbf{z}^T}{\partial \mathbf{y}}\right]^T \left[\left\{\frac{\partial \mathbf{y}^T}{\partial \mathbf{x}}\right\}^T \frac{d\mathbf{x}}{dt} + \frac{\partial \mathbf{y}}{\partial t}\right] + \left[\frac{\partial \mathbf{z}^T}{\partial \mathbf{x}}\right]^T \frac{d\mathbf{x}}{dt} + \frac{\partial \mathbf{z}}{\partial t}.$$

A.2-4
Taylor series expansion of a scalar function of the vector $\mathbf{x}$ *about* $\mathbf{x}_o$

In much of our work we have need for series expansions of scalar functions of vectors,

$$f(\mathbf{x}) = f(\mathbf{x}_o) + \left[\frac{\partial f(\mathbf{x}_o)}{\partial \mathbf{x}_o}\right]^T (\mathbf{x} - \mathbf{x}_o) + \frac{1}{2}(\mathbf{x} - \mathbf{x}_o)^T \left[\frac{\partial}{\partial \mathbf{x}_o}\left(\frac{\partial f(\mathbf{x}_o)}{\partial \mathbf{x}_o}\right)\right]^T (\mathbf{x} - \mathbf{x}_o) +$$
$$\frac{1}{6}(\mathbf{x} - \mathbf{x}_o)^T \left\{(\mathbf{x} - \mathbf{x}_o)^T \frac{\partial}{\partial \mathbf{x}_o}\left(\frac{\partial}{\partial \mathbf{x}_o}\left[\frac{\partial f(\mathbf{x}_o)}{\partial \mathbf{x}_o}\right]\right)^T\right\}(\mathbf{x} - \mathbf{x}_o) + \text{higher-order terms.}$$

The Taylor series for the expansion of a vector function of a vector $\mathbf{x}$ about $\mathbf{x}_o$ is defined in exactly the same fashion.

A.2-5
Trace of a matrix

We define the trace of $\mathbf{A}$, $\operatorname{tr}[\mathbf{A}]$, as being the sum of the principal diagonal elements. Thus

$$\operatorname{tr}[\mathbf{A}^T\mathbf{B}] = \operatorname{tr}[\mathbf{B}^T\mathbf{A}] = \operatorname{tr}[\mathbf{A}\mathbf{B}^T] = \operatorname{tr}[\mathbf{B}\mathbf{A}^T]$$

and

$$\operatorname{tr}[\mathbf{A} + \mathbf{B}] = \operatorname{tr}[\mathbf{A}] + \operatorname{tr}[\mathbf{B}].$$

Also

$$\text{tr}\,[\mathbf{A}] = \text{tr}\,[\mathbf{A}^T].$$

If

$$\mathbf{A} = \mathbf{x}\mathbf{y}^T,$$

then

$$\text{tr}\,[\mathbf{A}] = \mathbf{x}^T\mathbf{y}.$$

A.3
Linear vector differential equations

In many cases, a sufficiently accurate model of a differential system is provided by the following state variable equations where $\mathbf{x}$ is an n-vector, $\mathbf{u}$ is an m-vector, $\mathbf{z}$ is an r-vector, and $\mathbf{F}$, $\mathbf{G}$, and $\mathbf{H}$ are matrices of compatible orders,

$$\dot{\mathbf{x}} = \mathbf{F}(t)\mathbf{x} + \mathbf{G}(t)\mathbf{u}(t)$$

$$\mathbf{z}(t) = \mathbf{H}(t)\mathbf{x}(t).$$

If we assume that $\mathbf{F}$, $\mathbf{G}$, $\mathbf{H}$, and $\mathbf{u}$ are piecewise continuous functions of time, the system of equations will have a unique solution of the form

$$\mathbf{x}(t) = \boldsymbol{\varphi}(t, t_o)\mathbf{x}(t_o) + \int_{t_o}^{t} \boldsymbol{\varphi}(t, \tau)\mathbf{G}(\tau)\mathbf{u}(\tau)\,d\tau,$$

where the system transition matrix $\boldsymbol{\varphi}(t, \tau)$ satisfies the matrix differential equation (often called the fundamental matrix equation when $\boldsymbol{\varphi}(t, \tau)$ is called the fundamental matrix)

$$\frac{\partial \boldsymbol{\varphi}(t, \tau)}{\partial t} = \mathbf{F}(t)\boldsymbol{\varphi}(t, \tau)$$

for all t and τ. The initial condition matrix for the foregoing is

$$\boldsymbol{\varphi}(\tau, \tau) = \mathbf{I}, \qquad \forall \tau.$$

From these properties, it immediately follows that

$$\boldsymbol{\varphi}^{-1}(t, \tau) = \varphi(\tau, t), \qquad \forall\ \tau, t$$

$$\boldsymbol{\varphi}(\tau_3, \tau_2)\boldsymbol{\varphi}(\tau_2, \tau_1) = \boldsymbol{\varphi}(\tau_3, \tau_1), \qquad \forall\ \tau_1, \tau_2, \tau_3.$$

In the particular case for which the system is stationary $\mathbf{F}(t) = \mathbf{F}$, a constant, the solution for the transition matrix $\boldsymbol{\varphi}(t, \tau)$ can be immediately determined as

$$\boldsymbol{\varphi}(t, \tau) = e^{\mathbf{F}(t-\tau)},$$

where the matrix exponential is defined by

$$e^{\mathbf{A}t} = \mathbf{I} + \mathbf{A}t + \frac{\mathbf{A}^2t^2}{2!} + \frac{\mathbf{A}^3t^3}{3!} + \cdots.$$

Also, where $\mathbf{F}$ and $\mathbf{G}$ are constant, i.e., not functions of time, it is possible to write the vector Laplace transform of the system response as

$$\mathbf{x}(s) = (s\mathbf{I} - \mathbf{F})^{-1}\,[\mathbf{x}(t = 0) + \mathbf{G}\mathbf{u}(s)].$$

Thus for the case of a stationary system, the state transition matrix $\boldsymbol{\varphi}(t, \tau) = \boldsymbol{\varphi}(t - \tau)$ is seen to be the inverse transform of $[s\mathbf{I} - \mathbf{F}]^{-1}$.

A.4
Linear vector difference equations

For discrete systems which are linear, an appropriate difference equation to characterize a system is

$$\mathbf{x}(k + 1) = \mathbf{F}(k)\mathbf{x}(k) + \mathbf{G}(k)\mathbf{u}(k)$$

$$\mathbf{z}(k) = \mathbf{H}(k)\mathbf{x}(k), \qquad k = 0, 1, 2, 3, \ldots,$$

where $\mathbf{x}$, $\mathbf{u}$, and $\mathbf{z}$, are n-, m-, and r-vectors, respectively, and where the orders of the matrices are compatible; k denotes the stage of the process. Thus $\mathbf{x}(k + 1) = \mathbf{x}(t_{k+1})$ and $\mathbf{G}(k) = \mathbf{G}(t_k)$. The system of equations has a solution of the form

$$\mathbf{x}(k) = \boldsymbol{\varphi}(k, l)\mathbf{x}(l) + \sum_{j=l}^{k-1} \boldsymbol{\varphi}(k, j + 1)\mathbf{G}(j)\mathbf{u}(j),$$

where the discrete system transition matrix satisfies the matrix difference equation,

$$\boldsymbol{\varphi}(k + 1, l) = \mathbf{F}(k)\boldsymbol{\varphi}(k, l), \qquad \forall\ k, l$$

$$\boldsymbol{\varphi}(k, k) = \mathbf{I}$$

$$\boldsymbol{\varphi}(k, j) = \prod_{i=j}^{k-1} \mathbf{F}(i).$$

From these properties, it immediately follows that

$$\boldsymbol{\varphi}^{-1}(k, l) = \boldsymbol{\varphi}(l, k), \qquad \forall\ k, l$$

$$\boldsymbol{\varphi}(\alpha, \beta)\boldsymbol{\varphi}(\beta, \gamma) = \boldsymbol{\varphi}(\alpha, \gamma), \qquad \forall\ \alpha, \beta, \gamma.$$

In the particular case for which $\mathbf{F}$ is not a function of the stage, the solution to the foregoing equation is

$$\boldsymbol{\varphi}(k, l) = \mathbf{F}^{k-l}.$$

References

1. GANTMACHER, F.R., *Theory of Matrices*, Vol. I, II. Chelsea Publishing Co., New York, 1959.

2. PEASE, M.C., *Methods of Matrix Algebra*. Academic Press, New York, 1965.

3. KALMAN, R.E., "Mathematical Description of Linear Dynamical Systems." *SIAM J. Control*, Ser. A, **I**, *2*, (1963), 159–92.

4. PIPES, L.A., "Matrices in Engineering," E.F. Beckenbach, ed., *Modern Mathematics for the Engineer*. McGraw-Hill Book Co., New York, 1956, Chapter 13.

5. BELLMAN, R., *Introduction to Matrix Analysis*. McGraw-Hill Book Co., New York, 1960.

6. OGATA, K., *State Space Analysis of Control Systems*. Prentice-Hall, Inc., Englewood Cliffs, New Jersey, 1967.

7. GUPTA, S.C., *Transform and State Variable Methods in Linear Systems*. Wiley, New York, 1966.

8. NAYLOR, A.W., and SELL, G.R., *Linear Operator Theory in Engineering and Science*. Holt, New York, 1971.

9. ATHANS, M., and FALB, P.L., *Optimal Control*. McGraw-Hill Book Co., New York, 1966.

Abstract spaces

B

In this appendix we present several defining properties of certain abstract spaces that are useful for fuller appreciation of much of the material in this text. The references indicate where the interested reader can further explore the topics presented here. A single outstanding source for such a study is [1].

B.1 Functions and inverses

B.1-1 Functions

Let X and Y be two sets, and let f be a rule which assigns to each element in X one and only one element in Y. Then f is said to be a *function* which maps X into Y, which is often denoted as $f: X \longrightarrow Y$. Note that f or $f(\cdot)$ is a function while $f(x)$ is an element of Y.

B.1-2 Inverses

The function $g: Y \longrightarrow X$ is said to be the *inverse* of $f: X \longrightarrow Y$ if and only if $g[f(x)] = x$ for all $x \in X$ and $f[g(y)] = y$ for all $y \in Y$. It can be shown that g is unique and that f is the inverse of g. The inverse of f is typically denoted by f^{-1}. In Subsec. A.1-19 of Appendix A, we discuss a particular inverse, the inverse of linear operators, represented by matrices.

B.2 Topological structure

B.2-1 Metric space

A *metric space* (X, ρ) is a set X and a function $\rho: X \times X \longrightarrow R$, where R represents the real line, which satisfies the following axioms:

1. $\rho(x, y) \geq 0 \quad$ for all $\quad x, y \in X$
2. $\rho(x, y) = 0 \quad$ if and only if $\quad x = y$
3. $\rho(x, y) = \rho(y, x) \quad$ for all $\quad x, y \in X$
4. $\rho(x, y) \leq \rho(x, z) + \rho(z, y) \quad$ for all $\quad x, y, z \in X$.

We note that the metric satisfies many of our notions of a distance measure, where axiom 4 is referred to as the *triangle inequality*. *Examples:*

(a) Let $X = R^2$, and

$$\rho(x, y) = [(x_1 - y_1)^2 + (x_2 - y_2)^2]^{1/2}.$$

(b) Let X represent the set of all scalar-valued continuous functions on $[t_o, t_f]$, and

$$\rho(x, y) = \max_{t_o \leq t \leq t_f} |x(t) - y(t)|.$$

(c) Let $X = R^n$, and

$$\rho(x, y) = \left[\sum_{i=1}^{n} |x_i - y_i|^2\right]^{1/2},$$

(d) For any set $A \subset X$, define

$$d(x, A) = \min \{\rho(x, a): \quad a \in A\}.$$

The Hausdorff metric is defined as

$$\tilde{\rho}(A, B) = \tfrac{1}{2}[\max_{a \in A} d(a, B) + \max_{b \in B} d(b, A)].$$

B.2-2 Continuity

Let $f: X \longrightarrow Y$, where (X, ρ_x) and (X, ρ_y) are metric spaces. The function f is said to be continuous at $x_o \in X$ if for every $\epsilon > 0$, there exists a $\delta > 0$, $\delta = \delta(x_o, \epsilon)$, such that if for any $x \in X$, $\rho_x(x, x_o) < \epsilon$, then $\rho_y(f(x), f(x_o)) < \delta$. The function f is said to be *continuous* if it is continuous at all points in X. A continuous function is said to be *uniformly continuous* if for each $\epsilon > 0$, a δ can be determined which is independent of values in X; i.e., $\delta = \delta(\epsilon)$.

B.2-3 *Convergence*

A sequence $\{x_n\}$ of points in a metric space (X, ρ) is said to be convergent if there is a point $x_o \in X$ such that for any $\epsilon > 0$, there is an integer $N = N(\epsilon)$ such that if $n \geq N$, then $\rho(x_n, x_o) < \epsilon$. The point x_o is then called the limit of the sequence $\{x_n\}$, which is often denoted as $\lim_{n \to \infty} x_n = x_o$.

The notions of convergence and continuity are related by the following equivalent statements:

1. $f:(X, \rho_x) \longrightarrow (Y, \rho_y)$ is continuous at $x_o \in X$,
2. $\lim_{n \to \infty} f(x_n) = f(\lim_{n \to \infty} x_n)$ for any sequence $\{x_n\}$ such that $\lim_{n \to \infty} x_n = x_o$.

Therefore, continuity and the interchange of a function and the limit operator are equivalent notions.

B.2-4 *Closure*

A set A in a metric space (X, ρ) is said to be closed if and only if every convergent sequence $\{x_n\}$, $x_n \in A$ for all n, converges to a point in A.

B.2-5 *Completeness and Cauchy sequences*

A sequence $\{x_n\}$ in a metric space (X, ρ) is called a *Cauchy sequence* if for any given $\epsilon > 0$, there is an integer N, $N = N(\epsilon)$, such that if $m, n \geq N$, then $\rho(x_m, x_n) < \epsilon$.

It is easily shown from the triangle inequality that all convergent sequences are Cauchy sequences. However, the converse, is not, in general, true. If each Cauchy sequence in a metric space is a convergent sequence, then the metric space is said to be *complete*.

B.3 Algebraic structure

B.3-1 *Linear spaces and linear subspaces*

A *linear space* (on a field F, which will always be either the real or complex numbers here), is comprised of a nonempty set X and two operations, addition and scalar multiplication, which satisfy the following conditions:

1. $x_1 + x_2 = x_2 + x_1$ for all $x_1, x_2 \in X$
2. $(x_1 + x_2) + x_3 = x_1 + (x_2 + x_3)$ for all $x_1, x_2, x_3 \in X$
3. there exists a unique element in X, denoted by 0, such that $0 + x = x$ for all $x \in X$,
4. for every element $x \in X$, there is a unique element $-x \in X$ such that $x + (-x) = 0$
5. $1x = x$, for all $x \in X$,
6. $0x = 0$, for all $x \in X$,
7. $$\alpha(\beta x) = (\alpha\beta)x$$ $$(\alpha + \beta)x = \alpha x + \beta x$$ $$\alpha(x_1 + x_2) = \alpha x_1 + \alpha x_2$$
   for all $\alpha, \beta \in F$ and $x_1, x_2 \in X$.

When $F = R$, the real line, the linear space is called real; when $F = C$, the complex plane, the linear space is called complex.

A linear subspace is a subspace of a linear space which is also a linear space.

The linear subspace generated by the points x_i, $i = 1, \ldots, n$, (themselves elements of a linear space) is the set of all points of the form $\sum_{i=1}^{n} \alpha_i x_i$, for $\alpha_i \in R$, $i = 1, \ldots, n$.

B.3-2
Linear transformations

Let X and Y be two linear spaces. The function $f: X \longrightarrow Y$ is said to be *linear* if

$$f(\alpha x_1 + \beta x_2) = \alpha f(x_1) + \beta f(x_2),$$

for all $\alpha, \beta \in R$ and $x_1, x_2 \in X$.

Appendix A is concerned with linear transformations on finite dimensional spaces represented by matrices.

B.3-3
Linear independence and dependence

Let $x_i \in X$, $i = 1, \ldots, n$, where X is a linear space. Then the finite set $\{x_1, \ldots, x_n\}$ is said to be *linearly independent* if and only if the only scalar sequence $\{\alpha_1, \ldots, \alpha_n\}$ that satisfies $\sum_{i=1}^{n} \alpha_i x_i = 0$ is $\alpha_i = 0$, $i = 1, \ldots, n$. The set $\{x_1, \ldots, x_n\}$ is said to be linearly dependent if it is not linearly independent.

B.4
Combined topological and algebraic structure

B.4-1
Norms

Let X be a linear space. Then the function $||\cdot||: X \longrightarrow R$ is said to be a *norm* if the following conditions are satisfied:

1. $||x|| \geq 0$ for all $x \in X$
2. $||x|| = 0$ if and only if $x = 0$
3. $||x + y|| \leq ||x|| + ||y||,$ for all $x, y \in X$
4. $||\alpha x|| = |\alpha|\,||x||,$ for all $\alpha \in R, x \in X.$

It can be shown that the norm function $||\cdot||$ is continuous and obeys the inequality

$$||x|| - ||y|| \leq ||x - y||.$$

B.4-2
Normed linear spaces and Banach spaces

A *normed linear space* $(X, ||\cdot||)$ is a linear space X and a norm $||\cdot||: X \longrightarrow R$. The norm can be used to induce a metric. Define

$$\rho(x, y) = ||x - y||.$$

The normed linear space then becomes a metric space under the (usual) metric.

A *Banach space* is a complete (under the usual metric induced by the norm) normed linear space. *Examples:*

(a) l_p, $1 \leq p < \infty$, is the set of all scalar sequences $\{x_1, x_2, \ldots\}$ having the property

$$||\{x_n\}||_p = \sum_{i=1}^{\infty} |x_i|^p < \infty$$

(b) l_∞ is the set of all scalar sequences $\{x_1, x_2, \ldots\}$ having the property

$$||\{x_n\}||_\infty = \sup\{|x_i| : 1 \leq i < \infty\} < \infty$$

(c) $L_p[t_o, t_f]$, $1 \leq p < \infty$, is the set of all functions $f: R \longrightarrow [t_o, t_f]$ such that

$$||f||_p = \left[\int_{t_o}^{t_f} |f(\tau)|^p \, d\tau\right]^{1/p} < \infty,$$

where f is such that the integral is well-defined

(d) $L_\infty[t_o, t_f]$ is the set of all functions $f: R \longrightarrow [t_o, t_f]$ such that

$$||f||_\infty = \text{ess sup}\{|f(\tau)| : \tau \in [t_o, t_f]\} < \infty.$$

B.4-3
Inequalities

SCHWARTZ AND HOLDER INEQUALITIES: Let $1 < p < \infty$ and $1/p + 1/q = 1$. Then,

(a) if $\{x_n\} \in l_p$ and $\{y_n\} \in l_q$,

$$\sum_{i=1}^{\infty} |x_i y_i| \leq ||\{x_i\}||_p ||\{y_i\}||_q,$$

(b) if $f \in L_p[t_o, t_f]$ and $g \in L_q[t_o, t_f]$,

$$\int_{t_o}^{t_f} |f(\tau)g(\tau)| \, d\tau \leq ||f||_p ||g||_q.$$

MINKOWSKI INEQUALITY: Let $1 \leq p < \infty$. Then,

(a) if $\{x_i\}, \{y_i\} \in l_p$,

$$\left[\sum_{i=1}^{\infty} |x_i \pm y_i|^p\right]^{1/p} \leq ||\{x_i\}||_p + ||\{y_i\}||_p,$$

(b) if $f, g \in L_p[t_o, t_f]$,

$$\left[\int_{t_o}^{t_f} |f(\tau)g(\tau)|^p \, d\tau\right]^{1/p} < ||f||_p + ||g||_p.$$

B.4-4
Inner product spaces and Hilbert spaces

Let X be a complex linear space, and let an overbar ($\overline{}$) denote the complex conjugate. An inner product $(\cdot, \cdot): X \times X \rightarrow C$ satisfies the following conditions:

1. $(x_1 + x_2, x_3) = (x_1, x_3) + (x_2, x_3)$ for all $x_1, x_2, x_3 \in X$
2. $(\alpha x_1, x_2) = \alpha(x_1, x_2)$, for all $\alpha \in C, x_1, x_2 \in X$
3. $(x_1, x_2) = \overline{(x_2, x_1)}$ for all $x_1, x_2 \in X$
4. $(x, x) > 0$ when $x \neq 0$.

If X is a real linear space, the inner product is defined similarly.

An *inner product space* $(X, (\cdot, \cdot))$ is a (real or complex) linear space X and an inner product $(\cdot, \cdot)$.

The fact that $(x, x) \in R$ follows from condition 3 in the definition of an inner product. Note that

$$||x|| = (x, x)^{1/2}$$

defines a norm on X; thus, an inner product space generates a normed linear space (which in turn generates a metric space). A complete (with respect to the metric induced by the norm) inner product space is called a *Hilbert space*. *Examples:*

(a) l_2, having elements composed of sequences of complex-valued numbers, is a Hilbert space.
(b) $L_2[t_o, t_f]$ is a Hilbert space if it is composed of integrable complex valued functions.
The Cauchy inequality is:

$$|(x, y)| \leq ||x|| \, ||y||.$$

B.4-5
Orthogonality and orthonormal bases

Let $(X, (\cdot, \cdot))$ be an inner product space. Then $x, y \in X$ are said to be *orthogonal* if $(x, y) = 0$. Let M be a linear vector subspace of X. Then $x \in X$ is said to be *orthogonal to M* if $(x, m) = 0$ for all $m \in M$. A set of points $\{x_i\}$, $x_i \in X$, is said to be *orthonormal* if $(x_i, x_j) = 0$, $i \neq j$, and $(x_i, x_i) = 1$, for all i. The orthonormal set $\{x_i\}$ is said to be *maximal* in X, if $(x, x_i) = 0$ for all i implies that $x = 0$. A maximal orthonormal set of a Hilbert space is called an *orthonormal basis* for X.

B.4-6
The orthogonal projection theorem

Elementary geometry indicates that a perpendicular from a point to a line represents the shortest distance from the point to the line, a result easily generalized to determining minimum distance from a point to a plane. The central concept in this argument is the "perpendicularity" or orthogonality of the minimizing line segment from the point to the plane in relation to the plane. This concept can be generalized to a Hilbert space, which represents a particularly useful setting for the following result—the powerful and important *orthogonal projection theorem*. Proof of the theorem and the following lemma can be found in [1, 3] and elsewhere.

Theorem B.1
Let M be a closed linear subspace of a Hilbert space. Then, for any vector $\beta \in M$, there exists a unique (best estimator) vector $\hat{\beta} \in M$ such that $||\hat{\beta} - \beta|| \leq ||m - \beta||$ for all $m \in M$. Furthermore, the (minimum estimation error) vector $\beta - \hat{\beta}$ is necessarily and sufficiently orthogonal to all vectors in M.

We now present two particularly useful results of the projection theorem.

Lemma
Let $M \subset H$ be the linear subspace of H generated by the orthonormal basis $\{\alpha_i\}_{i=1}^m$. Then

$$\hat{\beta} = \sum_{i=1}^{m} (\beta, \alpha_i)\alpha_i. \tag{B.1}$$

Let $\alpha^* \in H$ be such that $(\alpha^*, y) = 0$ for all $y \in M$ and $(\alpha^*, \alpha^*) = 1$. Define M^* as the linear subspace generated by $\{\alpha_1, \ldots, \alpha_m, \alpha^*\}$, and let β^* satisfy Theorem B.1 with respect to M^*. Then

$$\beta^* = \hat{\beta} + (\beta, \alpha^*)\alpha^*. \tag{B.2}$$

Equation (B.1) describes the orthogonal projection $\hat{\beta}$ of β on M in terms of the orthonormal basis $\{\alpha_i\}_{i=1}^m$. Equation (B.2) states that recalculation of the orthogonal projection $\hat{\beta}$ of β on M is unnecessary when an additional element α^*, orthonormal to M, is allowed to be used in addition to $\{\alpha_i\}_{i=1}^m$ to better describe β. Both of these results, Eqs. (B.1) and (B.2), are of substantial value in the linear minimum variance estimation theory discussed in Chapter 8.

B.4-7
The Parseval equality

The Parseval equality is a result associated with Fourier series analysis and the Fourier series theorem [1, p. 307], which we now state.

Theorem B.2
Let $\{x_i\}$ be an orthonormal set in the Hilbert space X. Then the following statements are equivalent:

(a) $\{x_i\}$ is an orthonormal basis
(b) $x \in X$ implies

$$x = \sum_i (x, x_i)x_i$$

(c) $x, y \in X$ implies

$$(x, y) = \sum_i (x, x_i)\overline{(y, x_i)}$$

(d) $x \in X$ implies

$$\|x\|^2 = \sum_i |(x, x_i)|^2.$$

The coefficients (x, x_i) are called Fourier coefficients, (b) is called the Fourier series expansion, and (c) is Parseval's equality. A fifth equivalence condition is presented and further discussion can be found in [1].

References

1. NAYLOR, A.W., and SELL, G.R., *Linear Operator Theory in Engineering and Science*. Holt, New York, 1971.

2. PORTER, W.A., *Modern Foundations of Systems Engineering*. Macmillan, New York, 1966.

3. LUENBERGER, D.G., *Optimization by Vector Space Methods*. Wiley, New York, 1969.

4. HARDY, G.H., LITTLEWOOD, J.E., and POLYA, G., *Inequalities*. Cambridge University Press, New York, 1952.

5. GOFFMAN, C., and PEDRICK, G., *First Course in Functional Analysis*. Prentice-Hall, Inc., Englewood Cliffs, New Jersey, 1965.

6. HALMOS, P., *Finite Dimensional Vector Spaces*. Van Nostrand, Princeton, New Jersey, 1958.

7. HEWITT, E., and STROMBERG, K., *Real and Abstract Analysis*. Springer–Verlag, Berlin, 1965.

8. KOLMOGOROV, A.N., and FOMIN, S.V., *Elements of the Theory of Functions and Functional Analysis*, **I** and **II**. Graylock, Albany, New York, 1957 and 1961.

9. SIMMONS, G.F., *Introduction to Topology and Modern Analysis*. McGraw-Hill Book Co., New York, 1963.

10. ROYDEN, H.L., *Real Analysis*. Macmillan, New York, 1963.

11. RUBIN, W., *Principles of Mathematical Analysis*, 2nd ed. McGraw-Hill Book Co., New York, 1964.

Random variables and stochastic processes C

This appendix is intended to serve as background for the stochastic material presented in Chapters 8 and 9. Further discussion of these concepts can be found in the references. In particular, our outline for this appendix is, in part, based on [1].

C.1 Probability spaces

C.1-1 *σ–algebra*

A *σ–algebra* is a class of sets closed under complementation and countable union; i.e. if $\mathfrak{F}$ is a σ–algebra, then

1. $\mathcal{E} \in \mathfrak{F}$ implies $\mathcal{E}^c \in \mathfrak{F}$, where $\mathcal{E}^c$ is the complement of $\mathcal{E}$,
2. $\mathcal{E}_i \in \mathfrak{F}$, $i = 1, 2, \ldots$, implies $\bigcup_{i=1}^{\infty} \mathcal{E}_i \in \mathfrak{F}$.

C.1-2 *Probability measure*

A probability space $(S, \mathfrak{F}, P)$ consists of:

1. the sample space S
2. a σ–algebra of subsets or events of S

3. a function $P: \mathfrak{F} \longrightarrow R$ which is called a probability measure if it satisfies the following axioms:

   (a) $P(\mathcal{E}) \geq 0, \quad$ for all $\mathcal{E} \in \mathfrak{F}$
   (b) $P(S) = 1$
   (c) $P\left(\bigcup_{i=1}^{\infty} \mathcal{E}_i\right) = \sum_{i=1}^{\infty} P(\mathcal{E}_i), \quad$ for $\mathcal{E}_i \in \mathfrak{F}, \mathcal{E}_i \bigcap \mathcal{E}_j = \phi, i \neq j$, where ϕ is the null set.

C.1-3
Conditional probability

The conditional probability of the event $\mathcal{E}_1$ given event $\mathcal{E}_2$ is defined as

$$P(\mathcal{E}_1 \mid \mathcal{E}_2) = \frac{P(\mathcal{E}_1 \cap \mathcal{E}_2)}{P(\mathcal{E}_2)}, \qquad P(\mathcal{E}_2) \neq 0.$$

C.2 Random variables and distributions

C.2-1
Random vector

Let $(S, \mathfrak{F}, P)$ be a probability space, and let $\mathbf{x}: S \longrightarrow R^n$. If sets of the form $\{\xi \in S: \mathbf{X}(\xi) \leq \boldsymbol{\alpha}\} \in \mathfrak{F}$, for all $\boldsymbol{\alpha} \in R^n$, and if $P\{\xi \in S: \mathbf{x}(\xi) = \pm\infty\} = 0$, then $\mathbf{x}$ is said to be a *random vector*.

C.2-2
Distributions and densities

The *probability distribution function* $F_{\mathbf{x}}: R^n \longrightarrow R$ of the random vector $\mathbf{x}$ is defined as

$$F_{\mathbf{x}}(\boldsymbol{\alpha}) = P\{\xi \in S: \mathbf{x}(\xi) \leq \boldsymbol{\alpha}\}.$$

A *probability density function* $f_{\mathbf{x}}: R^n \longrightarrow R$ of the random vector $\mathbf{x}$ is any (possibly generalized) function that satisfies

$$F_{\mathbf{x}}(\boldsymbol{\alpha}) = \int_{-\infty}^{\boldsymbol{\alpha}} f_{\mathbf{x}}(\boldsymbol{\beta})\, d\boldsymbol{\beta}.$$

C.2-3
Joint and marginal distributions and densities

The *joint probability distribution function* $F_{\mathbf{x}_1\mathbf{x}_2}: R^n \times R^n \longrightarrow R$ of the random vectors $\mathbf{x}_1$ and $\mathbf{x}_2$ is defined as

$$\mathbf{F}_{\mathbf{x}_1\mathbf{x}_2}(\boldsymbol{\alpha}_1, \boldsymbol{\alpha}_2) = P(\mathbf{x}_1 \leq \boldsymbol{\alpha}_1, \mathbf{x}_2 \leq \boldsymbol{\alpha}_2).$$

The *joint probability density function* $f_{\mathbf{x}_1\mathbf{x}_2}: R^n \times R^n \longrightarrow R$ of the random vectors $\mathbf{x}_1$ and $\mathbf{x}_2$ is any (possibly generalized) function that satisfies

$$F_{\mathbf{x}_1\mathbf{x}_2}(\boldsymbol{\alpha}_1, \boldsymbol{\alpha}_2) = \int_{-\infty}^{\alpha_1}\int_{-\infty}^{\alpha_2} f_{\mathbf{x}_1\mathbf{x}_2}(\boldsymbol{\beta}_1, \boldsymbol{\beta}_2)\, d\boldsymbol{\beta}_1\, d\boldsymbol{\beta}_2.$$

It is easily shown that the *marginal probability distribution and density functions*, $F_{\mathbf{x}_1}$ and $f_{\mathbf{x}_1}$, satisfy

$$F_{\mathbf{x}_1}(\boldsymbol{\alpha}_1) = F_{\mathbf{x}_1\mathbf{x}_2}(\boldsymbol{\alpha}_1, \infty)$$

and

$$f_{\mathbf{x}_1}(\boldsymbol{\alpha}_1) = \int_{-\infty}^{+\infty} f_{\mathbf{x}_1\mathbf{x}_2}(\boldsymbol{\alpha}_1, \boldsymbol{\alpha}_2)\, d\boldsymbol{\alpha}_2.$$

C.2-4
Conditional distributions and densities

The conditional distribution function $F_{\mathbf{x}_1|\mathbf{x}_2}: R^n \times R^n \longrightarrow R$ of a random vector $\mathbf{x}_1$, given the event $\{\mathbf{x}_2 \leq \boldsymbol{\alpha}_2\}$, is defined as

$$F_{\mathbf{x}_1|\mathbf{x}_2}(\boldsymbol{\alpha}_1 \,|\, \boldsymbol{\alpha}_2) = \frac{F_{\mathbf{x}_1\mathbf{x}_2}(\boldsymbol{\alpha}_1, \boldsymbol{\alpha}_2)}{F_{\mathbf{x}_2}(\boldsymbol{\alpha}_2)}.$$

The conditional density function $f_{\mathbf{x}_1|\mathbf{x}_2}: R^n \times R^n \longrightarrow R$ of a random vector $\mathbf{x}_1$, given that $\mathbf{x}_2 = \boldsymbol{\alpha}_2$ is defined as

$$f_{\mathbf{x}_1|\mathbf{x}_2}(\boldsymbol{\alpha}_1 \,|\, \boldsymbol{\alpha}_2) = \frac{f_{\mathbf{x}_1\mathbf{x}_2}(\boldsymbol{\alpha}_1, \boldsymbol{\alpha}_2)}{f_{\mathbf{x}_2}(\boldsymbol{\alpha}_2)}.$$

C.2-5
Expection and the fundamental theorem of expectation

The expectation of a random vector $\mathbf{x}$ is defined as

$$\mathcal{E}(\mathbf{x}) = \int_{-\infty}^{+\infty} \boldsymbol{\alpha} f_{\mathbf{x}}(\boldsymbol{\alpha})\, d\boldsymbol{\alpha}.$$

The fundamental theorem of expectation is that for all functions $\mathbf{g}$, mapping random vectors into random vectors,

$$\mathcal{E}[\mathbf{g}(\mathbf{x})] = \int_{-\infty}^{+\infty} \mathbf{g}(\boldsymbol{\alpha}) f_{\mathbf{x}}(\boldsymbol{\alpha})\, d\boldsymbol{\alpha}.$$

C.2-6
The Hilbert space of random vectors

Let H be the set of all random vectors $\mathbf{x}$ having the property $(\mathbf{x}, \mathbf{x}) < \infty$, where

$$(\mathbf{x}, \mathbf{y}) = \mathcal{E}(\mathbf{x}^T\mathbf{y})$$

for any random vector $\mathbf{y}$. It is easily shown that $[H, (\cdot\,, \cdot)]$ is a Hilbert space.

C.2-7
Conditional expectation

The conditional expectation of a random vector $\mathbf{x}$, given the realization $\boldsymbol{\beta}$ of random vector $\mathbf{y}$, is defined as

$$\mathcal{E}(\mathbf{x} \mid \mathbf{y} = \boldsymbol{\beta}) = \int_{-\infty}^{+\infty} \boldsymbol{\alpha} f_{\mathbf{x}|\mathbf{y}}(\boldsymbol{\alpha} \mid \boldsymbol{\beta})\, d\boldsymbol{\alpha}.$$

C.2-8
Covariance

The covariance function of two random vectors $\mathbf{x}_1$ and $\mathbf{x}_2$ is

$$\operatorname{cov}(\mathbf{x}_1, \mathbf{x}_2) = \mathcal{E}\{[\mathbf{x}_1 - \mathcal{E}(\mathbf{x}_1)][\mathbf{x}_2 - \mathcal{E}(\mathbf{x}_2)]^T\}.$$

C.2-9
Independence, correlation, and orthogonality

Two events $\mathcal{E}_1$ and $\mathcal{E}_2$ are said to be *independent* if $P(\mathcal{E}_1 \cap \mathcal{E}_2) = P(\mathcal{E}_1)P(\mathcal{E}_2)$. Two random vectors $\mathbf{x}_1$ and $\mathbf{x}_2$ are said to be independent if the events $(\mathbf{x}_1 \leq \boldsymbol{\alpha}_1)$ and $(\mathbf{x}_2 \leq \boldsymbol{\alpha}_2)$ are independent for all $\boldsymbol{\alpha}_1, \boldsymbol{\alpha}_2$. Equivalent statements for two random vectors $\mathbf{x}_1$ and $\mathbf{x}_2$ to be independent are:

1. $$F_{\mathbf{x}_1\mathbf{x}_2}(\boldsymbol{\alpha}_1, \boldsymbol{\alpha}_2) = F_{\mathbf{x}_1}(\boldsymbol{\alpha}_1)F_{\mathbf{x}_2}(\boldsymbol{\alpha}_2)$$
2. $$f_{\mathbf{x}_1\mathbf{x}_2}(\boldsymbol{\alpha}_1, \boldsymbol{\alpha}_2) = f_{\mathbf{x}_1}(\boldsymbol{\alpha}_1)f_{\mathbf{x}_2}(\boldsymbol{\alpha}_2).$$

Two random vectors $\mathbf{x}_1$ and $\mathbf{x}_2$ are said to be *uncorrelated* if $\mathcal{E}(\mathbf{x}_1\mathbf{x}_2^T) = \mathcal{E}(\mathbf{x}_1)\mathcal{E}^T(\mathbf{x}_2)$.

Two random vectors $\mathbf{x}_1$ and $\mathbf{x}_2$ are said to be orthogonal if $\mathcal{E}(\mathbf{x}_1\mathbf{x}_2^T) = \mathbf{0}$.

It can be shown that independence implies uncorrelatedness but not conversely.

C.2-10
Gaussian random vectors

A random vector $\mathbf{x}$ is said to have a Gaussian distribution with mean $\mathcal{E}(\mathbf{x}) = \boldsymbol{\mu}$ and variance $\mathcal{E}[(\mathbf{x} - \boldsymbol{\mu})(\mathbf{x} - \boldsymbol{\mu})^T] = \mathbf{V}$ if the probability density function of x is

$$f_{\mathbf{x}}(\boldsymbol{\alpha}) = \frac{1}{[(2\pi)^n \det \mathbf{V}]^{1/2}} \exp\left[-\frac{1}{2}(\boldsymbol{\alpha} - \boldsymbol{\mu})^T\mathbf{V}^{-1}(\boldsymbol{\alpha} - \boldsymbol{\mu})\right].$$

It can be shown that uncorrelated Gaussian random vectors are independent.

C.3
Stochastic processes

A stochastic process $\{x(t), t \in T\}$ is simply a collection of random vectors defined on the same probability space, where t usually denotes time, and the index set T is either continuous or discrete. We have derived many expressions in this text for the propagation of the statistical properties of stochastic processes, particularly mean and variance propagation in linear systems.

C.3-1
Gaussian processes

An important class of stochastic processes is the *Gaussian process*. The Gaussian process approximates many random phenomena in physical systems and enjoys considerable analytical simplicity. We say that the stochastic process $\{x(t), t \in T\}$ is *Gaussian* if every finite linear combination $\sum_{i=1}^{n} a_i x(t_i)$, $t_i \in T$ for all i, is a Gaussian random variable. Let $j = \sqrt{-1}$. Then a necessary and sufficient condition for $\{x(t), t \in T\}$ to be Gaussian is for each $t \in T$, $\mathcal{E}[x^2(t)] < \infty$ and for each finite set $\{t_1, \ldots, t_n\}$, $t_i \in T$,

$$\mathcal{E}\left\{\exp\left[j \sum_{i=1}^{n} a_i x(t_i)\right]\right\} =$$

$$\exp\left\{j \sum_{i=1}^{n} a_i \mu(t_i) - \tfrac{1}{2} \sum_{i,k} a_i a_k \operatorname{cov}\,[x(t_i), x(t_k)]\right\},$$

where $\mu(t_i) = \mathcal{E}[x(t_i)]$[3]. It can be shown that all finite-dimensional distributions of a Gaussian process can be completely determined from knowledge of $\mu(t)$ and $\operatorname{cov}\,[x(t), x(s)]$.

A stochastic process $\{x(t), t \in T\}$ is said to have *independent increments* if for any $t_1 < \ldots < t_n$, $t_i \in T$, the random variables $x(t_1)$, $x(t_2) - x(t_1), \ldots, x(t_n) - x(t_{n-1})$ are mutually independent. An important class of processes having independent increments is a *Markov process*; $\{x(t), t \in T\}$ is said to be Markov if for any $t_1, \ldots, t_n$, $t_i \in T$,

$$\begin{aligned} P(x(t_n) \le \alpha_n \mid x(t_i) = \alpha_i, \quad i = 1, \ldots, n-1) \\ = P(x(t_n) \le \alpha_n \mid x(t_{n-1}) = \alpha_{n-1}). \end{aligned}$$

A particularly valuable descriptor of a stochastic process is the *Martingale property*; $\{x(t), t \in T\}$ is said to be a Martingale if (almost surely)

$$\mathcal{E}\{x(t) \mid x(s), s \le \tau\} = x(\tau)$$

for any $t \ge \tau$.

C.3-2
Wiener process

An important example of a Gaussian process that also has independent increments and is a Martingale is the *Wiener process* or *Brownian motion*, a particularly valuable model of the stochastic effects to which many physical systems are subject. The process $\{\omega(t), t \geq 0\}$ is said to be a *Wiener process* if it is a zero-mean, i.e., $\mu_\omega(t) = 0$ for all t, Gaussian process such that $\text{cov}\,[\omega(t), \omega(s)] = \min\,(t, s)$. The Wiener process has other properties of importance; it can be shown to have continuous sample paths (with probability one) but is nowhere differentiable and is of unbounded variation on every closed subinterval of $[0, \infty]$. Further we can show that the Wiener process is the time integral of a Gaussian white noise process [2]†. This allows the Wiener process to be intuitively interpreted as being such that (where $\sim$ denotes "is of the order of") $d\omega(t) \sim \sqrt{dt}$, a result useful in the determination of the stochastic Hamilton-Jacobi-Bellman equation, which we now derive in a more or less heuristic fashion.

C.3-3
Stochastic Hamilton-Jacobi-Bellman equation

Let the stochastic process $\{x(t), t \geq 0\}$ be described by the nonlinear scalar stochastic differential equation

$$dx(t) = f[x(t), u(t), t]\,dt + d\omega(t), \tag{C.1}$$

where $\{\omega(t), t \geq 0\}$ is a Wiener process. We wish to determine a stochastic version of the Hamilton-Jacobi-Bellman equation. We assume, as our cost function at time t, a conditional expectation of a function ϕ from time t to the terminal time T, where the conditioning variable is the present state $x(t)$. Thus we have

$$V[x(t), t] = \min_{u(\tau)} \mathcal{E}\left\{\int_t^T \phi[x(\lambda), u(\lambda), \lambda]\,d\lambda \,|\, x(t)\right\} \tag{C.2}$$

for $t \leq \tau \leq T$, where we assume the control $u(t)$ is a function of the state $x(t)$ for no time greater than t such that we have the feedback control law $u(t) = \gamma[x(t), t]$. Just as in our deterministic derivation of the Hamilton-Jacobi-Bellman equation in Chapter 4, we may easily show that

$$V[x(t), t] \cong \min_{u(t)} \mathcal{E}\{\phi[x(t), u(t), t]\Delta t + V[x(t + \Delta t), t + \Delta t] \,|\, x(t)\}. \tag{C.3}$$

---

†Comparison of this equation with the (formally identical) state equations presented in Chapters 8 and 9 implies that the derivative of a Wiener process is formally Gaussian white noise. Thus we could write $d\omega(t) = \omega(t)\,dt$ where $\omega(t)$ is white noise. Since a Wiener process is nowhere differentiable, further elaboration and substantiation is required. See [2, 3, 7] for further discussion.

A Taylor series expansion of V about $\Delta t = 0$ yields:

$$V[x(t+\Delta t), t+\Delta t] = V[x(t), t] + \frac{\partial V[x(t), t]}{\partial t}\Delta t + \frac{\partial V[x(t), t]}{\partial x(t)}\Delta x(t) +$$

$$\frac{1}{2}\frac{\partial^2 V[x(t), t]}{\partial t^2}(\Delta t)^2 + \frac{1}{2}\frac{\partial^2 V[x(t), t]}{\partial x^2(t)}[\Delta x(t)]^2 + \frac{\partial^2 V[x(t), t]}{\partial t\,\partial x(t)}\Delta x(t)\,\Delta t + \dots.$$

The Taylor series expansion has been taken out to the second order terms due to the fact that $\Delta\omega(t) \sim \sqrt{\Delta t}$. Note, however, that $\mathcal{E}[\Delta\omega(t)] = \mathcal{E}[\omega(t+\Delta t)] - \mathcal{E}[\omega(t)] = 0$. Thus, we obtain from Eq. (C.3), upon using Eqs. (C.2) and (C.4) where we suppress several time and state arguments for notational simplicity,

$$V[x(t), t] \cong \min_{u(t)} \mathcal{E}\left\{\phi\Delta t + V[x(t), t] + \frac{\partial V}{\partial t}\Delta t + \frac{\partial V}{\partial x}(f\Delta t + \Delta\omega) + \right.$$

$$\left. \frac{1}{2}\frac{\partial^2 V}{\partial x^2}(f\Delta t + \Delta\omega)^2 + \dots \,|\, x(t)\right\} \cong \min_{u(t)}\left\{\phi\,\Delta t + V[x(t), t] + \right.$$

$$\left.\frac{\partial V}{\partial t}\Delta t + \frac{\partial V}{\partial x}f\Delta t + \frac{1}{2}\frac{\partial^2 V}{\partial x^2}[f^2(\Delta t)^2 + 2f\Delta t\sqrt{\Delta t} + \Delta t] + \dots\right\}.$$

Subtracting $V[x(t), t]$ from both sides of the foregoing equation, dividing through by Δt and letting $\Delta t \longrightarrow 0$, we find that

$$0 = \min_{u(t)}\left\{\phi + \frac{\partial V}{\partial t} + \frac{\partial V}{\partial x}f + \frac{1}{2}\frac{\partial^2 V}{\partial x^2}\right\}, \tag{C.5}$$

which is the stochastic Hamilton-Jacobi-Bellman equation. From Eq. (C.2) we see that the terminal condition for this partial differential equation is $V[x(T), T] = 0$. For notational simplicity in this derivation, we have assumed that the state $x(t)$ is a scalar process. Equation (C.5) may be stated for the vector case as

$$\min_{\mathbf{u}(t)}\left\{\frac{\partial V[\mathbf{x}(t), t]}{\partial t} + \phi[\mathbf{x}(t)\mathbf{u}(t), t] + \left[\frac{\partial V[\mathbf{x}(t), t]}{\partial \mathbf{x}(t)}\right]^T \mathbf{f}[\mathbf{x}(t), \mathbf{u}(t), t] + \right.$$

$$\left.\frac{1}{2}\operatorname{tr}\mathbf{Q}(t)\frac{\partial^2 V[\mathbf{x}(t), t]}{\partial \mathbf{x}^2(t)}\right\} = 0 \tag{C.6}$$

with terminal condition

$$V[\mathbf{x}(T), T] = \theta[\mathbf{x}(T), T]. \tag{C.7}$$

This is the Hamilton-Jacobi-Bellman equation whose solution minimizes the cost function,

$$V[\mathbf{x}(t), t] = \min_{\mathbf{u}(t)} \mathcal{E}\left\{\theta[\mathbf{x}(T), T] + \int_t^T \phi[\mathbf{x}(\tau), \mathbf{u}(\tau), \tau]\,d\tau \,|\, \mathbf{x}(t)\right\}, \tag{C.8}$$

subject to the equality constraint

$$\mathbf{dx}(t) = \mathbf{f}[\mathbf{x}(t), \mathbf{u}(t), t]\,dt + \mathbf{d\omega}(t), \tag{C.9}$$

where $\boldsymbol{\omega}(t)$ is assumed a vector Wiener process with zero-mean and variance derivative $\mathbf{Q}(t)$.

References

1. MELSA, J.L., and SAGE, A.P., *An Introduction to Probability and Stochastic Processes.* Prentice-Hall Inc., Englewood Cliffs, New Jersey, 1973.
2. SAGE, A.P., and MELSA, J.L., *Estimation Theory With Applications to Communications and Control.* McGraw-Hill Book Co., New York, 1971.
3. WONG, E., *Stochastic Processes in Information and Dynamical Systems.* McGraw-Hill Book Co., New York, 1971.
4. HALMOS, P., *Measure Theory.* Van Nostrand, Princeton, New Jersey, 1950.
5. KOLMOGOROV, A.N., *Foundations of the Theory of Probability.* Chelsea Publishing Co., New York, 1950.
6. LOEVE, M., *Probability Theory.* Van Nostrand, Princeton, New Jersey, 1960.
7. ASTROM, K.J., *Introduction to Stochastic Control Theory.* Academic Press, New York, 1970.

Proof of the matrix inversion lemma D

The matrix inversion lemma can be stated as follows [1]:

If the matrices $\mathbf{P}_{n+1}$, $\mathbf{P}_n$, $\mathbf{H}_{n+1}$, and $\mathbf{R}_{n+1}$ satisfy the equation

$$\mathbf{P}_{n+1}^{-1} = \mathbf{P}_n^{-1} + \mathbf{H}_{n+1}^T \mathbf{R}_{n+1}^{-1} \mathbf{H}_{n+1}, \tag{D.1}$$

where $\mathbf{P}_{n+1}$, $\mathbf{P}_n$, $\mathbf{P}_{n+1}$, and $(\mathbf{H}_{n+1}\mathbf{P}_n\mathbf{H}_{n+1}^T + \mathbf{R})$ also exist. This requires that $\mathbf{P}_n^{-1}$, $\mathbf{P}_{n+1}^{-1}$, and $\mathbf{R}_{n+1}^{-1}$ be nonsingular, and $\mathbf{H}_{n+1}$ be of maximum rank.

Then $\mathbf{P}_{n+1}$ is given by

$$\mathbf{P}_{n+1} = \mathbf{P}_n - \mathbf{P}_n\mathbf{H}_{n+1}^T(\mathbf{H}_{n+1}\mathbf{P}_n\mathbf{H}_{n+1}^T + \mathbf{R}_{n+1})^{-1}\mathbf{H}_{n+1}\mathbf{P}_n. \tag{D.2}$$

We will now prove the matrix inversion lemma by manipulating the expressions in such a way that we can derive Eq. (D.2) directly from Eq. (D.1). Premultiplying Eq. (D.1) by $\mathbf{P}_{n+1}$ gives us

$$\mathbf{I} = \mathbf{P}_{n+1}\mathbf{P}_n^{-1} + \mathbf{P}_{n+1}\mathbf{H}_{n+1}^T\mathbf{R}_{n+1}^{-1}\mathbf{H}_{n+1}. \tag{D.3}$$

Postmultiplying Eq. (D.3) by $\mathbf{P}_n$ results in

$$\mathbf{P}_n = \mathbf{P}_{n+1} + \mathbf{P}_{n+1}\mathbf{H}_{n+1}^T\mathbf{R}_{n+1}^{-1}\mathbf{H}_{n+1}\mathbf{P}_n. \tag{D.4}$$

Postmultiplying Eq. (D.4) by $\mathbf{H}_{n+1}^T$, we obtain

$$\begin{aligned}\mathbf{P}_n\mathbf{H}_{n+1}^T &= \mathbf{P}_{n+1}\mathbf{H}_{n+1}^T + \mathbf{P}_{n+1}\mathbf{H}_{n+1}^T\mathbf{R}_{n+1}^{-1}\mathbf{H}_{n+1}\mathbf{P}_n\mathbf{H}_{n+1}^T \\ &= \mathbf{P}_{n+1}\mathbf{H}_{n+1}^T\mathbf{R}_{n+1}^{-1}[\mathbf{R}_{n+1} + \mathbf{H}_{n+1}\mathbf{P}_n\mathbf{H}_{n+1}^T].\end{aligned} \tag{D.5}$$

We postmultiply Eq. (D.5) by $[\mathbf{R}_{n+1} + \mathbf{H}_{n+1}\mathbf{P}_n\mathbf{H}_{n+1}^T]^{-1}$ to obtain

$$\mathbf{P}_{n+1}\mathbf{H}_{n+1}^T\mathbf{R}_{n+1}^{-1} = \mathbf{P}_n\mathbf{H}_{n+1}^T[\mathbf{H}_{n+1}\mathbf{P}_n\mathbf{H}_{n+1}^T + \mathbf{R}_{n+1}]^{-1}. \tag{D.6}$$

Postmultiplying this expression by $\mathbf{H}_{n+1}\mathbf{P}_n$, we see that

$$\mathbf{P}_{n+1}\mathbf{H}_{n+1}^T\mathbf{R}_{n+1}^{-1}\mathbf{H}_{n+1}\mathbf{P}_n = \mathbf{P}_n\mathbf{H}_{n+1}^T[\mathbf{H}_{n+1}\mathbf{P}_n\mathbf{H}_{n+1}^T + \mathbf{R}_{n+1}]^{-1}\mathbf{H}_{n+1}\mathbf{P}_n. \tag{D.7}$$

Subtracting Eq. (D.7) from $\mathbf{P}_n$ results in

$$\mathbf{P}_n - \mathbf{P}_{n+1}\mathbf{H}_{n+1}^T\mathbf{R}_{n+1}^{-1}\mathbf{H}_{n+1}\mathbf{P}_n = \mathbf{P}_n - \mathbf{P}_n\mathbf{H}_{n+1}^T[\mathbf{H}_{n+1}\mathbf{P}_n\mathbf{H}_{n+1}^T + \mathbf{R}_{n+1}]^{-1}\mathbf{H}_{n+1}\mathbf{P}_n. \tag{D.8}$$

From Eq. (D.4), we finally have

$$\mathbf{P}_{n+1} = \mathbf{P}_n - \mathbf{P}_n\mathbf{H}_{n+1}^T[\mathbf{H}_{n+1}\mathbf{P}_n\mathbf{H}_{n+1}^T + \mathbf{R}_{n+1}]^{-1}\mathbf{H}_{n+1}\mathbf{P}_n. \tag{D.9}$$

Equation (D.9) is identical to Eq. (D.2) as required; therefore, we have accomplished the desired task.

Reference

1. GRENANDER, U., and ROSENBLATT, M., *Statistical Analysis of Stationary Time Series*. Wiley, New York, 1957.

Index

T

U

V

W

Y